T0231792

Logic in Tehran

Lecture Notes in Logic

A Publication of

The Association for Symbolic Logic

LECTURE NOTES IN LOGIC 26

Logic in Tehran

Proceedings of the Workshop and Conference on Logic, Algebra and Arithmetic,
held October 18–22, 2003

Edited by

Ali Enayat
Department of Mathematics and Statistics
American University, Washington, D.C.

Iraj Kalantari
Department of Mathematics
Western Illinois University, Macomb

Mojtaba Moniri
Department of Mathematics
Tarbiat Modarres University, Tehran

CRC Press
Taylor & Francis Group
Boca Raton London New York

CRC Press is an imprint of the
Taylor & Francis Group, an informa business

Addresses of the Editors of Lecture Notes in Logic and a Statement of Editorial Policy may be found at the back of this book.

Sales and Customer Service:
A K Peters, Ltd.
888 Worcester Street, Suite 230
Wellesley, Massachusetts 02482, USA
http://www.akpeters.com/

Association for Symbolic Logic:
Sam Buss, Publisher
Department of Mathematics
University of California, San Diego
La Jolla, California 92093-0112, USA
http://www.aslonline.org/

Library of Congress Cataloging-in-Publication Data

Workshop and Conference on Logic, Algebra, and Arithmetic (2003 : Tehran, Iran)
    Logic in Tehran: proceedings of the Workshop and Conference on Logic, Algebra, and Arithmetic, held October 18-22, 2003 / edited by Ali Enayat, Iraj Kalantari, Mojtaba Moniri.
       p. cm. – (Lecture notes in logic ; 26)
    Includes bibliographical references.
    ISBN-13 978-1-56881-295-3 (acid-free paper)
    ISBN-10 1-56881-295-7 (acid-free paper)
    ISBN-13 978-1-56881-296-0 (pbk. : acid-free paper)
    ISBN-10 1-56881-296-5 (pbk. : acid-free paper)
    1. Logic, Symbolic and mathematical–Congresses. 2. Arithmetic–Congresses. I. Enayat, Ali, 1959- II. Kalantari, Iraj, 1947- III. Moniri, Mojtaba, 1961- IV. Title. V. Series.

QA9.A1W67 2003
511.3–dc22

2005058272

CRC Press
6000 Broken Sound Parkway, NW
Suite 300, Boca Raton, FL 33487
270 Madison Avenue
New York, NY 10016
2 Park Square, Milton Park
Abingdon, Oxon OX14 4RN, UK

Publisher's note: This book was typeset in LATEX, by the African Group for the ASL Typesetting Office, from electronic files produced by the authors, using the ASL document class asl.cls. The fonts are Monotype Times Roman. This book was printed by Friesens Corp, Altona, Manitoba, Canada, on acid-free paper. The cover design is by Richard Hannus, Hannus Design Associates, Boston, Massachusetts.

TABLE OF CONTENTS

Preface vii

Workshop and conference on logic, algebra, and arithmetic:
Tehran, October 18–22, 2003 ix

Foreword xi

Mohammad-Javad A. Larijani
    Mathematical logic in Iran: A perspective xiii

Seyed Masih Ayat and Mojtaba Moniri
    Real closed fields and IP-sensitivity 1

Seyed Mohammad Bagheri
    Categoricity and quantifier elimination for intuitionistic theories 23

Darko Biljakovic, Mikhail Kochetov, and Salma Kuhlmann
    Primes and irreducibles in truncation integer parts of real closed fields 42

Lou van den Dries
    On explicit definability in arithmetic 65

Ali Enayat
    From bounded arithmetic to second order arithmetic via
    automorphisms 87

Yuri L. Ershov
    Local-global principles and approximation theorems 114

Mehdi Ghasemi and Mojtaba Moniri
    Beatty sequences and the arithmetical hierarchy 126

Iraj Kalantari and Larry Welch
    Specker's theorem, cluster points, and computable quantum
    functions 134

Franz-Viktor Kuhlmann
    Additive polynomials and their role in the model theory of valued
    fields 160

Franz-Viktor Kuhlmann
  Dense subfields of henselian fields, and integer parts................ 204

Shahram Mohsenipour
  A recursive nonstandard model for open induction with GCD
  property and cofinal primes 227

Morteza Moniri
  Model theory of bounded arithmetic with applications to
  independence results 239

Zia Movahed
  Ibn-Sina's anticipation of the formulas of Buridan and Barcan 248

Anand Pillay
  Remarks on algebraic D-varieties and the model theory of
  differential fields... 256

Massoud Pourmahdian and Frank Wagner
  A simple positive Robinson theory with LSTP $\neq$ STP............... 270

Albert Visser
  Categories of theories and interpretations 284

PREFACE

We are pleased to present the proceedings volume of the invited and contributed talks delivered at the *Workshop and Conference on Logic, Algebra, and Arithmetic* that took place October 18-22, 2003 at the Institute for Studies in Theoretical Physics and Mathematics (IPM) in Tehran, Iran. The meeting provided a unique occasion for the exchange of ideas and methods in various frontiers of research in mathematical logic, and further strengthened the existing bonds of scientific cooperation between Iranian scholars and their colleagues abroad.

The meeting was made possible through the visionary leadership of the director of IPM, Professor M. J. A. Larijani, and the inspiring support of the head of the mathematics division at IPM, Professor G. B. Khosrovshahi. On behalf of all the participants, we also wish to acknowledge the industry and dedication of the staff at IPM, particularly Mr. M. Rahpeyma, Mr. M. Ashtiani, Mrs. N. Ramezani, and Ms. T. Parsa.

The papers that appear herein are in most cases revised and expanded versions of those that were originally presented at the meeting. We are grateful to the authors for their valuable contributions, patience, and cooperation during the refereeing and revision process. Furthermore, we offer a special note of gratitude to the numerous referees who assisted the authors and the editors in improving this volume.

Finally, we would like to extend our appreciation to the ASL for publishing this volume in the LNL series.

The Editors
Ali Enayat
Washington, DC
Iraj Kalantari
Macomb, Illinois
Mojtaba Moniri
Tehran

WORKSHOP AND CONFERENCE
ON
LOGIC, ALGEBRA, AND ARITHMETIC

TEHRAN, OCTOBER 18–22, 2003

- *Organizing committee*:
  M. Ardeshir, S. M. Bagheri, A. Enayat, I. Kalantari, M. J. A. Larijani, Mojtaba Moniri, Morteza Moniri (Coordinator), and M. Pourmahdian.

- *Invited speakers*:
  L. van den Dries, A. Enayat, Yu. L. Ershov, I. Kalantari, F.-V. Kuhlmann, S. Kuhlmann, A. Pillay, and A. Visser.

- *Registered participants* (*in addition to the above*):

M. Sh. Adib-Soltani	Sh. Etemad	A. Mohammadian
B. Afshari	A. Fallahi	Sh. Mohsenipour
M. Aghaei	H. G. Farahani	R. Moshtagh-Nazm
K. Aghigh	M. Ghari	Z. Movahed
S. Akbari	M. Ghasemi	L. Nabavi
M. Alizadeh	K. Ghayoom-Zadeh	P. Nasehpour
S. M. Amini	A. Haghany	M. R. Pournaki
S. Aqareb-Parast	H. Haghighi	N. Sadat-Rasooli
M. E. Arasteh-Rad	M. Hamideh	K. Reihani
S. M. Ayat	M. A. Hojjati	H. Sabzrou
M. Bargozini	S. A. Kalantari	P. Safari
M. Behzad	F. Karimi	M. Taherkhani
M. R. Changiz	M. Kashi	H. Talebi
A. R. Darabi	A. Madanshekaf	A. Taromi
F. Didehvar	M. Mahdavi-Hezavehi	M. Tousi
K. Divaani-Aazar	E. S. Mahmoodian	S. Yassemi
M. M. Ebrahimi	M. Mahmoudi	Rahim Zaare-Nahandi
J. S. Eivazloo	N. Mani	Rashid Zaare-Nahandi
B. Emamizadeh	S. H. Masoud	M. Zare
S. A. Esfahani	K. Mehrabadi	M. Zekavat

- *IPM support staff*:
  M. Ashtiani, N. Barati, A. Eslami, M. Hossein-zadeh Giv, M. Mashayekhi, T. Parsa, A. A. Rad, M. Rahpeyma, N. Ramezani, and A. Samie.

FOREWORD

Since the day of inception of the Institute for Studies in Theoretical Physics and Mathematics (IPM) in 1989, Mathematical Logic has been one of the main domains of activity at its School of Mathematics.

Briefly, through inviting a select number of prominent logicians from the republics of the former Soviet Union, activities in logic were initiated and the candle of interest was lit. This start was relatively successful in kindling interest in the subject in Iran. Furthermore, in those early days, an international congress and a summer school, both in logic, were organized and hosted by IPM, adding to the presence of the subject in Iran. Later, by establishing a Ph.D. program in Mathematical Logic at IPM, the cause was advanced with help from Iranian logicians educated abroad and the program in logic was consolidated. For more details, I refer you to the article in this volume by Professor Larijani, the Director of IPM.

Some three years ago, I proposed to our logic researchers at IPM the possibility of organizing another international workshop. This time our plan was focused on involving the Iranian logicians living abroad and getting help from them. We carried out this plan and it has been a complete success. Support from such logicians was a significant factor in the success of the workshop. I do not intend to add anything further about the workshop since I believe the list of its invited speakers and the list of its presentations provide sufficient manifestation of the success of the workshop.

But I would like to express, with great pleasure, my sincere gratitude to my friends, and IPM's fans, Iraj Kalantari and Ali Enayat. From distant places, they took time to participate in the planning and management of the workshop. Later, they continued in the same way and worked extensively with all, closing the distances through internet and email as if they were virtual officemates at IPM, to arrange for this volume to appear. I also wish to express my thanks to Mojtaba Moniri and Morteza Moniri who led the team of the local organizers.

A note about this proceedings: This is a truly significant contribution to the logic community in Iran and the rest of the world. All of the articles are bound to be of the highest caliber as they have been refereed with standards equal to those of the most prestigious journals. The three editors, Enayat,

Kalantari, and Mojtaba Moniri, indeed worked tirelessly to produce this wonderful volume. I thank ASL for arranging this proceedings to appear in the esteemed series of Lecture Notes in Logic.

Last but not least, I offer my deep appreciation to IPM and to other sponsors for financial support.

G. B. Khosrovshahi
Associate Director
and Head of School of Mathematics
IPM
July 2005

MATHEMATICAL LOGIC IN IRAN: A PERSPECTIVE

MOHAMMAD-JAVAD A. LARIJANI

§1. Introduction. As a member of the organizing committee for the Conference and Director of the Institute for Studies in Theoretical Physics and Mathematics (IPM), it is an honor for me to write this introductory essay. I am particularly pleased to witness the publication of the proceedings here in the LNL series. Although there were various research interests for Iranian logicians, we eventually decided to concentrate this conference on those areas of mathematical logic that significantly impact the fields of algebra, arithmetic, and analysis. The meeting was a reminder to all of us that, despite the tremendous increase in specialization and diversification, mathematical logic can still enjoy a vibrant sense of unity[1].

§2. Some local history.

2.1. A Revival. The history of philosophical logic in Iran dates back to the works of great thinkers such as Farabi (870-950 AD) and Ibn-Sina[2] (973-1037 AD). However, interest in *modern* logic is relatively recent. For example, the first monograph in mathematical logic published in Iran was Gholam-Hossein Mosaheb's rigorous text *Madkhalé Manteghé Soorat*[3] (Introduction to Formal Logic) in 1955. I am pleased to report that the last two decades have witnessed an extensive and dynamic growth of research interest in the field in Iran. I wish to put this progress in perspective by providing a sketch of its recent history (I apologize for any inadvertent omissions).

In the 1970s, the first generation of Iranians was schooled in mathematical logic: Iraj Kalantari (Cornell University, advisor: A. Nerode), A. Jalali-Naini (University of Oxford, advisor: R. Gandy), H. Lessan (Manchester University, advisor: G. Wilmers), and M. J. A. Larijani (University of California at Berkeley, advisor: R. Vaught). This list was extended in the 1980s and 1990s with E. Eslami (Iowa State University, advisor: A. Abian), Ali Enayat (University of Wisconsin, advisor: K. Kunen), Mojtaba Moniri (University of

---

[1]A particularly poignant example of this phenomenon is visible in the paper presented by Professor van den Dries, in which model theoretic methods are employed to establish an important result in complexity theory.

[2]Known as *Avicenna* in the West.

[3]Reviewed by L. A. Zadeh, J. Symbolic Logic, 22 (1957) pp. 354–355.

Minnesota, advisor: K. Prikry), M. Ardeshir (Marquette University, advisor: W. Ruitenburg), S. M. Bagheri (Université Claude Bernard Lyon-1, advisor: B. Poizat), and M. Pourmahdian (University of Oxford, advisor: F. Wagner). This list of mathematical logicians should also be supplemented with the names of M. Binaye-Motlagh (Göttingen, advisor: W. Hinz), Z. Movahed (University College, London, advisor: W. D. Hart), and H. Vahid (University of Oxford, advisor: D. Bostock) who received their schooling in modern philosophical logic.

The return of many of the aforementioned specialists from the US and Europe contributed to the development of standard coursework in mathematical logic, beginning with courses taught at Sharif University of Technology, and later at IPM, Tarbiat Modarres University, and Amir Kabir University of Technology. The first group of Master's students in mathematical logic completed their work in a diverse spectrum of topics under my supervision in the early to mid 1990s. This group consists of M. Ardeshir, F. Riazati, S. M. Bagheri, M. Pourmahdian, and F. Didehvar, all of whom later obtained their Ph.D. degrees in mathematical logic. By now, several dozen MS theses and half a dozen Ph.D.'s have been completed. The Ph.D. recipients in mathematical logic in Iran include M. Aghaei (advisor: M. Ardeshir), Morteza Moniri (advisor: Mojtaba Moniri), F. Didehvar (advisor: M. Ardeshir), M. Alizadeh (advisor: M. Ardeshir), S. Mohsenipour (advisor: A. Enayat), and J. S. Eivazloo (advisor: Mojtaba Moniri). Three Iranian logicians who have recently completed their Ph.D.'s abroad are F. Riazati (University of Florida, advisor: D. Cenzer), S. Salehi Pourmehr (University of Warsaw, advisor: Z. Adamowicz) and A. Togha (George Washington University, advisors: V. Harizanov and A. Enayat), all of whom received their initial schooling in mathematical logic in Iran.

Recent research work has included *philosophical logic* (the semantics of definite descriptions and proper names, theories of truth, nature of meaning in natural languages and mind, and philosophical problems of modal logic), *situational semantics* (the set theory it induces, and the model-theoretic view for natural languages that it offers), *general logic* (subsystems of classical logic, especially intuitionistic logic and basic logic, logic of natural languages, logic in computer science), *model theory* (stability theory, simple theories, model-theoretic algebra, models of arithmetic and set theory, o-minimality and its variants), *proof theory* (complexity of proofs, first order arithmetic, weak arithmetics and interactions with complexity theory, constructive mathematics, intuitionistic mathematics). There has also been research in computability theory, recursive analysis, set theory, and theoretical computer science (recently a School of Computer Science has been founded at IPM).

2.2. IPM's support of logic. The Institute for Studies in Theoretical Physics and Mathematics was founded in 1989, owing its success to a continuous

15-year period of campaigning and planning by many leading Iranian scientists, with the explicit goal of supporting research in mathematics and theoretical physics. Since the founding of IPM and its School of Mathematics, mathematical logic has been one of the main fields of interest. Beginning with 1992, IPM has supported research groups in mathematical logic, theoretical computer science, and philosophical logic, leading to the presentation of advanced courses and the supervision of doctoral dissertations, conference and workshop organization, and the publication of technical reports and lecture notes. Among the first series of courses in modern mathematical logic offered in Iran, one can mention those by Z. Movahed, M. J. A. Larijani, M. Ardeshir and H. Vahid. In the last decade and a half, several distinguished international scholars have visited Tehran to teach, present talks, and conduct collaborative research. In particular, IPM has hosted several logicians for long-term stays (Kanovei, Lyubetsky, Arslanov, Edalat, Enayat, Morozov, Goncharov, Palyutin) and many others for short-term stays (Baldwin, Poizat, van Dalen, M. van Atten, W. Veldman, W. Ruitenburg, P. Selinger). Moreover, IPM has helped organize three other international meetings[4] in logic. Mathematical logic is now an established discipline in Iran and our young logicians are teaching the subject at several institutes in the country. Their research papers are being published in high profile journals in the field and presented at international conferences. There have been over forty research papers supported by IPM research grants by the following authors: M. Aghaei, M. Alizadeh, M. Ardeshir, S. M. Ayat, S. M. Bagheri, F. Didehvar, J. S. Eivazloo, M. Ghasemi, S. Mohsenipour, Mojtaba Moniri, Morteza Moniri, M. Pourmahdian, S. Salehi Pourmehr, and H. Vahid. In summary, I am proud to witness IPM's role in the development and growth of the first *school* of mathematical logic in Iran.

Acknowledgments. I am grateful to my dear friend G. B. Khosrovshahi for his suggestion to write this introduction, and for his persistence and supervision. Thanks also to all the speakers and participants who helped us organize a successful meeting. External financial support for the meeting was provided by the Center for International Research and Collaboration (ISMO), and by the Institute for Research and Planning in Higher Education. I also wish to

---

[4]The other meetings were:

(a) The First Congress of Logic was held in 1990, its Proceedings was published by IPM in 1993 (for its table of contents, see http://www.ipm.ac.ir/IPM/publications/table_of_contents/logic.pdf).

(b) A two-week School on Set Theory was held by IPM at Shahid Beheshti University in 1991. This meeting was co-sponsored by ASL, and included several distinguished lecturers. See the report by A. Enayat, *Journal of Symbolic Logic*, vol. 58 (1993), p. 1476.

(c) The First International Seminar on Philosophy of Mathematics in Iran was held on October 17, 2001 at Shahid Beheshti University and was partially supported by IPM (invited speakers included C. Parsons, M. Detlefsen, M. van Atten and W. Veldman).

thank World Scientific publishers for supplying us with a number of books and sample copies of journal issues.

INSTITUTE FOR STUDIES IN THEORETICAL
PHYSICS AND MATHEMATICS
TEHRAN, IRAN
E-mail: larijani@ipm.ir

REAL CLOSED FIELDS AND IP-SENSITIVITY

SEYED MASIH AYAT AND MOJTABA MONIRI

Abstract. This work is concerned with the extent to which certain topological and algebraic phenomena, such as the mod one density of the square roots of positive integers, or the existence of irrationals and transcendentals, generalize from the standard setting of $\langle \mathbb{R}, \mathbb{Z} \rangle$ to structures of the form $\langle F, I \rangle$, where F is an ordered field and I is an integer part (IP) for F. Here are some of the highlights of the paper:

(i) Given an ordered field F, either all or none of the IP's for F (if any) are models of Open Induction, depending on whether F is dense in its real closure or not.

(ii) Density mod one of a subset is sensitive to a certain relaxation of the notion of an IP in non-Archimedean real closed fields with countable IP's. The weakened notion is that of (additive subgroup) *integer sets* in the sense of J. Schmerl.

(iii) There is a real closed field F with two IP's I_1 and I_2 such that for some $\alpha \in I_2$ which is an algebraic irrational with respect to I_1, the set $\{ u\alpha - \lfloor u\alpha \rfloor_{I_1} \mid u \in I_1 \}$ is dense in $[0, 1)_F$.

(iv) All real closed fields F of infinite transcendence degree have proper dense real closed subfields over which F can be of any positive transcendence degree, up to and including that of the transcendence degree of F (they are real closures of suitable proper dense subfields also). Together with a result of van den Dries on the existence of real closed fields F of infinite transcendence degree which possess certain IP's I with the peculiar property that every element of F is rational with respect to I, this demonstrates the IP-sensitivity of existence of transcendentals in F.

(v) Certain classes of real closed fields of infinite transcendence degree such as Puiseux series, and all real closed fields of finite transcendence degree, have algebraic irrationals relative to every IP. Also, Cauchy complete real closed fields have transcendentals relative to any IP.

(vi) For all $n \in \mathbb{N}^{\geq 1}$, there are real closed fields F and F' of transcendence degree n such that F is algebraic over all its proper dense subfields, but F' is algebraic over a proper dense subfield of itself and transcendental over another proper dense subfield of itself. This and the previous two issues are also dealt with in the second paper of F.-V. Kuhlmann in the present volume.

§1. Introduction. We begin with some background, motivation and connections to the literature.

1.1. Integer parts, models of OI, and integer sets. A discrete subring I of an ordered field F is an *integer part* (IP) for F if for any $x \in F$, there exists $a \in I$

2000 *Mathematics Subject Classification.* Primary 03H15, 12L15; Secondary 03C62, 11J99.

Key words and phrases. Real Closed Field, Integer Part, Open Induction, Dense Mod One, Integer Set, Proper Dense Subfield, Liouville's Theorem, Perfect Set of Irrationals.

Logic in Tehran
Edited by A. Enayat, I. Kalantari, and M. Moniri
Lecture Notes in Logic, 26

such that $a \leq x < a + 1$, this necessarily unique element a is called the *integer part* of x (with respect to I), written $a = \lfloor x \rfloor$ or $a = \lfloor x \rfloor_I$. Every real closed field has an integer part, see [MR]. The IP's for real closed fields are models of *Open Induction* (OI), a weak fragment of first order Peano Arithmetic in the language $\{+, \cdot, <, 0, 1\}$ in which the induction scheme is restricted to quantifier-free formulas with parameters. Conversely, models of OI are IP's for the real closures of their fraction fields. These facts were originally proved by J. Shepherdson in 1964, see [D]. On the other hand, there are p-real closed fields (for any prescribed p) without any IP's, see [B].

For F an ordered field, $RC(F)$ will denote the real closure of F. Consider the ordered field $P^a_{RC(\mathbb{Q})}(t)$ of Puiseux series with ascending exponents and real algebraic coefficients in terms of an infinitely small indeterminate t. This field is the union of Laurent series fields over the same field of coefficients in terms of $t^{\frac{1}{n}}$ for $n \in \mathbb{N}^{\geq 1}$. It is a real closed field of transcendence degree continuum. Here there are IP's included in $RC(\mathbb{Q}(t))$ and so of transcendence degree 1: Shepherdson's IP is defined as

$$S_t(\mathbb{Z}) = \left\{ \sum_{i=1}^{n} a_i t^{-r_i} + a_0 \mid n \in \mathbb{N}, a_i \in RC(\mathbb{Q}), a_0 \in \mathbb{Z}, r_i \in \mathbb{Q}^{>0} \right\}.$$

This is the model of OI obtained by applying Shepherdson's end-extension method, which we call S, to the standard model $\mathbb{Z}$ using the indeterminate t, see [D, Proposition 3.4]. If we extend with respect to an infinitesimal element z in $RC(\mathbb{Q}(t))$ instead of t itself, we get another IP denoted $S_z(\mathbb{Z})$ for $RC(\mathbb{Q}(t)) = RC(\mathbb{Q}(z))$, see [B, p. 326]. So there are infinitely many IP's for $RC(\mathbb{Q}(t))$, and all of them are also IP's for $P^a_{RC(\mathbb{Q})}(t)$.

We will also be considering Puiseux series fields with coefficients in other real closed fields (such series will always be with ascending exponents in this paper, so the indeterminate will be infinitely small with respect to the field of coefficients).

We will need the model for OI of transcendence degree $\aleph_0$ that van den Dries constructed in [D, p. 352]. It has a real closed fraction field (so there will be no irrationals there in the real closure with respect to that IP). He took the ω-th iteration of the Shepherdson end-extension method applied to $\mathbb{Z}$. Note that fraction field of the $(j + 1)$-st iteration contains real closure of (fraction field of) the j-th one, making the fraction field of the limit model real closed. We denote this model by $S^\omega(\mathbb{Z})$.

If S is a subset of F containing 0 that uniquely approximates each element of the field from below with an error less than 1, we say S is an *integer set* (IS) for F. As proved by J. Schmerl, every ordered field has an IS which is an additive subgroup, see [Sch, Proposition 1.2] (Schmerl's original definition took the nonnegative part of what we are calling an IS here).

1.2. IP-sensitivity: Motivation. The real field has the unique integer part $\mathbb{Z}$. Indeed $\mathbb{Z}$ is the unique IS for any Archimedean ordered field.[1] Non-Archimedean real closed fields, on the other hand, have many IP's. We study IP-sensitivity of some of the properties of $\langle \mathbb{R}, \mathbb{Z} \rangle$ in the case of a general $\langle F, I \rangle$ (the notation $\langle F, I \rangle$ is used throughout the paper to denote an ordered field F, most often real closed, equipped with an integer part I). The properties of interest will sometimes be first order expressible in the signature $\langle F, I \rangle$ and sometimes not so (even for the first order case, there could be sensitivity as the IP's are generally not elementary equivalent and if they are by any chance, still when paired with the field F, may form pairs $\langle F, I_1 \rangle$ and $\langle F, I_2 \rangle$ which are not elementary equivalent). We will be mainly concerned with whether a given set $S \subset F$ is dense mod one in subsection 2.1, and with the existence of irrationals and transcendentals in section 4. Let us say a little more on what we mean by the distinction that gives rise to the topic of the paper.

Given F, it will be nice to know whether any of the mentioned properties (or others) pertaining to a subset of F (including F itself) that holds with respect to a certain IP for F, will continue to hold with respect to any other IP for F. We call such a phenomenon *IP-absoluteness*. We will have notions such as *absolutely dense mod one*, for subsets S of F, and *absolute existence of irrationals*, for F itself. If IP-absoluteness fails given F (and if relevant, also S) for a property which depends on an IP for F, then we have an instance of the *IP-sensitivity* phenomenon.

Properties in terms of F could be IP-absolute in the positive (i.e., the same property holds in the reals with the integers) or negative (i.e. the negation holds in the standard case), or could be IP-sensitive[2]. The property $\mathrm{Frac}(I) \subseteq_{\mathrm{dense}} F$ is trivially IP-absolute in the positive. For a trivially IP-absolute property in the negative, consider existence of I-transcendental elements in $F = RC(\mathbb{Q}(x))$, with x infinitely small.

We will see more interesting instances in this paper, e.g., that of IP-absoluteness in the positive of existence of elements in $P^a_{RC(\mathbb{Q})}(t)$ which are I-algebraic-irrational. We also have interesting IP-sensitive instances: let F be the fraction field of the model constructed by van den Dries as defined above. Then the property of existence of irrationals is implied by our results to be IP-sensitive.

---

[1]The IS has 0, and the next number to the right cannot be less than 1 since otherwise sufficiently close points to the right of that second point would have more than one approximations from below within I in the IS. The second point cannot be more than 1 either since otherwise itself would have no such approximation. Similar observations apply to the negative elements of the IS. The same goes on cofinally in the Archimedean case and indeed for the Archimedean-limited part of any F.

[2]Of course one can also quantify, so to speak, on F in addition to on I. But as the title of the paper shows, we are not considering such variations here.

There seems to be a lot of interesting potential pursuits in this regard; here we wish to briefly discuss a particularly intriguing one. The issue is meant to provide more motivation for the general topic and will not be directly used or pursued further in this paper.

Beatty's theorem from the 1920's (rediscovered by others later, e.g., by Skolem and Bang in the 1950's) says that if α and β are positive reals, then the sets of integer parts of multiples of α and β by the positive integers (called the Beatty sequences corresponding to α and β) make a partition of $\mathbb{N}^{>0}$ if and only if α and β are irrational and $\frac{1}{\alpha} + \frac{1}{\beta} = 1$, see [N, Theorem 3.7]. It is not difficult to see that the *if* part of this theorem holds for general $\langle F, I \rangle$. We end this subsection with an interesting property sufficient to prove the *only if* part of the Beatty theorem for general $\langle F, I \rangle$ (yet the status of this property needs to be clarified). An important feature which is missing in the passage from the classical case to the general $\langle F, I \rangle$, is *representation in bases* (e.g., the decimal representation). A good alternative in lieu of base representations in general for an element α in $\langle F, I \rangle$ would be the Beatty *sequence* (or multiset, where repetitions count) of all $\lfloor n\alpha \rfloor_I$ for $n \in I$. Notice that classically the n-th digit (for $n \in \mathbb{N}^{\geq 1}$) of α in base say 10 is recoverable from the Beatty sequence of α, since it equals $\lfloor 10^n \alpha \rfloor - 10 \lfloor 10^{n-1} \alpha \rfloor$. It would be nice to figure out whether the Beatty multiset corresponding to α in $\langle F, I \rangle$ has all the information in α (to be distinguished in F) in general, i.e., whether different elements of F necessarily have different Beatty multisets.

1.3. Cauchy completions, generalized power series, and continued fractions.
Here we discuss more preliminaries that we will need. As shown in [Sco, Theorem 1], any ordered field F has a maximal ordered field extension in which F is dense, and this extension is unique up to an ordered field isomorphism which is the identity on F. Such an extension is called the *Cauchy completion* of F, denoted $CC(F)$ (the name *Scott completion* is also used in the literature for Cauchy completions of ordered fields). By [Sco, Theorem 2], $CC(F)$ is real closed if and only if F is dense in the $RC(F)$.

For any real closed field F and nonzero divisible ordered Abelian group G, the ordered field $[[F^G]]$ of generalized power series with exponents in G and coefficients in F is real closed and Cauchy complete (for this latter property see [ME, Theorem 3.2], on the other hand Puiseux series fields are never Cauchy complete, see [ME, Proposition 2.2(ii)]). The series in $[[F^G]]$ are essentially functions $G \to F$ whose supports are well ordered in G, with the obvious $+, \cdot,$ and $<$. There is also the corresponding formal sum representation for such series in terms of an indeterminate. Each series will have its infinitely large part with respect to F (terms with exponents in $G^{<0}$), its single term which has exponent 0 in G (which can be regarded as an element of F), and finally the part which is infinitely small with respect to F (terms with exponents in $G^{>0}$). Given an IP such as I for F, the corresponding Shepherdson IP for

$[[F^G]]$ will consist of series not having the third (F-infinitesimal) part and whose second (constant) parts are in I.

Some very basic terminology and facts regarding continued fractions in the real case will be used throughout the paper. This includes *convergents*, which provide us with cofinal quadratic rational approximations (convergent approximations of the form $|\alpha - \frac{p}{q}| < \frac{1}{q^2}$); see, e.g., [R, Section 7.1, Theorem 1.5]. One easily observes that such basic facts will still hold in the non-classical cases. Continued fractions have been considered in the literature for certain fields of power series, see, e.g., [L]. We will prove and use a general Liouville's Theorem, see [R, Section 8.1, Theorem 1.1] whose proof there is indeed valid for all real closed fields with an IP (certain facts such as the mean value theorem for polynomials being available).

1.4. The old friend DENSE IN RC matching a new dichotomy. The distinction between extensions of real closed fields which are dense and those that are not has been long known to affect related problems in model theoretic algebra. E.g., the theory of real closed fields with a distinguished proper dense real closed subfield is complete (as proved by A. Robinson in the 1950's and generalized by A. Macintyre in the 1960's), while that of real closed fields with a distinguished proper real closed subfield is undecidable (as proved by W. Baur in the 1980's). Here is another effect of the distinction where the smaller field is not necessarily real closed, a dichotomy for ordered fields with IP's:

PROPOSITION 1.1. *Suppose F is an ordered field which has IP's. Then we have the following*:

(a) *If F is dense in $RC(F)$, then all IP's for F are IP's for $RC(F)$ too. Therefore all IP's for F are models of OI.*

(b) *If F is not dense in $RC(F)$, then no IP's for F can be an IP for $RC(F)$; indeed no IP I for F can be an IP for $RC(\mathrm{Frac}(I))$ and therefore no IP's for F can be a model of OI.*

PROOF. (a) Clear.

(b) Let I be an IP for F. Then F, as a dense field extension of $\mathrm{Frac}(I)$, densely embeds in $CC(\mathrm{Frac}(I))$. Suppose for the purpose of a contradiction that I is an IP for $RC(\mathrm{Frac}(I))$. Then $\mathrm{Frac}(I)$ would be dense in $RC(\mathrm{Frac}(I))$. Therefore by [Sco, Theorem 2] that we mentioned in the above subsection, $CC(\mathrm{Frac}(I))$ would be real closed. In that case F, as a dense subfield of the real closed field $CC(\mathrm{Frac}(I))$, would clearly be dense in its real closure (the latter being embeddable in $CC(\mathrm{Frac}(I))$). ⊣

So for any F, either all or none of the IP's for F are models of OI (depending on whether F is dense in $RC(F)$ or not). We choose to concentrate in this paper on the study of real closed fields with an IP instead of say ordered fields dense in their real closure with an IP, although the latter would have still kept us with OI (some of the results may actually hold for this more general setting).[3]

1.5. Connections to a paper of F.-V. Kuhlmann in this volume. We point out the paper by Franz-Viktor Kuhlmann, [K2]. To some of the results in the Henselian case there, relate some of ours in Subsections 3.1 and 4.1. Here is a non-comprehensive list of connections:

- Regarding [K2, Theorem 1], we prove in Theorem 3.1 a similar statement when transcendence degree is infinite. There we show proper dense real closed subfields over which the field is of a prescribed transcendence degree at most that of itself. We show in Theorem 3.2 that for all $n \in \mathbb{N}^{\geq 1}$, there are real closed fields F and F' of transcendence degree n with F algebraic over all its proper dense subfields but F' algebraic over some but not all of its own. An easy minimality argument shows that every real closed field of finite transcendence degree has a dense real closed subfield L algebraic over *all* proper dense subfields of L. This minimal dense real closed subfield may or may not be proper. Compare with [K2, Proposition 24].

- Regarding [K2, Theorem 5], we prove in Corollary 4.2(i) that all Puiseux series fields over real closed fields enjoy what we call absolute existence of *algebraic* irrationals.

Also, Kuhlmann has mentioned a question elsewhere [K] which asks whether there is a (non-Archimedean) real closed field larger than, but still of the same cardinality as, fraction fields of all its IP's. Our results will show that the countable field $RC(\mathbb{Q}(x))$ (with x infinitely small) is as desired, as it has elements whose finite powers are cofinal, see Lemma 4.1.

§2. Dense Mod One subsets and variation of the IP.

In this section, we first define Dense Mod One (abbreviated DMO) subsets in the natural fashion. Then we define Absolutely DMO (abbreviated ADMO) subsets, and give some examples. We are unable to figure out whether the property ADMO is indeed stronger than DMO, however we prove a couple of partial results in this direction. We then give examples of pairs of IP's whose mutual differences, as well as their intersections, are cofinal and at the same time, their fraction fields coincide. We also give two distinct IP's for a certain F which intersect trivially, i.e., on the standard integers only.

2.1. Absolutely DMO subsets. The property of being uniformly distributed modulo one for certain sequences of reals is of much interest in Diophantine approximations and metric number theory. The same is true to some extent

---

[3]Incidentally, in [Mc, Theorem 5] an axiomatization for the class of ordered fields which are dense in their real closure was given (the instances of the scheme added to the ordered field axioms express that a 1-variable polynomial over F that changes sign on an interval in F, must come arbitrarily close to 0 on that interval in F).

for the weaker notion of certain merely countable subsets of the reals (which are not enumerated) being dense mod one. For example the sequences $\{\sqrt{n} \mid n \in \mathbb{N}\}$ and $\{n\alpha \mid n \in \mathbb{N}\}$ of the reals, where in the second sequence $\alpha \in \mathbb{R}$ is irrational, are uniformly distributed mod one and so their ranges are dense mod one. We will see that the conclusion for square roots easily generalizes to the context we are considering but the one for multiples, see [N, Theorem 3.2], is a deeper phenomenon.

A set $S \subset F$ is *dense mod one* in $\langle F, I \rangle$ (DMO for short) if the corresponding set of fractional parts $\{s - \lfloor s \rfloor_I \mid s \in S\}$, is dense in $[0, 1)_F$. For an irrational $\alpha > 0$ in $\langle F, I \rangle$, we say that $DMO(\alpha)$ holds whenever the set $I^{\geq 0}\alpha = \{n\alpha \mid n \in I^{\geq 0}\}$ is DMO (with respect to I). A set $S \subset F$ is *absolutely dense mod one* (ADMO for short) in F if it is DMO with respect to all IP's for F.

PROPOSITION 2.1. *For any $\langle F, I \rangle$ and $p \in \mathbb{N}^{\geq 2}$, the set $\{\sqrt[p]{u} \mid u \in I^{>0}\}$ is ADMO in F.*

PROOF. Let I_1 be any IP for F. Pick arbitrary $k, t \in I_1^{\geq 0}$ with $k < t$. We need to find $M \in I$ and $n \in I_1$ such that $\frac{k}{t} < \sqrt[p]{M} - n < \frac{k+1}{t}$, i.e. $(n + \frac{k}{t})^p < M < (n + \frac{k+1}{t})^p$. But $(n + \frac{k+1}{t})^p - (n + \frac{k}{t})^p > \frac{p}{t}(n + \frac{k}{t})^{p-1}$ and we can choose $n \in I_1$ such that $\frac{p}{t}(n + \frac{k}{t})^{p-1} > 1$ (say by taking $n > \sqrt[p-1]{\frac{t}{p}}$). Then any $M \in I$ in the interval above works (the existence of such an M follows from the fact that the length of the interval around it as above is greater than 1).   ⊣

PROPOSITION 2.2.   (i) *If S is a dense subset of an interval of F of length 1, then S is ADMO.*

(ii) *If S is DMO with respect to some IP and its elements are all pair-wise finitely apart, then S is ADMO.*

(iii) *Take a non-Archimedean $\langle F, I \rangle$. Recall that a $\mathbb{Z}$-chain in I is an equivalence class under being finitely apart. Fix a cofinal net $(C_\alpha)_{\alpha \in J}$ of $\mathbb{Z}$-chains in I and take $n_\alpha \in C_\alpha$. Then $S = \{\frac{p}{n_\alpha} \mid \alpha \in J, p \in I^{\geq 0}, \frac{p}{n_\alpha} \in [n_\alpha, n_\alpha + 1)\}$ is ADMO.*

PROOF. (i) Every IP will have a point α in such an interval. Then S and $S^{\geq \alpha} \cup (S^{<\alpha} + 1)$ (which coincide if α is the initial point of the interval) mod-one-reduce to the same dense set.

(ii) The set S is included in a galaxy A in F. Fix two IP's such as I_1 and I_2 for F and assume S is DMO with respect to I_1. All elements of $I_2 \cap A$ have the same fractional part $\lambda \in [0, 1)_F$ with respect to I_1. Those elements in S that contribute to being DMO with respect to I_1 in the interval $[0, \lambda)$ will do so for the interval $[1 - \lambda, 1)$ with respect to I_2. Also those elements in S that contribute to being DMO with respect to I_1 in the interval $[\lambda, 1)$ will do so for the interval $[0, 1 - \lambda)$ with respect to I_2. Therefore S will be DMO with respect to I_2.

(iii) Every IP will have a point m_α in such an interval $[n_\alpha, n_\alpha + 1)$. Then $S \cap [n_\alpha, n_\alpha + 1)$ and $(S \cap [n_\alpha, n_\alpha + 1))^{\geq m_\alpha} \cup ((S \cap [n_\alpha, n_\alpha + 1))^{< m_\alpha} + 1)$ (which coincide if $m_\alpha = n_\alpha$) mod-one-reduce to the same set which is *dense within* $\frac{1}{n_\alpha}$. We are in the non-Archimedean context and these get arbitrarily fine.   ⊣

We do not know whether it is possible for I_1 to be DMO with respect to I_2.[4] If so, it would in particular imply that ADMO is stronger than DMO. What we can say is that in such a case I_2 would be DMO with respect to I_1 too: For all $0 < a < b < 1$ in F, there exist $m \in I_1$ and $n \in I_2$ such that $1 - b < m - n < 1 - a$ and so $a < n - (m - 1) < b$. It is also noteworthy to observe that in the mentioned situation the set obtained by choosing just one element from every $\mathbb{Z}$-chain of I_1 will remain DMO with respect to I_2 and any two of its elements are infinitely apart.

Since we are unable to locate an example where two distinct IP's are DMO with respect to each other (if possible at all), we pursue a weakening: whether there are any (additive subgroup) IS's which are DMO with respect to an IP. We first observe that for any non-Archimedean $\langle F, I \rangle$, there exists an IS which is DMO (with respect to I). The reason goes as follows. There are as many $\mathbb{Z}$-chains in I as I-rational elements in $(0, 1)_F$ (they are both equipotent to I). Put these in a one to one correspondence. Then the set J obtained from I by (fixing the standard $\mathbb{Z}$-chain $\mathbb{Z}$ in I and) shifting every nonstandard $\mathbb{Z}$-chain there by the I-rational in $(0, 1)_F$ matching it, is as desired. Below we improve this when the IP is countable by showing such a DMO IS could even be an additive subgroup of the field:

THEOREM 2.3.   (i) *In all non-Archimedean real closed fields with a countable IP such as I, there exist additive subgroup IS's which are DMO with respect to I.*

(ii) *Suppose I is an IP for a non-Archimedean real closed field F, $\alpha \in F$ is I-irrational, and $DMO(\alpha)$ holds with respect to I. Then there exists an element $k \in I^{>0}$ and an additive subgroup IS such as J such that $I^{>0}(\alpha + k) \subseteq J$ (and so J is DMO with respect to I).*

(iii) *There are structures $\langle F, I_1, I_2 \rangle$, with F real closed, such that there is an element $\alpha \in I_2 \setminus \text{Frac}(I_1)$ for which $DMO(\alpha)$ holds with respect to I_1.*

PROOF. (i) Schmerl's result mentioned earlier in subsection 1.1 that there are always additive subgroup IS's follows from first observing that any maximal discrete (in the sense that 1 is the only element x such that $0 < x \leq 1$) additive subgroup of an arbitrary ordered field is closed, which means that it contains all elements of the additive divisible hull of the group inside the field not destroying discreteness. From the latter property one can then deduce that a maximal discrete additive subgroup is an IS by showing that if G is a closed discrete additive subgroup and $a \in F$ is such that for no $b \in G$

---

[4] We do not have examples of distinct IP's which fail to be DMO with respect to each other either.

do we have $a \leq b \leq a + 1$, then $G \cup \{a\}$ generates a discrete additive subgroup.

If we start with a set which is DMO with respect to an IP I and which generates a discrete additive subgroup, then the latter group could certainly be extended to a maximal discrete subgroup and we will have an additive subgroup IS which is DMO with respect to I. This will happen if we choose a set that is DMO with respect to I any pair of whose elements have all their standard integer multiples infinitely apart. For then any linear combination $a_1 x_1 + a_2 x_2 + \cdots + a_n x_n$ for $n \in \mathbb{N}^{\geq 2}$, $a_i \in \mathbb{Z} \setminus \{0\}$, and the x_i in the chosen set, will be infinite (positive or negative).

Enumerate the countable set of I-rational elements in $(0, 1)_F$ to get the sequence $(r_n)_{n \in \mathbb{N}}$. Take an infinite element $M \in I^{>0}$. Then the set $\{M^n + r_n \mid n \in \omega\}$ is as desired.

(ii) Take an element $k \in I^{>0}$ such that $\alpha + k$ is infinitely large in F. Any linear combination with standard integer coefficients of elements of the DMO set $I^{>0}(\alpha + k)$ is an element of that set which is either 0 or infinite. So the previous method applies again.

(iii) As mentioned more generally before, for $z = \sqrt{t^2 + t} \in RC(\mathbb{Q}(t))$ we have the IP $S_z(\mathbb{Z})$ for $RC(\mathbb{Q}(z)) = RC(\mathbb{Q}(t))$. Note that $z^{-1} = \frac{t^{-1}}{\sqrt{t^{-1}+1}} \in S_z(\mathbb{Z}) \setminus \operatorname{Frac}(S_t(\mathbb{Z}))$ (the irrationality assertion is proved more generally and independently in Proposition 3.3 below) and $\beta = \sqrt{t^{-1} + 1}$ has the continued fraction $\langle t^{-\frac{1}{2}}, 2t^{-\frac{1}{2}} \rangle$. This continued fraction tends to β in the scale of the field $RC(\mathbb{Q}(t))$ (as one easily checks that the denominators of the convergents are unbounded). Hence there exists an $S_t(\mathbb{Z})$-cofinal sequence $(Q_n)_{n \in \omega}$ in $S_t(\mathbb{Z})^{>0}$ such that for all n, there exists $P_n \in S_t(\mathbb{Z})^{>0}$ with $0 < P_n - Q_n\beta < \frac{1}{Q_n}$ and so $0 < P_n(\frac{t^{-1}}{\sqrt{t^{-1}+1}}) - t^{-1}Q_n < \frac{t^{-1}}{Q_n\beta}$. Given $\varepsilon > 0$ in the field, let $n \in \omega$ be such that $\frac{t^{-1}}{Q_n\beta} < \varepsilon$, $a = P_n$ and $b = t^{-1}Q_n$. Then $0 < a(\frac{t^{-1}}{\sqrt{t^{-1}+1}}) - b < \varepsilon$. Hence $DMO(z^{-1})$ holds with respect to $S_t(\mathbb{Z})$ since:

CLAIM. If $\alpha \in F$ is such that for all $\varepsilon \in F^{>0}$, we have $\{n\alpha - \lfloor n\alpha \rfloor_I \mid n \in I\} \cap (0, \varepsilon) \neq \emptyset$, then $DMO(\alpha)$ holds.

PROOF OF CLAIM. Given $0 < u < v < 1$ in F, for $\varepsilon = v - u$, let $n \in I^{>0}$ be such that $n\alpha - \lfloor n\alpha \rfloor_I < \varepsilon$. Then the interval $(\frac{u}{n\alpha - \lfloor n\alpha \rfloor_I}, \frac{v}{n\alpha - \lfloor n\alpha \rfloor_I})$ is of length greater than 1 and so contains some $k \in I^{>0}$. That gives us $0 < u < kn\alpha - k\lfloor n\alpha \rfloor_I < v < 1$ and from there $\lfloor kn\alpha \rfloor_I = k\lfloor n\alpha \rfloor_I$ and by putting $m = kn$, we have $u < m\alpha - \lfloor m\alpha \rfloor_I < v$. ⊣

2.2. Intersections of IP's and their fraction fields. In part (iii) of the proof just above, the fraction fields of the two IP's have cofinal intersection (consider, e.g., even powers of z). They also have cofinal difference sets. As we shall see now, two IP's themselves may have cofinal intersection and difference sets.

The situation where two IP's have nothing more in common than the standard integers is also achieved.[5]

THEOREM 2.4.   (i) *There is a real closed field with two IP's I_1 and I_2 such that the differences $I_1 \setminus I_2$ and $I_2 \setminus I_1$, as well as the intersection $I_1 \cap I_2$, are all cofinal and* $\mathrm{Frac}(I_1) = \mathrm{Frac}(I_2)$.

(ii) *There is a non-Archimedean real closed field with two IP's I_1 and I_2 such that $I_1 \cap I_2 = \mathbb{Z}$.*

PROOF. (i) First we give an easier example when two IP's for a real closed field simultaneously have cofinal intersection and cofinal mutual differences, not considering their fraction fields. Consider the second Shepherdson extensions $\mathcal{S}_x(\mathcal{S}_t(\mathbb{Z}))$ and $\mathcal{S}_x(\mathcal{S}_z(\mathbb{Z}))$, this time with respect to the same indeterminate x (z is as before). They both are IP's for the same real closed field $RC(\mathcal{S}_x(\mathcal{S}_t(\mathbb{Z}))) = RC(\mathcal{S}_x(\mathcal{S}_z(\mathbb{Z})))$. To see that they have a cofinal intersection, consider series with *constant* term 0 (series all whose terms involve x). On the other hand, they have cofinal differences, e.g., elements with *constant* term t^{-1} belong to the former and are cofinal there but they do not belong to the latter.

To get the stronger statement involving fraction fields, once again let $z = \sqrt{t^2 + t}$ and consider $\mathcal{S}_t(\mathbb{Z})$ and $\mathcal{S}_z(\mathbb{Z})$. Note that the ω-th Shepherdson's extensions $\mathcal{S}^\omega(\mathcal{S}_t(\mathbb{Z}))$ and $\mathcal{S}^\omega(\mathcal{S}_z(\mathbb{Z}))$ of these models of OI, with respect to the same sequence of indeterminates from that point on, continue to have cofinal intersection and also cofinal mutual differences. Those extensions have real closed fraction fields both containing t and z and this makes the fraction fields equal.

(ii) Recall that for an ordered field K, $P_K^a(x)$ is a real closed field if (and only if) K is so. If K is a real closed subfield of $\mathbb{R}$, then we have Shepherdson's IP, which could also be written as $K[x^{\mathbb{Q}^{<0}}] + \mathbb{Z}$, for $P_K^a(x)$. Now suppose $\{\pi_i : i \in \mathbb{N}\}$ is an algebraically independent set of distinct reals and put $L = RC(\mathbb{Q}(\{\pi_i : i \in \mathbb{N}\}))$. Let I_1 be Shepherdson's IP for $P_L^a(x)$. Let I_2 be the IP constructed by Berarducci and Otero for the same, see [BO] (disregarding recursiveness issues). Note that the constant terms of infinite elements of the latter model are nonzero polynomials with rational coefficients of the algebraically independent numbers above, so they are not in $\mathbb{Z}$. Hence I_1 and I_2 have no infinite elements in common.   ⊣

§3. **Existence of dense subfields.** There are ordered fields (necessarily of finite transcendence degree) with no proper dense subfields, e.g. $\mathbb{Q}(x)$, with x infinitely small (transcendence degrees will refer to those over $\mathbb{Q}$ unless stated otherwise). Every real closed field is the real closure of a proper subfield: Let F

---

[5]It is also possible that $\mathrm{Frac}(I_1) \subsetneq \mathrm{Frac}(I_2)$: let I_2 be van den Dries's model of OI, that is the ω-th Shepherson's extension of $\mathbb{Z}$ and I_1 an IP constructed as in Proposition 3.4 ahead for the real closed field $\mathrm{Frac}(I_2)$ of infinite transcendence degree.

be a real closed field with a transcendence basis B over $\mathbb{Q}$ and put $K = \mathbb{Q}(B)$. Clearly the real closure of K is F. Observe that K is not real closed (if $B \neq \emptyset$, then K is a purely transcendental extension of the rationals ordered as a subfield of F) and so proper in F. We prove a stronger statement for the case of infinite transcendence degree in the first subsection: all real closed fields of infinite transcendence degree have proper dense subfields where the extension is of any transcendence degree at most that of the field. Observing that purely transcendental extensions of finite degree of the rationals may have their transcendence bases ordered so that the resulting real closure does or does not have proper dense real closed subfields, we note that in the positive case some of the proper dense real closed subfields will be algebraic over all of their own proper dense subfields.

In the second subsection, we establish the not necessarily absolute existence of algebraic irrationals in real closed power series fields; and full size sets of elements which are algebraically independent over an IP, in arbitrary real closed fields of infinite transcendence degree.

3.1. Dense subfields and real closures. Let F be a real closed field and I an IP for F such that F is not algebraic over $\mathrm{Frac}(I)$. Then obviously there exists an extension (a purely transcendental one) of $\mathrm{Frac}(I)$ such as K which is a (proper dense) subfield of F and F is algebraic over K. If F is a proper algebraic extension of $\mathrm{Frac}(I)$, again F is the real closure of the proper dense subfield $\mathrm{Frac}(I)$. On the other hand, if $\mathrm{Frac}(I) = F$, then in this way we do not obtain F as the real closure of a proper dense subfield. Also, in the first case of the three, for any nonzero cardinal $\alpha \leq \mathrm{tr.deg.}\,[F : \mathrm{Frac}(I)]$, there exists a (proper) dense (indeed extending $\mathrm{Frac}(I)$) real closed subfield K of F such that $\mathrm{tr.deg.}\,[F : K] = \alpha$. We have the following couple of results when in issues similar to the above, transcendence degrees are considered over $\mathbb{Q}$.

THEOREM 3.1. *Let F be a real closed field of infinite transcendence degree, then*

(i) *F is the real closure of some of its proper dense subfields, and*
(ii) *For any nonzero cardinal $\alpha \leq \mathrm{tr.deg.}\,(F)$, F contains a dense real closed subfield K such that F is of transcendence degree α over K.*

PROOF. (i) It was shown in [EGH, Lemma 2.3] that for uncountable ordered fields F (equivalently those of uncountable transcendence degree), there are transcendence bases B dense in F. The same is true when the transcendence degree is countably infinite (a subclass of the countably infinite ordered fields). For convenience (and as we need it for part (ii) also) we bring the inductive argument here for the case $\mathrm{tr.deg.}\,(F) = \omega$, but it will be very similar to the general case. Before that however, observe that it will imply the assertion since then $\mathbb{Q}(B) \subsetneq_{\mathrm{dense}} RC(\mathbb{Q}(B)) = F$.

Enumerate the set of intervals (a, b) of F as $(I_j)_{j \in \omega}$. Choose an arbitrary element $t_0 \in I_0$. For any $n < \omega$ and given a set of elements $\{t_0, t_1, \ldots, t_n\}$ in F

such that for $0 \leq l \leq n$ we have $t_l \in I_l$ and which is algebraically independent over $\mathbb{Q}$, consider the set $E_n = \{x \in F \mid x \text{ is algebraic over } \mathbb{Q}(t_0, \ldots, t_n)\}$. Note that $E_n \subsetneq F$, since F is of infinite transcendence degree. So the subfield E_n does not cover any interval of F (proper subfields always have dense complements, see [EGH, Lemma 2.2]). Hence there exists $t_{n+1} \in I_{n+1}$ such that $\{t_0, \ldots, t_n, t_{n+1}\}$ is algebraically independent over $\mathbb{Q}$. So $\{t_n \mid n \in \omega\}$ is a set of algebraically independent elements dense in F and can be extended to a transcendence basis B over $\mathbb{Q}$ for F.

(ii) Similar to the construction of [EGH, Lemma 2.3] again, choose two points now (rather than one) in any interval when its turn comes up in the definition by transfinite induction. Consider two sets C and D by putting one of the points in any such pair in C and the other one in D. Pick a set E such that $C \cup D \cup E$ forms a transcendence basis for F. Let $A \subseteq D$ be a set of cardinality α and $B = C \cup (D \setminus A) \cup E$. Then the real closure of $\mathbb{Q}(B)$ inside F has the desired properties.   ⊣

THEOREM 3.2.   (i) *For every $n \in \mathbb{N}$, there is a real closed field F of transcendence degree n such that F has a proper dense subfield, but none of the proper dense subfields of F are real closed.*

(ii) *If $n \geq 1$ is an integer, then (in addition to the above kind of fields) there is a real closed field F' of transcendence degree n such that F' has proper dense subfields with both proper and improper real closures; and additionally, if $n \geq 2$, then F' can be also arranged to be non-Archimedean.*

(iii) *If a real closed field of finite transcendence degree has proper dense real closed subfields, then it has a proper dense real closed subfield L algebraic over all dense subfields of L.*

PROOF. (i) For $n = 0$, $RC(\mathbb{Q})$ is of transcendence degree 0 and has no proper real closed subfields. For $n = 1$, consider $RC(\mathbb{Q}(x))$ with x infinitely small. Dense subfields here have to be of transcendence degree 1, so their real closure would be improper. An example of a proper dense subfield is $\mathrm{Frac}(\mathcal{S}_x(\mathbb{Z}))$, see Corollary 4.2(iii) (which says that *every* IP for a real closed field of finite transcendence degree has its fraction field proper (it is dense as always)).

Now fix $n \geq 2$ and consider $F = RC(\mathbb{Q}(x_1, \ldots, x_n))$ inside say $[[RC(\mathbb{Q})^{\mathbb{R}}]]$, where $x_j = t^{r_j}$ (t is the indeterminate used to represent elements of the power series field) and $r_j \in \mathbb{R}^{>0}$, $j = 1, \ldots, n$, are picked $\mathbb{Q}$-linearly independent. Then F is algebraic over all its dense subfields, since if a subfield of F misses x_j, for some j, then it cannot have any elements in the interval $(x_j, 2x_j)$ of F (by the linear independence of r_j for $j = 1, \ldots, n$, over $\mathbb{Q}$).

As examples of proper dense subfields of F, we may still refer to Corollary 4.2(iii). Here is another way to establish existence of irrationals with respect to the familiar IP for F, the argument would not cover the case $n = 1$, it is good for $n \geq 2$. One checks easily that if p is a positive infinite prime in an IP such that the elements p^n for $n \in \mathbb{N}$ form a cofinal sequence, then the element $\sqrt{p}$

in the real closure of the fraction field of the IP is irrational with respect to the same IP. This is applicable here since in the multi-variate (number of variables at least two) Shepherdson extension $S_{x_1,\dots,x_n}(\mathbb{Z})$ (with $n \geq 2$) of the integers (which forms an IP for F), we have the infinite prime $p = x_1^{-1} + \cdots + x_n^{-1} + 1$, see the theorem at the end of [Mo].

(ii) For $n \geq 1$, let $\{\pi_1, \dots, \pi_n\}$ be an algebraically independent set of reals. Consider $F' = RC(\mathbb{Q}(\pi_1, \dots, \pi_n))$. We have the proper dense real closed subfield of real algebraic numbers and the proper dense subfield $\mathbb{Q}(\pi_1, \dots, \pi_n)$ whose real closure is F'. Now let $n \geq 2$, then we have the following non-Archimedean example also. Consider the Cauchy completion K (which is of transcendence degree continuum) of $L = RC(\mathbb{Q}(x))$, with x infinitely small. Adjoin $n - 1$ algebraically independent elements in K transcendental over L to the latter to get a field A and take real closure of A to get F'. Then F' is of transcendence degree n and has L as a proper real closed subfield. Now L is dense in F' because K is real closed and so F' can be taken as a subfield of K (in which L is dense). On the other hand A is a proper dense subfield of F' and has its real closure equal to F'.

(iii) Let F_1 be a proper dense real closed subfield of the real closed field F of finite transcendence degree. If F_1 does not have any proper dense real closed subfields, then that is the field we wanted. Otherwise, let F_2 be a proper dense real closed subfield of F_1. Continue this process and note that in each step, transcendence degree is strictly reduced. Let $L = F_j$ be the field achieved when this process stops. It will be algebraic over all its dense subfields. ⊣

3.2. Existence of algebraic irrationals and transcendentals.

PROPOSITION 3.3. (*Suggested by S. Kuhlmann*) *For any real closed field K and nonzero divisible ordered Abelian group G, the ordered field $[[K^G]]$ possesses IP's whose fraction fields leave out some algebraic irrationals.*

PROOF. (Communicated to us by L. van den Dries) Let t be the positive K-infinitesimal indeterminate used to represent elements of $[[K^G]]$. Fix an arbitrary element $g \in G^{<0}$. We show that the elements $\sqrt{t^g \pm 1} \in [[K^G]]$ do not belong to $\mathrm{Frac}(K[G])$. Here $K[G]$ is the ring of elements in $[[K^G]]$ all whose exponents are non-positive. Now since the Shepherdson-like IP, for any fixed IP for K itself, is included in the latter, we will be done. Take a basis $B \subset G^{<0}$ containing g of the $\mathbb{Q}$-vector space G. For any positive integer n, let $G_n = \sum_{b \in B} \frac{b\mathbb{Z}}{n}$ and $K_n = \mathrm{Frac}(K[G_n])$, the latter being the fraction field of the domain of power series in $[[K^G]]$ all whose exponents belong to $G_n^{\leq 0}$. As $\mathrm{Frac}(K[G]) = \bigcup_{n=1}^{\infty} K_n$, it is enough to show $\sqrt{t^g \pm 1} \notin K_n$ for any fixed n. The elements $(t^{\frac{b}{n}})_{b \in B}$ in $K[G_n]$ are algebraically independent over K. So the elements $t^g \pm 1$ in $K[G_n]$ are not squares there (since the n-th power ± 1 is not the squaring function!). Therefore the polynomial $Y^2 - (t^g \pm 1)$ is irreducible in $K[G_n][Y]$. Now as $K[G_n]$ is a unique factorization domain, by (a consequence of) Gauss' lemma (see [H, Chapter 3, Lemma 6.13] for the statement we use) $Y^2 - (t^g \pm 1)$ is irreducible in $\text{Frac}(K[G_n])[Y] = K_n[Y]$ also, hence the result. ⊣

We show in the next section that if G is Archimedean, then with respect to all IP's for $[[F^G]]$ we have existence of algebraic irrationals.

PROPOSITION 3.4. *Real closed fields of infinite transcendence degree have IP's over which they are of the same transcendence degree as over* $\mathbb{Q}$.

PROOF. Using Theorem 3.1(ii), take a proper dense real closed subfield over which the original field is of transcendence degree the same as over the rationals. Any IP for such a field is also one for F and with respect to which transcendence degree is the same as over the rationals. ⊣

We will see below that with the stronger condition of Cauchy completeness (instead of being of infinite transcendence degree), transcendentals will exist with respect to *every* IP.

§4. **Absolute existence of irrationals.** In this section we show that for real closed fields which are either of finite transcendence degree, or are a field of Puiseaux series, *there always exist algebraic irrational elements no matter what IP one picks.* We call this property *Absolute Existence of Algebraic Irrationals*, abbreviated AEAI. Note that AEAI implies absolute existence of a dense set of algebraic irrationals. We show that Cauchy complete real closed fields have Absolute Existence of Transcendentals AET and obtain some results about the existence of perfect sets of irrationals.

4.1. Absolute existence of algebraic irrationals. As mentioned in the preliminaries, we use the basic terminology and notation for continued fractions, see [R, Section 7.1]. The proof below indeed works for 2-real closed, instead of real closed, ordered fields having IP's as well:

LEMMA 4.1. *If there is* $M \in F^{\geq 0}$ *with* $\{M^n\}_{n \in \mathbb{N}}$ *cofinal in* F, *then* F *has AEAI.*

PROOF. Suppose I is an IP for F and put $N = \lfloor M \rfloor_I + 1$. The sequence $\{N^n\}_{n \in \mathbb{N}}$ dominates $\{M^n\}_{n \in \mathbb{N}}$ and therefore will also be cofinal in F. Hence we may as well assume $M \in I$.

Consider the element $\alpha = \sqrt{M^2 + 1} \in F$. The corresponding continued fraction with respect to I is $\langle M, \overline{2M} \rangle$. The convergents $c_n = \frac{P_n}{Q_n}$ of this continued fraction form a Cauchy sequence in F. This is because $\{M^n\}_{n \in \mathbb{N}}$ is cofinal in F and therefore the same holds for the sequence of denominators Q_n. As usual, every other convergent is to the left, respectively to the right of α. Therefore, the above continued fraction indeed tends in F to α. For every $n \in \mathbb{N}$ we have, as in the standard case, $0 < |\alpha - \frac{P_n}{Q_n}| < \frac{1}{Q_n^2}$. If it were the case that $\alpha = \frac{p}{q}$ for some $p, q \in I^{>0}$, then for all n, we had

$|\frac{p}{q} - \frac{P_n}{Q_n}| = \frac{|pQ_n - qP_n|}{qQ_n} \geq \frac{1}{qQ_n}$ (the latter inequality being true since the numerator of its left hand side which is an element of I is nonzero due to the continued fraction being non-stabilizing). But this contradicts the fact that we may take n large enough so that $Q_n > q$ (this being possible by the cofinality assumption again). ⊣

COROLLARY 4.2. *The following three families of fields all satisfy AEAI:*

(i) *Puiseux series fields over real closed fields,*
(ii) *Generalized power series fields $[[F^G]]$ with G Archimedean, and*
(iii) *2-real closed fields of finite transcendence degree.*

PROOF. (i, ii) In such ordered fields, all positive infinite elements have cofinal standard powers.

(iii) Suppose that the finite set $\{\alpha_1, \ldots, \alpha_n\}$ is a $\mathbb{Q}$-transcendence basis for F and without loss of generality assume all the α_j's are greater than 1. Then there must be a j such that α_j has cofinal standard powers in F. ⊣

4.2. Absolute existence of transcendentals. The Cauchy completion of any real closed field of an at most countable transcendence degree has transcendentals with respect to some IP's. The reason is that transcendence degree of any Cauchy complete ordered field is at least continuum, see [ME, Proposition 2.1 and the Claim therein]. We see in this subsection that even without the above restriction on the transcendence degree, those fields indeed have AET. The transcendental elements we construct will have the property DMO (for their IP-multiples), providing additive subgroup IS's which are DMO with respect to an IP.

In the classical setting, Liouville's theorem states that approximation exponents for algebraic irrationals are less than degrees of algebraicity (and of course, there is the improvement by Roth). As mentioned in [D, p. 347], a similar statement also holds more generally (below, F need not be assumed real closed):

PROPOSITION 4.3. *Fix an algebraic element of degree d with respect to I such as α in F. Then there is a constant $c(\alpha) \in F^{>0}$ depending on α only such that for every $\frac{p}{q} \in \mathrm{Frac}(I)$ distinct from α, we have $|\alpha - \frac{p}{q}| > \frac{c(\alpha)}{q^d}$.*

PROOF. It goes without saying that the case $d = 1$ is true with $C = 1$ if approximations are excluded, as in the statement above, from being exact. Let $P(X)$ be a polynomial of degree $d \in \mathbb{N}^{>1}$ over I having α as a root. Let $C(\alpha)$ be the reciprocal of $2\sum_{i=1}^{d} \frac{|P^i(\alpha)|}{i!}$, where the upper superscript denotes the order of derivatives (note that the last term, and so the sum, is nonzero). Then for any $\frac{p}{q} \in \mathrm{Frac}(I)$ with $|\frac{p}{q} - \alpha| < 1$ (the case one needs to consider), we have $|P(\frac{p}{q})| = |\frac{p}{q} - \alpha| \cdot |\sum_{i=1}^{d} (\frac{p}{q} - \alpha)^{i-1} \frac{P^i(\alpha)}{i!}| < |\frac{p}{q} - \alpha| \cdot \frac{1}{c(\alpha)}$. Now since $P(\frac{p}{q})$ in $\mathrm{Frac}(I)$ is nonzero and can be expressed with denominator q, we have $|P(\frac{p}{q})| \geq \frac{1}{q^d}$ and the result is proved. ⊣

Liouville's method of constructing transcendental numbers in $\mathbb{R}$, as in [R, Section 8.1] for instance, can be used in our setting as follows.

THEOREM 4.4. *Every Cauchy complete real closed ordered field has AET.*

PROOF. Obviously one needs to consider non-Archimedean such fields only. Let I be an IP for F.

Case I. Suppose F is of countable cofinality. Let the ascending sequence $\{M_j\}_{j\in\omega}$ of elements of $I^{>0}$ be cofinal in I. For all $n \in \mathbb{N}$, let $N_n = \prod_{j=0}^{n} M_j$ and then for any $k \in \mathbb{N}$, let $C_k = \sum_{n=0}^{k} N_n^{-n!}$. For $l > k$, we have $C_l - C_k = \frac{1}{N_l^{(k+1)!}} + \cdots + \frac{1}{N_l^{l!}} \le \frac{l-k}{N_{k+1}^{(k+1)!}}$. So, $(C_k)_{k\in\mathbb{N}}$ is a Cauchy sequence in F. Let $\alpha = \lim_{k\to\infty} C_k \in F$. We show that α is an I-transcendental element. For any $m \in \mathbb{N}^{\ge 1}$ (a to-fail candidate for the degree of algebraicity of α) and $k > m$, with $p_k \in I$ being such that $\sum_{n=0}^{k} N_n^{-n!} = \frac{p_k}{N_k^{k!}}$ (note that if $l < k$, then $N_l | N_k$), we have $0 < \alpha - \frac{p_k}{N_k^{k!}} = \alpha - \sum_{n=0}^{k} N_n^{-n!} = \sum_{n=k+1}^{\infty} N_n^{-n!} < \sum_{j=1}^{\infty} \frac{1}{N_{k+1}^{j(k+1)!}} = \frac{1}{N_{k+1}^{(k+1)!}-1} < \frac{1}{N_k^{(k+1)!}}$. So if it were the case that α were an I-algebraic element of degree m, then by Liouville's theorem there would exist $C \in F^{>0}$ such that $\frac{C}{(N_k^{k!})^m} < \frac{1}{N_k^{(k+1)!}}$. But this is impossible as there could be no nonzero elements in an ordered field infinitesimal in the scale of that very field.

Case II. Suppose F is of cofinality $cf(F) = \delta > \omega$. Take a cofinal sequence $\{M_\beta\}_{\beta<\delta}$ in $I^{>0}$ such that $(\forall\beta < \delta)(\forall n \in \mathbb{N})(M_\beta^n < M_{\beta+1})$. Fix an arbitrary element $a_0 \in I^{>0}$ and let I_0 be the interval $[\frac{a_0}{M_0} - \frac{2}{M_1}, \frac{a_0}{M_0}]$. For ordinal β with $0 < \beta < \delta$, pick $a_\beta \in I^{>0}$ such that for all $\eta < \beta$, we have $I_\beta := [\frac{a_\beta}{M_\beta} - \frac{2}{M_{\beta+1}}, \frac{a_\beta}{M_\beta}] \subsetneq I_\eta$. Note that in the limit case, the already constructed intervals are of length greater than $\frac{2}{M_\beta}$ and so it is possible to choose an interval of length $\frac{2}{M_{\beta+1}}$ with the above kind of end points. By the density of Frac(I) in F and ongoing refinement of the intervals, there is a never-stabilizing δ-sequence $(x_\beta)_{\beta<\delta}$ with $x_\beta \in I_\beta \cap$ Frac(I). Note that for all $\beta_1 < \beta_2 < \delta$, we have $|x_{\beta_1} - x_{\beta_2}| < \frac{2}{M_{\beta_1+1}}$ and so $(x_\beta)_{\beta<\delta}$ is Cauchy in Frac$(I) \subseteq F$. Therefore there is an element $\rho = \lim_{\beta<\delta} x_\beta \in F$. For all $\beta < \delta$ and $d \in \mathbb{N}$, we have $|\frac{a_\beta}{M_\beta} - \rho| < \frac{2}{M_{\beta+1}} < \frac{1}{M_\beta^{d+1}}$. So, once again since there could be no nonzero K-infinitesimals in an ordered field K, by Liouville's theorem ρ is not algebraic of degree d over I. $\dashv$

We proved the existence of additive subgroup DMO IS's in non-Archimedean real closed fields with either a countable IP, or when there is an irrational with the DMO property (for its multiples) in Theorem 2.3(i, ii). Now we can state:

COROLLARY 4.5. *For any IP such as I for an arbitrary non-Archimedean Cauchy complete real closed field, there are additive subgroup IS's which are DMO with respect to I.*

PROOF. Note that for the elements α in Case I and ρ in case II in the proof of Theorem above, we have $DMO(\alpha)$ and $DMO(\rho)$ respectively. For α, the reason is that the proof shows $N_k^{k!}\alpha - p_k$ could be made arbitrarily small in that field and we can use the Claim at the end of our argument for Theorem 2.3(iii). Now consider ρ and fix $\varepsilon \in F^{>0}$. Let $\beta < \delta$ be such that $\frac{1}{M_\beta} < \varepsilon$. Then we have $|a_\beta - M_\beta\rho| < \frac{1}{M_\beta} < \varepsilon$. Using the same Claim again, this proves $DMO(\rho)$.   ⊣

COROLLARY 4.6. *There is a real closed field with an IP with respect to which there exist irrationals but none of which are algebraic.*

PROOF. Take the Cauchy completion of the real closed fraction field of van den Dries' model $\mathcal{S}^\omega(\mathbb{Z})$ (that we defined in subsection 1.1), with the same IP.   ⊣

4.3. Perfect sets of irrationals and transcendentals. A perfect set in a topological space is a closed subset none of whose points is isolated. We now deal with nonempty perfect sets which avoid all rational or even all algebraic elements in real closed fields (with the order topology and) equipped with an IP. In the real case $\langle \mathbb{R}, \mathbb{Z} \rangle$, there are perfect sets with such properties.[6] These are special cases of the next Lemma which enables us to conclude similar but more general statements. It turns out that not all that standard phenomena such as the fact that nonempty perfect subsets of Polish spaces (in particular of the real field) are of cardinality continuum, continue to hold in our general setting.

LEMMA 4.7. *Every dense subset A of an ordered field F whose complement in F has cardinality at most $cf(F)$, contains a non-empty perfect subset (with respect to the order topology of F).*

PROOF. This is trivial if A contains an open interval of F. Let $F \setminus A$ be of cardinality α with an enumeration $(r_\beta)_{\beta<\alpha}$. Inductively define an α-sequence of open intervals in F as follows. For any $\beta < \alpha$, if the intervals I_δ for $\delta < \beta$ are already defined, find the least $\gamma < \alpha$ such that $r_\gamma \notin \cup_{\delta<\beta}I_\delta$. Choose an open interval I_β containing r_γ with endpoints in A (using the density of A in F) whose closure does not intersect union of closures of the intervals already defined (this is possible since $\beta < \alpha \leq cf(F)$). The complement P with respect to F of the union of the open intervals constructed this way is a closed subset of F and contains endpoints of all the intervals I_β, for $\beta < \alpha$, and so is non-empty. It is obviously included in A.

To see that P is perfect, fix $a \in P$ and $\varepsilon \in F^{>0}$. Assume a is not the left end point of any of the open intervals above (the other case, when it's not the right end point of any, is similar). Let $b \in F \setminus A$ be such that $a < b < a + \varepsilon$.

---

[6]There are more striking results such as the fact that each perfect subset of $\mathbb{R}$ contains a perfect set of algebraically independent numbers, see [My].

Assume β is such that $b \in I_\beta$ and let the latter interval have left end point c. Then $c \in P \setminus \{a\}$ is within ε of a.   ⊣

When card(F) is regular, this shows that any dense subset of F, and in particular complement of any proper subfield of F, contains non-empty sets perfect in F. Under the same condition the transcendence bases constructed before for F's of infinite transcendence degree, show the existence of a non-empty set perfect in F of elements algebraically independent over the rationals.[7] If F is of infinite transcendence degree even over the fraction field of an IP such as I, then similar to what we did over $\mathbb{Q}$, we get a dense transcendence basis over Frac(I).

COROLLARY 4.8. *In* $\langle F, I \rangle$ *with* card(F) *regular,*

(i) *if there are any irrationals, then there is a non-empty set of irrationals perfect in* F.

(ii) *if the transcendence degree of F over an IP is infinite, then there exists a nonempty perfect subset of F that is algebraically independent over* Frac(I).

§5. Concluding remarks and some remaining questions.

5.1. DMO vs. ADMO.
We proved in this paper that:

(a) For subsets of a general real closed field F, the property of being DMO is sensitive with respect to additive subgroup IS's for F. Working around whether the same property is IP-sensitive, we showed that there are two IP's with just trivial intersection (i.e. intersecting on the standard integers only) and if an IP is DMO with respect to another, then so is the other way around.

(b) There are integer parts I_1 and I_2 of the same real closed field F for which there exists $\alpha \in I_2 \setminus$ Frac(I_1) such that $DMO(\alpha)$ holds in $\langle F, I_1 \rangle$.

In this connection, we would like to raise the following:

QUESTION 5.1. (a) For subsets of an arbitrary real closed field, is being DMO an IP-sensitive property? E.g., if two IP's intersect only trivially, will they be DMO with respect to each other?

(b) Does the property (for all irrationals α) ($DMO(\alpha)$) hold in $\langle F, I \rangle$ in general? If not, does there exist an example of a mutually irrational $\alpha \in F \setminus [\text{Frac}(I_1) \cup \text{Frac}(I_2)]$ such that $DMO(\alpha)$ holds in $\langle F, I_1 \rangle$, but not in $\langle F, I_2 \rangle$?

Regarding part (b), we mention the following:

REMARK 5.2. (i) Note that this property is a first-order sentence in structures $\langle F, +, \cdot, <, I \rangle$. By the Downward Löwenheim-Skolem Theorem, there

---

[7]We are not sure whether one can improve this to the existence of a perfect set which forms a transcendence basis over $\mathbb{Q}$.

exist countable elementary substructures. Hence, if any easier, it would suffice to find out the answer for countable structures of that form. A similar observation applies to the second question in part (b) considered for structures of the form $\langle F, I_1, I_2 \rangle$.

(ii) We saw earlier in the paper that if $\alpha \in F$ is such that for all $\varepsilon \in F^{>0}$, we have $\{n\alpha - \lfloor n\alpha \rfloor_I \mid n \in I\} \cap (0, \varepsilon) \neq \emptyset$ (let us call this property $P(\alpha)$), then we have $\mathrm{DMO}(\alpha)$. Arguments of a similar nature show that $\mathrm{DMO}(\alpha)$ implies $(\forall m \in I^{>0})(\forall k \in I^{\geq 0})(\lfloor mI^{>0}\alpha \rfloor \cap (mI^{>0} + k) \neq \emptyset)$, and that $(\forall m \in I^{>0})(\lfloor mI^{>0}\alpha \rfloor \cap mI^{>0} \neq \emptyset)$ implies $P(\alpha)$. So they are all equivalent. It is not clear whether these could be of any use in resolving the problem though.

(iii) The standard method of proving the statement in the classical setting is via cofinal quadratic rational approximations. Such approximations are obtained from the continued fractions associated to irrationals (they converge to them). Sufficiency of having a convergent (in the scale of the field) continued fraction for existence of cofinal quadratic rational approximations (itself implying IP-multiples being DMO) holds in general but its necessity fails. This happens, e.g., for a 'rational' shift of the element $\sqrt{t^{-1} + 1}$ that we considered earlier, namely $\sqrt{t^{-1} + 1} + \sqrt{2}$ in the Puiseux series field with real algebraic coefficients equipped with Shepherdson's IP. It can easily be seen that the continued fraction of the fractional part of this infinitely large irrational element is the same as the continued fraction for $\sqrt{2} - 1$. Therefore the differences between the convergents and the fractional part of $\sqrt{t^{-1} + 1} + \sqrt{2}$ remain greater than all infinitesimals in the field, preventing the convergents from actually converging. Nevertheless, we can shift a cofinal quadratic rational approximation for $\sqrt{t^{-1} + 1}$ (see the proof of Theorem 2.3(iii)) by the 'rational' $\sqrt{2} = \frac{\sqrt{2}t^{-1/5}}{t^{-1/5}}$ to get a cofinal quadratic rational approximation for $\sqrt{t^{-1} + 1} + \sqrt{2}$.

5.2. AE vs. EA. In this paper we presented results on Absolute Existence of Irrationals and Absolute Existence of Transcendentals. Let us now consider Existence of Absolute Irrationals and Existence of Absolute Transcendentals. We say an element $\alpha \in F$ is an Absolute Algebraic Irrational (AAI for short) if for each IP I for F, α is algebraic over $\mathrm{Frac}(I)$ but not an element of the latter. Absolute Transcendentals (abbreviated AT) are defined similarly by being transcendental with respect to (fraction field of) all IP's. The question, of course, is whether such elements exist (that is whether EAAI or EAT are possible).

The following four types of elements can be made rational (among other possibilities in some cases) by taking appropriate IP's: (1) infinitely large elements, (2) standard rationals plus an infinitesimal, (3) real algebraic numbers, and (4) standard transcendentals which happen to belong to the field. So no such element can be an AAI or AT.

A standard algebraic irrational r plus a nonzero infinitesimal ε can be made algebraic with respect to a suitable IP by making ε rational again.[8] The elements of the form $r + \varepsilon$ as above therefore cannot be AT's. So let us try to figure out whether any such elements can be an AAI.

In the case of real closed fields F of infinite transcendence degree, $r + \varepsilon$ can indeed also become transcendental (and so it will not be AAI either): We can produce a transcendence basis $(t_i)_{i \in J}$ for F as in Theorem 3.1(i) such that $t_1 = r + \varepsilon$. Then $RC(\mathbb{Q}(t_i : 1 \neq i \in J)) \subsetneq_{\text{dense}} F$ and $RC(\mathbb{Q}(t_i : 1 \neq i \in J))$ has an IP which is an IP for F too and with respect to which $r + \varepsilon$ is transcendental. In fact the argument just given works when the field has a proper dense real closed subfield missing the infinitesimal ε: any element in the difference of the two fields becomes transcendental with respect to any IP for the smaller (and hence the larger) field.

What about the case when there are even no proper dense real closed sub-fields (as we know from Theorem 3.2(ii) this happens for some real closed fields of finite transcendence degree), let alone any such subfield that misses particular infinitesimals? For instance consider $RC(\mathbb{Q}(x))$, with x infinitely small. No elements, and in particular no infinitesimals, can become transcendental with respect to any IP. Therefore an element such as $\sqrt{2} + \varepsilon$ will always be algebraic in $RC(\mathbb{Q}(x))$, no matter what IP one picks.

On the other hand, and again in the infinite transcendence degree case, any element of the form of a standard transcendental number plus a nonzero infinitesimal can similarly be made transcendental. One can see that the same conclusion holds in the case of finite transcendence degree as standard transcendentals in the field require no transcendence from the IP (that much of the approximability within 1 is for free) and so there remains room for the above kind of elements to become transcendental. No such element therefore is an AAI.

These arguments show in particular that a class of real closed fields, each of which satisfies AEAI as we saw earlier, fail the stronger property:

COROLLARY 5.3. *In all Puiseux series fields over real closed fields, the property EAAI fails.*

But some questions remain unanswered, for example:

QUESTION 5.4. (a) Is there an infinitesimal ε (necessarily nonzero) in RC $(\mathbb{Q}(x))$ such that $\sqrt{2} + \varepsilon$ is irrational with respect to all the IP's for $RC(\mathbb{Q}(x))$? (In that case $\sqrt{2} + \varepsilon$ will be an AAI.)
(b) Is there any (necessarily nonzero) infinitesimal $\varepsilon \in [[\mathbb{R}^{\mathbb{Q}}]]$ such that $\pi + \varepsilon$ is an AT?[9] How about a similar question for the case $\varepsilon \in RC(\mathbb{Q}(\pi, x))$ (with x infinitely small) in the latter real closed field?

---

[8] Observe that the class of IP's making r rational may possibly be disjoint from the class of IP's making ε rational, we have no reason to rule that out.

[9] Similar to the previous footnote, notice that π (as an element of the field of coefficients of the generalized power series field) and ε can become rational separately, that is with respect to two

Acknowledgment. This work was supported in part by a grant (No. 82030211) from Institute for Studies in Theoretical Physics and Mathematics (IPM). It grew out of our two presentations at the wonderful and memorable WCLAA-2003 meeting. We are indebted to A. Enayat for making the results lucid and the presentation of the paper a whole lot better. We thank the referee for comments improving the paper; S. M. Bagheri, S. Boughattas, L. van den Dries, F.-V. Kuhlmann, S. Kuhlmann, and J.-P. Ressayre for useful discussions; and A. Enayat, I. Kalantari, and G. B. Khosrovshahi for their continued understanding and encouragement.

REFERENCES

[BO] A. BERARDUCCI and M. OTERO, *A recursive nonstandard model of normal open induction*, **The Journal of Symbolic Logic**, vol. 61 (1996), no. 4, pp. 1228–1241.

[B] S. BOUGHATTAS, *Résultats optimaux sur l'existence d'une partie entière dans les corps ordonnés*, **The Journal of Symbolic Logic**, vol. 58 (1993), no. 1, pp. 326–333.

[D] L. VAN DEN DRIES, *Some model theory and number theory for models of weak systems of arithmetic*, **Model Theory of Algebra and Arithmetic (Proc. Conf., Karpacz, 1979)** (L. Pacholski, J. Wierzejewski, and A. J. Wilkie, editors), Lecture Notes in Mathematics, vol. 834, Springer, Berlin, 1980, pp. 346–362.

[EGH] P. ERDÖS, L. GILLMAN, and M. HENRIKSEN, *An isomorphism theorem for real-closed fields*, **Annals of Mathematics. Second Series**, vol. 61 (1955), pp. 542–554.

[H] T. W. HUNGERFORD, *Algebra*, Graduate Texts in Mathematics, vol. 73, Springer, New York, 1974, (5th printing 1989).

[K] FRANZ-VIKTOR KUHLMANN, *Dense Subfields and Integer Parts*, abstract of a talk at International Congress M.ARI.AN. 2004 (see http://math.usask.ca/~fvk/Fvktalks.html).

[K2] ———, *Dense Subfields of Henselian Fields, and Integer Parts*, **Logic in Tehran** (A. Enayat, I. Kalantari, and M. Moniri, editors), Lecture Notes in Logic, vol. 26, ASL and AK Peters, 2006, this volume, pp. 204–226.

[L] A. LASJAUNIAS, *A survey of Diophantine approximation in fields of power series*, **Monatshefte für Mathematik**, vol. 130 (2000), no. 3, pp. 211–229.

[Mc] K. MCKENNA, *New facts about Hilbert's seventeenth problem*, **Model Theory and Algebra (A Memorial Tribute to Abraham Robinson)** (D. H. Saracino and V. B. Weispfenning, editors), Lecture Notes in Mathematics, vol. 498, Springer, Berlin, 1975, pp. 220–230.

[Mo] M. MONIRI, *Recursive models of open induction of prescribed finite transcendence degree > 1 with cofinal twin primes*, **Comptes Rendus de l'Académie des Sciences. Série I. Mathématique**, vol. 319 (1994), no. 9, pp. 903–908.

[ME] M. MONIRI and J. S. EIVAZLOO, *Relatively complete ordered fields without integer parts*, **Fundamenta Mathematicae**, vol. 179 (2003), no. 1, pp. 17–25.

[MR] M.-H. MOURGUES and J.-P. RESSAYRE, *Every real closed field has an integer part*, **The Journal of Symbolic Logic**, vol. 58 (1993), no. 2, pp. 641–647.

[My] J. MYCIELSKI, *Algebraic independence and measure*, **Fundamenta Mathematicae**, vol. 61 (1967), pp. 165–169.

[N] I. NIVEN, *Diophantine Approximations*, Interscience Publishers, a division of John Wiley & Sons, New York, 1963.

---

perhaps necessarily distinct IP's (the former by taking the Shepherdson IP). A similar observation applies to the second question in part (b) and to $\sqrt{2}$ and ε in part (a).

[R] H. E. ROSE, *A Course in Number Theory*, Oxford Science Publications, The Clarendon Press Oxford University Press, New York, 1994.

[Sch] J. H. SCHMERL, *Models of Peano arithmetic and a question of Sikorski on ordered fields*, *Israel Journal of Mathematics*, vol. 50 (1985), no. 1-2, pp. 145–159.

[Sco] D. SCOTT, *On completing ordered fields, Applications of Model Theory to Algebra, Analysis, and Probability (Internat. Sympos., Pasadena, Calif., 1967)* (W. A. J. Luxemburg, editor), Holt, Rinehart and Winston, New York, 1969, pp. 274–278.

INSTITUTE FOR STUDIES IN THEORETICAL
  PHYSICS AND MATHEMATICS
    TEHRAN, IRAN
and
  DEPARTMENT OF MATHEMATICS
    TARBIAT MODARRES UNIVERSITY
      TEHRAN, IRAN
E-mail: smasih@ipm.ir
E-mail: mojmon@ipm.ir

CATEGORICITY AND QUANTIFIER ELIMINATION
FOR INTUITIONISTIC THEORIES

SEYED MOHAMMAD BAGHERI

Abstract. We study a class of Kripke models with a naturally and well-behaved embedding relation. After proving a completeness theorem, we generalize some usual classical theorems of model theory into this framework. As an application, we give examples of non-classical ω-categorical theories which admit quantifier elimination.

§1. Semi-classical logic. Usually, Kripke models are used for proof-theoretic reasons, to provide a semantics for Intuitionistic logic. But, some of these generalized models are very similar to the usual models of classical first order logic and we may wish to do some sort of model theory with them. This can help us to better understand the expressive power of intuitionistic theories. A concrete question in this respect is the existence of a proper intuitionistic ω-categorical theory (in some sense!). Also, some general questions may be answered by the intervention of model-theoretic techniques. An example is the existence of a proper intuitionistic theory which admit quantifier elimination. Another motivation is the new tendencies in stability theory where first order theories are replaced by abstract elementary classes (see [Baldwin]). Roughly speaking, an abstract elementary class is a category (see [Ben]) which satisfies the elementary chain property and Löwenheim-Skolem theorem. What is obtained in this paper is in fact an abstract elementary class with some constructive flavor. Nevertheless, our ambition is rather to follow the natural approach of first order model theory with classical models replaced by Kripke models. In other words, we try to bring Kripke models into the realm of algebra.

One of the most crucial notions in model theory is that of "substructure" or more generally "embedding". In [V], A. Visser introduces various options of this notion for the class of Kripke models and proves some interesting preservation theorems. In [BM], we give some results with a particular embedding relation defined for the class of Kripke models over a fixed frame. However,

---

2000 *Mathematics Subject Classification.* 03B55, 03C35, 03C90.

Logic in Tehran
Edited by A. Enayat, I. Kalantari, and M. Moniri
Lecture Notes in Logic, 26

it seems that defining a fully satisfactory embedding relation for the whole class of Kripke models is impossible and it is better to restrict ourselves to a particular class. Here, we choose the class of models having a linear frame and a constant domain. In fact, we put some additional conditions and call them *semi-classical*. In section 2 we review some general model-theoretic properties for these models. We show that with a naturally defined embedding relation, some basic notions and results in classical models theory extend to this class. For example, any non-empty subset of a semi-classical model is contained in a unique smallest submodel (so we may speak of the generated submodel). This is not true for an arbitrary linear frame constant domain Kripke model. Also, the notion of submodel has a syntactic characterization by means of diagrams. In particular, each model is identified by its diagram. One may wonder these models still are of interest in the constructive realm. Model theory is essentially the study of definable sets. We believe that here, the constructive aspects of intuitionistic logic reduces to the level of definability. Imagine a semi-classical model observed in first order classical logic. Then there are sets defined constructively (those which are definable in semi-classical logic) and definable sets which are not constructive.

To start the matter, we prepare a logical framework for semi-classical models, i.e. we give some axioms and prove a completeness theorem. We note that there are various completeness theorems in the literature but, the corresponding class of Kripke models does not match this definition of embedding. Nevertheless, some results of this paper remain valid for arbitrary linear frame constant domain Kripke models. First, we recall some basic definitions in intuitionistic logic. Further details can be found in [D] and [TD]. We use $\wedge, \vee, \Rightarrow, \exists$ and $\forall$ as primitive logical symbols. $\top$ and $\perp$ are considered as both logical symbols and atomic sentences. Let $(k, \preceq, 0)$ be a linearly ordered set with minimum 0. Elements of k will be called nodes. Let L be a fixed language, that we assume to be finite for some technical reasons. What we call here a classical model is an L-structure in the ordinary sense. A (classical) model M is a weak submodel of N if it is a subset of N and any atomic sentence (with parameters in M) which is satisfied in M, is also satisfied in N. Semi-classical models defined below are special kinds of Kripke models.

DEFINITION 1.1. A semi-classical model over the frame k is a family of classical models $(M_\alpha)_{\alpha \in k}$ over the same base set M such that whenever $\alpha \preceq \beta$, M_α is a weak submodel of M_β and in addition:

$C1$) for each atomic formula $\theta(\overline{x})$ and $\overline{a} \in M$, if there is an α such that $M_\alpha \models \theta[\overline{a}]$ then there is a least such α.

$C2$) for each node α there is an atomic formula $\theta(\overline{x})$ and $\overline{a}$ such that α is the least node at which $\theta(\overline{a})$ is satisfied.

Thus, a semi-classical model has a constant domain and a linear frame. Moreover, each node in its frame corresponds to the least occurrence of an atomic formula, and conversely each atomic formula, if it is forced at some node, should appear at a least node. In other words, each node is identified by some relation and each relation has a commencement. The effect of these conditions on (elementary) embeddings will be cleared later. Note that any ordinary classical model corresponds to a semi-classical model as defined above when the frame is reduced to a point. Satisfaction relation is replaced by forcing relation defined below:

DEFINITION 1.2 (*Forcing*). The forcing relation $\Vdash$ is defined inductively as follows (where φ, ψ are $L(M_0)$-sentences):

- For atomic φ, $M_\alpha \Vdash \varphi$ if and only if $M_\alpha \models \varphi$, and also, $M_\alpha \Vdash \top$ and $M_\alpha \not\Vdash \bot$;
- $M_\alpha \Vdash \varphi \vee \psi$ if and only if $M_\alpha \Vdash \varphi$ or $M_\alpha \Vdash \psi$;
- $M_\alpha \Vdash \varphi \wedge \psi$ if and only if $M_\alpha \Vdash \varphi$ and $M_\alpha \Vdash \psi$;
- $M_\alpha \Vdash \varphi \Rightarrow \psi$ if and only if for all $\beta \succeq \alpha$, $M_\beta \Vdash \varphi$ implies $M_\beta \Vdash \psi$;
- $M_\alpha \Vdash \forall x \varphi(x)$ if and only if for all $a \in M$, $M_\alpha \Vdash \varphi(a)$;
- $M_\alpha \Vdash \exists x \varphi(x)$ if and only if there exists $a \in M$ such that $M_\alpha \Vdash \varphi(a)$.

Forcing is preserved upward, i.e. if φ is forced in M_α it is also forced in M_β for any $\beta \succeq \alpha$. Intuitively, $\varphi \Rightarrow \psi$ holds if ψ is forced sooner than φ, i.e. at a smaller node. In particular, we define $\neg \varphi = (\varphi \Rightarrow \bot)$ as a shorthand for $\varphi \Rightarrow \bot$. Its interpretation sounds more "never" than the classical "not". Similarly, $\neg\neg\varphi$ is interpreted as "finally φ", i.e. it holds at some node. Semi-classical models may be considered as realistic description of concrete situations, where the knowledge of partial information on the structure increases with the passing of the time, as exemplified in what follows.

EXAMPLE 1. To compare two real numbers, we usually look at the corresponding digits in their (say binary) representation. Consider the sequence of classical structures $(\mathbb{R}_n, <_n)_{n \in \omega}$ in the language $L = \{<\}$ where $\mathbb{R}_n = \mathbb{R}$ and the relation $x <_n y$ is defined by $y - x \geq 2^{-n}$. Then, if $a \neq b$, there exists n such that either $\mathbb{R}_n \models a < b$ or $\mathbb{R}_n \models b < a$. If we regard $(\mathbb{R}_n, <_n)_{n \in \omega}$ as a semi-classical model, this means that $\mathbb{R}_0 \Vdash \neg\neg(a < b \vee b < a)$, i.e. finally a and b can be compared. This semi-classical model may be called a *semi-linearly ordered set*. It forces the usual axioms of linearly ordered sets except that points are cofinally comparable instead of being immediately comparable. In fact, the sentence $\neg \forall xy (x = y \vee x < y \vee y < x)$ is forced in $\mathbb{R}_0$. If we endow $(\mathbb{R}_n, <_n)$ with the usual addition operation, we may call $(\mathbb{R}_n, +, <_n)_{n \in \omega}$ a semi-linearly ordered group as $\mathbb{R}_0$ forces the sentence $\forall xyz(x < y \Rightarrow x + z < y + z)$. Note however that $(\mathbb{R}_r, <_r)_{r \in \mathbb{Q}^+}$ where $x <_r y$ if $y - x \geq 2^{-r}$ is not semi-classical because $C1, C2$ are not satisfied.

NOTATION 1.3. Semi-classical models will be simply denoted by M, N, etc. We will write $M \Vdash \varphi$ instead of $M_0 \Vdash \varphi$. We usually prefer to use the word model rather than the phrase semi-classical model.

For any semi-classical model M, the set

$$\{\theta[\overline{a}] : \theta \text{ is atomic, } \overline{a} \in M, M \Vdash \neg\neg\theta[\overline{a}]\}$$

is linearly pre-ordered by the relation $\Rightarrow$. The quotient of this set by the equivalence relation $\Leftrightarrow$ is a linearly ordered set with minimum $[\top]$ which, regarding $C1, C2$, is isomorphic to the frame of M. This is the main difference of a semi-classical model with other linear frame constant domain Kripke models. This in particular makes possible to define a semi-classical model without any reference to its nodes and just by giving its base set M and determining for every atomic θ, η and $\overline{a} \in M$ whether or not $\theta(\overline{a}) \Rightarrow \eta(\overline{a})$ holds. If R is a n-arc relation symbol, a classical mind may call the interpretation of $R(\overline{x}) \Rightarrow R(\overline{y})$ a n-ary *semi-relation*. In other words, one may imagine that a semi-classical model is a classical model augmented by a number of semi-relations and that, a semi-relation is relation if it is decidable i.e., for each $\overline{a}$, either $R(\overline{a})$ or $\neg R(\overline{a})$ is forced.

NOTATION 1.4. $\varphi \rightarrow \psi$ is an abbreviation for the formula $(\psi \Rightarrow \varphi) \Rightarrow \psi$.

When ψ is not forced in the root model, $\varphi \rightarrow \psi$ may be read as: ψ is forced strictly before (relative to the ordering of k) than φ. We may view this as a weak form of the true negation for $\psi \Rightarrow \varphi$. In what follows, by a *semi-atomic formula* is meant a formula of the form $\varphi \Rightarrow \psi$ or $\varphi \rightarrow \psi$ where φ and ψ are atomic. *Semi-classical logic* formalized in the finite language L is the intermediate logic consisting of the following list of rules and axioms:

I. Modus Ponens.
II. Usual axioms of intuitionistic first order logic.
III. Universal closure of the following formulas (φ, ψ are arbitrary formulas of L):

III0. $x = y \lor x \neq y$

III1. $(\varphi \Rightarrow \psi) \lor (\psi \Rightarrow \varphi)$

III2. $\forall x(\varphi \lor \psi(x)) \Leftrightarrow (\varphi \lor \forall x \psi(x))$ where x is not free in φ

III3. $\exists x(\theta \Rightarrow \psi(x)) \Leftrightarrow (\theta \Rightarrow \exists x \psi(x))$ where θ is atomic and x is not free in θ

III4. $(\varphi \rightarrow \psi) \Rightarrow \exists \theta \exists \overline{\imath}[\varphi \rightarrow \theta(\overline{\imath}) \land \theta(\overline{\imath}) \Rightarrow \psi]$, where $\exists \theta$ means "there is an atomic θ" (we deal with a finite disjunction of formulas).

The last two axioms correspond to the conditions $C1$ and $C2$. Obviously, any semi-classical model satisfies all these axioms (*soundness*). The finiteness assumption for the language is to make the assertion $\exists \theta$ first order. A natural option for infinite languages could be "there is an atomic subformula of $\varphi \land \psi$" but, although this make forcing to be more intrinsic (i.e. makes $\Vdash \varphi$ to depend just on the subformulas of φ) and so compatible with reductions, it is problematic with the soundness. We could also consider arbitrary languages containing a finite number of special relation symbols called "semi-relation symbols". Some easy consequences of the above axioms are listed below (some of them are true in intuitionistic logic).

(0) $(\varphi \Rightarrow \psi) \vee (\psi \to \varphi)$

(1) $(\varphi \Leftrightarrow \psi) \vee (\varphi \to \psi) \vee (\psi \to \varphi)$

(2) $(\varphi \Rightarrow \psi) \equiv [(\varphi \Leftrightarrow \psi) \vee (\varphi \to \psi)]$

(3) $[(\varphi \to \psi) \wedge (\psi \Rightarrow \varphi)] \equiv (\varphi \wedge \psi)$.

(4) $[(\varphi_1 \vee \varphi_2) \Rightarrow \varphi_3] \equiv [(\varphi_1 \Rightarrow \varphi_3) \wedge (\varphi_2 \Rightarrow \varphi_3)]$

(5) $[(\varphi_1 \wedge \varphi_2) \Rightarrow \varphi_3] \equiv [(\varphi_1 \Rightarrow \varphi_3) \vee (\varphi_2 \Rightarrow \varphi_3)]$

(6) $[\varphi_1 \Rightarrow (\varphi_2 \vee \varphi_3)] \equiv [(\varphi_1 \Rightarrow \varphi_2) \vee (\varphi_1 \Rightarrow \varphi_3)]$

(7) $[\varphi_1 \Rightarrow (\varphi_2 \wedge \varphi_3)] \equiv [(\varphi_1 \Rightarrow \varphi_2) \wedge (\varphi_1 \Rightarrow \varphi_3)]$

(8) $[\varphi_1 \Rightarrow (\varphi_2 \Rightarrow \varphi_3)] \equiv [(\varphi_1 \wedge \varphi_2) \Rightarrow \varphi_3]$

(9) $[(\varphi_1 \Rightarrow \varphi_2) \Rightarrow \varphi_3] \equiv [((\varphi_2 \to \varphi_1) \wedge (\varphi_2 \Rightarrow \varphi_3)) \vee \varphi_3]$

(10) $[\exists x \varphi \Rightarrow \psi] \equiv \forall x[\varphi \Rightarrow \psi]$ where x is not free in ψ

(11) $[\varphi \Rightarrow \forall x \psi] \equiv \forall x[\varphi \Rightarrow \psi]$ where x is not free in φ

(12) $[\exists x \varphi \to \theta] \equiv \forall x[\varphi \to \theta]$ where θ is atomic and x is not free in θ.

The notion of proof is defined as usual. A theory is any set of sentences which is consistent in semi-classical logic i.e. it does not prove $\perp$. In this paper, a theory T will be called prime if $T \vdash \sigma \vee \eta$ implies $T \vdash \sigma$ or $T \vdash \eta$. Also, T is complete if it proves just one of σ and $\neg\sigma$ for each σ. Any complete theory is prime. The theory of a model M is the collection of all sentences forced at M_0 and this is a prime but not necessarily complete theory. A consistent pair is a pair (T, Δ) of sets of sentences such that $\perp \in \Delta$ and for any $\sigma_1, \ldots, \sigma_n \in T$ and $\eta_1, \ldots, \eta_m \in \Delta$, $\wedge_i \sigma_i \not\vdash \vee_i \eta_i$. We usually assume that T is closed under deduction and Δ is closed under disjunction. A complete pair is a pair for which every sentence belongs to either T or Δ. A model of the pair (T, Δ) is a model M of T which does not force any sentence in Δ. Below, by $\Delta \vee \theta$ we mean $\Delta \cup \{\eta \vee \theta : \eta \in \Delta\}$.

THEOREM 1.5 (*Completeness*). *Any consistent pair* (T, Δ) *has a semi-classical model.*

PROOF. To simplify the proof, we assume the language is countable. Let C be a countable set of new constant symbols and $\{X, Y, Z\}$ a partition of ω into infinite sets. Enumerate all $L(C)$-sentences of the form $\exists x \varphi(x)$ by elements of X and all $L(C)$-sentences of the form $\forall x \varphi(x)$ by elements of Y. Let also $\{\sigma_i : i \in Z\}$ be an enumeration of all sentences of $L(C)$. We assume that each $\exists x \varphi(x)$ has an infinite number of occurrences in the enumeration. First, we construct an increasing sequence (T_i, Δ_i) of consistent pairs such that Δ_i is closed under disjunction. Put and $T_0 = T$ and $\Delta_0 = \vee\Delta$, the collection of all finite disjunctions of sentences of Δ. Given (T_i, Δ_i), we define (T_{i+1}, Δ_{i+1}) depending on the case as follows:

1. $i \in X$: If $T_i \not\vdash \exists x \varphi_i$, we put $T_{i+1} = T_i$ and $\Delta_{i+1} = \Delta_i$. Let $T_i \vdash \exists x \varphi_i(x)$ and b be a new constant symbol not appeared in $T_i \cup \Delta_i \cup \{\varphi_i\}$. If $T_i, \varphi_i(b) \vdash \delta$ for some $\delta \in \Delta_i$, then $T_i, \exists x \varphi_i(x) \vdash \delta$ and so $T_i \vdash \delta$. This contradicts the

induction hypothesis. Thus $T_i \cup \{\varphi_i(b)\}$ cannot prove any sentence of Δ_i and we may put $\Delta_{i+1} = \Delta_i$ and $T_{i+1} = T_i \cup \{\varphi_i(b)\}$.

2. $i \in Y$: If $T_i \vdash \forall x \varphi_i \vee \delta$ for some $\delta \in \Delta$, we put $T_{i+1} = T_i$ and $\Delta_{i+1} = \Delta_i$. Let $T_i \nvdash \forall x \varphi_i(x) \vee \delta$ for any $\delta \in \Delta_i$ and b be a new constant symbol which is not used in $T_i \cup \Delta_i \cup \{\varphi_i\}$. Put $T_{i+1} = T_i$ and $\Delta_{i+1} = \Delta_i \vee \varphi_i(b)$. We have only to show that T_i does not prove any sentence of Δ_{i+1}. Let $T_i \vdash \delta \vee \varphi_i(b)$ for some $\delta \in \Delta_i$. Since b is a new constant symbol, we have $T_i \vdash \forall x(\delta \vee \varphi_i(x))$ and by III2, we have $T_i \vdash \delta \vee \forall x \varphi_i(x)$. This is a contradiction.

3. $i \in Z$: If $T_i \cup \{\sigma_i\} \nvdash \Delta$ put $T_{i+1} = T_i \cup \{\sigma_i\}$ and $\Delta_{i+1} = \Delta_i$. Otherwise, $T_i \nvdash \Delta_i \vee \sigma_i$ and we can put $T_{i+1} = T_i$ and $\Delta_{i+1} = \Delta_i \vee \sigma_i$.

Now put $\overline{T} = \cup_i T_i$ and $\overline{\Delta} = \cup_i \Delta_i$. Then we have:

(i) $(\overline{T}, \overline{\Delta})$ is a complete consistent pair.
(ii) for each $\varphi(x)$ there is a b such that $\overline{T} \vdash \exists x \varphi(x)$ implies $\overline{T} \vdash \varphi(b)$.
(iii) for each $\psi(x)$, there is a b such that $\overline{T} \nvdash \forall x \varphi(x)$ implies $\varphi(b) \in \overline{\Delta}$. In particular, $\overline{T} \vdash \forall x \varphi(x)$ if and only if $\overline{T} \Vdash \varphi(b)$ for every constant symbol b in $L(C)$.

Finally, we construct a model for $\overline{T}$. Let $\overline{a} = a_0, a_1, \ldots$ be an enumeration of all constant symbols of $L(C)$. For simplicity, we omit the parameter set $\overline{a}$ and write sentences of $L(C)$ in the form φ, ψ, etc.. As underlying set of nodes k, take all atomic sentences θ for which $\overline{T} \vdash \neg\neg\theta$ modulo the equivalence relation $\overline{T} \vdash \theta \Leftrightarrow \theta'$. Put a relation on k by defining $[\theta] \succeq [\theta']$ if $\overline{T} \vdash \theta \Rightarrow \theta'$. Then k is a linearly ordered set whose minimum is $0 = [\top]$. Now we construct the intended semi-classical model. For $a, b \in D = \{a_0, a_1, \ldots\}$ define $a \sim b$ if $\overline{T} \vdash a = b$. For each θ let $M = M_\theta = \frac{D}{\sim}$. We don't distinguish between a_i and its class (likewise for θ). Put a structure on M_θ by defining, for each atomic formula $\eta(\overline{x})$, $M_\theta \models \eta[\overline{a}]$ if and only if $\overline{T} + \theta \vdash \eta$. This does not depend on the choice of representatives. Clearly, $M = (M_\alpha)_{\alpha \in k}$ is a semi-classical model. We prove by induction on the complexity of φ that for each atomic θ:

$$M_\theta \Vdash \varphi \quad \text{if and only if} \quad \overline{T} + \theta \vdash \varphi.$$

1. φ is atomic: by definition.
2. $\varphi_1 \wedge \varphi_2$: clear.
3. $\varphi_1 \vee \varphi_2$: by primeness of $\overline{T}$.
4. $\varphi_1 \Rightarrow \varphi_2$: if: clear.

only if: Assume that $M_\theta \Vdash \varphi_1 \Rightarrow \varphi_2$. By the induction hypothesis, for each $\theta' \succeq \theta$, $\overline{T} + \theta' \vdash \varphi_1$ implies $\overline{T} + \theta' \vdash \varphi_2$. We have nothing to prove if $\overline{T} + \theta \vdash \varphi_2$. Assume that $\overline{T} + \theta \nvdash \varphi_2$. In this case, we claim that $\overline{T} \vdash \varphi_1 \Rightarrow \varphi_2$. Assume not. Then $\overline{T} \vdash \varphi_2 \rightarrow \varphi_1$. By (ii) and axiom III4, there is an atomic η such that

(*) $$\overline{T} \vdash (\varphi_2 \longrightarrow \eta) \wedge (\eta \Longrightarrow \varphi_1)$$

By the induction hypothesis, $M_\eta \Vdash \varphi_1$. On the other hand, $\overline{T} \vdash \eta \Rightarrow \varphi_2$ by (*) leads to $\overline{T} \vdash \varphi_2$. Thus, $M_\eta \nVdash \varphi_2$. In particular, we cannot have $\eta \succeq \theta$. Thus, $M_\theta \Vdash \varphi_1$ which by the assumption $\overline{T} + \theta \nvdash \varphi_2$ leads to a contradiction.

5. $\exists x\varphi(x)$: $M_\theta \Vdash \exists x\varphi(x)$ if and only if there is a $b \in M$ such that $M_\theta \Vdash \varphi(b)$. By the induction hypothesis, this is equivalent to saying that there is a b with $\overline{T} + \theta \vdash \varphi(b)$. By (ii) and III3, this is equivalent to $\overline{T} + \theta \vdash \exists x\varphi(x)$.

6. $\forall x\varphi(x)$: if: clear.

only if: assume that $\overline{T} \nvdash \theta \Rightarrow \forall x\varphi(x) \equiv \forall x(\theta \Rightarrow \varphi(x))$. Then, there is a constant symbol b such that $\overline{T} \nvdash (\theta \Rightarrow \varphi(b))$. By the induction hypothesis, $M_\theta \nVdash \varphi(b)$.

Finally, we have $M_0 \Vdash \varphi$ if and only if $\overline{T} \Vdash \varphi$. In particular, M is a model of the pair (T, Δ). Note also that M is a semi-classical model.   ⊣

Every prime theory T has a model whose theory is exactly T. It is natural to consider prime theories as semi-classical counterparts of complete theories in classical model theory.

EXAMPLE 2. By the height of a model we mean the height of its frame. The assertion M is of height at least n is expressible negatively. Assume the language consists only of one relation symbol S. Then the sentence

$$\forall \overline{x}_0 \cdots \overline{x}_n \left(\bigvee_{i \neq j} S\overline{x}_i \iff S\overline{x}_j \right)$$

is forced in M if and only if M is of height at most n. Then by the completeness theorem, a theory which has models of arbitrarily large heights has a model of infinite height.

Using (only) the axiom III1 we can prove a normal form theorem for quantifier-free formulas.

PROPOSITION 1.6 (*Normal forms*). *Every quantifier-free formula is equivalent to a normal disjunctive formula i.e. a formula of the form* $\bigvee_i \bigwedge_j \varphi_{ij}$ *where each* φ_{ij} *is semi-atomic. Similarly, any quantifier-free formula is equivalent to a normal conjunctive formula.*

PROOF. We prove the claim by induction on the complexity of formulas. There is nothing to prove for atomic formulas. The cases $\varphi \vee \psi$ and $\varphi \wedge \psi$ are clear. Let us consider the case $\varphi \Rightarrow \psi$. An easy use of the rules $(0) - (12)$ proves the claim for all possible combinations for φ or ψ. The only non-trivial case is the formula $(\varphi_1 \Rightarrow \varphi_2) \Rightarrow \psi$. For this case, first represent $\varphi_1 \Rightarrow \varphi_2$ and ψ in normal disjunctive forms. Then use De Morgan laws to represent $(\varphi_1 \Rightarrow \varphi_2) \Rightarrow \psi$ in the form $\bigvee_i \bigwedge_j \psi_{ij}$ where each ψ_{ij} has one of the forms: semi-atomic, $(\theta_1 \Rightarrow \theta_2) \Rightarrow \theta_3$ or $[(\theta_1 \Rightarrow \theta_2) \Rightarrow \theta_3] \Rightarrow \theta_4$ (each θ_i atomic). We have only to show that the last two formulas have equivalent normal disjunctive forms. This is true $(\theta_1 \Rightarrow \theta_2) \Rightarrow \theta_3$ by (9). Finally, an application of the same rule for the part $(\theta_1 \Rightarrow \theta_2) \Rightarrow \theta_3$ of the third form shows that only $[\theta_2 \rightarrow \theta_1] \Rightarrow \theta_4 = [(\theta_1 \Rightarrow \theta_2) \Rightarrow \theta_1] \Rightarrow \theta_4$ should be put in normal disjunctive form. Another application of the same rule for the latter one leads us to $\theta_1 \rightarrow (\theta_1 \Rightarrow \theta_2)$ which is easily seen to be equivalent to $\theta_1 \Rightarrow \theta_2$. A similar proof works for normal conjunctive forms.   ⊣

§2. Some basic model theory.

DEFINITION 2.1. Let M and N be semi-classical models. An injective function $f : M \rightarrow N$ is called embedding if for each quantifier-free formula $\varphi(\overline{x})$ and $\overline{a} \in M$:

$$M \Vdash \varphi[\overline{a}] \quad \text{if and only if} \quad N \Vdash \varphi[\overline{a}].$$

In other words, f preserves all quantifier-free formulas. An elementary embedding is an embedding which preserves all formulas. If M is a subset of N, it is a submodel (resp. elementary submodel) of N denoted by $M \subseteq N$ (resp. $M \leq N$), if the inclusion map is an embedding (elementary embedding). A surjective elementary embedding is called isomorphism.

Note that there is no reference to the nodes in this definition. Regarding the normal form theorem, for f to be an embedding, it is sufficient to preserve semi-atomic formulas. In fact, it easy to see that f is an embedding if and only if it preserves simple semi-atomic formulas i.e. formulas of the form $\theta \Rightarrow \eta$ where θ and η are atomic. More interesting is the following proposition which is due to the fact that each node corresponds to a new atomic relation.

PROPOSITION 2.2. *Any surjective embedding is an isomorphism.*

PROOF. Let $f : M \rightarrow N$ be a surjective embedding. We prove by induction on the complexity of formulas that for each formula φ each $\overline{a} \in M$:

$$M \Vdash \varphi[\overline{a}] \quad \text{if and only if} \quad N \Vdash \varphi[f(\overline{a})].$$

For simplicity, we omit $\overline{a}$ and $f(\overline{a})$ from the formulas. The claim holds for semi-atomic formulas by the assumption. Regarding $(0) - (12)$, the only non-trivial case is when the formula is of the form $\varphi \Rightarrow \psi$. In fact, we have only to consider two cases: $\forall x \varphi'(x) \Rightarrow \psi$ and $\varphi \Rightarrow \exists x \psi'(x)$: We prove the claim for the first case. The second case is similar.

Let $M \Vdash \forall x \varphi'(x) \Rightarrow \psi$. If $N \nVdash \forall x \varphi'(x) \Rightarrow \psi$ then there is a node $\alpha \in k_N$ such that $N_\alpha \Vdash \forall x \varphi'(x)$ and $N_\alpha \nVdash \psi$. Let $\theta(f(\overline{c}))$ be an atomic sentence which represents α, i.e. α is the least node such that $N_\alpha \Vdash \theta(f(\overline{c}))$. Then we have $N_0 \Vdash \theta(f(\overline{c})) \Rightarrow \forall x \varphi'(x)$ and $N_0 \nVdash \theta(f(\overline{c})) \Rightarrow \psi$. Thus, by the induction hypothesis, we have $M_0 \Vdash \theta(\overline{c}) \Rightarrow \forall x \varphi'(x)$ (the complexity can be easily reduced here) and $M_0 \nVdash \theta(\overline{c}) \Rightarrow \psi$. This is a contradiction. The converse is similar.   ⊣

As we noted in the proof, if θ represents the node α, $M_\alpha \Vdash \varphi$ may be replaced by $M_0 \Vdash \theta \Rightarrow \varphi$. This explains the crucial role of the conditions $\mathcal{C}_1, \mathcal{C}_2$. The following example shows that these constraints are not superfluous.

EXAMPLE 3. Consider the Kripke models $M = (\mathbb{Q}, <_r)_{r \in \mathbb{Q}^+}$ and $N = (\mathbb{Q}, \prec_r)_{r \in \mathbb{Q}^+}$ where $x <_r y$ if $y - x \geq \frac{1}{1+r}$ and $x \prec_r y$ if $y - x \geq e^{-r}$. The first one is a semi-classical model while the second one is not, because it does not satisfy $\mathcal{C}1, \mathcal{C}2$. The identity function preserves the formula $x > y \Rightarrow z > t$, hence it is a surjective embedding. However, it is not an isomorphism because N forces $0 < \frac{1}{2} \Rightarrow \exists x (0 < x \wedge \neg\neg x < \frac{1}{2})$ while M does not (use the fact that $\ln 2$ is irrational).

Given a non-empty subset $X \subseteq M$, let N be the smallest set containing X and interpretations of constant symbols, and closed under the operations of M. Then for each atomic formulas θ, θ' and $\overline{a} \in N$ put $N \Vdash \theta(\overline{a}) \Rightarrow \theta'(\overline{a})$ if and only if $M \Vdash \theta(\overline{a}) \Rightarrow \theta'(\overline{a})$. This defines a model N which is the smallest submodel of M containing X. Regarding the above counterexample, if the ambient model M is not semi-classical, several structures on the same subset of M may be considered as submodels of M. In particular, the notion of generated submodel does not make sense in general. An easy description of the notion of elementary submodel is:

PROPOSITION 2.3 (*Tarski test*). *Let $M \subseteq N$. Then M is an elementary submodel of N if and only if for each formula $\varphi(t, \overline{x})$ and $\overline{a} \in M$ the following conditions hold:*
– *if there is a $b \in N$ such that $N \Vdash \varphi[b, \overline{a}]$ then b can also be found in M.*
– *if there is a $b \in N$ such that $N \nVdash \varphi[b, \overline{a}]$ then b can also be found in M.*

PROOF. $\Rightarrow$: obvious.

$\Leftarrow$: We prove by induction on the complexity of formulas that for each formula $\varphi(\overline{x})$ and any $\overline{a} \in M$, $M \Vdash \varphi[\overline{a}]$ if and only if $N \Vdash \varphi[\overline{a}]$. For semi-atomic ones we have nothing to do. The cases $\varphi \vee \psi$ and $\varphi \wedge \psi$ are obvious. Assume the claim be proved for $\varphi(t, \overline{x})$. If $N \Vdash \forall t \varphi[t, \overline{a}]$ then clearly $M \Vdash \forall t \varphi[t, \overline{a}]$. Assume $N \nVdash \forall t \varphi[t, \overline{a}]$. Then there is a $b \in N$ such that $N \nVdash \varphi[b, \overline{a}]$. By the assumption, b can also be found in M. Thus, by the induction hypothesis, $M \nVdash \forall t \varphi[b, \overline{a}]$. The case $\exists t \varphi(t, \overline{x})$ is similar. The only non-trivial case is the formula $\varphi \Rightarrow \psi$. By the rules $(0) - (12)$, if φ or ψ has one of the forms $\varphi_1 \wedge \varphi_2$, $\varphi_1 \vee \varphi_2$ and $\varphi_1 \Rightarrow \varphi_2$ then the problem reduces to formulas of smaller lengths. The same thing is true if φ has the form $\exists x \varphi'(x)$ or ψ has the form $\forall x \psi'(x)$. It remains the following two cases:

$\varphi \Rightarrow \exists x \psi(x)$: Let $M \Vdash \varphi \Rightarrow \exists x \psi(x)$ but $N \nVdash \varphi \Rightarrow \exists x \psi(x)$. In particular, we must have $M, N \nVdash \exists x \psi$. By rule (0) and axiom III4 there is an atomic formula θ and $\overline{b} \in N$ such that

$(**)$ $\qquad N \Vdash \exists x \psi(x) \longrightarrow \theta(\overline{b}) \wedge \theta(\overline{b}) \Longrightarrow \varphi.$

By the first condition of the proposition (or rather by its generalization), we may assume that $\overline{b} \in M$. By the induction hypothesis, $M \Vdash \theta(\overline{b}) \Rightarrow \varphi$. If we also show that $M \Vdash \exists x \psi(x) \rightarrow \theta(\overline{b})$, then we can deduce that $M \Vdash \exists x \psi(x) \rightarrow \varphi$ which by (3) contradicts the assumptions. Assume not. Then $M \Vdash \theta(\overline{b}) \Rightarrow \exists x \psi(x)$ and $M \nVdash \theta(\overline{b})$. So, by axiom III3 and the induction hypothesis we must have $N \Vdash \theta(\overline{b}) \Rightarrow \exists x \psi(x)$ which by (**) implies $N \Vdash \exists x \psi$. This is a contradiction.

Conversely assume that $\varphi \Rightarrow \exists x \psi(x)$ is forced in N but not forced in M. In particular, we must have $M, N \nVdash \exists x \psi$. There is an atomic formula θ and $\overline{b} \in M$ such that

$$(***) \qquad\qquad M \Vdash \exists x \psi(x) \longrightarrow \theta(\overline{b}) \wedge \theta(\overline{b}) \Longrightarrow \varphi.$$

So by the induction hypothesis, $N \Vdash \theta(\overline{b}) \Rightarrow \varphi$ and, as above, we have only to show that $N \Vdash \exists x \psi(x) \rightarrow \theta(\overline{b})$. Assume not. Then $N \Vdash \theta(\overline{b}) \Rightarrow \exists x \psi(x)$ but $N \nVdash \theta(\overline{b})$. So, by axiom III3, the first condition and the induction hypothesis we must have $M \Vdash \theta(\overline{b}) \Rightarrow \exists x \psi(x)$ which by (***) implies $M \Vdash \exists x \psi$.

$\forall x \varphi(x) \Rightarrow \psi$: Similar. $\dashv$

Clearly, the union of a chain of submodels of M is again a submodel. This helps us to prove the downward Löwenheim-Skolem theorem.

COROLLARY 2.4. *Let $\aleph_0 \leq \kappa \leq |M|$ and X be a subset of M of cardinality at most κ. Then there is an elementary submodel N of M containing X whose cardinality is κ.*

The upward one is a simple consequence of the compactness theorem. With a little more effort, we can prove the stronger version where an elementary extension of a given model (with a prescribed cardinality) is found.

PROPOSITION 2.5 (*elementary chain property*). *Let I be a linearly ordered set and $\{A^i\}_{i \in I}$ an elementary chain of models. Then there exists a model $A = \cup_i A^i$ such that $A^i \leq A$ for each i.*

PROOF. Let $A = \cup_i A^i$ and $\theta \Rightarrow \eta$ be a simple semi-atomic formula. For each $\overline{a} \in A$, let $A \Vdash \theta \Rightarrow \eta[\overline{a}]$ if there exists A^i containing $\overline{a}$ such that, $A^i \Vdash \theta \Rightarrow \eta[\overline{a}]$. In fact, the frame of A is the union of the frames of A^i's. Each A^i is a submodel of A. We prove by induction on the complexity of formulas that for each formula φ, $i \in I$ and $\overline{a} \in A^i$:

$$A^i \Vdash \varphi[\overline{a}] \quad \text{if and only if} \quad A \Vdash \varphi[\overline{a}].$$

To be brief, we treat only the cases $\forall$ and $\Rightarrow$. Clearly $A \Vdash \forall x \varphi[x, \overline{a}]$ implies $A^i \Vdash \forall x \varphi[x, \overline{a}]$. Let $A^i \Vdash \forall x \varphi[x, \overline{a}]$. Then for any j and any $b \in A^j$ we have $A^j \Vdash [b, \overline{a}]$. Thus, by the induction hypothesis, for any $b \in A$ we have $A \Vdash [b, \overline{a}]$ which means that $A \Vdash \forall x \varphi[x, \overline{a}]$.

Let consider the formula $\varphi \Rightarrow \psi$ where at least one of φ and ψ is not atomic. By the rules (0)–(12) we have only to consider the following cases:

$\forall x \varphi' \Rightarrow \psi$: Let $A^i \Vdash \forall x \varphi' \Rightarrow \psi$ but $A \nVdash \forall x \varphi' \Rightarrow \psi$. In particular, A and A^i do not force ψ. There is an atomic θ, $j \geq i$ and $\overline{b} \in A^j$ such that

$$A \Vdash \psi \longrightarrow \theta(\overline{b}) \wedge \theta(\overline{b}) \Longrightarrow \forall x \varphi'.$$

Using the induction hypothesis we can easily deduce that $A^j \Vdash \psi \rightarrow \forall x \varphi'$. Since $A^i \leq A^j$ we have $A^i \Vdash \psi \rightarrow \forall x \varphi'$ which by the assumption implies $A^i \Vdash \psi$. This is a contradiction. The converse is similar.

$\varphi \Rightarrow \exists x \psi'(x)$: Similar.   ⊣

Usually, preservation theorems explain the relation between embeddings and theories. Here we prove Łoś-Tarski theorem, some more preservation results may be found in [Bagheri]. Let $\mathrm{diag}^+(M)$ (resp. $\mathrm{diag}^-(M)$) denote the set of all quantifier-free sentences with parameters in M which are forced in M (resp. not forced in M). It is clear that if M is a subset of N, then $M \subseteq N$ if and only if N is a model of the pair $(\mathrm{diag}^+(M), \mathrm{diag}^-(M))$.

PROPOSITION 2.6. *A theory T is universal if and only the class of models of T is closed under taking submodels.*

PROOF. Obviously, the class of models of any universal theory is closed under taking submodels. Assume that the condition is satisfied for T. Let Γ be the collection of all universal consequences of T. We have only to show that any model of Γ is also a model of T. Let N be a model of Γ. We show that $T \cup \mathrm{diag}^+(N)$ does not prove any sentence in $\mathrm{diag}^-(N)$. Assume that $T, \varphi(\overline{c}) \vdash \psi(\overline{c})$ where $\varphi(\overline{c}) \in \mathrm{diag}^+(N)$ and $\psi(\overline{c}) \in \mathrm{diag}^-(N)$. Then $\forall \overline{x}(\varphi(\overline{x}) \Rightarrow \psi(\overline{x}))$ belongs to Γ and so it is forced in N. This is a contradiction since $N \nVdash \varphi(\overline{c}) \Rightarrow \psi(\overline{c})$. By the completeness theorem, there exists a model M of $T \cup \mathrm{diag}^+(N)$ which does not force any sentence in $\mathrm{diag}^-(N)$. In particular, N is a submodel of M and thus $N \Vdash T$.   ⊣

The following proposition provides us a suitable criterion for quantifier elimination.

PROPOSITION 2.7. *Let T be a theory and $\varphi(\overline{x})$ be a formula. Then the following assertions are equivalent:*

(i) *There is a quantifier-free formula $\psi(\overline{x})$ such that $T \Vdash \varphi \Leftrightarrow \psi$.*
(ii) *If M and N are models of T and A is a submodel of both M and N, then for all $\overline{a} \in A$, $M \Vdash \varphi[\overline{a}]$ if and only if $N \Vdash \varphi[\overline{a}]$.*

PROOF. (i)⇒(ii): Obvious.
(ii)⇒(i): Let

$$\Gamma = \{\psi(\overline{x}) : \psi \text{ is quantifier-free and } T \Vdash \varphi \Longrightarrow \psi\}.$$

Let $\overline{d}$ be a sequence of new constants. We have only to show that $T \cup \Gamma(\overline{d}) \vdash \varphi(\overline{d})$. Assume that $T \cup \Gamma(\overline{d}) \nvdash \varphi(\overline{d})$. Then there is a model $M \Vdash T \cup \Gamma(\overline{d})$ such that $M \nVdash \varphi(\overline{d})$. Let A be the submodel of M generated by $\overline{d}$. We claim that

$$T \cup \text{diag}^+(A) \cup \varphi(\overline{d}) \not\vdash \text{diag}^-(A).$$

Otherwise, there are $\psi^+(\overline{d}) \in \text{diag}^+(A)$ and $\psi^-(\overline{d}) \in \text{diag}^-(A)$ such that $T \cup \{\varphi(\overline{d}), \psi^+(\overline{d})\} \vdash \psi^-(\overline{d})$. Thus $A \Vdash \psi^+(\overline{d}) \Rightarrow \psi^-(\overline{d})$. But, since $A \Vdash \psi^+(\overline{d})$, we obtain $A \Vdash \psi^-(\overline{d})$ which is a contradiction. Now, by completeness, there is a model N such that $A \subseteq N$ and $N \Vdash T + \varphi(\overline{d})$. In particular, $A \Vdash \varphi(\overline{d})$. This is a contradiction.   ⊣

In practice, it is sufficient to verify quantifier elimination for a special class of formulas.

COROLLARY 2.8. *To show that a theory T has quantifier elimination, it is sufficient to verify that for each quantifier-free formula $\varphi(x, \overline{y})$ and $\overline{a} \in A \subseteq M, N \Vdash T$:*

- *if there is a $b \in M$ such that $M \Vdash \varphi[b, \overline{a}]$ then there is a $c \in N$ such that $N \Vdash \varphi[c, \overline{a}]$.*
- *if there is a $b \in M$ such that $M \not\Vdash \varphi[b, \overline{a}]$ then there is a $c \in N$ such that $N \not\Vdash \varphi[c, \overline{a}]$.*

At the end of this section we mention two sporadic results which, to some extent, is related to the expressive power of semi-classical logic. The first one is that, finite elementarily equivalent models are isomorphic. This is obtained by a simple modification of the classical proof. More generally, if the theory of a finite model M is contained in the theory of a model N then there is a node α in the frame of M such that $(M_\beta)_{\beta \succeq \alpha}$ is isomorphic to N. The second result concerns indiscernible sequences. In classical model theory, an indiscernible sequence in M is a sequence $\{a_i\}_{i \in \kappa}$ of elements of M such that for each formula $\varphi(\overline{x})$ and each finite increasing sequences $a_{i_1}, \ldots, a_{i_n}$ and $a_{j_1}, \ldots, a_{j_n}$,

$$M \Vdash \varphi[a_{i_1}, \ldots, a_{i_n}] \quad \text{iff} \quad M \Vdash \varphi[a_{j_1}, \ldots, a_{j_n}].$$

This last assertion may be replaced by

$$M \Vdash \varphi[a_{i_1}, \ldots, a_{i_n}] \Longleftrightarrow \varphi[a_{j_1}, \ldots, a_{j_n}].$$

In semi-classical setting, the second assertion is stronger than the first one. Below, we prove the existence of indiscernible sequences in the weaker sense and leave open the existence of such sequences in the stronger sense. Like before, $\text{ediag}^+(M)$ (resp. $\text{ediag}^-(M)$) denotes the collection of all sentences with parameters in M which are forced (resp. not forced) in M.

PROPOSITION 2.9. *Every infinite model M has elementary extensions with arbitrary large indiscernible sequences.*

PROOF. For simplicity assume that $M = \{a_0, a_1, \ldots\}$ is countable. Let $\varphi_0, \varphi_1, \ldots$ be an enumeration of all formulas of the language L. Consider a set $C = \{c_0, c_1, \dots\}$ of new constant symbols. We construct a sequence C_n of subsets of C, a sequence (T_n, Δ_n) of consistent pairs in the languages $L(C_n)$ and a function v which assigns to each L-formula either 0 or 1.

Interpret each symbol c_i by a_i. We start by putting $C_{-1} = C$, $T_{-1} = \text{ediag}^+(M)$ and $\Delta_{-1} = \text{ediag}^-(M)$. Let (T_{n-1}, Δ_{n-1}) and C_{n-1} be constructed and v be defined over $\varphi_0, \dots, \varphi_{n-1}$. Let ℓ be the number of free variables in φ_n and X be the collection of all sets $\{c_{i_1}, \dots, c_{i_\ell}\} \subseteq C_{n-1}$ with $i_1 < \cdots < i_\ell$ such that $M \Vdash \varphi_n(c_{i_1}, \dots, c_{i_\ell})$. Likewise let Y be the set of those sets $\{c_{i_1}, \dots, c_{i_\ell}\} \subseteq C_{n-1}$ such that $M \nVdash \varphi_n(c_{i_1}, \dots, c_{i_\ell})$. By Ramsey theorem there is a homogeneous set C_n for the partition $\{X, Y\}$. If it is a homogeneous set for X, put $\Delta_n = \Delta_{n-1}$,

$$T_n = (T_{n-1} \downarrow C_n) \cup \{\varphi_n(c_{i_1}, \dots, c_{i_\ell}) : i_1, \dots, i_\ell \text{ is an increasing sequence in } C_n\}$$

and $v(\varphi_n) = 1$ where $T_{n-1} \downarrow C_n$ denotes the restriction of T_{n-1} to the language $L(C_n)$. Otherwise, C_n is a homogeneous for Y. Then put $T_n = T_{n-1}$,

$$\Delta_n = (\Delta_{n-1} \downarrow C_n) \cup \{\varphi_n(c_{i_1}, \dots, c_{i_\ell}) : i_1, \dots, i_\ell \text{ is an increasing sequence in } C_n\}$$

and $v(\varphi_n) = 0$. In either case (T_n, Δ_n) is a consistent pair and for every increasing sequences $c_{i_1}, \dots, c_{i_\ell}$ and $c_{j_1}, \dots, c_{j_\ell}$ in C_{n-1} we have $M \Vdash \varphi_n[c_{i_1}, \dots, c_{i_\ell}]$ if and only if $M \Vdash \varphi_n[c_{j_1}, \dots, c_{j_\ell}]$.

Now, let $\{d_i\}_{i < \kappa}$ be an arbitrary set of new constant symbols distinct from those of C. Put

$$T^\infty = \text{ediag}^+(M) \cup \{\varphi_n(d_{i_1}, \dots, d_{i_\ell}) : n \in \omega, \, i_1 < \cdots < i_\ell, \, v(\varphi_n) = 1\}$$

$$\Delta^\infty = \text{ediag}^-(M) \cup \{\varphi_n(d_{i_1}, \dots, d_{i_\ell}) : n \in \omega, \, i_1 < \cdots < i_\ell, \, v(\varphi_n) = 0\}.$$

Then, $(T^\infty, \Delta^\infty)$ is a consistent pair because each finite part (U, V) of $(T^\infty, \Delta^\infty)$ can be injected in some (T_n, Δ_n). By the compactness theorem, there exists a model N for $(T^\infty, \Delta^\infty)$. Thus M is an elementary submodel of N and N contains an indiscernible sequence of cardinality κ.   ⊣

§3. ω-**Categoricity.** In this section we modify two simple classical theories and obtain examples of non-classical ω-categorical theories which admit quantifier elimination. Our goal is just to prove the existence of such theories and to clarify the situation for further study. One may wish to construct better examples such as the theory of semi-classical random graphs or even to examine the method of Fraisse.

DEFINITION 3.1. A semi-classical theory T is ω-categorical if for all countable models M and N, $Th(M) = Th(N) = T$ implies that M is isomorphic to N.

Any ω-categorical theory is prime but not necessarily complete. It is complete if all its countable models are isomorphic.

THE THEORY OF BLACK-WHITE POINTS:

Let $L = \{P\}$ where P is a unary relation symbol. As is known, the classical theory whose axioms state that P and its complement are infinite, is ω-categorical and has quantifier elimination. We modify this theory and obtain a non-classical ω-categorical theory T_P which admits quantifier elimination.

Axioms of T_P:

1. P and $\neg P$ are infinite.
2$_m$. $\forall \bar{t} x \exists y (Py \Leftrightarrow Px \wedge \bigwedge_i y \neq t_i]$ where m is the length of $\bar{t}$.
3$_m$. $\forall \bar{t} x y [(Px \rightarrow Py) \Rightarrow \exists t ((Px \rightarrow Pt) \wedge (Pt \rightarrow Py) \wedge \bigwedge_i^m t \neq t_i)]$ where $m = |\bar{t}|$.
4. $\neg \forall t (Pt \vee \neg Pt)$.
5. $\forall x [\forall t (Pt \vee (Pt \Rightarrow Px)) \Rightarrow Px]$.

Axiom 4 states that at each level there are points which are neither black nor white. So, it makes the theory to be non-classical. Axioms 3 and 5 state that every model of T_P has a dense frame (in fact 5 implies 3). In particular, there is no model of T_P in which 0 has a successor. A model M for this theory is obtained as follows: take an infinite set A and for each $r \in \mathbb{Q}^+$ put $M_r = (A \times \mathbb{Q}^+, P_r)$ where $P_r = A \times [1, 2 + r]$. It is easy to see that axioms 1-5 are forced in M_0.

Let us recall that a formula θ is decidable in T if $T \Vdash \forall \bar{x} (\theta(\bar{x}) \vee \neg \theta(\bar{x}))$. Below, we will need the following fact.

FACT 3.2. If θ or θ' is decidable in T then:

(i) $\theta \Rightarrow \theta' \equiv_T \neg \theta \vee \theta'$,
(ii) $\theta \rightarrow \theta' \equiv_T (\neg \theta \wedge \neg \neg \theta') \vee \theta'$.

PROPOSITION 3.3. T_P has quantifier elimination.

PROOF. By the normal form theorem, we have to show that any formula of the form $\exists t [\varphi_1(t, \bar{x}) \wedge \cdots \wedge \varphi_n(t, \bar{x})]$ or $\forall t [\varphi_1(t, \bar{x}) \vee \cdots \vee \varphi_n(t, \bar{x})]$, where each φ_i is semi-atomic, is equivalent to a quantifier-free formula.

Existential case: We have nothing to do if one of φ_i's is of the form $t = x$. Thus, regarding fact 3.2, we have only to show that the following formula is equivalent to a quantifier-free formula:

$$\varphi = \exists t \left[\left(\bigwedge_{i=1}^k t \neq t_i \right) \wedge \left(\bigwedge_{i=1}^{\ell} Pt \Longleftrightarrow \theta_i^1 \right) \wedge \left(\bigwedge_{i=1}^m \theta_i^2 \longrightarrow Pt \right) \wedge \left(\bigwedge_{i=1}^n Pt \longrightarrow \theta_i^3 \right) \right]$$

where, each θ_j^i is an atomic formula in which t does not appear and $k, \ell, m, n \geq 0$. By axiom 1, we may assume $\ell + m + n \neq 0$. We have two cases:

• $\ell \geq 1$: We replace all occurrences of Pt in φ (except one) by θ_1^1. Then, φ is equivalent to a formula of the form $\exists t [t \neq t_1 \wedge \cdots \wedge t \neq t_k \wedge (Pt \Leftrightarrow \theta_1^1)] \wedge \psi$ where t does not appear in ψ. This is equivalent to ψ by axioms 1 and 2.

- $\ell = 0$: If $m = 0$, φ is equivalent to $\bigwedge_i \neg\neg\theta_i^3$. If $n = 0$, φ is equivalent to $\top$. Otherwise, by axiom 3, φ is equivalent to the formula $\bigwedge_{ij}(\theta_i^2 \to \theta_j^3)$.

Universal case: Like the existential case, we can simplify the formula and assume that:

$$\varphi = \forall t \left[t = t_1 \vee \cdots \vee t = t_k \vee \right.$$

$$\left(\bigvee_{i=1}^{\ell} Pt \Longrightarrow \theta_i^1 \right) \vee \left(\bigvee_{i=1}^{h} Pt \longrightarrow \theta_i^2 \right) \vee \left(\bigvee_{i=1}^{m} \theta_i^3 \Longrightarrow Pt \right) \vee$$

$$\left. \left(\bigvee_{i=1}^{n} \theta_i^4 \longrightarrow Pt \right) \right]$$

where, each θ_i^j is an atomic formula in which t does not appear and $k, \ell, h, n, m \geq 0$. By axiom 1, we may assume that $\ell + h + m + n > 0$. We have various cases:

- $m = n = 0$: φ is equivalent to $(\bigvee_i \theta_i^1) \vee (\bigvee_i \theta_i^2)$.
- $\ell = h = m = 0$: By axiom 1, φ is equivalent to $\perp$.
- $\ell = h = 0$, $m \neq 0$: φ is equivalent to $(\bigvee_i \neg\theta_i^3)$.
- $m + n \neq 0$, $\ell + h \neq 0$: In this case, by all the axioms, φ is equivalent to the formula:

$$\left(\bigvee_{ij} \theta_i^3 \Longrightarrow \theta_j^1 \right) \vee \left(\bigvee_{ij} \theta_i^4 \Longrightarrow \theta_j^1 \right) \vee \left(\bigvee_{ij} \theta_i^3 \Longrightarrow \theta_j^2 \right) \vee \left(\bigvee_{ij} \theta_i^4 \longrightarrow \theta_j^2 \right).$$

Let for example check this for the case $h = n = 0$ and $\ell = m = 1$. It is clear that $\theta_1^3 \Rightarrow \theta_1^1$ implies φ. For the converse, assume that for some model M of T_P^i and some $\bar{a} \in M_0$ we have $M \Vdash \varphi[\bar{a}]$ but $M_0 \nVdash (\theta_1^3 \Rightarrow \theta_1^1)[\bar{a}]$. Then $M_0 \Vdash (\theta_1^1 \to \theta_1^3)[\bar{a}]$ and either $M_0 \nVdash \theta_1^3$ or $M_0 \Vdash \theta_1^3$ but $M_0 \nVdash \theta_1^1$. The first case is impossible by axiom 3 and the second case is impossible by axiom 5. Thus, φ is equivalent to $\theta_1^3 \Rightarrow \theta_1^1$. ⊣

THEOREM 3.4. T_P is ω-categorical and complete.

PROOF. Let M and N be two arbitrary countable models of T_P. We use the back and forth method and show that they are isomorphic. Let $f : a_i \mapsto b_i$, $i = 1, \ldots, n$, be a partial isomorphism from M into N. So we have:

$$M \Vdash Pa_i \quad \text{iff} \quad N \Vdash Pb_i,$$
$$M \Vdash \neg Pa_i \quad \text{iff} \quad N \Vdash \neg Pb_i,$$
$$M \Vdash Pa_i \Longrightarrow Pa_j \quad \text{iff} \quad N \Vdash Pb_i \Longrightarrow Pb_j$$

where, $1 \leq i, j \leq n$. Given a new $a \in M$, we will find a new $b \in N$ such that

$f \cup \{(a, b)\}$ be a partial isomorphism. We have 6 cases:

1. $M \Vdash Pa$: By axiom 1, choose a $b \in N$ different from $b_1, \ldots, b_n$ such that $N \Vdash Pb$.
2. $M \Vdash \neg Pa$: By axiom 1, choose a $b \in N$ different from $b_1, \ldots, b_n$ such that $N \Vdash \neg Pb$.
3. $Pa \leftrightarrow Pa_{i_0}$ for some i_0: Use axiom 2 to find a $b \in N$ different from $b_1, \ldots, b_n$ such that $Pb \leftrightarrow Pb_{i_0}$.
4. $(\bot \to Pa) \wedge (Pa \to Pa_i)$ for all i: Apply axiom 4 to find a $b \in N$ such that $(\bot \to Pb) \wedge (Pb \to Pb_i)$ for all i.
5. $Pa_i \to Pa$ for all i but, $M \nVdash Pa$: Apply axiom 5 to find a b such that $Pb_i \to Pb$ for all i but $N \nVdash Pb$.
6. Neither of the above 5 cases hold: In this case, $\{1, \ldots, n\}$ can be partitioned into A, B such that whenever $i \in A$ and $j \in B$ then $(Pa_i \to Pa) \wedge (Pa \to Pa_j)$. Then, use axioms 3 or 5 (depending on whether $M \Vdash Pa_j$ for each $j \in B$ or not), such that for all $i \in A$ and $j \in B$, $N \Vdash (Pb_i \to Pb) \wedge (Pb \to Pb_j)$ but $N \nVdash Pb$.

It is clear that $f \cup \{(a, b)\}$ is a partial isomorphism. The back argument can be done Likewise. Now, let $M = \{a_0, a_1, \ldots\}$ and $N = \{b_0, b_1, \ldots\}$. Construct a sequence $f_0 \subseteq f_1 \subseteq \ldots$ of partial isomorphisms such that for each n, a_n is in the domain of f_{2n} and b_n is in the range of f_{2n+1}. Then, $\cup_n f_n$ is an isomorphism.   ⊣

Using the quantifier elimination, we are able to characterize all parametrically definable subsets of $M = (A \times \mathbb{Q}^+, P_r)_{r \in \mathbb{Q}^+}$. By the definition, $P(M) = A \times [1, 2]$, $\neg P(M) = A \times [0, 1)$ and $\neg\neg P(M) = A \times [1, \infty)$. Other $\emptyset$-definable subsets are those obtained by intersections and unions of these subsets. If $2 < r \leq s < \infty$, then $A \times [r, s]$ is definable by the formula $P(a, s) \Rightarrow Px \Rightarrow P(a, r)$ where $a \in A$ is fixed. Likewise, for any $2 < r < s \leq \infty$, the sets $A \times [r, s)$, $A \times (r, s]$ and $A \times (r, s)$ are definable. Also, for any $2 \leq r \leq \infty$, $A \times [1, r]$ and $A \times [1, r)$ are definable. Note that the family consisting of these sets (including $\emptyset$-definable sets) is closed under finite intersections. Any other parametrically definable subset of M is equal to a finite union of sets of these forms. Note however that while $A \times [0, 2]$ is definable, $A \times (2, \infty)$ is not definable. From the constructive point of view, a definable set is a collection of points which can be defined constructively i.e. by a positive property. In other words, "not satisfying P" is not positive property. Another property of M is that while the classical theory of back-white points is ω-stable, T_P has the order property: simply fix $a \in A$ and consider the sequence $(a, 2), (a, 3), \ldots$; then $Py \Rightarrow Px$ defines an order on this sequence. In fact, any non-classical complete theory interprets an infinite linear order (note that the language is finite).

It may happen that an ω-categorical theory be incomplete. Let $c \in M = (A \times \mathbb{Q}^+, P_r)_{r \in \mathbb{Q}^+}$ be such that $M_s \Vdash Pc$ if and only if $s \geq 3$. Then the theory of (M, c) is not complete since neither $M \Vdash Pc$ nor $M \Vdash \neg Pc$. However, if (N, d) is countable and elementarily equivalent to (M, c), then the partial function $f_0 : c \mapsto d$ can be extended (by back and forth) to an isomorphism from (M, c) to (N, d). Note also that, for each n, both M^n and $(M, c)^n$ have a finite number of definable subsets. This put forth the question whether the Ryll-Nardzewski theorem holds.

THE THEORY OF AN INCREASING EQUIVALENCE RELATION:

Let $L = \{E\}$ where E denotes a binary relation symbol for equivalence relation. To simplify description of T_E, we write $\varphi \to \psi \to \eta$ for $(\varphi \to \psi) \wedge (\psi \to \eta)$ and use English phrases in place of the exact expression of axioms.

Axioms of T_E:

1. $\forall xyz[xEx \wedge (xEy \Rightarrow yEx) \wedge ((xEy \wedge yEz) \Rightarrow xEz)]$

2_n. $\exists t_1, \ldots, t_n (\bigwedge_{i \neq j} \neg t_i E t_j)$.

3. For each atomic $\theta(\overline{x})$ and distinct $y_1, \ldots, y_k$ if $\theta(\overline{x}) \Rightarrow (\bigwedge_{i \neq j} y_i E y_j)$ then there are infinitely many t such that $\theta(\overline{x}) \Leftrightarrow (\bigwedge_i t E y_i)$.

4. For each atomic $\theta(\overline{x}), \eta(\overline{x})$ and distinct $y_1, \ldots, y_k$ if $\theta(\overline{x}) \to [\eta \wedge (\bigwedge_{i \neq j} y_i E y_j)]$ then there are infinitely many t such that $\theta(\overline{x}) \to (\bigwedge_i t E y_i) \to \eta(\overline{x})$.

5. $\forall y \neg \forall x (xEy \vee \neg xEy)$.

6. $\forall xyz[\forall t(tEz \vee (tEz \Rightarrow xEy)) \Rightarrow xEy]$.

By axiom 2 there are infinitely many classes at each node. Taking $\theta = \top$ and $k = 1$ in axiom 3, we see that each such class must be infinite. Axiom 5, makes models of T_E to be non-classical. A consequence of axioms 4 and 6 is that each model has a dense frame. To find a model for T_E, take an infinite set A and let M be the family of all functions f from $\mathbb{Q}^+$ into A for which there is a c such that $f(r) = c$ except on a finite set. Denote elements of M by $(a_t), (b_t)$ etc.. For each $r \in \mathbb{Q}^+$ define an equivalence relation E_r by $(a_t) E_r (b_t)$ if and only if $x_t = y_t$ for all $t \geq r$. It is not hard to show that $(M, E_r)_{r \in \mathbb{Q}^+}$ is a countable model of T_E. By a complicated (although not hard) argument similar to that of T_P, we can show that T_E admits quantifier elimination. Below, we only prove that it is ω-categorical.

REMARK 3.5. T_E proves $\forall xyz[(xEy \to xEz) \Rightarrow (xEy \Leftrightarrow yEz)]$.

THEOREM 3.6. T_E is ω-categorical and complete.

PROOF. Let M and N be models of T_E and $f : a_i \mapsto b_i$ $i = 1, \ldots, n$ be a partial isomorphism. This means that for every $1 \leq i, j, r, s \leq n$:

$$M \Vdash a_i E a_j \quad \text{if and only if} \quad N \Vdash b_i E b_j,$$

$$M \Vdash \neg a_i E a_j \quad \text{if and only if} \quad N \Vdash \neg b_i E b_j \quad \text{and}$$

$$M \Vdash a_i E a_j \Longrightarrow a_r E a_s \quad \text{if and only if} \quad N \Vdash b_i E b_j \Longrightarrow b_r E b_s.$$

Given a new element $a \in M$ we will find a $b \in N$ such that $f \cup \{(a, b)\}$ be a partial isomorphism. To be economic, we write $\varphi(\overline{a})$ for $M \Vdash \varphi[\overline{a}]$ (likewise for $\varphi(\overline{b})$). By axiom 2, we may assume that some a_i is clinally equivalent to a. In fact, without loss of generality, we can assume that each a_i is clinally equivalent to a. There is nothing to prove for $n = 0$. The case $n = 1$ is easily obtained by using axioms 3, 4 and 5. Let $n \geq 2$ and without loss of generality, assume that k is the largest integer such that:

$$aEa_n \Longrightarrow \cdots \Longrightarrow aEa_{k+1} \longrightarrow aEa_k \Longleftrightarrow \cdots \Longleftrightarrow aEa_1.$$

Let also $\theta_1, \ldots, \theta_m$ be the list of all atomic sentences with parameters chosen from $a_1, \ldots, a_n$ and $\theta'_1, \ldots, \theta'_m$ be the corresponding atomic sentences with parameters chosen from $b_1, \ldots, b_n$. We have two cases:

(a) $aEa_1 \Leftrightarrow \theta_{i_0}$ for some i_0:

Then whenever $i < j \leq k$ we have $\theta_{i_0} \Rightarrow a_iEa_j$. Since f is a partial isomorphism, we have $\theta'_{i_0} \Rightarrow b_iEb_j$ for each $i < j \leq k$. By axiom 3, there is a $b \in N_0$ such that for each $i \leq k$, $bEb_i \Leftrightarrow \theta'$. Then for $i \leq k$ we have $aEa_i \Leftrightarrow \theta$ and $bEb_i \Leftrightarrow \theta'$. For $i > k$ we have $aEa_i \rightarrow aEa_1$ and so (by remark 3.5), $aEa_i \Leftrightarrow a_1Ea_i$. On the other hand, $bEa_i \rightarrow bEb_1$ since otherwise we have $bEb_1 \Rightarrow bEa_i$ which implies $bEb_1 \Rightarrow b_1Ea_i$ and this is a contradiction. Therefore, we have $bEb_i \Leftrightarrow b_1Eb_i$.

Now we show that $f \cup \{(a, b)\}$ is a partial isomorphism. To see this, let $\eta \Rightarrow \zeta$ be a semi-atomic sentence with parameters in $a, a_1, \ldots, a_n$. We can assume that at most one of η and ζ contains the parameter a. Let for example $\eta = aEa_i$. By the above discussion, if $i \leq k$ the assertion $aEa_i \Rightarrow \zeta$ is equivalent to $\theta \Rightarrow \zeta$ and since f is a partial isomorphism, this is equivalent to $\theta' \Rightarrow \zeta'$ which is finally equivalent to $bEb_i \Rightarrow \zeta'$. Similarly, if $i > k$, $aEa_i \Rightarrow \zeta$ is equivalent to $a_1Ea_i \Rightarrow \zeta$ which is also equivalent to $b_1Eb_i \Rightarrow \zeta'$ and finally this is equivalent to $bEb_i \Rightarrow \zeta'$.

(b) $M \nVdash aEa_1 \Leftrightarrow \theta_i$ for each $i = 1, \ldots, m$:

Without loss of generality, assume that

$$\theta_1 \longrightarrow \cdots \longrightarrow \theta_\ell \longrightarrow aEa_1 \longrightarrow \theta_{\ell+1} \longrightarrow \cdots \longrightarrow \theta_m.$$

Use one of the axioms 5, 6, 4 (depending on the cases $\ell = 0$, $\ell = m$ and $\ell \neq 0, m$) to find a b such that

$$\theta'_1 \longrightarrow \cdots \longrightarrow \theta'_\ell \longrightarrow bEb_1 \longrightarrow \theta'_{\ell+1} \longrightarrow \cdots \longrightarrow \theta'_m$$

and $N_0 \nVdash bEb_1 \Leftrightarrow \theta'_i$ for any i. We show that $f \cup \{(a, b)\}$ is a partial isomorphism. Like the previous case, we have only to find a suitable substitution for aEa_i. If $i = 1$, we have nothing to do by the choice of b. Assume $i \neq 1$. Then if $aEa_1 \rightarrow a_1Ea_i$ we have also $bEb_1 \rightarrow b_1Eb_i$. Thus $aEa_i \Leftrightarrow aEa_1$ and $bEb_i \Leftrightarrow bEb_1$. Similarly if $a_1Ea_i \rightarrow aEa_1$, then $b_1Eb_i \rightarrow bEb_1$. So we have $aEa_i \Leftrightarrow a_1Ea_i$ and $bEb_i \Leftrightarrow b_1Eb_i$.

By the same argument, if we were given $b \in N$, we could find an $a \in M$ such that $f \cup \{(a, b)\}$ be a partial isomorphism. Now let $M = \{a_0, a_1, \ldots\}$ and $N = \{b_0, b_1, \ldots\}$. Construct a sequence $f_0 \subseteq f_1 \subseteq \cdots$ of partial isomorphisms from M to N such that for each n, a_n is in the domain of f_{2n} and b_n is in the range of f_{2n+1}. Obviously, $\cup_n f_n$ is an isomorphism. ⊣

Let us conclude the paper with some natural questions. Regarding the examples, one may ask whether the Ryll-Nardzewski theorem holds in semi-classical setting. One direction of this theorem requires the omitting types theorem. Also, we restricted ourselves to semi-classical models to prevent the confliction between the frames and embeddings. In fact, the presence of the nodes in an arbitrary Kripke model is too strong. Semi-classical models smoother the effect of this presence a lot (though there are still some anomalies: reduction of a model to a sublanguage changes the notion of forcing). Now, the question is to find a suitable soundness-completeness theorem for infinite languages.

The comments by the two referees were very constructive and helped the improvement of the paper. Included in those was this interesting question: does the Craig interpolation property holds in semi-classical setting? How about the Beth definability theorem?

Acknowledgements. Une partie de ce travail a été fait pendant mon séjour à l'Institut Girard Desargues, Université Lyon1 en tant que ATER. Je tiens à remercier Bruno Poizat pour son très chaleureux accueil et ses commentaires fructueux.

REFERENCES

[Bagheri] S. M. BAGHERI, *Some preservation theorems in an intermediate logic*, to appear in Mathematical Logic Quarterly.

[BM] S. M. BAGHERI and MORTEZA MONIRI, *Some results on Kripke models over an arbitrary fixed frame*, **MLQ. Mathematical Logic Quarterly**, vol. 49 (2003), no. 5, pp. 479–484.

[Baldwin] J. BALDWIN, *Abstract Elementary Classes*, draft of Monograph.

[Ben] ITAY BEN-YAACOV, *Positive model theory and compact abstract theories*, **Journal of Mathematical Logic**, vol. 3 (2003), no. 1, pp. 85–118.

[TD] A. S. TROELSTRA and D. VAN DALEN, *Constructivism in Mathematics. Vol. I*, North-Holland Publishing Co., Amsterdam, 1988.

[D] D. VAN DALEN, *Logic and Structure*, Springer, 1997.

[V] A. VISSER, *Submodels of Kripke models*, **Archive for Mathematical Logic**, vol. 40 (2001), no. 4, pp. 277–295.

INSTITUTE FOR STUDIES IN THEORETICAL
PHYSICS AND MATHEMATICS (IPM)
P. O. BOX 19395-5746
TEHRAN, IRAN
and
DEPARTMENT OF MATHEMATICS
TARBIAT MODARRES UNIVERSITY
TEHRAN, IRAN
E-mail: Bagheri@ipm.ir

PRIMES AND IRREDUCIBLES IN TRUNCATION INTEGER
PARTS OF REAL CLOSED FIELDS

DARKO BILJAKOVIC, MIKHAIL KOCHETOV, AND SALMA KUHLMANN

Abstract. Berarducci (2000) studied irreducible elements of the ring $k((G^{<0})) \oplus \mathbb{Z}$, which is an integer part of the power series field $k((G))$ where G is an ordered divisible abelian group and k is an ordered field. Pitteloud (2001) proved that some of the irreducible elements constructed by Berarducci are actually prime. Both authors mainly concentrated on the case of archimedean G. In this paper, we study *truncation integer parts* of any (non-archimedean) real closed field and generalize results of Berarducci and Pitteloud. To this end, we study the canonical integer part $\mathrm{Neg}(F) \oplus \mathbb{Z}$ of *any truncation closed subfield* F of $k((G))$, where $\mathrm{Neg}(F) := F \cap k((G^{<0}))$, and work out in detail how the general case can be reduced to the case of archimedean G. In particular, we prove that $k((G^{<0})) \oplus \mathbb{Z}$ has (cofinally many) prime elements for any ordered divisible abelian group G. Addressing a question in the paper of Berarducci, we show that every truncation integer part of a non-archimedean exponential field has a cofinal set of irreducible elements. Finally, we apply our results to two important classes of exponential fields: exponential algebraic power series and exponential-logarithmic power series.

§1. Introduction. An *integer part* (IP for short) Z of an ordered field K is a discretely ordered subring, with 1 as the least positive element, and such that for every $x \in K$, there is a $z \in Z$ such that $z \leq x < z + 1$. It follows that the ring of integers $\mathbb{Z}$ is a convex subring of Z. If K is archimedean, then $\mathbb{Z}$ is the only IP of K, so we will be interested in the case of non-archimedean K.

Shepherdson [S] showed that IP's of real closed fields are precisely the models of a fragment of Peano Arithmetic called Open Induction (OI for short). OI is the first-order theory, in the language $L := \{+, \cdot, <, 0, 1\}$, of discretely ordered commutative rings with 1 whose set of non-negative elements satisfies, for each quantifier-free formula $\Phi(x, y)$, the associated

---

2000 *Mathematics Subject Classification.* Primary 06F25; Secondary 13A16, 03H15, 03E10, 12J25, 13A05.

Second author supported by an NSERC Postdoctoral Fellowship, third author partially supported by an NSERC Discovery Grant. This paper was written while the third author was on sabbatical leave at Université Paris 7. The authors wish to thank the Équipe de Logique de Paris 7 for its support and hospitality.

Logic in Tehran
Edited by A. Enayat, I. Kalantari, and M. Moniri
Lecture Notes in Logic, 26
© 2006, Association for Symbolic Logic

induction axiom $I(\Phi)$:

$$\forall y \left[\Phi(0, y) \text{ and } \forall x \left[\Phi(x, y) \longrightarrow \Phi(x + 1, y)\right] \longrightarrow \forall x \Phi(x, y)\right].$$

This correspondence led Shepherdson to investigate the arithmetic properties of IP's of real closed fields. Given a field k and an ordered abelian group G, let us denote by $k((G))$ the field of *generalized power series* with exponents in G and coefficients in k (see Section 2). We write $k(G) := k(t^g \mid g \in G)$ for the subfield generated by k and by the monomials t^g. This subfield is the quotient field of the *group ring* $k[G]$ (which is a domain by the same argument as for the ring of polynomials). If k is an ordered field, then $k((G))$ can be ordered lexicographically. Shepherdson considered the countable recursive "algebraic Puiseux series field" $\mathbb{Q}(t^g \mid g \in \mathbb{Q})^r$, which can be viewed as a subfield of the field $\mathbb{Q}^r((\mathbb{Q}))$ of power series with coefficients in the field of real algebraic numbers $\mathbb{Q}^r$ and exponents in the additive group of rational numbers $\mathbb{Q}$. (The superscript r denotes real closure.) He constructed an IP of this field in which the only irreducible elements are those in $\mathbb{Z}$. In particular, the set of primes is not cofinal in this model of OI. Thus the "cofinality of primes" is not provable from Open Induction. On the other hand, subsequent to the work of Shepherdson, several authors (e.g. [M], [B–O], [Bi]) constructed various models of OI with (a cofinal set of) infinite primes.

In [M–R], Mourgues and Ressayre establish the existence of an IP for any real closed field K as follows. Let V be the natural valuation on K. Denote by k the residue field and by G the value group of K (see Section 2). [M–R] show that there is an order preserving embedding φ of K into the field of generalized power series $k((G))$ such that $\varphi(K)$ is a truncation closed subfield (see Section 3 for details). They observe that for the field $k((G))$, an integer part is given by $k((G^{<0})) \oplus \mathbb{Z}$, where $k((G^{<0}))$ is the (non-unital) ring of power series with negative support. It follows easily (see Proposition 3.1) that for any truncation closed subfield F of $k((G))$, an integer part is given by $Z_F = \text{Neg}(F) \oplus \mathbb{Z}$, where $\text{Neg}(F) := k((G^{<0})) \cap F$. We shall call Z_F the *canonical integer part* of F. Finally $\varphi^{-1}(Z_F)$ is an integer part of K if we take $F = \varphi(K)$. An integer part Z of K obtained in this way from a truncation closed embedding shall be called a *truncation integer part* of K.

In [R], a proof for an exponential analogue of the main result of [M–R] is sketched: every exponential field (see Section 6) has an *exponential* integer part (EIP for short). An EIP is an IP that satisfies some closure conditions under the exponential function (see Section 6).

In [Bo2], Boughattas considers the following extension of OI in the language $L \cup \{2^x\}$ containing a symbol for the exponential function. He defines $OI(2^x)$ to be the first-order theory of discretely ordered commutative rings with 1 whose set of non-negative elements satisfies, for each quantifier-free formula $\Phi(x, y)$ of $L \cup \{2^x\}$, the associated induction axiom $I(\Phi)$, and also the following basic axioms for 2^x:

$$2^0 = 1, \qquad 2^1 = 2, \qquad 2^{x+y} = 2^x 2^y, \qquad x < 2^x.$$

Analogously to Shepherdson's result, Boughattas constructs a model of $OI(2^x)$ in which the only irreducible elements are those in $\mathbb{Z}$, and thus the set of primes is not cofinal in this model of $OI(2^x)$. We note that, unlike the case of OI, an algebraic description of models of $OI(2^x)$ is not known. In particular, the relationship between models of $OI(2^x)$ on the one hand and EIP's of exponential fields on the other hand remains unclear. In [B, Concluding Remarks], Berarducci asks (and attributes the question to Ressayre) for an explicit axiomatisation of the class of EIP's of exponential fields. We do not consider this question in this paper.

Using a new kind of valuation whose values are ordinal numbers, Berarducci [B] studies irreducible elements in the ring $k((G^{<0})) \oplus \mathbb{Z}$, that is, in the canonical IP of the power series field $k((G))$, focusing mainly on the case when G is *archimedean*. He gives a test for irreducibility based only on the order type of the support of a series. It is not known if every irreducible element of $k((G^{\leq 0}))$ is prime (that is, generates a prime ideal). Refining the methods of [B], Pitteloud [P] shows that some of the irreducible series constructed in [B] are actually prime, in the case when G is *archimedean* or *contains a maximal proper convex subgroup*. In particular, [P, Theorem 4.1] says that, in the case of archimedean divisible group G, if $s \in k((G^{<0}))$ is such that support s has order type ω and least upper bound 0, then $s + 1$ is prime in $k((G^{<0})) \oplus \mathbb{Z}$.

In this paper, we extend the results of [B] and [P] to the canonical integer part $\mathrm{Neg}(F) \oplus \mathbb{Z}$ of any *truncation closed subfield* F,

$$k(G) \subset F \subset k((G)),$$

for an *arbitrary* divisible ordered abelian group $G \neq 0$. We shall denote by $k[G^{<0}]$ (respectively, by $k[G^{\leq 0}]$) the *semigroup ring* consisting of power series with negative (respectively, non-positive) and *finite* support. Note that $k[G^{<0}] \subset \mathrm{Neg}(F)$ since $k(G) \subset F$. [B, Theorem 11.2] says that, in the case of archimedean G, all irreducible elements of $k[G^{<0}] \oplus \mathbb{Z}$ remain irreducible in $k((G^{<0})) \oplus \mathbb{Z}$. We study, for an arbitrary G, the behaviour of primes and irreducibles under the ring extensions

$$(1) \qquad k[G^{<0}] \oplus \mathbb{Z} \subset \mathrm{Neg}(F) \oplus \mathbb{Z} \subset k((G^{<0})) \oplus \mathbb{Z}.$$

In [B, Concluding Remarks], the author asks whether every EIP of an exponential field contains a cofinal set of irreducible elements. We give a partial answer to this question: applying our results to truncation EIP's of non-archimedean exponential fields, we show that these EIP's indeed contain a cofinal set of irreducible elements. Note that in this case, the rank of G (see Section 2) is a dense linearly ordered set without endpoints

[K, Corollary 1.23], in particular G cannot be archimedean, nor can it contain a maximal proper convex subgroup.

The structure of this paper is the following. In Section 2, we fix the notation and review the necessary background concerning ordered groups and fields. In Section 3, we establish some straightforward facts used in the subsequent sections, in particular, to study the extensions (1). The main results of Section 4 are Proposition 4.8 and Corollary 4.10 which provides (under mild conditions) cofinal sets of irreducibles (=primes) in $k[G^{<0}] \oplus \mathbb{Z}$.

In Section 5, we generalize [B, Theorem 11.2] (see Corollaries 5.4 and 5.5) using the reduction to the case of an archimedean group G given in our Theorem 5.2 and Corollary 5.3 (some special cases of Corollary 5.3 already appeared in [B] and [P]). As a consequence, we extend Corollary 4.10 to the ring $k((G^{<0})) \oplus \mathbb{Z}$ obtaining cofinal sets of *irreducibles* with finite support in $\mathrm{Neg}(F) \oplus \mathbb{Z}$ (see Corollary 5.10). In the special case when F is the *field of algebraic power series*, that is, $F = k(G)^r$, we can improve the result to obtain cofinal sets of *primes* with finite support in $\mathrm{Neg}(F) \oplus \mathbb{Z}$ (see Corollary 5.11). Using our generalization of [P, Theorem 4.2] (see Theorem 5.12), we show that $k((G^{<0})) \oplus \mathbb{Z}$ *has a cofinal set of primes* with infinite support (see Corollary 5.16).

In Section 6, we apply the results of Section 5 to study EIP's of exponential fields. The main application is given in Theorem 6.3: we establish that *every exponential field has an EIP with a cofinal set of irreducible elements*. We work out two examples in detail. In Example 6.4, we consider the countable "Algebraic Power Series Fields with Exponentiation" described in [K, Example 1.45]. These fields are of the form $E(t^g \mid g \in G)^r$, where E is a countable exponentially closed subfield of the reals, and G is the lexicographic sum, taken over the rationals, of copies of the additive group of E. These fields may be viewed as truncation closed subfields of $E((G))$. We start by studying the canonical IP of such a field and show that this canonical IP is an EIP. We establish that this canonical EIP has a cofinal set of *prime elements with finite support*. In Example 6.5 we study the canonical EIP of the "Exponential-Logarithmic Power Series Fields" introduced in [K, Chapter 5, p. 79]. We show that this EIP has cofinally many *primes with infinite support*.

§2. **Preliminaries.** Let G be an ordered abelian group. Set $|g| := \max\{g, -g\}$ for $g \in G$. For non-zero $g_1, g_2 \in G$ we say that g_1 is *archimedean equivalent to* g_2 if there exists $r \in \mathbb{N}$ such that

(2) $$r|g_1| \geq |g_2| \quad \text{and} \quad r|g_2| \geq |g_1|.$$

We write $g_1 \ll g_2$ if $r|g_1| < |g_2|$ for all $r \in \mathbb{N}$. Denote by $[g]$ the equivalence class of $g \neq 0$, and by v the *natural valuation* on G, that is, $v(g) := [g]$ for $g \neq 0$, and $v(0) := \infty$.

The *rank* of G is defined to be $\Gamma := v(G \setminus \{0\})$. The relation $\ll$ on G induces a linear order on the set Γ by setting $v(g_1) > v(g_2)$ if $g_1 \ll g_2$ (note the reversed order). For $g_1, g_2 \in G$, we have $v(g_1 - g_2) \geq \min\{v(g_1), v(g_2)\}$. For each $\gamma \in \Gamma$, fix $g_\gamma \in G$ such that $v(g_\gamma) = \gamma$. Let C_γ and D_γ denote, respectively, the smallest convex subgroup containing g_γ and the largest convex subgroup not containing g_γ. Note that C_γ and D_γ are independent from the choice of the representative g_γ. In fact

$$C_\gamma = \{g \in G \mid v(g) \geq \gamma\} \quad \text{and} \quad D_\gamma = \{g \in G \mid v(g) > \gamma\}.$$

Note also that D_γ is a maximal proper convex subgroup of C_γ and thus the quotient

$$A_\gamma := C_\gamma / D_\gamma$$

is archimedean. We call A_γ an *archimedean component* of G.

Let Q be an archimedean field. We now recall some general definitions and facts about ordered Q-vector spaces. Clearly, if G is an ordered Q-vector space, then for $g_1, g_2 \in G$, g_1 is archimedean equivalent to g_2 if and only if there exists $r \in Q$ such that (2) holds. A subset $B \subset G \setminus \{0\}$ is *strongly independent* if B consists of pairwise archimedean inequivalent elements. We say that $B = \{b_i \mid i \in I\}$ is *Q-valuation independent* if for all $r_i \in Q$ such that $r_i = 0$ for all but finitely many $i \in I$,

$$v\left(\sum_{i \in I} r_i b_i\right) = \min_{\{i \in I \mid r_i \neq 0\}} \{v(b_i)\}.$$

A *Q-valuation basis* is a Q-basis which is Q-valuation independent.

If G is a Q-vector space, then C_γ and D_γ are Q-subspaces, so A_γ is a Q-vector space, isomorphic to any maximal archimedean *subspace* of G containing g_γ, and to any Q-vector space complement to D_γ in C_γ.

REMARK 2.1. For $Q = \mathbb{Q}$, A_γ is isomorphic to any maximal archimedean subgroup of G containing g_γ. This is because if A is a subgroup of G, then A is archimedean if and only if the divisible hull $\langle A \rangle_\mathbb{Q}$ of A is archimedean, which implies that a maximal archimedean subgroup of G is necessarily a $\mathbb{Q}$-subspace.

We let π_γ denote the natural homomorphism $C_\gamma \rightarrow A_\gamma = C_\gamma / D_\gamma$. The following characterization of valuation independence is useful and easy to prove (see [K]).

PROPOSITION 2.2. *Let $B \subset G \setminus \{0\}$. Then B is Q-valuation independent if and only if the following holds: for all $n \in \mathbb{N}$ and distinct $b_1, \ldots, b_n \in B$ with $v(b_1) = \cdots = v(b_n) = \gamma$, the elements $\pi_\gamma(b_1), \ldots, \pi_\gamma(b_n)$ in A_γ are Q-linearly independent.*

It follows that a Q-valuation basis is a maximal Q-valuation independent set. (But in general, a maximal Q-valuation independent set need *not* be a basis.) We also have

COROLLARY 2.3. *Let* $B \subset G$ *be Q-linearly independent and consist of archimedean equivalent elements. Then* $\langle B \rangle_Q$ *is archimedean if and only if* B *is Q-valuation independent.*

If all archimedean components of G have dimension 1 over Q, then $B \subset G$ is Q-valuation independent if and only if B is strongly independent. (In particular, if $Q = \mathbb{R}$, then B is Q-valuation independent if and only if B is strongly independent, because an archimedean $\mathbb{R}$-vector space has necessarily dimension 1 over $\mathbb{R}$.)

In [Br] it is shown that every ordered Q-vector space of countable dimension has a Q-valuation basis. (In particular, every ordered $\mathbb{R}$-vector space of countable dimension has a strongly independent basis.)

We also need to recall some facts about valued fields. (In this paper, we mainly deal with fields of characterisitc 0.) Let K be a field, G an ordered abelian group and ∞ an element greater than every element of G.

A surjective map $w : K \to G \cup \{\infty\}$ is a *valuation* on K if for all $a, b \in K$ (i) $w(a) = \infty$ iff $a = 0$, (ii) $w(ab) = w(a) + w(b)$, (iii) $w(a - b) \geq \min\{w(a), w(b)\}$.

We say that (K, w) is a *valued field*. The *value group* of (K, w) is $w(K) := G$. The *valuation ring* of w is $\mathcal{O}_w := \{a \mid a \in K$ and $w(a) \geq 0\}$ and the *valuation ideal* is $\mathcal{M}_w := \{a \mid a \in K$ and $w(a) > 0\}$. The field $\mathcal{O}_w/\mathcal{M}_w$, denoted by Kw, is the *residue field*. For $b \in \mathcal{O}_w$, bw is its image under the residue map.

A valued field (K, w) is *henselian* if given a polynomial $p(x) \in \mathcal{O}_w[x]$, and $a \in Kw$ a simple root of the reduced polynomial $p(x)w \in Kw[x]$, we can find a root $b \in K$ of $p(x)$ such that $bw = a$.

Let $K \subset L$ be an extension of valued fields. By abuse of notation, denote by w the valuation on both K and L. Then K is said to be *w-dense* in L if for every $a \in L$ and $\alpha \in w(L)$ there is some $b \in K$ such that $w(a - b) > \alpha$.

If $(K, +, \cdot, 0, 1, <)$ is an ordered field, we denote by V its natural valuation, that is, the natural valuation V on the ordered abelian group $(K, +, 0, <)$. (The set of archimedean classes becomes an ordered abelian group by setting $[x] + [y] := [xy]$.) For the natural valuation, we shall use the notation $\mathcal{O}_K := \{x \in K \mid V(x) \geq 0\}$ and $\mathcal{M}_K := \{x \in K \mid V(x) > 0\}$, respectively, for the valuation ring and valuation ideal. We shall also often write k for the residue field $\mathcal{O}_K/\mathcal{M}_K$ and G for the value group $V(K)$. Note that k is an archimedean ordered field, and that V is *compatible* with the order, that is, has a convex valuation ring. We denote by $\mathcal{U}_K^{>0}$ the multiplicative group of positive units (invertible elements) of $\mathcal{O}_K$. The subgroup $1 + \mathcal{M}_K$ of $\mathcal{U}_K^{>0}$ is called the group of 1-units.

For ordered fields, there is another notion of density that we would like to review briefly. For more details see [K, Chapter 1, Section 6]. Let $K \subset L$ be an extension of ordered fields. K is *order dense* in L if for all $a, c \in L$ with $a < c$ there is $b \in K$ such that $a \leq b \leq c$. Let $K \subset L$ be an extension of ordered fields, and let w be a compatible valuation on both of them. Assume that L is non-archimedean. Then K is order dense in L if and only if K is w-dense in L.

We need further well-known facts about ordered fields endowed with a compatible valuation. Let K be an ordered field and w a compatible valuation on K. Then K is real closed if and only if (i) (K, w) is henselian (ii) its residue field Kw is real closed field and (iii) its value group $w(K)$ is divisible.

If K is real closed, then the residue field Kw embeds in K and K admits a cross-section, that is, an embedding η of the value group G into $(K^{>0}, \cdot)$ such that $w(\eta(g)) = g$ for all $g \in G$ (see [PC]).

Therefore, whenever needed, we can assume without loss of generality that $k(G) \subset K$. (More precisely, we identify the residue field k with a maximal subfield of $\mathcal{O}_K$ through the residue map, and G with $\eta(G)$.)

Recall from the Introduction that for G an ordered abelian group, k a field, $k((G))$ denotes the field of power series with coefficients in k and exponents in G. Every series $s \in k((G))$ is of the form $\sum_{g \in G} s_g t^g$ with $s_g \in k$ and well-ordered support $\{g \in G \mid s_g \neq 0\}$. Addition is pointwise, multiplication is given by the usual formula for multiplying power series:

$$\left(\sum_{g \in G} r_g t^g\right)\left(\sum_{g \in G} s_g t^g\right) = \sum_{g \in G} \left(\sum_{g' \in G} r_{g'} s_{g-g'}\right) t^g.$$

The *canonical valuation* on $k((G))$ is given by $V_{\min}(s) := \min(\text{support } s)$ for any series $s \in k((G))$. If k is an ordered field, we can endow $k((G))$ with the lexicographic order: a series is positive if its least nonzero coefficient is positive. With this order, $k((G))$ is an ordered field, and $V_{\min}$ is compatible with this order. If k is archimedean, then $V_{\min}$ coincides with the natural valuation V. Clearly, the value group of $(k((G)), V_{\min})$ is (isomorphic to) G and the residue field is (isomorphic to) k. The valuation ring $k((G^{\geq 0}))$ consists of the series with non-negative exponents, and the valuation ideal $k((G^{>0}))$ of the series with positive exponents. The *constant term* of a series $\sum_{g \in G} s_g t^g$ is the coefficient s_0. The units of $k((G^{\geq 0}))$ are the series in $k((G^{\geq 0}))$ with a non-zero constant term. Every series $s \in k((G))$ can be written as $s = s_{<0} + s_0 + s_{>0}$ where s_0 is the constant term of s and $s_{<0}$, resp. $s_{>0}$, denotes the restriction of s to $G^{<0}$, resp. $G^{>0}$. Thus the (non-unital) ring $k((G^{<0}))$ of generalized power series with negative support is a complement in $(k((G)), +)$ to the valuation ring. Note that it is in fact a k-algebra. We shall denote by $k((G^{\leq 0}))$ the ring of generalized power series with non-positive support.

Given $s \in k((G))^{>0}$, we can factor out the monomial of smallest exponent $g \in G$ and write $s = t^g u$ with u a unit with a positive constant term. Thus the multiplicative subgroup $\text{Mon}\, k((G)) := \{t^g \mid g \in G\}$ consisting of the (monic) monomials t^g is a complement in $(k((G))^{>0}, \cdot)$ to the subgroup of positive units.

§3. **Truncation integer parts.** A subfield F of $k((G))$ is *truncation closed* if whenever $s = \sum_{g \in G} s_g t^g \in F$ and $a \in G$, the restriction $s_{<a} = \sum_{g \in G^{<a}} s_g t^g$ of s to the initial segment $G^{<a}$ of G also belongs to F. As mentioned in the Introduction, [M–R] show that given a real closed field K with residue field k and value group G (for the natural valuation V on K), with $k(G) \subset K$, there is a *truncation closed embedding* of K into $k((G))$ over $k(G)$, that is, an embedding φ such that $F := \varphi(K)$ is truncation closed. Note that since the restriction of φ to $k(G)$ is the identity, φ is in particular an embedding of k-vector spaces.

PROPOSITION 3.1. *Let* $F \subset k((G))$ *be a truncation closed subfield that contains* $k(G)$. *Then* $Z_F := \mathrm{Neg}(F) \oplus \mathbb{Z}$ *is an integer part of* F (*that contains* $k[G^{<0}] \oplus \mathbb{Z}$).

PROOF. Clearly, Z_F is a discrete subring of F. Let $s \in k((G))$. Let $\lfloor s_0 \rfloor \in \mathbb{Z}$ be the integer part of $s_0 \in k$. Define

$$z_s = \begin{cases} s_{<0} + s_0 - 1 & \text{if } s_0 \in \mathbb{Z} \text{ and } s_{>0} < 0, \\ s_{<0} + \lfloor s_0 \rfloor & \text{otherwise.} \end{cases}$$

Clearly, $z_s \leq s < z_s + 1$. From truncation closedness of F it follows that if $s \in F$, then $z_s \in Z_F$.   ⊣

It follows that $\varphi^{-1}(Z_F)$ is an IP of K. Recall that we refer to IP's obtained in this way via a truncation closed embedding as truncation IP's.

We now state some easy facts about IP's of ordered fields in general. We note that all IP's of a given ordered field are *isomorphic as ordered sets* (if Z_1, Z_2 are IP's, an isomorphism is obtained by mapping $s \in Z_1$ to its integer part z_s with respect to Z_2), but in general, they need not be isomorphic as ordered rings, not even elementarily equivalent (see Remark 3.3). If Z is an IP of K, then the fraction field of Z is an order-dense subfield of K. If L is an order-dense subfield of K, then every IP of L is an IP of K. Conversely, if $L \subset K$ and there exists an IP Z of L that remains an IP of K, then L is order-dense in K.

REMARK 3.2. In [Bo] examples of ordered fields without IP's are given. We conjectured that every *henselian* ordered field admits a truncation closed embedding into a field of power series and thus, admits an IP. This conjecture is being studied in [F]. We also asked for a direct proof of the existence of IP's for henselian fields (without arguing via truncation closed embeddings in fields of power series). This question is addressed in [KF].

REMARK 3.3. Truncation IP's of non-archimedean real closed fields are very peculiar models of OI. It would be interesting to investigate the algebraic and model-theoretic properties of this class.

(i) They admit $\mathbb{Z}$ as a direct summand. This is not the case for an arbitrary model of OI. For example, as observed by D. Marker (unpublished), $\mathbb{Z}$ cannot be a direct summand of a non-standard model of Peano Arithmetic.

This remark implies in particular that *not every IP of a real closed field is a truncation IP.*

(ii) They are intimately related to complements of the valuation ring (see [KF]). A truncation IP Z of K decomposes as $Z = A \oplus \mathbb{Z}$ (lexicographic sum), where the summand A is a k-algebra. It is easily verified that A is an additive complement to $\mathcal{O}_K$. Since additive complements to $\mathcal{O}_K$ are unique up to isomorphism of ordered groups, all truncation IP's of K are *isomorphic as ordered groups.* Note that if an additive complement A to $\mathcal{O}_K$ is closed under multiplication, then $Z(A) := A \oplus \mathbb{Z}$ is an IP of K. However, A need not be a k-algebra, so $Z(A)$ need not be a truncation IP. It is not known to us whether every IP of the form $Z(A)$, with A a k-algebra, is a truncation IP.

(iii) They are never normal (that is, never integrally closed in their field of fractions). This is because $\mathrm{Neg}(F)$ is a k-vector space, so $\sqrt{2}t^g \in Z_F$ for $g \in G^{<0}$ and hence $\sqrt{2}$ is rational over Z_F: $\sqrt{2} = \frac{\sqrt{2}t^g}{t^g}$. Since the *normal* model obtained in [B–O] is an IP of a real closed field, which also has a *non-normal* IP, we see that the IP's of a given real closed field need not be elementarily equivalent. This motivates the following

Open Question: Does every real closed field have a *normal* integer part?

Let $F = k(G)$ and F' its real closure. By [M–R], F and F' are truncation closed subfields of $k((G))$. We already noted that $k[G^{<0}] \subset \mathrm{Neg}(F)$. We are interested in understanding when $k[G^{<0}] = \mathrm{Neg}(F)$. The following fact was observed by F.-V. Kuhlmann.

PROPOSITION 3.4. *Assume that G is archimedean and divisible, and that k is a real closed field. Then $\mathrm{Neg}(F) = \mathrm{Neg}(F') = k[G^{<0}]$.*

PROOF. We first show that $\mathrm{Neg}(F) = k[G^{<0}]$. Let $f \in F = k(G)$. So $f = p/q$ with $p = \sum_{i=0}^{l} a_i t^{g_i}$, $q = \sum_{i=0}^{l} a_i' t^{g_i'} \in k[G]$. By factoring out the monomial of least exponent in q, say $a_0' t^{g_0'}$, we may rewrite

$$f = \sum_{i=0}^{l} c_i t^{h_i}(1+\varepsilon)^{-1} = \sum_{i=0}^{l} c_i t^{h_i} \sum_{n=0}^{\infty} (-1)^n \varepsilon^n$$

where $c_i = a_i/a_0'$, $h_i = g_i - g_0'$, and $\varepsilon \in k[G^{>0}]$. Since G is archimedean, we may choose N large enough so that $N V_{\min}(\varepsilon) \geq -h_i$ for all $i = 0, \ldots, l$. We now rewrite:

$$f = \sum_{i=0}^{l} c_i t^{h_i} \sum_{n=0}^{N-1} (-1)^n \varepsilon^n + \sum_{i=0}^{l} c_i t^{h_i} \sum_{n=N}^{\infty} (-1)^n \varepsilon^n.$$

By our choice of N, we have that $V_{\min}(\sum_{i=0}^{l} c_i t^{h_i} \sum_{n=N}^{\infty} (-1)^n \varepsilon^n) \geq 0$. It follows that if $f \in \mathrm{Neg}(F)$, then $f = \sum_{i=0}^{l} c_i t^{h_i} \sum_{n=0}^{N-1} (-1)^n \varepsilon^n$, so $f \in k[G]$.

We now show that $\mathrm{Neg}(F) = \mathrm{Neg}(F')$. By assumptions on k and G, we have $F^r = F^h :=$ the henselization of F with respect to $V_{\min}$. Since G is archimedean, F is dense in F^h (see [Ri]). Thus $\text{Neg}(F) = \text{Neg}(F^h)$ (see [K, Lemma 1.32]). ⊣

REMARK 3.5. If G is not archimedean, then $k[G^{<0}] \neq \text{Neg}(F)$: choose $g_1, g_2 \in G^{<0}$ with $v(g_1) < v(g_2)$, then $t^{g_1}(1 - t^{-g_2})^{-1} \in F$ and $t^{g_1}(1 - t^{-g_2})^{-1} = \sum_{n=0}^{\infty} t^{g_1 - ng_2}$ has *infinite* negative support. Also, $\text{Neg}(F') \neq \text{Neg}(F)$: consider $t^{g_1}(1 + t^{-g_2})^{1/2} \in F'$. Since $t^{g_1}(1 + t^{-g_2})^{1/2} = t^{g_1} + \frac{1}{2}t^{g_1 - g_2} - \frac{1}{8}t^{g_1 - 2g_2} + \cdots$ has negative support, it belongs to $\text{Neg}(F')$. On the other hand, $(1 + t^{-g_2})^{1/2} \notin F$ so $t^{g_1}(1 + t^{-g_2})^{1/2} \notin F$.

We will frequently use the following two lemmas. If R is a ring, we denote by $U(R)$ its group of units.

LEMMA 3.6. *Let G be an ordered abelian group, K a field. Let R be a subring of K and Neg a (non-unital) K-subalgebra of $K((G^{<0}))$. Let $r \in R$. Then r is prime (resp., irreducible) in R iff r is prime (resp., irreducible) in $\text{Neg} \oplus R$.*

PROOF. Clearly, if $a, b \in \text{Neg} \oplus R$ and $ab \in R$, then $a, b \in R$. It follows that $U(\text{Neg} \oplus R) = U(R)$ and that r is irreducible in R iff r is irreducible in $\text{Neg} \oplus R$. Now let $a \in \text{Neg} \oplus R$ and let a_0 be the constant term of a. Clearly, r divides a in $\text{Neg} \oplus R$ iff r divides a_0 in R. It follows that r is prime in R iff r is prime in $\text{Neg} \oplus R$. ⊣

LEMMA 3.7. *Let G be an ordered abelian group, K a field. Let R be a subring of K and Neg a (non-unital) K-subalgebra of $K((G^{<0}))$. Let f be a non-constant element of $\text{Neg} \oplus R$ with constant term 1. Then f is prime (resp., irreducible) in $\text{Neg} \oplus R$ iff f is prime (resp., irreducible) in $\text{Neg} \oplus K$.*

PROOF. ($\Rightarrow$) Suppose f is irreducible in $\text{Neg} \oplus R$ and assume that $f = ab$ for some non-invertible $a, b \in \text{Neg} \oplus K$. Let a_0, b_0 be the constant terms of a, b. Then $a \neq a_0$, $b \neq b_0$, and $a_0 b_0 = 1$. Let $\tilde{a} = b_0 a$ and $\tilde{b} = a_0 b$. Then $\tilde{a}, \tilde{b}$ are non-invertible elements of $\text{Neg} \oplus R$ and $f = \tilde{a}\tilde{b}$, a contradiction. Now suppose f is prime in $\text{Neg} \oplus R$, $a, b \in \text{Neg} \oplus K$, and f divides ab in $\text{Neg} \oplus K$. Let $\tilde{a} = a/a_0$ if $a_0 \neq 0$ and $\tilde{a} = a$ otherwise. Similarly, let $\tilde{b} = b/b_0$ if $b_0 \neq 0$ and $\tilde{b} = b$ otherwise. Then f divides $\tilde{a}\tilde{b}$ in $\text{Neg} \oplus R$. Hence f divides $\tilde{a}$ or $\tilde{b}$ in $\text{Neg} \oplus R$, which implies that f divides a or b in $\text{Neg} \oplus K$.

($\Leftarrow$) Suppose f is irreducible in $\text{Neg} \oplus K$ and assume that $f = ab$ for some non-invertible $a, b \in \text{Neg} \oplus R$. Then $a_0 b_0 = 1$ and, therefore, $a \neq a_0$ and $b \neq b_0$. But then a, b are not invertible in $\text{Neg} \oplus K$ and $f = ab$, a contradiction. Now suppose f is prime in $\text{Neg} \oplus K$, $a, b \in \text{Neg} \oplus R$, and f divides ab in $\text{Neg} \oplus R$. Then f divides ab in $\text{Neg} \oplus K$. Hence f divides a or b in $\text{Neg} \oplus K$, which implies that f divides a or b in $\text{Neg} \oplus R$ since f has constant term 1. ⊣

We want to study prime and irreducible elements of the rings $k[G^{<0}] \oplus \mathbb{Z}$ and $k((G^{<0})) \oplus \mathbb{Z}$, but it is often more convenient to work with $k[G^{\leq 0}]$ and $k((G^{\leq 0}))$ instead. It is easy to see how this change affects primality and irreducibility.

If f is irreducible in $k[G^{<0}] \oplus \mathbb{Z}$, then either $f \in \mathbb{Z}$ or $f \notin \mathbb{Z}$ and the constant term $f_0 = \pm 1$. Indeed, if $f \notin \mathbb{Z}$ and $f_0 \neq \pm 1$, then we can factor f in $k[G^{<0}] \oplus \mathbb{Z}$ as follows: $f = f_0(f/f_0)$ if $f_0 \neq 0$ and $f = 2(f/2)$ if $f_0 = 0$. Using Lemmas 3.6 and 3.7, we conclude that prime (resp., irreducible) elements of $k[G^{<0}] \oplus \mathbb{Z}$ are of two types: 1) primes in $\mathbb{Z}$ and 2) prime (resp., irreducible) elements of $k[G^{\leq 0}]$ with constant term ± 1. We will show in Section 4 that in fact all irreducible elements of $k[G^{\leq 0}]$ are prime.

By similar considerations, primes (resp., irreducibles) of $k((G^{<0})) \oplus \mathbb{Z}$ are of two types: 1) primes in $\mathbb{Z}$ and 2) primes (resp., irreducibles) of $k((G^{\leq 0}))$ with constant term ± 1. We do not know whether or not all irreducibles of $k((G^{\leq 0}))$ are prime.

The following two lemmas are standard. We omit the proofs.

LEMMA 3.8. *Let $(I, \leq)$ be a directed set (that is, $\leq$ is a partial order and, for any $i_1, i_2 \in I$, there exists $i_0 \in I$ such that $i_0 > i_1, i_2$). Let R be an integral domain and assume that, for every $i \in I$, we assigned a subring $R_i \subset R$ in such a way that R is the direct limit of R_i (that is, $R_{i_1} \subset R_{i_2}$ for $i_1 \leq i_2$ and $R = \cup_{i \in I} R_i$). Let $r \in R$. If there exists $i_0 \in I$ such that r is prime (resp., irreducible) as an element of R_i for $i \geq i_0$, then r is prime (resp., irreducible) in R.*

LEMMA 3.9. *Let R be an integral domain and $T \subset R \setminus \{0\}$ a multiplicative subsemigroup. Denote by R_T the corresponding ring of fractions. Let $r \in R$.*

1) *If r is prime in R and $r \nmid t$ in R for all $t \in T$, then r is prime in R_T.*
2) *If r is prime in R_T and, for all $t \in T$ and $a \in R$, $r \mid ta$ in $R \Rightarrow r \mid a$ in R, then r is prime in R.*

As an immediate application of Lemma 3.9 (with $R := k[G^{\leq 0}]$ and $T := \{t^g \mid g \in G^{\leq 0}\}$), we obtain

COROLLARY 3.10. *A non-monomial $f \in k[G^{\leq 0}]$ is prime in $k[G^{\leq 0}]$ if and only if it is prime in $k[G]$.*

§4. **Primes and irreducibles in $k[G^{\leq 0}]$.** Let G be a torsion-free abelian group and k a field. Then G is orderable, so the group algebra $k[G]$ is a domain. First, we want to show that all irreducibles of $k[G]$ or $k[G^{\leq 0}]$ are prime.

PROPOSITION 4.1. *Let G be a torsion-free abelian group, k a field. Then every irreducible element of $k[G]$ is prime.*

PROOF. Since G is orderable, $R := k[G]$ is a domain and the only invertible elements of R are of the form λt^g where $g \in G$, $0 \neq \lambda \in k$. Let I be the set of finitely generated subgroups of G, partially ordered by inclusion. For each $A \in I$, set $R_A = k[A]$. Then R is the direct limit of R_A. Fix an irreducible element $f \in R$. Suppose that $f \in R_A$ for some $A \in I$. Since $U(R_A) = U(R) \cap R_A$, f is irreducible in R_A. Pick a $\mathbb{Z}$-basis $\{g_1, \ldots, g_n\}$ of A. Then $R_A \cong k[x_1^{\pm 1}, \ldots, x_n^{\pm 1}]$, the ring of Laurent polynomials, where $x_l = t^{g_l}$,

$l = 1, \ldots, n$, so R_A is a UFD. It follows that f is prime in R_A. Applying Lemma 3.8, we conclude that f is prime in R.   ⊣

COROLLARY 4.2. *Let G be an ordered abelian group, k a field. Then every irreducible element of $k[G^{\leq 0}]$ is prime.*

PROOF. Let $f \in k[G^{\leq 0}]$ be irreducible. Then either f has a nonzero constant term or else $f = \lambda t^{g_0}$ where $0 \neq \lambda \in k$ and $g_0 = \max G^{<0}$. In the second case, $k[G^{\leq 0}]/(f) \cong k$, so f is prime. In the first case, using the fact that every element $0 \neq s \in k[G]$ can be written uniquely in the form $\tilde{s} t^g$ where $g \in G$ and $\tilde{s} \in k[G^{\leq 0}]$ with nonzero constant term, one checks that f is irreducible in $k[G]$. Then by Proposition 4.1, f is prime in $k[G]$. Applying Corollary 3.10, we conclude that f is prime in $k[G^{\leq 0}]$.   ⊣

REMARK 4.3. Note, however, that $k[G^{\leq 0}]$ and $k[G]$ are rarely UFD's. For instance, if $g \in G^{<0}$ is divisible in G (that is, the equation $nx = g$ has a solution for every $n \in \mathbb{N}$), then obviously the (non-invertible) element t^g cannot have a factorization into primes in $k[G^{\leq 0}]$. As to the group ring $k[G]$, if $g \neq 0$ is divisible in G and k is real closed, then the (non-invertible) element $t^g + 1$ does not have a factorization into primes in $k[G]$. Indeed, assume $t^g + 1 = p_1 \cdots p_l$ is such a factorization. Let $H = \langle g \rangle_{\mathbb{Q}}$, the divisible hull of $\langle g \rangle$, and pick a complement H' for H in G. Then $k[G] = R[H']$ where $R = k[H]$. Since $t^g + 1 \in R$, all $p_i \in R$ (otherwise the product $p_1 \cdots p_l$ would contain a term of the form $r t^h$ where $0 \neq r \in R$ and $0 \neq h \in H'$ and thus $p_1 \cdots p_l \notin R$). Since $U(R) = U(k[G]) \cap R$, p_i are irreducible(=prime) in R. But this is a contradiction by Proposition 4.5, below.

The following simple observation will be useful:

LEMMA 4.4. *Let $H \subset G$ be a divisible subgroup and $f \in k[G]$ (resp., $f \in k[G^{\leq 0}]$). Assume that $\operatorname{support} f \subset H$. Then f is prime in $k[G]$ (resp., in $k[G^{\leq 0}]$) iff f is prime in $k[H]$ (resp., in $k[H^{\leq 0}]$).*

PROOF. Pick a complement H' for H in G. Let $R = k[H]/(f)$. Then $k[G]/(f) \cong R[H']$. Obviously, $R[H']$ is a domain iff R is a domain. This proves that f is prime in $k[G]$ iff it is prime in $k[H]$. The case of $k[G^{\leq 0}]$ and $k[H^{\leq 0}]$ follows from here by Corollary 3.10. Indeed, since H is divisible, $H^{<0}$ does not have a greatest element. Hence any $f \in k[H^{\leq 0}]$ that is prime in $k[H^{\leq 0}]$ or in $k[G^{\leq 0}]$ must have nonzero constant term. Then by Corollary 3.10, f is prime in $k[H^{\leq 0}]$, resp. $k[G^{\leq 0}]$, iff f is prime in $k[H]$, resp. $k[G]$.   ⊣

From now on, we assume that G is divisible, that is, a $\mathbb{Q}$-vector space.

PROPOSITION 4.5. *Let G be a divisible ordered abelian group, k a real closed field. If $f \in k[G^{\leq 0}]$ is a non-constant prime, then $\operatorname{support} f$ does not lie in a 1-dimensional $\mathbb{Q}$-subspace of G.*

PROOF. Let $\operatorname{support} f = \{g_1, \ldots, g_n, 0\}$. Write $f = a_1 t^{g_1} + \cdots + a_n t^{g_n} + a_0$. Assume that $g_i = \frac{m_i}{n_i} g_1$ for $m_i, n_i \in \mathbb{N}$. Set $N = \operatorname{lcm}\{n_i\}$, $g = \frac{1}{3N} g_1$, and

$x = t^g$. Rewriting f in terms of x, we obtain the polynomial $f(x) = a_1 x^{N_1} + \cdots + a_n x^{N_n} + a_0$ (where $N_i = 3m_i M_i$ and $N = M_i n_i$). Since $f(x) \in k[x]$ has degree ≥ 3, it factors in $k[x]$, which provides a factorization of f in $k[G^{\leq 0}]$. $\dashv$

REMARK 4.6. In [B, p. 555] it is wrongly asserted that the support must be a $\mathbb{Q}$-linearly independent set. A counterexample is constructed as follows. Let $g_1, g_2 \in G^{<0}$ be $\mathbb{Q}$-linearly independent, then the element $f = t^{2g_1} + t^{g_1 + g_2} + t^{g_1} + t^{g_2} + 1$ is prime in $k[G]$ by an argument similar to the proof of Proposition 4.8, below. Indeed, fix any $m \in \mathbb{N}$ and set set $x = t^{\frac{1}{m}g_1}$, $y = t^{\frac{1}{m}g_2}$, then rewriting f in terms of x and y, we obtain an irreducible polynomial $f(x, y) = x^{2m} + (y^m + 1)x^m + (y^m + 1)$. This implies that f is prime in $k[G]$ and, consequently, in $k[G^{\leq 0}]$.

REMARK 4.7. Shepherdson's model mentioned at the beginning of the Introduction is the canonical integer part of $\mathbb{Q}(t^g \mid g \in \mathbb{Q})^r$. By Proposition 3.4, this is just $\mathbb{Q}^r[\mathbb{Q}^{<0}] \oplus \mathbb{Z}$. By Proposition 4.5, a non-constant element cannot be prime. This explains why Shepherdson's model has no non-standard primes.

Next, we want to construct prime elements in $k[G^{\leq 0}]$. We first establish the following result:

PROPOSITION 4.8. *Let G be a divisible torsion-free abelian group (resp., divisible ordered abelian group), k a field of characteristic zero. Let f be an element of $k[G]$ (resp., $k[G^{\leq 0}]$). Assume that support $f = \{g_1, \ldots, g_n, 0\}$ where $g_1, \ldots, g_n$ are linearly independent over $\mathbb{Q}$ and $n \geq 2$. Then f is prime in $k[G]$ (resp., in $k[G^{\leq 0}]$).*

PROOF. Write $f = a_1 t^{g_1} + \cdots + a_n t^{g_n} + a_0$ where $0 \neq a_i \in k$. Let $H = \langle g_1, \ldots, g_n \rangle_{\mathbb{Q}}$. First we show that f is prime in $k[H]$. For $m \in \mathbb{N}$, let $H_m = \langle \frac{1}{m} g_1, \ldots, \frac{1}{m} g_n \rangle_{\mathbb{Z}}$. Then $k[H]$ is the direct limit of $k[H_m]$. By Lemma 3.8, it suffices to show that f is prime in $k[H_m]$ for sufficiently large m. Clearly, $k[H_m] \cong k[x_1^{\pm 1}, \ldots, x_n^{\pm 1}]$ where $x_i = t^{\frac{1}{m}g_i}$. Under this isomorphism f corresponds to the polynomial $f_m(x_1, \ldots, x_n) = a_1 x_1^m + \cdots + a_n x_n^m + a_0$. This polynomial is irreducible in $k[x_1, \ldots, x_n]$. One can show this e.g. by induction on $n \geq 2$ applying Eisenstein's criterion to $f_m(x_1, \ldots, x_n) = a_1 x_1^m + (a_2 x_2^m + \cdots + a_n x_n^m + a_0)$ viewed as an element of $R[x_1]$ where $R = \bar{k}[x_2, \ldots, x_n]$ and $\bar{k}$ is the algebraic closure of k. For $n = 2$, take $p = x_2 - \xi \in R$ where ξ is an m-th root of $-a_0/a_2$, then p satisfies the conditions of Eisenstein's criterion for f_m. For $n > 2$, take $p = a_2 x_2^m + \cdots + a_n x_n^m + a_0$, which is prime in R by induction hypothesis and satisfies the conditions of Eisenstein's criterion for f_m.

Now by Lemma 3.9, 1) (with $R := k$ and T is generated by $x_1, \ldots, x_n$), f_m is prime in $k[x_1^{\pm 1}, \cdots, x_n^{\pm 1}]$, so f is prime in $k[H_m]$ for any m. Thus we have proved that f is prime in $k[H]$. By Lemma 4.4, we conclude that f is prime in $k[G]$. If $f \in k[G^{\leq 0}]$, then by Corollary 3.10, f is also prime in $k[G^{\leq 0}]$. $\dashv$

EXAMPLE 4.9. Let G be a divisible ordered abelian group, k a real closed field. Consider $f = a_1 t^{g_1} + a_2 t^{g_2} + 1$ where $g_1, g_2 \in G^{<0}$, $0 \neq a_1, a_2 \in k$. Then f is prime in $k[G^{\leq 0}]$ iff g_1 and g_2 are linearly independent over $\mathbb{Q}$.

PROOF. Follows from Propositions 4.5 and 4.8.   ⊣

COROLLARY 4.10. Let G be a divisible ordered abelian group, k an ordered field. If $\dim_{\mathbb{Q}} G \geq 2$, then the ordered ring $k[G^{<0}] \oplus \mathbb{Z}$ has a cofinal set of primes.

PROOF. Let $s \in k[G^{<0}] \oplus \mathbb{Z}$ and let $g = \min(\text{support } s) \in G^{\leq 0}$. Then we can find $\mathbb{Q}$-linearly independent $g_1, g_2 \in G^{<0}$ such that $g_1 < g$. Let $f = t^{g_1} + t^{g_2} + 1$. Then $f > s$ and f is prime by Proposition 4.8.   ⊣

§5. **Primes and irreducibles in $k((G^{\leq 0}))$.** Throughout this section, G is a divisible ordered abelian group, with rank Γ, k a field of characteristic zero, and F is a truncation closed subfield of $k((G))$ that contains $k(G)$. Note that it follows that the restriction of any series in F to any convex subset of G also belongs to F (because the restriction of the series to the convex subset can be written as the difference of two initial segments of the series). We want to investigate when a prime element of $k[G^{\leq 0}]$ remains prime or at least irreducible through the ring extensions

$$k[G^{\leq 0}] \subset \text{Neg}(F) \oplus k \subset k((G^{\leq 0})).$$

First we consider the following example to see how things can go wrong.

EXAMPLE 5.1. Suppose we have $g_1, g_2 \in G^{<0}$ such that $v(g_1) < v(g_2)$. Then g_1 and g_2 are linearly independent over $\mathbb{Q}$, so by Proposition 4.8, $f = t^{g_1} + t^{g_2} + 1$ is prime in $k[G^{\leq 0}]$. However, f can be factored in $k((G^{\leq 0}))$ as follows: $f = (t^{g_2} + 1)(\frac{t^{g_1}}{t^{g_2}+1} + 1) = (t^{g_2} + 1)(t^{g_1 - g_2}(1 + t^{-g_2})^{-1} + 1) = (t^{g_2} + 1)(\sum_{n=1}^{\infty} (-1)^{n-1} t^{g_1 - n g_2} + 1)$. Thus f does not remain irreducible in $k((G^{\leq 0}))$.

Generalizing this trick, we can show that if f (non-constant) is irreducible in $\text{Neg } F \oplus k$, then all nonzero elements of support f must be archimedean equivalent, that is,

$$\Gamma_f := v(\text{support } f \setminus \{0\})$$

must be a singleton (this observation is due to Gonshor [G]). Indeed, let $\gamma = \min \Gamma_f$ and let $f_\gamma = \sum_{g:v(g)=\gamma} f_g t^g$. Since F is truncation closed, $f_\gamma \in \text{Neg}(F)$. Let $\varepsilon = f - f_\gamma$. If Γ_f is not a singleton, then $\varepsilon \notin k$. Clearly, for all $g \in \text{support } \varepsilon$, $v(g) > \gamma$. It follows that $f = \varepsilon(\frac{f_\gamma}{\varepsilon} + 1)$ is a nontrivial factorization of f in $\text{Neg}(F) \oplus k$.

This phenomenon suggests that the problems of irreducibility and primality in $\text{Neg}(F) \oplus k$ should essentially reduce to the case when G is archimedean. Let f be an element of $\text{Neg}(F) \oplus k$ such that $\Gamma_f = \{\gamma\}$. Fix a complement for

D_γ in C_γ so that we can view the archimedean component A_γ as a subgroup of C_γ. Then $C_\gamma = A_\gamma \oplus D_\gamma$ (lexicographic sum). Consequently, $k((C_\gamma)) = k((D_\gamma))((A_\gamma))$ and

$$(3) \qquad k((C_\gamma^{<0})) = k((D_\gamma))((A_\gamma^{<0})) \oplus k((D_\gamma^{<0})).$$

Set $k_\gamma = k((D_\gamma)) \cap F$ and $F_\gamma = k((C_\gamma)) \cap F$. Then $f \in F_\gamma$. Also from truncation closedness of F it follows that $F_\gamma = k_\gamma((A_\gamma)) \cap F$, since each coefficient that appears when we represent a series from F_γ as an element of $k((D_\gamma))((A_\gamma))$, is the restriction of the series to a convex subset of C_γ, divided by a monomial, and thus also belongs to F. So F_γ is a truncation closed subfield of $k_\gamma((A_\gamma))$ containing $k_\gamma(A_\gamma)$. Viewing the elements of k_γ as constants, define $\mathrm{Neg}_{k_\gamma}(F_\gamma) = F_\gamma \cap k_\gamma((A_\gamma^{<0}))$. Since we are changing the field of constants, we have to indicate by a subscript over which ground field each Neg is taken. In particular, what was previously denoted by $\mathrm{Neg}(F)$ now will be written as $\mathrm{Neg}_k(F)$.

THEOREM 5.2. *Let f be an element of* $\mathrm{Neg}_k(F) \oplus k$ *with constant term $f_0 \in k$. Assume that $\Gamma_f = \{\gamma\}$. Then f is prime (resp., irreducible) in $\mathrm{Neg}_k(F) \oplus k$ iff the following two conditions hold: 1) f is prime (resp., irreducible) in $\mathrm{Neg}_{k_\gamma}(F_\gamma) \oplus k_\gamma$ and 2) either $f_0 \neq 0$ or $\gamma = \max \Gamma$.*

PROOF. First we observe that if condition 2) does not hold, that is, $f_0 = 0$ and there exists $g_0 \in G^{<0}$ such that $v(g_0) > \gamma$, then f can be factored in $\mathrm{Neg}_k(F) \oplus k$: $f = t^{g_0}(f t^{-g_0})$. So from now on we assume that condition 2) holds. The proof that f is prime (resp., irreducible) in $\mathrm{Neg}_k(F) \oplus k$ iff f is prime (resp., irreducible) in $\mathrm{Neg}_{k_\gamma}(F_\gamma) \oplus k_\gamma$ will proceed in two steps.

Step 1. We claim that f is prime (resp., irreducible) in $\mathrm{Neg}_{k_\gamma}(F_\gamma) \oplus k_\gamma$ iff f is prime (resp., irreducible) in $\mathrm{Neg}_k(F_\gamma) \oplus k$.

If $\gamma = \max \Gamma$, then $k_\gamma = k$ and the claim is trivial. So assume $f_0 \neq 0$. Replacing f by f/f_0, we can assume that f has constant term 1 as an element of $k((G))$. Since $\Gamma_f = \{\gamma\}$, f still has constant term 1 when viewed as an element of $k_\gamma((A_\gamma))$. By Lemma 3.7 with $G := A_\gamma$, $K := k_\gamma$, $\mathrm{Neg} := \mathrm{Neg}_{k_\gamma}(F_\gamma)$, and $R := \mathrm{Neg}_k(k_\gamma) \oplus k$, f is prime (resp., irreducible) in $\mathrm{Neg} \oplus K = \mathrm{Neg}_{k_\gamma}(F_\gamma) \oplus k_\gamma$ iff f is prime (resp., irreducible) in $\mathrm{Neg} \oplus R = \mathrm{Neg}_{k_\gamma}(F_\gamma) \oplus \mathrm{Neg}_k(k_\gamma) \oplus k$. It remains to observe that decomposition (3) implies that $\mathrm{Neg}_k(F_\gamma) = \mathrm{Neg}_{k_\gamma}(F_\gamma) \oplus \mathrm{Neg}_k(k_\gamma)$.

Step 2. We claim that f is prime (resp., irreducible) in $\mathrm{Neg}_k(F_\gamma) \oplus k$ iff f is prime (resp., irreducible) in $\mathrm{Neg}_k(F) \oplus k$.

Let E_γ be a complement for C_γ in G. Then $G = E_\gamma \oplus C_\gamma$ (lexicographic sum). Consequently, $k((G)) = k((C_\gamma))((E_\gamma))$ and

$$(4) \qquad k((G^{<0})) = k((C_\gamma))((E_\gamma^{<0})) \oplus k((C_\gamma^{<0})).$$

Since F is truncation closed, $F \subset F_\gamma((E_\gamma))$. It remains to apply Lemma 3.6 with $G := E_\gamma$, $K := F_\gamma$, Neg $:= \mathrm{Neg}_{F_\gamma}(F)$, and $R := \mathrm{Neg}_k(F_\gamma) \oplus k$ and observe that $\mathrm{Neg} \oplus R = \mathrm{Neg}_k(F) \oplus k$ since $\mathrm{Neg}_k(F) = \mathrm{Neg}_{F_\gamma}(F) \oplus \mathrm{Neg}_k(F_\gamma)$ by (4). $\dashv$

Taking $F = k((G))$, we obtain the following important

COROLLARY 5.3. *Let f be an element of $k((G^{\leq 0}))$ with constant term f_0. Assume that $\Gamma_f = \{\gamma\}$. Then f is prime (resp., irreducible) in $k((G^{\leq 0}))$ iff the following two conditions hold*: 1) f *is prime (resp., irreducible) in $k_\gamma((A_\gamma^{\leq 0}))$ where $k_\gamma = k((D_\gamma))$ and* 2) *either $f_0 \neq 0$ or $\gamma = \max \Gamma$.*

As an application of Corollary 5.3, we prove the following generalizations of [B, Theorem 11.2]. First we introduce some notation. Let f be an arbitrary element of $\mathrm{Neg}(F) \oplus k$ and let $\gamma \in \Gamma_f$. As before, set $f_\gamma = \sum_{g:v(g)=\gamma} f_g t^g$. Then $f_\gamma \in F_\gamma = k((C_\gamma)) \cap F$. Set

$$S_f(\gamma) := \pi_\gamma(\text{support } f_\gamma) \subset A_\gamma.$$

Once a complement for D_γ in C_γ is fixed, we can view F_γ as a subfield of $k_\gamma((A_\gamma))$, where $k_\gamma = k((D_\gamma)) \cap F$ as before. Then $S_f(\gamma)$ is just the support of f_γ viewed as an element of $k_\gamma((A_\gamma))$.

COROLLARY 5.4. *Assume that $\Gamma_f = \{\gamma\}$, $S_f(\gamma)$ is finite, and f is prime as an element of $k_\gamma[A_\gamma^{\leq 0}]$. Then f is irreducible in $k((G^{\leq 0}))$.*

PROOF. Since A_γ is archimedean, f remains irreducible in $k_\gamma((A_\gamma^{\leq 0}))$ by Berarducci's result [B, Theorem 11.2]. Now apply Corollary 5.3. $\dashv$

COROLLARY 5.5. *Let f be prime in $k[G^{\leq 0}]$ and suppose that support f generates an archimedean subgroup of G. Then f remains irreducible in $k((G^{\leq 0}))$ and, consequently, in $\mathrm{Neg}(F) \oplus k$, for any truncation closed F, $k(G) \subset F \subset k((G))$.*

PROOF. Let H be the subgroup generated by support f. Since H is archimedean, all nonzero elements of H have the same valuation, say γ. Moreover, we can pick a complement A_γ for D_γ in C_γ in such a way that $H \subset A_\gamma$. Since f is prime in $k[G^{\leq 0}]$, the constant term $f_0 \neq 0$. By Corollary 5.4, it suffices to show that f is prime in $k_\gamma[A_\gamma^{\leq 0}]$. By the choice of A_γ, $f \in k[A_\gamma^{\leq 0}]$. By Lemma 4.4, f is prime in $k[A_\gamma^{\leq 0}]$, so $R := k[A_\gamma^{\leq 0}]/(f)$ is a domain. It follows that $R((D_\gamma))$ is also a domain. But $R((D_\gamma)) = (k[A_\gamma^{\leq 0}]/(f))((D_\gamma)) = k[A_\gamma^{\leq 0}]((D_\gamma))/(f) = k((D_\gamma))[A_\gamma^{\leq 0}]/(f) = k_\gamma[A_\gamma^{\leq 0}]/(f)$. We conclude that f is prime in $k_\gamma[A_\gamma^{\leq 0}]$. $\dashv$

EXAMPLE 5.6. Consider $f = a_1 t^{g_1} + a_2 t^{g_2} + 1$ where $g_1, g_2 \in G^{<0}$ are $\mathbb{Q}$-linearly independent and $0 \neq a_1, a_2 \in k$. As we know, f is prime in $k[G^{\leq 0}]$. It remains irreducible in $k((G^{\leq 0}))$ iff $v(g_1) = v(g_2)$ and g_1, g_2 are $\mathbb{Q}$-valuation independent.

PROOF. If the conditions on g_1, g_2 are satisfied, then the subgroup $\langle g_1, g_2 \rangle$ is archimedean by Corollary 2.3, so f remains irreducible in $k((G^{\leq 0}))$ by Corollary 5.5. Now if $v(g_1) \neq v(g_2)$, then by the same argument as in Example 5.1, f does not remain irreducible in $k((G^{\leq 0}))$. If $v(g_1) = v(g_2) = \gamma$, but g_1, g_2 are not $\mathbb{Q}$-valuation independent, then there exist $m, n \in \mathbb{N}$ such that $v(g) > \gamma$ where $g = ng_1 - mg_2$. In this case, let $x = t^{\frac{1}{3m}g_1}$, then $f = a_1 x^{3m} + a_2' x^{3n} + 1$ where $a_2' = a_2 t^{-\frac{1}{m}g} \in k_\gamma = k((D_\gamma))$. Clearly, if k is real closed, this polynomial factors in $k_\gamma[x]$, so f does not remain irreducible in $k((G^{\leq 0}))$.   ⊣

This example shows that in Corollary 5.5 the condition that support f generates an archimedean subgroup cannot be weakened to say just that Γ_f is a singleton. On the other hand, consider $f' = t^{2g_1} + t^{g_1 + g_2} + t^{g_1 - g} + t^{g_2 + g} + 1$ where $g_1, g_2 \in G^{<0}$ are $\mathbb{Q}$-valuation independent and $v(g_1) = v(g_2) = \gamma < v(g)$. If $g \neq 0$, then the subgroup generated by support f' is not archimedean, but f' is still irreducible in $k((G^{\leq 0}))$. Indeed, choose A_γ such that $g_1, g_2 \in A_\gamma$. Then f' is prime in $k_\gamma[A_{\overline{\gamma}}^{\leq 0}]$: $f' = t^{2g_1} + \alpha^{-1}(\alpha t^{g_2} + 1)t^{g_1} + (\alpha t^{g_2} + 1)$ where $\alpha = t^g \in k_\gamma$, so by rewriting f' in terms of $x := t^{\frac{1}{n}g_1}$ and $y := t^{\frac{1}{m}g_2}$ ($m \in \mathbb{N}$), we obtain an irreducible polynomial $x^{2m} + \alpha^{-1}(\alpha y^m + 1)x^m + (\alpha y^m + 1)$ (cf. Remark 4.6). Therefore, by Corollary 5.4, f' is irreducible in $k((G^{\leq 0}))$.

We do not know in general in Corollary 5.5 if f actually remains prime in $\text{Neg}(F) \oplus k$, but we can show this in the case when the field F is "small".

THEOREM 5.7. *Assume that for any $f \in \text{Neg}(F) \oplus k$ and any $\gamma \in \Gamma_f$, the set $S_f(\gamma)$ is finite. Then every irreducible element of $\text{Neg}(F) \oplus k$ is prime.*

PROOF. Let f be irreducible in $\text{Neg}(F) \oplus k$. Then $\Gamma_f = \{\gamma\}$ for some $\gamma \in \Gamma$. By Theorem 5.2, f is irreducible in $\text{Neg}_{k_\gamma}(F_\gamma) \oplus k_\gamma$. By our assumption on F, $\text{Neg}_{k_\gamma}(F_\gamma) \oplus k_\gamma = k_\gamma[A_{\overline{\gamma}}^{\leq 0}]$. By Corollary 4.2, f is prime in $k_\gamma[A_{\overline{\gamma}}^{\leq 0}]$. Hence by Theorem 5.2, f is prime in $\text{Neg}(F) \oplus k$.   ⊣

One important case when the sets $S_f(\gamma)$ are finite is $F = k(G)^r$, the field of algebraic power series. (Recall that by [M–R], $k(G)^r$ is a truncation closed subfield of $k((G))$.) In the following corollaries we assume that k is a real closed field.

COROLLARY 5.8. *Let $F = k(G)^r$. Then every irreducible element of $\text{Neg}(F) \oplus k$ is prime.*

PROOF. By Theorem 5.7, it suffices to show that for any $f \in k((G^{\leq 0}))$ which is algebraic over $k(G)$ and any $\gamma \in \Gamma_f$, the set $S_f(\gamma)$ is finite. (As a matter of fact, the set Γ_f is also finite, but we do not need this property here.) By truncation closedness of F, we can assume without loss of generality that $f = f_\gamma$.

We claim that f is algebraic over $k(C_\gamma)$. Indeed, since f is algebraic over $k(G)$, there exists a nonzero polynomial $p(x) \in k[G][x]$ such that $p(f) = 0$.

Choose a complement E_γ for C_γ in G. Then $k[G] = k[C_\gamma][E_\gamma]$ and so we can write $p(x) = \sum_{g \in E_\gamma} p_g(x)t^g$ (finite sum) where $p_g(x) \in k[C_\gamma][x]$. Since $p(x)$ is nonzero, there exists $g_0 \in E_\gamma$ such that $p_{g_0}(x)$ is nonzero. Since $p(f) = 0$ and the set $\{t^g \mid g \in E_\gamma\}$ is linearly independent over $k((C_\gamma))$, it follows that $p_{g_0}(f) = 0$, which proves the claim.

Now let $K = k((D_\gamma))$. Then $f \in K((A_\gamma^{\leq 0}))$. Since $k(C_\gamma) \subset K(A_\gamma)$, f is algebraic over $K(A_\gamma)$. But A_γ is archimedean, so by Proposition 3.4, the support of f, viewed as an element of $K((A_\gamma))$, is finite. As noted earlier, this support is precisely $S_f(\gamma)$. $\dashv$

EXAMPLE 5.9. Let $F = k(G)^r$. Let $g_1, g_2 \in G^{<0}$ be $\mathbb{Q}$-valuation independent and $v(g_1) = v(g_2)$. Consider $f = a_1 t^{g_1} + a_2 t^{g_2} + 1$ where $0 \neq a_1, a_2 \in k$. Then f is prime in $\text{Neg}(F) \oplus k$.

PROOF. Combine Corollaries 5.5 and 5.8. $\dashv$

COROLLARY 5.10. *Assume that for every $\gamma \in \Gamma$ we have $\dim_{\mathbb{Q}} A_\gamma \geq 2$. Then for any truncation closed F, $k(G) \subset F \subset k((G))$, the ordered ring $\text{Neg}(F) \oplus \mathbb{Z}$ has a cofinal set of irreducible elements with finite support.*

PROOF. Let $s \in \text{Neg}(F) \oplus \mathbb{Z}$ and let $g = \min(\text{support } s) \in G^{\leq 0}$. Take $g_1 < g$ and set $\gamma = v(g_1)$. Choose A_γ such that $g_1 \in A_\gamma$. By hypothesis, we can find $g_2 \in A_\gamma$ linearly independent from g_1. Without loss of generality, $g_2 < 0$. Then $f = t^{g_1} + t^{g_2} + 1$ is irreducible in $\text{Neg}(F) \oplus \mathbb{Z}$ by Corollary 5.5. Clearly, $f > s$. $\dashv$

COROLLARY 5.11. *Assume that for every $\gamma \in \Gamma$ we have $\dim_{\mathbb{Q}} A_\gamma \geq 2$. Let $F = k(G)^r$. Then the ordered ring $\text{Neg}(F) \oplus \mathbb{Z}$ has a cofinal set of primes with finite support.*

PROOF. Combine Corollaries 5.10 and 5.8. $\dashv$

As another application of Corollary 5.3, we obtain the following improvement of [P, Theorem 4.1] and [P, Theorem 4.2].

THEOREM 5.12. *Let $s \in k((G^{<0}))$ be such that all elements of support s have the same valuation $\gamma \in \Gamma$. Assume also that the order type of the set $\pi_\gamma(\text{support } s) \subset A_\gamma^{\leq 0}$ is ω and the least upper bound is 0. Then $s + 1$ is prime in $k((G^{\leq 0}))$.*

PROOF. By Corollary 5.3, it suffices to show that $f := s + 1$ is prime in $k_\gamma((A_\gamma^{\leq 0}))$ where $k_\gamma = k((D_\gamma))$. Since A_γ is archimedean and the set $S_f(\gamma) = \pi_\gamma(\text{support } s)$ is precisely the support of s viewed as an element of $k_\gamma((A_\gamma))$, we can apply Pitteloud's result [P, Theorem 4.1] to conclude that f is indeed prime in $k_\gamma((A_\gamma^{\leq 0}))$. $\dashv$

COROLLARY 5.13. *Let $s \in k((G^{<0}))$ be such that the support of s is of order type ω and is contained in an archimedean subgroup A of G.*

Assume that s is not divisible by any monomial t^a with $a \in A^{<0}$. Then $s + 1$ is prime in $k((G^{\leq 0}))$.

EXAMPLE 5.14. For any $g \in G^{<0}$, the series

$$p_g := \sum_{n=1}^{\infty} t^{\frac{1}{n}g} + 1$$

is prime in $k((G^{\leq 0}))$.

EXAMPLE 5.15. Let $g, g' \in G^{<0}$ be such that $v(g') > v(g)$. Then the series

$$p_{g,g'} := \sum_{n=1}^{\infty} \sum_{m=1}^{\infty} t^{\frac{1}{n}g + \frac{1}{m}g'} + 1$$

is prime in $k((G^{\leq 0}))$. Note that the support of $f := p_{g,g'} - 1$ has order type ω^2, but $S_f(\gamma)$, where $\gamma = v(g)$, has order type ω, so Theorem 5.12 applies.

COROLLARY 5.16. *The ordered ring $k((G^{<0})) \oplus \mathbb{Z}$ has a cofinal set of prime elements with infinite support.*

PROOF. Let $s \in k((G^{<0})) \oplus \mathbb{Z}$ and let $g = \min(\text{support } s) \in G^{\leq 0}$. Pick $g' < g$, then $p_{g'}$ is prime by Example 5.14 and $p_{g'} > s$.   ⊣

§6. **Exponential integer parts of exponential fields.** In this Section, we apply the results of the previous sections to study EIP's of exponential fields. Let $(K, +, \cdot, 0, 1, <)$ be a real closed field (RCF). We say that K is an *exponential field* if there exists an exponential on K, that is, a map exp such that

(EXP) exp : $(K, +, 0, <) \simeq (K^{>0}, \cdot, 1, <)$ (that is, exp is an isomorphism of ordered groups).

We shall further consider exponentials that satisfy the *growth axiom scheme*:

(GA) for all $x \in K$, $x \geq n^2 \Longrightarrow \exp(x) > x^n$ $(n \geq 1)$.

Thus $(K, \exp)$ is a model for the fragment

$$T_{EXP} = RCF + EXP + GA$$

of $\text{Th}(\mathbb{R}, \exp) :=$ the elementary theory of the ordered field of the reals with exponentiation.

A *logarithm* on K is the compositional inverse $\log = \exp^{-1}$ of an exponential.

As before, we consider the residue field k embedded into K. By appropriately modifying exp (see [K, Lemma 1.18]), we can always assume that

$$\exp(\mathcal{O}_K) = \mathcal{U}_K^{>0} \quad \text{and} \quad \exp(\mathcal{M}_K) = 1 + \mathcal{M}_K \quad \text{and} \quad \exp(k) = k^{>0}.$$

THEOREM 6.1. *Every truncation integer part of an exponential field has a cofinal set of irreducibles.*

PROOF. We can apply Corollary 5.10 because the condition on G is always fulfilled when G is the value group of an exponential field (see [K, Proposition 1.22]).   ⊣

An *exponential integer part* (EIP) of K is an integer part Z such that for any $z \in Z^{>0}$, we have $2^z \in Z^{>0}$ (where by definition $2^x := \exp(x \log 2)$ for all $x \in K$). The following easy proposition tells us when a truncation IP is actually an EIP.

PROPOSITION 6.2. *Let F be a truncation closed subfield of $k((G))$ that contains $k(G)$. Assume that F admits an exponential $\exp$ such that $\exp(\mathrm{Neg}(F)) \subset \mathrm{Mon}(F)$ where $\mathrm{Mon}(F) := \mathrm{Mon}\, k((G))$. Then the canonical integer part $Z_F = \mathrm{Neg}(F) \oplus \mathbb{Z}$ is an exponential integer part.*

PROOF. Let $z = a + z_0 \in Z_F^{\geq 0}$, $a \in \mathrm{Neg}(F)$, $z_0 \in \mathbb{Z}$. If $a = 0$ then $z_0 > 0$ and $2^{z_0} \in \mathbb{Z}$. If not, then $a > 0$ and $2^{a+z_0} = 2^{z_0} 2^a = 2^{z_0} \exp(a \log 2)$. Since $\log 2 \in k$ and $\mathrm{Neg}(F)$ is a k-vector space, $a \log 2 \in \mathrm{Neg}(F)$. So $\exp(a \log 2) = t^g$ for some $g \in G$. Since $a \log 2 > 0$, $t^g > 1$, so $g < 0$. Therefore, $2^{a+z_0} = 2^{z_0} t^g \in \mathrm{Neg}(F) \subset Z_F$ as required. ⊣

THEOREM 6.3. *Every exponential field has an exponential integer part with a cofinal set of irreducibles.*

PROOF. In [R] it is shown that given an exponential field K, there is a truncation closed embedding φ of K into $k((G))$ such that $\exp(\mathrm{Neg}(F)) \subset \mathrm{Mon}(F)$ where $F := \varphi(K)$, so $\mathrm{Neg}(F) \oplus \mathbb{Z}$ is an EIP of F. It follows that K has a truncation IP which is an EIP, thus the assertion follows by Theorem 6.1. ⊣

In the following example, we describe a procedure to construct countable non-archimedean models of T_{EXP} which are algebraic power series fields, and investigate their canonical IP's. We recall that a real closed field K is an ordered vector space over the residue field k, and all archimedean components of $(K, +, 0, <)$ have dimension 1 over k.

EXAMPLE 6.4. *Countable Fields of Algebraic Power Series with Exponentiation.* Let E be a countable exponential subfield of $(\mathbb{R}, \exp)$ (for example, take E to be the smallest real closed, exp-closed and log-closed subfield of $\mathbb{R}$). Set $G = \oplus_{\mathbb{Q}} E$ the lexicographic sum over $\mathbb{Q}$ of copies of $(E, +, 0, <)$, that is,

$$G := \left\{ \sum_{q \in \mathbb{Q}} g_q 1_q \mid g_q \in E, g_q = 0 \text{ for all but finitely many } q \in \mathbb{Q} \right\}$$

endowed with pointwise addition and lexicographic order (here, 1_q denotes the characteristic function on the singleton $\{q\}$). Set $E = E(G)^r$, then E is countable and real closed. Further E is a truncation closed subfield of $E((G))$ [M–R].

We want an exponential $\exp$ (or equivalently, a logarithm $\log$) on E such that $(E, \exp)$ is a model of T_{EXP}, and such that the canonical IP $Z_E = \mathrm{Neg}(E) \oplus \mathbb{Z}$ is an EIP.

Observe first that

$$(5) \qquad\qquad (E, +, 0, <) = \mathrm{Neg}(E) \oplus \mathcal{O}_E.$$

$$(6) \qquad\qquad (E^{>0}, \cdot, 1, <) = \mathrm{Mon}(E) \times \mathcal{U}_E^{\geq 0}.$$

We shall define log on $E^{>0}$ as follows. We first define log on $\mathcal{U}_E^{\geq 0} = E^{>0} \times (1 + \mathcal{M}_E)$. By [K, Theorem 2.22] there exists a $\mathbb{Q}$-valuation basis B of $\mathcal{M}_E$ such that the set $\{1 + b \mid b \in B\}$ is a $\mathbb{Q}$-valuation basis of $1 + \mathcal{M}_E$. Define $\log(1 + b) := b$ and extend log to $1 + \mathcal{M}_E$ by linearity. One proves that log is an isomorphism of ordered groups with $\log(1 + \mathcal{M}_E) = \mathcal{M}_E$ [K, Lemma 2.21]. For $u \in \mathcal{U}_E^{\geq 0}$, write $u = ry$ where $r \in E^{>0}$, $y \in 1 + \mathcal{M}_E$, and define $\log u := \log r + \log y$ (where $\log r \in \mathbb{R}$ is the natural logarithm). Using these definitions, it is easily seen that log is an isomorphism of ordered groups with $\log(\mathcal{U}_E^{\geq 0}) = \mathcal{O}_E$.

Now we define log on $\mathrm{Mon}(E)$. By [K, Proposition 2.16]; fix an isomorphism of chains $s : \mathbb{Q} \to G^{<0}$ such that for all $g \in G^{<0}$

$$(7) \qquad\qquad s(v(g)) > g.$$

Note that $\{1_q \mid q \in \mathbb{Q}\}$ is a strongly independent E-basis of G. Now $\mathrm{Neg}(E)$ is a countable ordered E-vector space with 1-dimensional archimedean components, so it admits a strongly independent E-basis.

Fix such a basis $\{b_{s(q)} \mid q \in \mathbb{Q}\}$ with $V(b_{s(q)}) = s(q)$ and $b_{s(q)} < 0$ for all $q \in \mathbb{Q}$. Define $\log(t^{1_q}) := b_{s(q)}$ for all $q \in \mathbb{Q}$, and extend log to $\mathrm{Mon}(E)$ by linearity:

$$\log(t^g) = \sum_{q \in \mathbb{Q}} g_q b_{s(q)} \quad \text{for } g = \sum_{q \in \mathbb{Q}} g_q 1_q.$$

It is easily verified that log is an isomorphism of ordered groups with $\log(\mathrm{Mon}(E)) = \mathrm{Neg}(E)$.

We extend log to $E^{>0}$ using the above definitions via the decomposition (6). The inequality (7) implies that $\exp := \log^{-1}$ satisfies (GA) [K, Theorem 2.5]. Thus $(E, \exp)$ is a model of T_{EXP}. By Proposition 6.2, $Z_E = \mathrm{Neg}(E) \oplus \mathbb{Z}$ is an EIP. By Corollary 5.11, it has a cofinal set of primes (with finite support).

EXAMPLE 6.5. *Exponential-Logarithmic Power Series Fields.* We describe the canonical integer part of the Exponential-Logarithmic Power Series Fields (EL-series fields for short). The EL-series fields are models of $\mathrm{Th}(\mathbb{R}, \exp)$ constructed in [K, Chapter 5 p. 79].

The EL-series field $\mathbb{R}((\Gamma))^{EL}$ is a countable increasing union of power series fields with exponents in properly chosen value groups G_n:

$$\mathbb{R}((\Gamma))^{EL} = \bigcup_{n \in \mathbb{N}} \mathbb{R}((G_n)).$$

The field $\mathbb{R}((\Gamma))^{EL}$ is a proper subfield of $\mathbb{R}((G_\omega))$ where

$$G_\omega := \bigcup_{n \in \mathbb{N}} G_n.$$

In fact, $\mathbb{R}((\Gamma))^{EL}$ is truncation closed and contains $\mathbb{R}(G_\omega)$ (so its value group is G_ω). Furthermore,

$$\mathrm{Neg}(\mathbb{R}((\Gamma))^{EL}) = \bigcup_{n \in \mathbb{N}} \mathrm{Neg}\,\mathbb{R}((G_n)).$$

Therefore, the truncation integer part of $\mathbb{R}((\Gamma))^{EL}$ is

$$Z_{EL} = \bigcup_{n \in \mathbb{N}} Z_n$$

where $Z_n := \mathbb{R}((G_n^{<0})) \oplus \mathbb{Z}$ is the canonical integer part of $\mathbb{R}((G_n))$. The construction of $\mathbb{R}((\Gamma))^{EL}$ guarantees that

$$\exp\left(\mathrm{Neg}\left(\mathbb{R}((\Gamma))^{EL}\right)\right) = \mathrm{Mon}\left(\mathbb{R}((\Gamma))^{EL}\right).$$

By Proposition 6.2 we have that Z_{EL} is indeed an EIP of $\mathbb{R}((\Gamma))^{EL}$. We now show that Z_{EL} has a cofinal set of primes (with infinite support):

PROPOSITION 6.6. *The canonical exponential integer part Z_{EL} of $\mathbb{R}((\Gamma))^{EL}$ has a cofinal set of primes with infinite support.*

PROOF. Let $g \in G_\omega^{<0}$. Then $g \in G_n$ for some $n \in \mathbb{N}$ and the series $p_g = \sum_{l=1}^{\infty} t^{\frac{g}{l}} + 1 \in \mathbb{R}((G_n))$ is prime in Z_m for all $m \geq n$ by Example 5.14. Therefore, by Lemma 3.8, p_g remains prime in $Z_{EL} = \bigcup_{n \in \mathbb{N}} Z_n$. The set $\{p_g \mid g \in G_\omega^{<0}\}$ is clearly cofinal in $\mathbb{R}((G_\omega))$ and thus in $\mathbb{R}((\Gamma))^{EL}$. ⊣

REFERENCES

[B] A. BERARDUCCI, *Factorization in generalized power series*, **Transactions of the American Mathematical Society**, vol. 352 (2000), no. 2, pp. 553–577.

[B–O] A. BERARDUCCI and M. OTERO, *A recursive nonstandard model of normal open induction*, **The Journal of Symbolic Logic**, vol. 61 (1996), no. 4, pp. 1228–1241.

[Bi] D. BILJAKOVIC, *Recursive Models of Open Induction With Infinite Primes*, preprint, 1996.

[Bo] S. BOUGHATTAS, *Résultats optimaux sur l'existence d'une partie entière dans les corps ordonnés*, **The Journal of Symbolic Logic**, vol. 58 (1993), no. 1, pp. 326–333.

[Bo2] ——, *Trois théorèmes sur l'induction pour les formules ouvertes munies de l'exponentielle*, **The Journal of Symbolic Logic**, vol. 65 (2000), no. 1, pp. 111–154.

[Br] R. BROWN, *Valued vector spaces of countable dimension*, **Publicationes Mathematicae Debrecen**, vol. 18 (1971), pp. 149–151.

[F] A. FORNASIERO, *Embedding Henselian fields into power series*, submitted (2005) http://www.dm.unipi.it/~fornasiero/ressayre.pdf.

[G] H. GONSHOR, *An Introduction to The Theory of Surreal Numbers*, Cambridge University Press, Cambridge, 1986.

[KF] F.-V. KUHLMANN, *Dense Subfields of Henselian Fields, and Integer Parts*, this volume.

64   DARKO BILJAKOVIC, MIKHAIL KOCHETOV, AND SALMA KUHLMANN

[K] S. KUHLMANN, *Ordered Exponential Fields*, Fields Institute Monographs, vol. 12, American Mathematical Society, Providence, RI, 2000.

[M] M. MONIRI, *Recursive models of open induction of prescribed finite transcendence degree > 1 with cofinal twin primes*, **Comptes Rendus de l'Académie des Sciences. Série I. Mathématique**, vol. 319 (1994), no. 9, pp. 903–908.

[M–R] M.-H. MOURGUES and J. P. RESSAYRE, *Every real closed field has an integer part*, **The Journal of Symbolic Logic**, vol. 58 (1993), no. 2, pp. 641–647.

[P] D. PITTELOUD, *Existence of prime elements in rings of generalized power series*, **The Journal of Symbolic Logic**, vol. 66 (2001), no. 3, pp. 1206–1216.

[PC] S. PRIEß-CRAMPE, *Angeordnete Strukturen. Gruppen, Körper, projektive Ebenen*, Ergebnisse der Mathematik und ihrer Grenzgebiete, vol. 98, Springer-Verlag, Berlin, 1983.

[R] J.-P. RESSAYRE, *Integer parts of real closed exponential fields (extended abstract)*, Arithmetic, Proof Theory, and Computational Complexity (Prague, 1991) (P. Clote and J. Krajicek, editors), Oxford Logic Guides, vol. 23, Oxford Univ. Press, New York, 1993, pp. 278–288.

[Ri] P. RIBENBOIM, *Théorie des Valuations*, vol. 1964, Les Presses de l'Université de Montréal, Montreal, Que., 1968.

[S] J. C. SHEPHERDSON, *A non-standard model for a free variable fragment of number theory*, **Bulletin de l'Académie Polonaise des Sciences. Série des Sciences Mathématiques, Astronomiques et Physiques**, vol. XII (1964), pp. 79–86.

FACULTY OF AGRICULTURE
UNIVERSITY OF ZAGREB
CROATIA
E-mail: biljakovic@agr.hr

DEPARTMENT OF MATHEMATICS
CARLETON UNIVERSITY
OTTAWA, CANADA
E-mail: kochetov@math.carleton.ca

RESEARCH UNIT ALGEBRA AND LOGIC
UNIVERSITY OF SASKATCHEWAN
SASKATOON, CANADA
E-mail: skuhlman@math.usask.ca

ON EXPLICIT DEFINABILITY IN ARITHMETIC

LOU VAN DEN DRIES

Abstract. We characterize the arithmetic functions in *one* variable that are *explicitly definable* from the familiar arithmetic operations. This characterization is deduced from algebraic properties of some non-standard rings of integers. Elementary model theory is also used to show that the greatest common divisor function and related functions are *not* explicitly definable from the usual arithmetic operations. It turns out that explicit definability is equivalent to *computability with bounded complexity*, for the recursive algorithms of Y. Moschovakis and with respect to a certain natural cost function.

§1. Introduction.

Motivating Question: Which functions are explicitly definable in the arithmetic structures

$$\mathbb{Z}(1) := (\mathbb{Z}, 0, 1, +, -, \mathrm{iq}, \mathrm{rem}, <)$$

$$\text{and } \mathbb{Z}(2) := (\mathbb{Z}, 0, 1, +, -, \cdot, \mathrm{iq}, \mathrm{rem}, <)?$$

Here iq ("integer quotient") and rem ("remainder") describe

integer division with remainder,

that is, $\mathrm{iq} : \mathbb{Z}^2 \to \mathbb{Z}$ and $\mathrm{rem} : \mathbb{Z}^2 \to \mathbb{Z}$ are the functions such that for all $a, b \in \mathbb{Z}$ we have $a = \mathrm{iq}(a,b)b + \mathrm{rem}(a,b)$, with $0 \leq \mathrm{rem}(a,b) < b$ for $b > 0$, $b < \mathrm{rem}(a,b) \leq 0$ for $b < 0$, and $\mathrm{iq}(a,0) = \mathrm{rem}(a,0) = a$.

We now specify what we mean by *explicitly definable*. Let $\mathcal{M} = (M, 0, 1, \dots)$ be an L-structure with distinct marked elements $0, 1 \in M$. A function $f : M^m \to M$ is said to be *explicitly definable in $\mathcal{M}$* if there are L-terms $t_1(x), \dots, t_n(x)$ with $x = (x_1, \dots, x_m)$, such that each set

$$\{a \in M^m : f(a) = t_i(a)\} \quad (i = 1, \dots, n)$$

is definable in $\mathcal{M}$ by a quantifier-free L-formula, and these sets together cover M^m. (Throughout this paper, m and n range over $\mathbb{N} = \{0, 1, 2, \dots\}$.)

An n-ary relation $R \subseteq M^n$ is said to be *explicitly definable in $\mathcal{M}$* if its characteristic function $\chi_R : M^n \to \{0, 1\} \subseteq M$ is explicitly definable in $\mathcal{M}$.

Partially supported by NSF grant DMS 01-00979

Logic in Tehran
Edited by A. Enayat, I. Kalantari, and M. Moniri
Lecture Notes in Logic, 26

Remarks on explicit definability.

(1) *Connection to computation*: The *recursive programs on* $\mathcal{M}$ from [7, 8] have the property that explicit definability in $\mathcal{M}$ is equivalent to computability by a recursive program on $\mathcal{M}$ in a uniformly bounded number of steps, as measured by a certain cost function. A precise statement to this effect can be found in [8].

(2) *Connection to quantifier-free definability*: If a function $f : M^m \to M$ is explicitly definable in $\mathcal{M}$, then its graph is definable in $\mathcal{M}$ by a quantifier-free L-formula; the converse of this implication is in general false. A *relation* $R \subseteq M^n$ is explicitly definable in $\mathcal{M}$ if and only if R is definable in $\mathcal{M}$ by a quantifier-free L-formula.

(3) *Robustness*: If $f : M^m \to M$ and $g_1, \ldots, g_m : M^n \to M$ are explicitly definable in $\mathcal{M}$, so is $f(g_1, \ldots, g_m) : M^n \to M$. If $\mathcal{M}'$ is an expansion of $\mathcal{M}$ by relations and functions that are explicitly definable in $\mathcal{M}$, then explicit definability in $\mathcal{M}'$ is equivalent to explicit definability in $\mathcal{M}$, and definability in $\mathcal{M}'$ by a quantifier-free L'-formula is equivalent to definability in $\mathcal{M}$ by a quantifier-free L-formula.

Throughout this paper we use elementary model theory. A simple version of a result proved in this way in Section 2 is the following:

PROPOSITION 1.1. *Let* $S \subseteq \mathbb{Z}$ *and* $f : \mathbb{Z} \to \mathbb{Z}$. *Then*

(1) *the unary relation* S *is explicitly definable in* $\mathbb{Z}(2)$ *if and only if* S *is a finite union of arithmetic progressions*;

(2) *the unary function* f *is explicitly definable in* $\mathbb{Z}(2)$ *if and only if there are polynomials* $f_1(T), \ldots, f_k(T)$ *in* $\mathbb{Q}[T]$ *and a partition of* $\mathbb{Z}$ *into arithmetic progressions* $S_1, \ldots, S_k$ *such that* $f(s) = f_i(s)$ *on* S_i *for* $i = 1, \ldots, k$.

Less formally: f is explicitly definable in $\mathbb{Z}(2)$ iff f is *piecewise-polynomial*, the finitely many pieces being arithmetic progressions. In this paper, an *arithmetic progression* is a set of the form

$$\{a \in \mathbb{Z} : a \equiv i \mod m, \ \alpha < a < \beta\}$$

where $0 \le i < m$, $\alpha < \beta$ with $\alpha, \beta \in \mathbb{Z} \cup \{-\infty, +\infty\}$. So an arithmetic progression can be finite, or infinite in one or both directions. Note that the finite unions of arithmetic progressions are exactly the subsets of $\mathbb{Z}$ that are definable in $(\mathbb{Z}, 0, 1, +, -, <)$, by Presburger, see [5]. The easy "if"-direction in both parts of the proposition is left to the reader. In Section 2 below we shall prove the more interesting "only if"-direction.

We obtain a stronger result by allowing also the greatest common divisor function among the primitives. To state these improvements accurately, define the *greatest common divisor* of the integers a, b to be the unique nonnegative integer $c =: \gcd(a, b)$ such that $c\mathbb{Z} = a\mathbb{Z} + b\mathbb{Z}$. (So $\gcd(a, b)$ is indeed the largest integer dividing both a and b, if $a \ne 0$ or $b \ne 0$.) We also introduce operations that express $\gcd(a, b)$ as an integer linear combination of a and b. We define such operations $\sigma, \tau : \mathbb{Z}^2 \to \mathbb{Z}$ by requiring that

$$\gcd(a, b) = \sigma(a, b)a + \tau(a, b)b$$

subject to the conventions

$$a \neq 0, \ b \neq 0 \implies 0 \leq \sigma(a, b) < |b|$$
$$a = 0 \implies \sigma(a, b) = 0, \ \tau(a, b) = \operatorname{sign} b$$
$$b = 0 \implies \sigma(a, b) = \operatorname{sign} a, \ \tau(a, b) = 0.$$

PROPOSITION 1.2. *Let $S \subseteq \mathbb{Z}$ and $f : \mathbb{Z} \to \mathbb{Z}$. Then*

(1) *the unary relation S is explicitly definable in $(\mathbb{Z}(2), \gcd, \sigma, \tau)$ if and only if S is a finite union of arithmetic progressions;*

(2) *the unary function f is explicitly definable in $(\mathbb{Z}(2), \gcd, \sigma, \tau)$ if and only if there are polynomials $f_1(T), \ldots, f_k(T)$ in $\mathbb{Q}[T]$ and a partition of $\mathbb{Z}$ into arithmetic progressions $S_1, \ldots, S_k$ such that $f(s) = f_i(s)$ on S_i for $i = 1, \ldots, k$.*

The degrees of the polynomials in (2) can be bounded by the *degree* of the terms that explicitly define f, as shown at the end of Section 2.

It is unclear if this result on unary functions has a good binary analogue. For binary functions our work goes rather in an opposite direction: in Section 3 we show for example that the greatest common divisor function and the coprimeness relation are not explicitly definable in $\mathbb{Z}(1)$, and in Section 4 we extend this to $\mathbb{Z}(2)$. (The *coprimeness relation* is the binary relation $\perp$ on $\mathbb{Z}$ defined by: $x \perp y \Leftrightarrow \gcd(x, y) = 1$.) That coprimeness is not explicitly definable in $\mathbb{Z}(2)$ follows from earlier work by Mansour, Schieber and Tiwari [3], and our goal is just to give another treatment of questions of this nature: our model-theoretic approach highlights a connection to irrationality and transcendence that may be suggestive to some readers. This model-theoretic approach was initiated in [6], but in Sections 3–5 we improve on [6] in various ways.

CONJECTURE. σ is not explicitly definable in $(\mathbb{Z}(2), \gcd)$.

In Section 6 we consider in more detail certain rings of non-standard integers that play a key role in Section 2, and pose some problems that may interest those who study algebraic fragments of first-order Peano arithmetic.

NOTATIONS. We fix an $\aleph_1$-saturated elementary extension $^*\mathbb{Z}$ of the ordered ring $\mathbb{Z}$ of integers, and let $^*\mathbb{Q}$ denote its ordered fraction field. For the usual ring operations $+, -,$ and $\cdot$ on $\mathbb{Z}$ and $\mathbb{Q}$, and the relations $<, \leq, >, \geq$ on $\mathbb{Z}$ and $\mathbb{Q}$, their natural extensions to $^*\mathbb{Z}$ and $^*\mathbb{Q}$ are indicated by the same symbols. The same convention applies to the operations iq, rem, gcd, σ and τ on $\mathbb{Z}$, the integer part function $x \mapsto \lfloor x \rfloor : \mathbb{Q} \to \mathbb{Z}$, and the (binary) divisibility relation $|$ on $\mathbb{Z}$ defined by: $x|y \Leftrightarrow kx = y$ for some $k \in \mathbb{Z}$.

In Sections 3 and 4, $^*\mathbb{Z}$ is part of an $\aleph_1$-saturated elementary extension $(^*\mathbb{R}, {}^*\mathbb{Z})$ of the structure $(\mathbb{R}, \mathbb{Z})$, the ordered field of real numbers with a predicate for the set of integers; $^*\mathbb{Q}$ is then the fraction field of $^*\mathbb{Z}$ inside the ambient field $^*\mathbb{R}$. We use the same symbol to denote a standard operation or relation on $\mathbb{R}$ that is definable in $(\mathbb{R}, \mathbb{Z})$, and its natural extension to $^*\mathbb{R}$. We say that $\varepsilon \in {}^*\mathbb{R}$ is *infinitesimal* if $|\varepsilon| < c$ for all real $c > 0$.

As usual, we assume the ordering of an *ordered abelian group* to be linear. A *ring* is always commutative with 1, and a subring has the same 1 as its ambient ring. Let A be an ordered abelian (additively written) group. For $x, y \in A \setminus \{0\}$ we write $x \ll y$ (or $y \gg x$, or $x = o(y)$) to mean that $n|x| < |y|$ for all n, and $x \asymp y$ to mean that $|x| < n|y|$ and $|y| < n|x|$ for some n. So $\asymp$ is an equivalence relation on $A \setminus \{0\}$; its equivalence classes are called *archimedean* classes, and the set of archimedean classes is linearly ordered by $[x] < [y] \Leftrightarrow x \ll y$, where $[x] :=$ archimedean class of x.

I thank Yiannis Moschovakis for many email exchanges that inspired several results of this paper. Indeed, the proof of Proposition 3.3 is due to him, and its way of constructing non-trivial isomorphisms became essential in our joint work [7], [8], where one can find other connections to the present paper.

§2. One-variable functions.
The characterization of explicitly definable univariate functions will be obtained from some purely algebraic facts about the structures $\mathbb{Z}\langle a \rangle$ and $\mathbb{Z}^\times\langle a \rangle$, defined as follows for $a \in {}^*\mathbb{Z}$:

$$\mathbb{Z}\langle a \rangle := (\mathbb{Q} + \mathbb{Q}a) \cap {}^*\mathbb{Z},$$

$$\mathbb{Z}^\times\langle a \rangle := \mathbb{Q}(a) \cap {}^*\mathbb{Z}.$$

We first state the relevant fact for the structures $\mathbb{Z}\langle a \rangle$ and show how it yields the corresponding result about explicit definability in $(\mathbb{Z}(1), \gcd, \sigma, \tau)$.

LEMMA 2.1. *The additive group* $\mathbb{Z}\langle a \rangle$ *is closed under* iq, rem, gcd, *and* σ *and* τ. *If in addition* $a' \in {}^*\mathbb{Z}$ *and for all positive integers* N,

$$a \equiv a' \mod N, \qquad \text{sign}(a - N) = \text{sign}(a' - N),$$

then there is an isomorphism

$$\imath : \mathbb{Z}\langle a \rangle \longrightarrow \mathbb{Z}\langle a' \rangle, \quad \imath(a) = a',$$

of ordered abelian groups such that $\imath(\phi(x, y)) = \phi(\imath(x), \imath(y))$ *for all* $x, y \in \mathbb{Z}\langle a \rangle$ *and* $\phi \in \{$iq, rem, gcd, $\sigma, \tau\}$.

We postpone the proof, and show first how this lemma yields the result on explicit definability that interests us. The derivation uses some elementary model theory that we have assembled in Section 7 as an appendix to this paper.

COROLLARY 2.2. *For any* $f : \mathbb{Z} \longrightarrow \mathbb{Z}$ *the following are equivalent*:

(1) f *is explicitly definable in* $(\mathbb{Z}(1), \gcd, \sigma, \tau)$;

(2) $\mathbb{Z}$ *has a partition into finitely many arithmetic progressions* $S_1, \ldots, S_k$ *and there are polynomials* $f_1(T), \ldots, f_k(T)$ *in* $\mathbb{Q}[T]$ *of degree at most* 1 *such that* $f(s) = f_i(s)$ *on* S_i *for* $i = 1, \ldots, k$.

PROOF. Put $\mathcal{M} := (\mathbb{Z}(1), \gcd, \sigma, \tau)$, and let L denote the language of this structure. Suppose f is explicitly definable in $\mathcal{M}$. For each $a \in {}^*\mathbb{Z}$ the subgroup $\mathbb{Z}\langle a \rangle$ of ${}^*\mathbb{Z}$ is closed under iq, rem, gcd, σ and τ, by lemma 2.1. Hence by lemma 7.1 in the Appendix there are polynomials $f_1(T), \ldots, f_k(T) \in \mathbb{Q}[T]$ of degree at most 1 such that $f(s) \in \{f_1(s), \ldots, f_k(s)\}$ for all $s \in \mathbb{Z}$. It remains to show that for $i = 1, \ldots, k$ the set $\{s \in \mathbb{Z} : f(s) = f_i(s)\}$ is a finite union of arithmetic progressions. This set is quantifier-free definable in $\mathcal{M}$, which allows us to use Proposition 7.2 in the Appendix as follows.

Given any $a, a' \in {}^*\mathbb{Z}$, there is an isomorphism $\mathbb{Z}\langle a \rangle \cong \mathbb{Z}\langle a' \rangle$ of substructures of ${}^*\mathcal{M}$ sending a to a' if and only if $\mathrm{rem}(a, N) = \mathrm{rem}(a', N)$ and $\mathrm{sign}(a - N) = \mathrm{sign}(a' - N) \in \{-1, 0, 1\}$ for all $N \in \mathbb{N}^{>0}$. In other words, the quantifier-free L-type of any $a \in {}^*\mathbb{Z}$ over $\mathbb{Z}$ is completely determined by the formulas $\mathrm{rem}(x, N) = i$, $N < x$, $N = x$, and $x < N$ that it satisfies, where N ranges over $\mathbb{N}^{>0}$ and $0 \le i < N$. Using Proposition 7.2 of the Appendix it follows that each subset of $\mathbb{Z}$ which is quantifier-free definable in $\mathcal{M}$ is a finite union of arithmetic progressions.

This proves the direction (1) $\Rightarrow$ (2). The converse is straightforward.   ⊣

We now continue with the analogue of the above for $\mathbb{Z}(2)$, starting with some purely algebraic results. In a remark following Lemma 2.7 we indicate how the proofs of these algebraic results also yield Lemma 2.1.

Below we let R be a subring of ${}^*\mathbb{Z}$ with fraction field $F \subseteq {}^*\mathbb{Q}$. We say that R *preserves divisibility* if $F \cap {}^*\mathbb{Z} = R$, equivalently, whenever $x, y \in R$ and $x|y$ in ${}^*\mathbb{Z}$, $x \ne 0$, then $\frac{y}{x} \in R$. For $a \in {}^*\mathbb{Z}$ we put $R^\times \langle a \rangle := F(a) \cap {}^*\mathbb{Z}$. Note that $\mathbb{Z}$ preserves divisibility. For our results on explicit definability we need the lemmas below only for $R = \mathbb{Z}$, but it may be of some interest to pin down exactly what it is about $\mathbb{Z}$ that makes things work.

LEMMA 2.3. *Suppose R preserves divisibility, $a \in {}^*\mathbb{Z}$ and $a > R$. Then $R^\times \langle a \rangle \subseteq F[a]$, $R^\times \langle a \rangle$ preserves divisibility, and R is convex in $R^\times \langle a \rangle$.*

PROOF. Let $f(X), g(X) \in R[X]$ be such that $f(X) \ne 0$ and $f(a)|g(a)$ in ${}^*\mathbb{Z}$. We show that $\frac{g(a)}{f(a)} \in F[a]$. We can assume that $\deg f(X) > 0$. Write

$$g(X) = q(X)f(X) + r(X)$$

with $q(X), r(X) \in F[X]$, $\deg r(X) < \deg f(X)$.

Then $\frac{g(a)}{f(a)} = q(a) + \frac{r(a)}{f(a)} \in {}^*\mathbb{Z}$; after multiplying $g(X)$, $q(X)$, $r(X)$ with some positive constant from R we can assume that $q(X), r(X) \in R[X]$, hence

$q(a) \in {}^*\mathbb{Z}$, and thus $\frac{r(a)}{f(a)} \in {}^*\mathbb{Z}$. In view of $|r(a)| < |f(a)|$ this implies $r(a) = 0$, hence $r(X) = 0$, so $\frac{g(a)}{f(a)} = q(a) \in F[a]$ as promised. $\quad\dashv$

Note that R is closed under iq if and only if R preserves divisibility and is closed under rem.

LEMMA 2.4. *Suppose R is closed under* iq, $a \in {}^*\mathbb{Z}$, $a > R$, *and* $\mathrm{rem}(a, r) \in R$ *for all non-zero* $r \in R$. *Then* $R^\times \langle a \rangle$ *is closed under* iq.

PROOF. First we show that $\lfloor f(a) \rfloor$ belongs to $R^\times \langle a \rangle$ for $f(X) \in F[X]$. To see this, write $f(X) = \frac{g(X)}{r}$ with $g(X) \in R[X]$ and $0 \neq r \in R$. Let $\mathrm{rem}(a, r) = s \in R$, so $a = qr + s$ with $q \in R^\times \langle a \rangle$. This gives $f(a) = b + f(s)$ where $b \in R[q] \subseteq R^\times \langle a \rangle$. Since $f(s) \in F$ this yields $\lfloor f(a) \rfloor = b + \lfloor f(s) \rfloor \in R^\times \langle a \rangle$.

Let $f(X), g(X) \in F[X]$, $f(X) \neq 0$. We finish the proof of the lemma by showing that then $\lfloor \frac{g(a)}{f(a)} \rfloor$ lies in $R^\times \langle a \rangle$. The case that $\deg f(X) = 0$ falls under the result stated at the beginning of the proof. So we can assume that $\deg f(X) > 0$. Write $g(X) = q(X)f(X) + r(X)$ with $q(X), r(X) \in F[X]$ and $\deg r(X) < \deg f(X)$. Then $\frac{g(a)}{f(a)} = q(a) + \frac{r(a)}{f(a)} \in {}^*\mathbb{Q}$. In view of $|r(a)| < |f(a)|$ this yields

$$\lfloor g(a)/f(a) \rfloor \in \{\lfloor q(a) \rfloor - 1, \lfloor q(a) \rfloor, \lfloor q(a) \rfloor + 1\} \subseteq R^\times \langle a \rangle$$

where the last inclusion follows by the result in the beginning of the proof. $\quad\dashv$

The hypothesis of Lemma 2.4 is satisfied for $R = \mathbb{Z}$ and any $a > \mathbb{Z}$ in ${}^*\mathbb{Z}$. This is also the case for the hypotheses of Lemmas 2.5 and 2.7 below.

LEMMA 2.5. *Suppose R is closed under* iq *and* gcd, $a \in {}^*\mathbb{Z}$, $a > R$ *and* $\mathrm{rem}(a, r) \in R$ *for all non-zero* $r \in R$. *Then* $R^\times \langle a \rangle$ *is closed under* gcd.

PROOF. Let $\phi \in R^\times \langle a \rangle$ with $\phi \neq 0$, and $r \in R$, $r \neq 0$. Then $\gcd(\phi, r) \in R$: after multiplying ϕ and r by a common non-zero element of R we can assume $\phi = f(a)$ with $f(X) \in R[X]$. With $s := \mathrm{rem}(a, r)$ we obtain $\phi = f(a) \equiv f(s) \bmod r$, so $\gcd(\phi, r) = \gcd(f(s), r) \in R$.

Next, let $\phi, \psi \in R^\times \langle a \rangle$ with $\phi, \psi \neq 0$. Writing $\phi = f(a)$ and $\psi = g(a)$ where $f(X), g(X) \in F[X]$, let $d(X) \in R[X]$ generate the ideal of $F[X]$ generated by $f(X)$ and $g(X)$, and put $\delta = d(a) \in R$. Then $\phi = \alpha\delta$, $\psi = \beta\delta$, $\delta = \lambda\phi + \mu\psi$ with $\alpha, \beta, \lambda, \mu \in F[a]$. Take non-zero $r \in R$ such that $r\alpha, r\beta, r\lambda, r\mu \in R[a]$. Then $r\phi = (r\alpha)\delta$ and $r\psi = (r\beta)\delta$, so $\delta | r\phi$ and $\delta | r\psi$. Put $\phi' := \frac{r\phi}{\delta}$ and $\psi' := \frac{r\psi}{\delta}$, so $\phi', \psi' \in R^\times \langle a \rangle$. From $r\delta = (r\lambda)\phi + (r\mu)\psi$ we obtain $\gcd(\phi, \psi) | r\delta$, so $\gcd(\phi', \psi') | r^2$. In view of the result in the beginning of the proof this yields

$$g := \gcd(\phi', \psi') = \gcd\left(\gcd(\phi', r^2), \gcd(\psi', r^2)\right) \in R.$$

Hence $g\delta = \gcd(r\phi, r\psi)$, in particular, $r | g\delta$, hence $\gcd(\phi, \psi) = \frac{g\delta}{r} \in R^\times \langle a \rangle$.
$\quad\dashv$

A *Bezout domain* is an integral domain A such that for all $a, b \in A$ there exists $c \in A$ with $Aa + Ab = Ac$. Of course, $\mathbb{Z}$ and $^*\mathbb{Z}$ are Bezout domains, and below we construct further subrings of $^*\mathbb{Z}$ that are Bezout domains.

LEMMA 2.6. *R is a Bezout domain closed under* rem *if and only if R preserves divisibility and is closed under σ and τ. In that case, R is also closed under* iq *and* gcd.

PROOF. Suppose R is a Bezout domain closed under rem. Let $a, b \in R$, $a, b \neq 0$. Take nonnegative $c \in R$ and $s, t, u, v \in R$ such that $sa + tb = c$, $a = uc$, and $b = vc$. It follows that $c = \gcd(a, b)$. So R is closed under gcd. If also $b \mid a$ in $^*\mathbb{Z}$, then $c = |b|$, hence $a/b = \pm u \in R$. It follows that R preserves divisibility, and hence is also closed under iq. Therefore $s = qb + r$ with $q, r \in R$, $0 \leq r < b$, so $c = ra + (qa + t)b$, hence $\sigma(a, b) = r \in R$ and $\tau(a, b) = qa + t \in R$. The other direction of the equivalence is left to the reader as an exercise. ⊣

It would be desirable to have an example of an R that is closed under iq and gcd but is not a Bezout domain. The existence of such an example would imply that the function σ cannot be explicitly defined in $(\mathbb{Z}(2), \gcd)$. (This non-explicit-definability is highly plausible but I have no proof.)

LEMMA 2.7. *Suppose R is a Bezout domain closed under* rem. *Let $a \in {}^*\mathbb{Z}$ be such that $a > R$ and $\mathrm{rem}(a, r) \in R$ for all non-zero $r \in R$. Then $R^\times \langle a \rangle$ is a Bezout domain.*

PROOF. We already know that R is also closed under iq and gcd. Let $\phi, \psi \in R^\times \langle a \rangle$, $\phi \neq 0$, $\psi \neq 0$ and $d = \gcd(\phi, \psi)$; it suffices to show that $d = \alpha\phi + \beta\psi$ for suitable $\alpha, \beta \in R^\times \langle a \rangle$. We first consider the case that $\psi \in R$. Multiplying ϕ, ψ and d with a suitable non-zero element of R we can assume that $\phi \in R[a]$. Then with $a \equiv r \bmod \psi$ with $r \in R$ we have $\phi \equiv s \bmod \psi$ with $s \in R$, so $d = \gcd(\phi, \psi) = \gcd(s, \psi) \in R$.

Next we consider the general case. Dividing ϕ and ψ by d we can assume $d = 1$. By the proof of the previous lemma we know that then $\alpha\phi + \beta\psi = \gamma$ with $\alpha, \beta \in R^\times \langle a \rangle$ and $0 \neq \gamma \in R$. Now write $\gcd(\phi, \gamma) = u\phi + v\gamma$ with $u, v \in R^\times \langle a \rangle$, and write $\gcd(\psi, \gamma) = u'\psi + v'\gamma$ with $u', v' \in R^\times \langle a \rangle$. We clearly have $x \cdot \gcd(\phi, \gamma) + y \cdot \gcd(\psi, \gamma) = 1$ with $x, y \in R$, so

$$1 = xu\phi + yu'\psi + (xv + yv')\gamma$$
$$= (xu + \alpha xv + \alpha yv')\phi + (yu' + \beta xv + \beta yv')\psi.$$

Now use that $xu + \alpha xv + \alpha yv'$ and $yu' + \beta xv + \beta yv'$ lie in $R^\times \langle a \rangle$. ⊣

REMARKS. Let R be a Bezout domain closed under rem, and let $a \in {}^*\mathbb{Z}$ be such that $a > R$ and $\mathrm{rem}(a, s) \in R$ for all $s \in R^{>0}$. A careful look at the proof of the lemmas above yields (1) and (2) below:

(1) If $f(X), g(X) \in F[X]$ have degree $\leq d \in \mathbb{N}$ and $f(a), g(a) \in R^\times \langle a \rangle$, then $\mathrm{iq}(f(a), g(a))$, $\mathrm{rem}(f(a), g(a))$, $\gcd(f(a), g(a))$, $\sigma(f(a), g(a))$,

and $\tau(f(a), g(a))$ are all of the form $h(a)$ with $h(X) \in F[X]$ of degree $\leq d$.

(2) If $a' \in {}^*\mathbb{Z}$, $a' > R$ and $a' \equiv a \bmod s$ for all $s \in R^{>0}$, then we have an isomorphism $\imath : R^\times \langle a \rangle \cong R^\times \langle a' \rangle$ of ordered rings that is the identity on R such that $\imath(a) = a'$ and $\imath(\phi(x, y)) = \phi(\imath(x), \imath(y))$ for all $x, y \in R^\times \langle a \rangle$ and $\phi = $ iq, rem, gcd, σ, τ.

(3) Applying remark (1) to the case $R = \mathbb{Z}$ and $d = 1$ yields the first part of Lemma 2.1, and remark (2) yields its second part.

COROLLARY 2.8. *Let* $a \in {}^*\mathbb{Z}$. *Then* $\mathbb{Z}^\times \langle a \rangle$ *is closed under* iq *and* σ *(hence under* rem, gcd *and* τ*). If in addition* $a' \in {}^*\mathbb{Z}$ *is such that* $a \equiv a' \bmod N$ *and* $\mathrm{sign}(a - N) = \mathrm{sign}(a' - N)$ *for all positive integers* N, *then there is an isomorphism* $\imath : \mathbb{Z}^\times \langle a \rangle \to \mathbb{Z}^\times \langle a' \rangle$ *of ordered rings such that* $\imath(a) = a'$, *and* $\imath(\phi(x, y)) = \phi(\imath(x), \imath(y))$ *for all* $x, y \in \mathbb{Z}\langle a \rangle$ *and* $\phi \in \{$iq, rem, gcd, $\sigma, \tau\}$.

This algebraic result now yields Proposition 1.2 in the same way as Corollary 2.2 followed from Lemma 2.1, using the Appendix with

$$\mathcal{M} := (\mathbb{Z}(2), \gcd, \sigma, \tau).$$

The degree bound of Remark (1) above yields a more informative version of Proposition 1.2. To formulate this, let x be a single variable; then the *degree* $\deg(t(x))$ of a term $t(x)$ in the language of $(\mathbb{Z}(2), \gcd, \sigma, \tau)$ is the natural number defined inductively as follows: if $t(x)$ does not contain the variable x, then $\deg(t(x)) = 0$; if $t(x)$ is the variable x, then $\deg(t(x)) = 1$; if $t(x) = \phi(t_1(x), t_2(x))$ with $\phi \in \{+, -, \text{iq}, \text{rem}, \gcd, \sigma, \tau\}$, then $\deg(t(x)) = \max(\deg(t_1(x)), \deg(t_2(x)))$; if $t(x) = t_1(x) \cdot t_2(x)$, then $\deg(t(x)) = \deg(t_1(x)) + \deg(t_2(x))$.

A function $f : \mathbb{Z} \to \mathbb{Z}$ is said to be *explicitly definable of degree* $\leq d$ in $(\mathbb{Z}(2), \gcd, \sigma, \tau)$ (where $d \in \mathbb{N}$) if there are terms $t_1(x), \ldots, t_n(x)$ in the language of $(\mathbb{Z}(2), \gcd, \sigma, \tau)$ of degree $\leq d$, such that the sets

$$\{a \in \mathbb{Z} : f(a) = t_i(a)\} \quad (i = 1, \ldots, n)$$

cover $\mathbb{Z}$ and are definable in $(\mathbb{Z}\langle 2 \rangle, \gcd, \sigma, \tau)$ by a quantifier-free formula. For example, if $g : \mathbb{Z}^{m+1} \to \mathbb{Z}$ is explicitly definable in $(\mathbb{Z}\langle 2 \rangle, \gcd, \sigma, \tau)$, then there exists a $d \in \mathbb{N}$ such that every function $g(a, -) : \mathbb{Z} \to \mathbb{Z}$ $(a \in \mathbb{Z}^m)$ is explicitly definable of degree $\leq d$ in $(\mathbb{Z}(2), \gcd, \sigma, \tau)$.

COROLLARY 2.9. *Suppose* $f : \mathbb{Z} \to \mathbb{Z}$ *is explicitly definable of degree* $\leq d$ *in* $\langle \mathbb{Z}(2), \gcd, \sigma, \tau \rangle$. *Then there are polynomials* $f_1(T), \ldots, f_k(T) \in \mathbb{Q}[T]$ *of degree* $\leq d$ *and a partition of* $\mathbb{Z}$ *into arithmetic progressions* $S_1, \ldots, S_k$ *such that* $f(s) = f_i(s)$ *on* S_i *for* $i = 1, \ldots, k$.

REMARK. Throughout we assumed that ${}^*\mathbb{Z}$ is an $\aleph_1$-saturated elementary extension of the ring $\mathbb{Z}$ of integers. The algebraic lemmas actually hold under purely algebraic assumptions on ${}^*\mathbb{Z}$ and this may be of interest to those who study algebraic fragments of first-order Peano arithmetic. Indeed,

Lemmas 2.3, 2.4, 2.5, 2.6, and 2.7 go through for subrings R of $^*\mathbb{Z}$ if we assume about $^*\mathbb{Z}$ only that it is a discretely ordered Bezout domain with a map $a \mapsto \lfloor a \rfloor : {}^*\mathbb{Q} \to {}^*\mathbb{Z}$ on its fraction field $^*\mathbb{Q}$ satisfying $\lfloor a \rfloor \leq a < \lfloor a \rfloor + 1$, with the usual identification of $\mathbb{Z}$ with a subring of $^*\mathbb{Z}$, and where iq, rem, gcd, σ and τ are defined on $^*\mathbb{Z}$ in the same way as on $\mathbb{Z}$.

§3. **The greatest common divisor and irrationality.** A byproduct of this section is that gcd and $\perp$ are not explicitly definable in $\mathbb{Z}(1)$, but the main goal is to improve a result of [6] as well as its exposition, and to highlight once again the connection to irrationality.

Let $a, b \in \mathbb{Z}$. Define a sequence $(G_n(a, b))$ of finite subsets of $\mathbb{Z}$:

$$G_0(a, b) \subseteq G_1(a, b) \subseteq G_2(a, b) \subseteq \cdots$$

$$G_0(a, b) := \{0, 1, a, b\}$$

$$G_{n+1}(a, b) := G_n(a, b) \cup \{x + y, x - y, \mathrm{iq}(x, y), \mathrm{rem}(x, y) : x, y \in G_n(a, b)\}.$$

Put

$$g(a, b) := \text{least } n \text{ such that } \gcd(a, b) \in G_n(a, b).$$

(Think of $g(a, b)$ as the number of steps needed to obtain $\gcd(a, b)$ from $a, b, 0, 1$ where at each step we apply $+$, $-$, iq and rem to all integers obtained at earlier steps.) By well-known results on the euclidean algorithm we have $g(a, b) \leq 2 \log_2 b$ for $a > b > 0$. The next result provides sequences of inputs along which g tends to infinity.

PROPOSITION 3.1. *Let $t > 0$ be any irrational number, and let (a_n), (b_n) and (κ_n) be any sequences of positive integers such that $a_n/b_n \to t$, and $\kappa_n \to \infty$. Then $g(\kappa_n a_n, \kappa_n b_n) \to \infty$.*

EXAMPLE. Let $F_n := n^{\text{th}}$ Fibonacci number. Then $F_n/F_{n+1} \to 2/(1 + \sqrt{5})$, so $g(nF_n, nF_{n+1}) \to \infty$ by the proposition.

An immediate consequence is that gcd is not explicitly definable in $\mathbb{Z}(1)$ in the following strong form: for each term $t(x)$ in the language of $\mathbb{Z}(1)$ and any sequences (a_n), (b_n) and (κ_n) as in the proposition, there are only finitely many n such that $\gcd(\kappa_n a_n, \kappa_n b_n) = t(\kappa_n a_n, \kappa_n b_n)$. In particular, gcd is not explicitly definable in any expansion of $\mathbb{Z}(1)$ by *relations*.

We shall work in an $\aleph_1$-saturated elementary extension $(^*\mathbb{R}, {}^*\mathbb{Z})$ of the structure $(\mathbb{R}, \mathbb{Z})$, the ordered field of real numbers with a predicate for the set of integers.

For $a, b \in {}^*\mathbb{Z}$ we define the following additive subgroup of $^*\mathbb{Z}$:

$$\mathbb{Z}\langle a, b \rangle := (\mathbb{Q} + \mathbb{Q}a + \mathbb{Q}b) \cap {}^*\mathbb{Z}.$$

LEMMA 3.2. *Suppose $t > 0$ is an irrational real number, and let $a, b \in {}^*\mathbb{Z}$ be such that $b > \mathbb{N}$ and $\frac{a}{b} - t$ is infinitesimal. Then $\mathbb{Z}\langle a, b \rangle$ is closed under iq and rem.*

PROOF. The ratio a/b fills the same Dedekind cut in $\mathbb{Q}$ as the irrational number t. By the Eudoxos-Dedekind theory of proportions and real numbers we have therefore an embedding of ordered abelian groups

$$pa + qb \longmapsto pt + q \ : \ \mathbb{Q}a + \mathbb{Q}b \longrightarrow \mathbb{R}, \quad (p, q \in \mathbb{Q}).$$

Hence all non-zero elements of $\mathbb{Q}a + \mathbb{Q}b$ have the same archimedean class. In view of $b > \mathbb{N}$ it follows easily that all non-zero elements of $\mathbb{Q} + \mathbb{Q}a + \mathbb{Q}b$ have the same archimedean class as b or as 1.

Let $x, y \in \mathbb{Z}\langle a, b \rangle$, $y \neq 0$, and write $x = qy + r$ with $q, r \in {}^*\mathbb{Z}$, $q = \mathrm{iq}(x, y)$ and $r = \mathrm{rem}(x, y)$. We have to show that $q, r \in \mathbb{Z}\langle a, b \rangle$.

CASE 1. $y \notin \mathbb{Z}$. Then $|x/y| < n$ for some n, so $q \in \mathbb{Z}$, and hence $r = x - qy \in \mathbb{Z}\langle a, b \rangle$.

CASE 2. $y \in \mathbb{Z}$. Then $|r| < |y|$ yields $r \in \mathbb{Z}$, hence

$$q = (x - r)/y \in (\mathbb{Q} + \mathbb{Q}a + \mathbb{Q}b) \cap {}^*\mathbb{Z} = \mathbb{Z}\langle a, b \rangle. \qquad \dashv$$

Proposition 3.1 is now easy to derive:

PROOF. The idea is to smooth out the problem by replacing the sequences $(\kappa_n a_n)$ and $(\kappa_n b_n)$ by non-standard elements κa and κb. Indeed, let $a, b, \kappa \in {}^*\mathbb{Z}$ be positive infinite such that $\frac{a}{b} - t$ is infinitesimal (so a and b serve in effect as *incommensurable* integers). It suffices to show that then $g(\kappa a, \kappa b)$ is infinite. Put $\alpha := \kappa a$, $\beta := \kappa b$, so $\frac{\alpha}{\beta} - t = \frac{a}{b} - t$ is infinitesimal.

By the previous lemma, with α and β instead of a and b, we have $G_n(\alpha, \beta) \subseteq \mathbb{Z}\langle \alpha, \beta \rangle$ for all n, so it only remains to show that $\gcd(\alpha, \beta) \notin \mathbb{Z}\langle \alpha, \beta \rangle$. We have $\gcd(\alpha, \beta) \geq \kappa > \mathbb{N}$. From the incommensurability of α and β we get $\gcd(\alpha, \beta) \ll \alpha$. By the proof of Lemma 3.2, $\mathbb{Z}\langle \alpha, \beta \rangle \setminus \{0\}$ has only two archimedean classes, hence $\gcd(\alpha, \beta) \notin \mathbb{Z}\langle \alpha, \beta \rangle$. $\qquad \dashv$

Proposition 3.1 improves [6], Proposition 2.1, which has $\kappa_n = n!$. In [6], this choice of multipliers made α and β divisible by all positive integers, resulting in the nice identity $\mathbb{Z}\langle \alpha, \beta \rangle = \mathbb{Z} + \mathbb{Q}\alpha + \mathbb{Q}\beta$, but thereby missing the correct level of generality.

Proposition 3.1 says nothing about explicit definability of coprimeness. The next proposition does; the result is not new, since it follows from [3], but it is the *proof* that is of interest.

PROPOSITION 3.3. *The relation $\perp$ on $\mathbb{Z}$ is not explicitly definable in $\mathbb{Z}(1)$.*

PROOF. Take any irrational $t > 0$ and sequences (a_n) and (b_n) of positive integers such that $a_n/b_n \to t$ and $a_n \perp b_n$ for all n. Put $\kappa_n := n! + 1$, and $\alpha_n := \kappa_n a_n$ and $\beta_n := \kappa_n b_n$, so $\alpha_n/\beta_n = a_n/b_n$, but $\alpha_n \not\perp \beta_n$.

CLAIM. Let $\phi(x, y)$ be a quantifier-free formula in the language of $\mathbb{Z}(1)$. Then for all but finitely many n we have

$$\mathbb{Z}(1) \models \phi(a_n, b_n) \longleftrightarrow \phi(\alpha_n, \beta_n).$$

The proposition follows immediately from this claim.

To prove the claim, we take $a, b, \kappa, \alpha, \beta$ as in the proof of Proposition 3.1; it suffices to show that then ${}^*\mathbb{Z}(1) \models \phi(a, b) \leftrightarrow \phi(\alpha, \beta)$. By Lemma 3.2 we can view $\mathbb{Z}\langle a, b\rangle$ and $\mathbb{Z}\langle \alpha, \beta\rangle$ as substructures of ${}^*\mathbb{Z}(1)$, and thus

$$
{}^*\mathbb{Z}(1) \models \phi(a, b) \Longleftrightarrow \mathbb{Z}\langle a, b\rangle \models \phi(a, b)
$$
$$
{}^*\mathbb{Z}(1) \models \phi(\alpha, \beta) \Longleftrightarrow \mathbb{Z}\langle \alpha, \beta\rangle \models \phi(\alpha, \beta).
$$

Using the fact that $\alpha \equiv a \bmod n$ and $\beta \equiv b \bmod n$ for all $n > 0$ it is easy to check that we have an isomorphism

$$
p + qa + rb \mapsto p + q\alpha + r\beta : \mathbb{Z}\langle a, b\rangle \to \mathbb{Z}\langle \alpha, \beta\rangle, \quad (p, q, r \in \mathbb{Q})
$$

between substructures of ${}^*\mathbb{Z}(1)$. Hence

$$
\mathbb{Z}\langle a, b\rangle \models \phi(a, b) \Longleftrightarrow \mathbb{Z}\langle \alpha, \beta\rangle \models \phi(\alpha, \beta).
$$

In view of the earlier equivalences this finishes the proof.   ⊣

The proof yields the following stronger result: *Let $t > 0$ be irrational, and let (a_n), (b_n) be sequences of positive integers such that $a_n/b_n \to t$. Then, given any quantifier-free formula $\phi(x, y)$ in the language of $\mathbb{Z}(1)$, we have $\mathbb{Z}(1) \models \phi(a_n, b_n) \leftrightarrow \phi((n! + 1)a_n, (n! + 1)b_n)$ for all sufficiently large n.*

§4. **The greatest common divisor and transcendence.** We are now going to consider the effect of having multiplication as an extra primitive. Proposition 4.1 below is from [6] and implies that gcd is not explicitly definable in any expansion of $\mathbb{Z}(2)$ by *relations*. The main goal of this section, however, is to improve a certain quadruple logarithmic lower bound in [6] to a triple logarithmic lower bound. The proof of this better lower bound uses [6] as back ground knowledge, and can be skipped without harm to understanding the rest of the paper.

Let $a, b \in \mathbb{Z}$. We define an increasing sequence of finite subsets of $\mathbb{Z}$,

$$
G_0^\times(a, b) \subseteq G_1^\times(a, b) \subseteq G_2^\times(a, b) \subseteq \cdots
$$

in the same way as we defined the $G_n(a, b)$ in Section 4, with $G_0^\times(a, b) = \{0, 1, a, b\}$, except that $G_{n+1}^\times(a, b)$ contains also all products xy with $x, y \in G_n^\times(a, b)$. Next, we put

$$
g^\times(a, b) := \text{least } n \text{ such that } \gcd(a, b) \in G_n^\times(a, b).
$$

We have a crude upperbound: $g^\times(a, b) \le 2 \lg \lg b$ for $a \ge b \ge 4$, see [6], Section 4. (But note that $g^\times$ differs slightly from the function $g^\times$ in [6].)

The function $g^\times$ is unbounded:

PROPOSITION 4.1. *Suppose $r > 0$ is transcendental, and (a_n) and (b_n) are sequences of positive integers such that $a_n/b_n \to r$. Then there are integers $\kappa_n > 0$ such that $g^\times(\kappa_n a_n, \kappa_n b_n) \to \infty$.*

The conclusion holds for any integer multipliers $\kappa_n > 0$ such that for each homogeneous non-zero polynomial $p(X, Y) \in \mathbb{Z}[X, Y]$ we have: $p(a_n, b_n)|\kappa_n$ for all sufficiently large n.

Here we just summarize the proof from [6] in a way that adds to the analogy with the proof of Proposition 3.1. We work in $^*\mathbb{Z}$, $^*\mathbb{Q}$ and $^*\mathbb{R}$, and let a and b be any positive infinite elements of $^*\mathbb{Z}$ such that $\frac{a}{b} - r$ is infinitesismal. Next, take any positive $\kappa \in {}^*\mathbb{Z}$ such that $p(a, b)|\kappa$ for all non-zero homogeneous polynomials $p(X, Y) \in \mathbb{Z}[X, Y]$. Put

$$\alpha := \kappa a, \qquad \beta := \kappa b, \qquad \mathbb{Z}^\times \langle \alpha, \beta \rangle := \mathbb{Q}(\alpha, \beta) \cap {}^*\mathbb{Z}.$$

The ring $\mathbb{Z}^\times \langle \alpha, \beta \rangle$ is analogous to the additive group $\mathbb{Z}\langle \alpha, \beta \rangle$ in the proof of Proposition 3.1. Clearly $\alpha, \beta \in \mathbb{Z}^\times \langle \alpha, \beta \rangle$. One can show that $\mathbb{Z}^\times \langle \alpha, \beta \rangle$ is closed under iq (hence under rem), and that $\gcd(\alpha, \beta) \notin \mathbb{Z}^\times \langle \alpha, \beta \rangle$. Hence $g^\times(\alpha, \beta) > \mathbb{N}$, which yields the desired result.

The proof that $\mathbb{Z}^\times \langle \alpha, \beta \rangle$ is closed under iq is given in Section 4 of [6], where $\mathbb{Z}^\times \langle \alpha, \beta \rangle$ occurs in the guise of the ring R, defined as follows: put $t := a/b = \alpha/\beta$, $K := \mathbb{Q}(t)$ (an archimedean subfield of $^*\mathbb{Q}$), and

$$R := \{f(\beta) : f(U) \in K[U], \; f(0) \in \mathbb{Z}\}.$$

The particular choice of κ guarantees that $R \subseteq {}^*\mathbb{Z}$.

It is not hard to show that $R = \mathbb{Z}^\times \langle \alpha, \beta \rangle$, but we will leave this to the interested reader. (The above definition of R is adapted to showing that R is closed under iq.) Once we know that $R \subseteq {}^*\mathbb{Z}$ and R is closed under iq, the equality $R = \mathbb{Z}^\times \langle \alpha, \beta \rangle$ is also clear from the following obvious fact:

If A is a subring of $^*\mathbb{Z}$ closed under iq, *then* $A = F \cap {}^*\mathbb{Z}$ *where* $F \subseteq {}^*\mathbb{Q}$ *is the fraction field of A.*

A triple logarithmic lower bound. Consider the rational approximations to the transcendental number e given by

$$\frac{a_n}{b_n} = \sum_{i=0}^{n} \frac{1}{i!}, \quad n > 0, \; b_n := n!.$$

Note that $a_n/b_n \to e$.

THEOREM 4.2. *There are positive integers* $\kappa_n < \exp(\exp((\log n)^3))$ *such that for* $\alpha_n := \kappa_n a_n$ *and* $\beta_n := \kappa_n b_n$ *we have*

$$g^\times(\alpha_n, \beta_n) > \frac{1}{3}\sqrt[3]{\log \log \log \beta_n}$$

for all sufficiently large n.

The κ_n in this theorem grow much slower than the κ_n specified in Section 5 of [6], and this is related to the improvement of the *quadruple* log bound there to a *triple* log bound here. Theorem 4.2 is an effective version of a special case of Proposition 4.1. Accordingly, in the proof we can expect the use of an effective *measure of transcendence* of e. As in [6], such a transcendence measure is provided by a theorem of Cijsouw [2].

We can copy most of the proof in Section 5 of [6], and we only indicate in detail the two points to be modified. First some notation from [6] that we need in order to discuss these points.

We write $l_2(x)$ and $l_3(x)$ for $\log\log x$ and $\log\log\log x$, respectively. For a polynomial p with real coeffients (in possibly more than one indeterminate) we let $\|p\|_\infty$ be the maximum of the absolute values of the coefficients of p. We work in $^*\mathbb{N}$, $^*\mathbb{Z}$, $^*\mathbb{R}$, $^*(\mathbb{Z}[X, Y])$, and so on, but now these are viewed as part of a non-standard universe as in Section 5 of [6]; see also [1]. We consider the natural extensions of the sequences (a_n) and (b_n) to sequences (a_ν) and (b_ν) with ν ranging over $^*\mathbb{N}^{>0}$, so $b_\nu = \nu!$. Theorem 4.2 is clearly implied by the following:

(*)   *For each $M \in {}^*\mathbb{N}$ with $M > \mathbb{N}$, there is $\kappa \in {}^*\mathbb{N}^{>0}$ such that*

$$\kappa < \exp\left(\exp\left((\log M)^3\right)\right) \text{ and } g^\times(\alpha_M, \beta_M) > \frac{1}{3}\sqrt[3]{l_3(\beta_M)}$$

where $\alpha_M := \kappa a_M$ and $\beta_M := \kappa b_M$

To prove (*), let $M \in {}^*\mathbb{N}$ with $M > \mathbb{N}$, put $a := a_M$, $b := b_M$ and $t := a/b$, so $t \in {}^*\mathbb{R}$, $t < e$ and $e - t$ is infinitesimal. We also put

$$Z := \{x \in {}^*\mathbb{Z} : |x| < (\log b)^n \text{ for some } n\}.$$

The value $\kappa = \lfloor b^{l_2(b)l_3(b)} \rfloor!$ in [6] was chosen such that Lemma 5.5 there holds. This lemma says:

Let $p(X, Y) \in {}^*(\mathbb{Z}[X, Y])$ be non-zero and homogeneous of total degree $O(l_2(b))$ with coefficients in Z. Then $p(a, b)|\kappa$.

The main difference with [6] is the choice of a much smaller value for κ such that this Lemma 5.5 of [6] goes through. We now indicate how to make this choice.

Let $d \in {}^*\mathbb{N}$ and $1 \le H \in {}^*\mathbb{R}$. The internal set

$$\{p(X, Y) \in {}^*(\mathbb{Z}[X, Y]) : p \text{ is homogeneous of degree } d, \text{ and } \|p\|_\infty \le H\}$$

has internal cardinality $(2\lfloor H \rfloor + 1)^{d+1}$, and for $p(X, Y)$ in this set we have

$$|p(a, b)| \le Hb^d\left(1 + t + \cdots + t^d\right) \le 2H(eb)^d.$$

It follows that there is a $\kappa(d, H) \in {}^*\mathbb{N}^{>0}$ with $\kappa(d, H) \le (2H(eb)^d)^{(2H+1)^{d+1}}$ such that $q(a, b)|\kappa(d, H)$ for each homogeneous $q(X, Y) \in {}^*(\mathbb{Z}[X, Y])$ of degree $\le d$ with $\|q\|_\infty \le H$ and $q(a, b) \ne 0$. (Here we use the fact that for each such q there is a $p = X^k q$, $k \in {}^*\mathbb{N}$, such that p is of exact degree d, so p lies in the set displayed above.)

Let now $n > 0$, $d = \lfloor nl_2(b) \rfloor$, $H = (\log b)^n$, and choose $\kappa(n) := \kappa(d, H)$ with the properties above. Then $\kappa(n) > b^n$ and

$$\log \kappa(n) \le (2H + 1)^{d+1}(d + \log 2H + d \log b) \le (2H + 1)^{d+1}(d + 1) \log b,$$

hence

$$l_2(\kappa(n)) \le (d + 1) \log(2H + 1) + \log(d + 1) + l_2(b) \le (n^2 + n + 1)(l_2(b))^2,$$

and thus $l_3(\kappa(n)) < (2 + \varepsilon) l_3(b)$ for each real $\varepsilon > 0$. An easy saturation argument yields a $\kappa \in {}^*\mathbb{N}$ with $\kappa \ge b^n$ for all n and $l_3(\kappa) < (5/2) l_3(b)$, such that Lemma 5.5 in [6] (stated above) goes through. For the rest of this subsection κ has this value.

One checks easily, using $b \le M^M$, that $\kappa < \exp(\exp((\log M)^3))$. Next we put $\alpha := \kappa a$ and $\beta := \kappa b$. It follows easily that $l_3(\beta) < 3 l_3(b)$. It remains to prove the part of (*) saying that $g^\times(\alpha, \beta) > \frac{1}{3}\sqrt[3]{l_3(\beta)}$.

Section 5 in [6] following Lemma 5.5 now goes through without change, except for the very last paragraph (the last three sentences), which can be replaced by the following: "Thus $g^\times(\alpha, \beta) > \frac{1}{2}\sqrt[3]{l_3(b)} > \frac{1}{2}\sqrt[3]{\frac{1}{3}l_3(\beta)} > \frac{1}{3}\sqrt[3]{l_3(\beta)}$. This finishes the proof of Theorem 4.2."

§5. Coprimeness is not explicitly definable in $\mathbb{Z}(2)$.

The goal of this section is to give a model-theoretic proof of:

PROPOSITION 5.1. *The relation $\perp$ is not explictly definable in $\mathbb{Z}(2)$.*

We shall construct iq-closed subrings R and S of ${}^*\mathbb{Z}$ and an ordered ring isomorphism $\phi : R \cong S$ with the following properties: $\phi(\mathrm{iq}(x, y)) = \mathrm{iq}(\phi x, \phi y)$ for all $x, y \in R$, the ring R has elements α_0, α_1 with $\alpha_0 \perp \alpha_1$, and S has elements β_0, β_1 with $\beta_0 \not\perp \beta_1$, and $\phi(\alpha_i) = \beta_i$ for $i = 0, 1$. (The ordering on R and S is induced by the ordering of ${}^*\mathbb{Z}$.) It is clear that Proposition 5.1 is a consequence of such a construction.

Some familiarity with standard facts about ordered fields will be assumed; a good reference for this material is [4].

LEMMA 5.2. *Let A be a subgroup of ${}^*\mathbb{Z}$ containing $\mathbb{Z}$, let $f, g \in A$ and let N be a positive integer such that $\mathrm{iq}(Nf, g) \in A$. Then $\mathrm{iq}(f, g)$ and $\mathrm{rem}(f, g)$ are of the form a/N with $a \in A$.*

PROOF. We can assume $g \ne 0$. Write $Nf = qg + r$ with $q = \mathrm{iq}(Nf, g)$ and $r = \mathrm{rem}(Nf, g)$. Put $j := \mathrm{rem}(q, N) \in \{0, \ldots, N - 1\}$. One easily checks that then $\mathrm{iq}(f, g) = \frac{q-j}{N}$, and $\mathrm{rem}(f, g) = \frac{r+jg}{N}$. ⊣

LEMMA 5.3. *Let $a, b \in {}^*\mathbb{Z}$ such that $a > \mathbb{Z}$ and $b > a^e$ for all $e \in \mathbb{N}$. Let $f \in \mathbb{Z}[a, b]$ and $g \in \mathbb{Z}[b]$. Then $\mathrm{iq}(f, g) \in \mathbb{Q}[a, b]$.*

PROOF. We can assume $g \ne 0$. Write $f = F(a)$ with $F(X) = \sum_{i=0}^d \beta_i X^i \in \mathbb{Z}[b][X]$ with all $\beta_i \in \mathbb{Z}[b]$. Write $g = G(b)$ where $G(X) \in \mathbb{Z}[X]$. We take a positive integer N such that $N\beta_i = \gamma_i g + \delta_i$ for $i = 0, \ldots, d$, where $\gamma_i, \delta_i \in \mathbb{Z}[b]$

and $\delta_i = \Delta_i(b)$, with $\Delta_i(X) \in \mathbb{Z}[X]$ of lower degree than $G(X)$. Hence

$$Nf = \left(\sum_i \gamma_i a^i\right)g + \sum_i \delta_i a^i, \quad \left|\sum_i \delta_i a^i\right| < |g|.$$

Thus $\text{iq}(Nf, g) \in \{-1+\sum \gamma_i a^i, \sum \gamma_i a^i, 1+\sum \gamma_i a^i\}$, in particular $\text{iq}(Nf, g) \in \mathbb{Z}[a, b]$. Now apply the previous lemma. ⊣

LEMMA 5.4. *Let $n > 0$ and $a_0, \ldots, a_n \in {}^*\mathbb{Z}$ be such that $a_0 > \mathbb{Z}$, $a_{i+1} > a_i^e$ for $0 \le i < n$ and all $e \in \mathbb{N}$, and $a_{i+1} \equiv a_{i-1} \bmod f$ for all non-zero $f \in \mathbb{Z}[a_0, \ldots, a_i]$ and $0 < i < n$. Then $\text{iq}(f, g) \in \mathbb{Q}(a_0, \ldots, a_n)$ for $f \in \mathbb{Z}[a_0, \ldots, a_n]$ and $g \in \mathbb{Z}[a_1, \ldots, a_n]$. (Hence $\text{rem}(f, g) \in \mathbb{Q}(a_0, \ldots, a_n)$ for $f \in \mathbb{Z}[a_0, \ldots, a_n]$ and $g \in \mathbb{Z}[a_1, \ldots, a_n]$.)*

PROOF. We proceed by induction on n. The case $n = 1$ follows from the previous lemma.

Suppose $a_0, \ldots, a_n$ satisfy the hypothesis and conclusion of the lemma, and that in addition $a_{n+1} \in {}^*\mathbb{Z}$ is such that $a_{n+1} > a_n^e$ for all $e \in \mathbb{N}$ and $a_{n+1} \equiv a_{n-1} \bmod f$ for all non-zero $f \in \mathbb{Z}[a_1, \ldots, a_n]$. Let $f \in \mathbb{Z}[a_0, \ldots, a_{n+1}]$ and $0 \ne g \in \mathbb{Z}[a_1, \ldots, a_{n+1}]$. We need to show that $\text{iq}(f, g) \in \mathbb{Q}(a_0, \ldots, a_{n+1})$. Write $f = F(a_{n+1})$ and $g = G(a_{n+1})$ with $F(X) \in \mathbb{Z}[a_0, \ldots, a_n][X]$ and $G(X) \in \mathbb{Z}[a_1, \ldots, a_n][X]$. Division with remainder in the polynomial ring $\mathbb{Q}(a_0, \ldots, a_n)[X]$ and clearing denominators yields $qf = q'g + r$ with $0 \ne q \in \mathbb{Z}[a_1, \ldots, a_n]$, $q' \in \mathbb{Z}[a_0, \ldots, a_{n+1}]$ and $r = R(a_{n+1})$ with $R(X) \in \mathbb{Z}[a_0, \ldots, a_n][X]$ of lower degree in X than $G(X)$. (For q we can take a power of the leading coefficient of $G(X)$.) Then

$$\frac{f}{g} = \frac{q'}{q} + \frac{r}{qg}, \quad \left|\frac{r}{qg}\right| < 1.$$

So it suffices to show that $\text{iq}(q', q) \in \mathbb{Q}(a_0, \ldots, a_{n+1})$. Write $q' = Q'(a_{n+1})$ where $Q'(X) \in \mathbb{Z}[a_0, \ldots, a_n][X]$. Let $q'' := Q'(a_{n-1}) \in \mathbb{Z}[a_0, \ldots, a_n]$. Then $q' \equiv q'' \bmod q$ and $\text{iq}(q', q) = \frac{q'-q''}{q} + \text{iq}(q'', q)$. Now use that by the inductive assumption we have $\text{iq}(q'', q) \in \mathbb{Q}(a_0, \ldots, a_n)$. ⊣

REMARKS. (1) A closer look shows that $\text{iq}(f, g) = \frac{A}{B}$ where

$$A \in \mathbb{Z}[a_0, \ldots, a_n] \quad \text{and} \quad 0 \ne B \in \mathbb{Z}[a_1, \ldots, a_{n-1}].$$

(2) In addition to $a_0, \ldots, a_n$ as in Lemma 5.4, let $b_0, \ldots, b_n \in {}^*\mathbb{Z}$ be such that $b_0 > \mathbb{Z}$, $b_{i+1} > b_i^e$ for $0 \le i < n$ and all $e \in \mathbb{N}$, and $b_{i+1} \equiv b_{i-1} \bmod f$ for all non-zero $f \in \mathbb{Z}[b_1, \ldots, b_i]$ and $0 < i < n$. Assume also that $a_0 \equiv b_0 \bmod N$ and $a_1 \equiv b_1 \bmod N$ for all positive integers N. Let $\phi : \mathbb{Q}(a_0, \ldots, a_n) \to \mathbb{Q}(b_0, \ldots, b_n)$ be the ordered field isomorphism with $\phi(a_i) = b_i$ for $i = 0, \ldots, n$. Then ϕ also preserves iq in the sense that $\phi(\text{iq}(f, g)) = \text{iq}(\phi f, \phi g)$ for all $f \in \mathbb{Z}[a_0, \ldots, a_n]$ and $g \in \mathbb{Z}[a_1, \ldots, a_n]$.

This follows by a close look at the proofs of the three lemmas above.

We now take an $a_0 \in {}^*\mathbb{Z}$ such that $a_0 > \mathbb{Z}$ and $a_0 \equiv 1 \bmod N$ for each positive integer N. Next we take an $a_1 \in {}^*\mathbb{Z}$ such that $a_1 > a_0^e$ for all $e \in \mathbb{N}$, $a_1 \equiv 1 \bmod N$ for each positive integer N and $a_1 \equiv 1 \bmod a_0$. By induction on n and using saturation we obtain for each $n > 1$ a sequence $a_0, a_1, \ldots, a_n$ in ${}^*\mathbb{Z}$ such that $a_{i+1} > a_i^e$ for all $e \in \mathbb{N}$, and $a_{i+1} \equiv a_{i-1} \bmod f$ for all non-zero $f \in \mathbb{Z}[a_0, \ldots, a_i]$ $(0 < i < n)$. In particular $\mathrm{rem}(a_{i+1}, a_i) = a_{i-1}$ $(0 < i < n)$, so $\gcd(a_n, a_{n-1}) = \gcd(a_1, a_0) = 1$.

For later convenience we reverse the indexing: Put $\alpha_i := a_{n-i}$. Then we have $\alpha_0, \ldots, \alpha_n \in {}^*\mathbb{Z}$, $\alpha_0, \ldots, \alpha_n$ are positive infinite, $\alpha_{i-1} > \alpha_i^e$ for all $e \in \mathbb{N}$, and $\alpha_{i-1} \equiv \alpha_{i+1} \bmod f$ for all non-zero $f \in \mathbb{Z}[\alpha_i, \alpha_{i+1}, \ldots, \alpha_n]$, $(0 < i < n)$. Note also that $\alpha_i \equiv 1 \bmod N$ for all positive integers N, and that $\gcd(\alpha_0, \alpha_1) = \gcd(a_0, a_1) = 1$.

By a further saturation argument it follows that there exists an infinite sequence $(\alpha_i)_{i \in \mathbb{N}}$ of positive infinite elements of ${}^*\mathbb{Z}$ such that $\gcd(\alpha_0, \alpha_1) = 1$, $\alpha_0 \equiv 1 \bmod N$ for all positive integers N, and for all $i > 0$: $\alpha_{i-1} > \alpha_i^e$ for all $e \in \mathbb{N}$, and $\alpha_{i-1} \equiv \alpha_{i+1} \bmod f$ for all non-zero $f \in \mathbb{Z}[\alpha_j \mid j \geq i]$. In particular, $\alpha_i \equiv 1 \bmod N$ for all i and all positive integers N. Let $R := \mathbb{Q}(\alpha_i : i \in \mathbb{N}) \cap {}^*\mathbb{Z}$. It follows easily from Lemma 5.4 that the ring R is closed under iq.

Parallel to a_0 and a_1 we construct b_0 and b_1 as follows: take $b_0 \in {}^*\mathbb{Z}$ such that $b_0 > \mathbb{Z}$ and $b_0 \equiv 1 \bmod N$ for each positive integer N. Next we take a $b_1 \in {}^*\mathbb{Z}$ such that $b_1 > b_0^e$ for all $e \in \mathbb{N}$, $b_1 \equiv 1 \bmod N$ for each positive integer N and $b_1 \equiv 0 \bmod b_0$. (This last congruence is the key difference with the a's.) By induction on n and using saturation we obtain for each $n > 1$ a sequence $b_0, b_1, \ldots, b_n$ in ${}^*\mathbb{Z}$ such that $b_{i+1} > b_i^e$ for all $e \in \mathbb{N}$, and $b_{i+1} \equiv b_{i-1} \bmod f$ for all non-zero $f \in \mathbb{Z}[b_0, \ldots, b_i]$ $(0 < i < n)$. In particular $\mathrm{rem}(b_{i+1}, b_i) = b_{i-1}$ $(0 < i < n)$, so $\gcd(b_n, b_{n-1}) = \gcd(b_1, b_0) = b_0$.

Reversing indices and using saturation this yields as before an infinite sequence $(\beta_i)_{i \in \mathbb{N}}$ of positive infinite elements of ${}^*\mathbb{Z}$ such that $\gcd(\beta_0, \beta_1) = b_0$, $\beta_0 \equiv 1 \bmod N$ for all positive integers N, and for all $i > 0$: $\beta_{i-1} > \beta_i^e$ for all $e \in \mathbb{N}$, and $\beta_{i-1} \equiv \beta_{i+1} \bmod f$ for all non-zero $f \in \mathbb{Z}[\beta_j \mid j \geq i]$. In particular, $\beta_i \equiv 1 \bmod N$ for all i and all positive integers N. Let $S := \mathbb{Q}(\beta_i : i \in \mathbb{N}) \cap {}^*\mathbb{Z}$, so the ring S is closed under iq.

It follows from the second remark following the proof of Lemma 5.4 that the ordered field isomorphism $\mathbb{Q}(\alpha_i \mid i \in \mathbb{N}) \cong \mathbb{Q}(\beta_i \mid i \in \mathbb{N})$ that sends each α_i to β_i maps R onto S. Since $\gcd(\alpha_0, \alpha_1) = 1$ and $\gcd(\beta_0, \beta_1) = b_0 \neq 1$, it follows that coprimeness is not quantifier-free definable in $\mathbb{Z}(2)$.

§6. More on the rings $\mathbb{Z}^\times \langle a \rangle$.

A key fact established in Section 2 is that the rings $\mathbb{Z}^\times \langle a \rangle$ with $a \in {}^*\mathbb{Z}$ are Bezout domains, since they are closed under iq, gcd, and σ. In this section we consider these rings a bit closer, mainly out of curiosity, not with particular applications in mind. We shall ask some questions to which it would be nice to know the answer.

Algebraic extensions of $\mathbb{Z}^\times \langle a \rangle$. For a subring R of $^*\mathbb{Z}$ with fraction field $F \subseteq {}^*\mathbb{Q}$, put

$$R^{\text{cl}} := \{x \in {}^*\mathbb{Q} : x \text{ is algebraic over } F\} \cap {}^*\mathbb{Z}.$$

In particular, $\mathbb{Z}^{\text{cl}} = \mathbb{Z}$. By a saturation argument there exist $a \in {}^*\mathbb{Z} \setminus \mathbb{Z}$ such that $\mathbb{Z}^\times \langle a \rangle^{\text{cl}} = \mathbb{Z}^\times \langle a \rangle$. More interesting is the case where the ring $\mathbb{Z}^\times \langle a \rangle^{\text{cl}}$ properly extends $\mathbb{Z}^\times \langle a \rangle$. Is $\mathbb{Z}^\times \langle a \rangle^{\text{cl}}$ always closed under iq, gcd and σ (and hence a Bezout domain)?

Take the case of positive infinite $a, b \in {}^*\mathbb{Z}$ with $a^2 - 2b^2 = 1$. (Such a, b exist since the Pell equation $x^2 - 2y^2 = 1$ has infinitely many solutions in positive integers. See for example [5] for this fact, and for other results on Pell equations used below.) Then $\mathbb{Q}(a, b)$ is a quadratic extension of the fraction field $\mathbb{Q}(a)$ of $\mathbb{Z}^\times \langle a \rangle$, so $\mathbb{Z}^\times \langle a, b \rangle := \mathbb{Q}(a, b) \cap {}^*\mathbb{Z}$ is a subring of $\mathbb{Z}^\times \langle a \rangle^{\text{cl}}$. From a lemma in [8] it follows that

$$\mathbb{Z}^\times \langle a, b \rangle \subseteq \mathbb{Q}[b] + \mathbb{Q}[b]a = \mathbb{Q}[a, b],$$

and that $\mathbb{Z}^\times \langle a, b \rangle$ is closed under iq and rem. The ring $\mathbb{Z}^\times \langle a, b \rangle$, however, is not closed under gcd: by another lemma in [8], either $a + 1$ is a square in $^*\mathbb{Z}$ and $\gcd(a+1, b) = \sqrt{a+1}$, or $2(a+1)$ is a square in $^*\mathbb{Z}$ and $\gcd(a+1, b) = \sqrt{2(a+1)}$. In both cases, $\gcd(a+1, b) \notin \mathbb{Z}^\times \langle a, b \rangle$, but $\gcd(a+1, b) \in \mathbb{Z}^\times \langle a \rangle^{\text{cl}}$.

The solution sets in $\mathbb{Z}^2$ and in $^*\mathbb{Z}^2$ of the Pell equation $x^2 - 2y^2 = 1$ are commutative groups G and $^*G$ with group operation given by

$$(x_1, y_1) \cdot (x_2, y_2) = (x_1 x_2 + 2y_1 y_2, x_1 y_2 + x_2 y_1).$$

Consider the algebraic extension field $K \subseteq {}^*\mathbb{Q}$ of $\mathbb{Q}(a)$ obtained by adjoining to $\mathbb{Q}(a)$ the (coordinates of the points of the) divisible hull of the subgroup of $^*G$ generated by G and (a, b). Then K has infinite degree over $\mathbb{Q}(a)$, and $K \cap {}^*\mathbb{Z} \subseteq \mathbb{Z}^\times \langle a \rangle^{\text{cl}}$. Is $K \cap {}^*\mathbb{Z}$ closed under iq, gcd, and σ? Do we have $K \cap {}^*\mathbb{Z} = \mathbb{Z}^\times \langle a \rangle^{\text{cl}}$? (Perhaps the answer depends on the particular solution (a, b) of $x^2 - 2y^2 = 1$.)

The prime ideal structure of $\mathbb{Z}^\times \langle a \rangle$. Here we assume $a \in {}^*\mathbb{Z}$, $a \notin \mathbb{Z}$. For each $n > 0$ the unique ring morphism $\mathbb{Z} \to \mathbb{Z}^\times \langle a \rangle / n\mathbb{Z}^\times \langle a \rangle$ is surjective with kernel $n\mathbb{Z}$, and thus induces a ring isomorphism

$$\mathbb{Z}/n\mathbb{Z} \cong \mathbb{Z}^\times \langle a \rangle / n\mathbb{Z}^\times \langle a \rangle.$$

It follows that the prime ideals of $\mathbb{Z}^\times \langle a \rangle$ that intersect $\mathbb{Z}$ non-trivially are exactly the ideals $p\mathbb{Z}^\times \langle a \rangle$ with p a prime number, and these prime ideals are maximal ideals. The above also shows that for each prime number p there is a unique ring morphism $\phi_p : \mathbb{Z}^\times \langle a \rangle \to \mathbb{Z}_p$ (with $\mathbb{Z}_p$ the ring of p-adic integers), and that $\ker \phi_p = \bigcap_n p^n \mathbb{Z}^\times \langle a \rangle$ is a prime ideal properly contained in $p\mathbb{Z}^\times \langle a \rangle$. We shall see below that in case $\ker \phi_p$ is non-trivial, it is the only non-trivial prime ideal of $\mathbb{Z}^\times \langle a \rangle$ that is properly contained in $p\mathbb{Z}^\times \langle a \rangle$. To get a better grasp of the prime ideals of the domain $\mathbb{Z}^\times \langle a \rangle$ it may be instructive to look at an example.

EXAMPLE. Suppose $a \equiv 0 \bmod n$ for all $n > 0$. Then $\mathbb{Z}^\times \langle a \rangle = \mathbb{Z} + a\mathbb{Q}[a]$. Clearly $a\mathbb{Q}[a]$ is an ideal of $\mathbb{Z}^\times \langle a \rangle$ with residue ring isomorphic to $\mathbb{Z}$, so $a\mathbb{Q}[a]$ is a non-trivial prime ideal but not a maximal ideal; the maximal ideals containing $a\mathbb{Q}[a]$ are exactly the $p\mathbb{Z}^\times \langle a \rangle$ with p a prime number. One verifies easily that $a\mathbb{Q}[a]$ is the only prime ideal that contains a and intersects $\mathbb{Z}$ trivially. Note that $a\mathbb{Q}[a]$ is generated as an ideal of $\mathbb{Z}^\times \langle a \rangle$ by its subset $\{a/n : n > 0\}$, but that no finite subset of $\{a/n : n > 0\}$ generates $a\mathbb{Q}[a]$; in particular, $\mathbb{Z}^\times \langle a \rangle$ is not noetherian.

The ideals $(a + 1)\mathbb{Z}^\times \langle a \rangle$ and $(a^2 + 1)\mathbb{Z}^\times \langle a \rangle$ are another kind of prime ideals, namely maximal ideals with residue fields isomorphic to $\mathbb{Q}$ and $\mathbb{Q}(i)$, respectively, where $i^2 = -1$.

This example suggests the following general description of prime ideals. Let f be a non-constant irreducible polynomial in $\mathbb{Z}[X]$ with positive leading coefficient, and put

$$\mathcal{D}_f := \mathcal{D}_{a,f} := \{n > 0 : n | f(a) \text{ in } {}^*\mathbb{Z}\}.$$

This set of positive integers contains 1 and is closed under taking divisors and taking least common multiples; in particular, if $\mathcal{D}_f$ is finite, then

$$\mathcal{D}_f = \{n > 0 : n | D_f\}$$

for a unique positive integer D_f. Put

$$\mathfrak{p}_f := \text{ideal generated in } \mathbb{Z}^\times \langle a \rangle \text{ by the } \frac{f(a)}{n} \text{ with } n \in \mathcal{D}_f.$$

It is easy to check that $\mathfrak{p}_f$ is the kernel of the ring morphism

$$\mathbb{Z}^\times \langle a \rangle \hookrightarrow \mathbb{Q}[a] \longrightarrow \mathbb{Q}[\alpha]$$

that sends a to α, where $f(\alpha) = 0$, $\alpha \in \mathbb{C}$. Thus $\mathfrak{p}_f$ is a prime ideal of $\mathbb{Z}^\times \langle a \rangle$ that intersects $\mathbb{Z}$ trivially. For every non-trivial prime ideal $\mathfrak{p}$ of $\mathbb{Z}^\times \langle a \rangle$ that intersects $\mathbb{Z}$ trivially there is a unique non-constant irreducible polynomial f in $\mathbb{Z}[X]$ with positive leading coefficient such that $\mathfrak{p} = \mathfrak{p}_f$: take f to be the non-constant polynomial in $\mathbb{Z}[X]$ of least degree and least positive leading coefficient in that degree with $f(a) \in \mathfrak{p}$. The next result summarizes some basic facts about these prime ideals.

PROPOSITION 6.1. *For each prime number p, we have*

$$\mathfrak{p}_f \subseteq p\mathbb{Z}^\times \langle a \rangle \iff p^n \in \mathcal{D}_f \text{ for all } n.$$

The prime ideal $\mathfrak{p}_f$ is maximal if and only if for each prime number p there is a largest n such that $p^n \in \mathcal{D}_f$. The ideal $\mathfrak{p}_f$ is finitely generated if and only if $\mathcal{D}_f$ is finite; in that case, $\mathfrak{p}_f$ is maximal and generated by a single element $f(a)/D_f$.

The forward part of the first equivalence is clear by induction on n, and the rest follows easily. In case $\mathfrak{p}_f$ is maximal, the above ring morphism $\mathbb{Z}^\times\langle a\rangle \to \mathbb{Q}[\alpha]$ induces a field isomorphism

$$\mathbb{Z}^\times\langle a\rangle/\mathfrak{p}_f \cong \mathbb{Q}[\alpha].$$

If the prime ideal $\mathfrak{p}_f$ is not maximal, then the induced ring embedding $\mathbb{Z}^\times\langle a\rangle/\mathfrak{p}_f \to \mathbb{Q}[\alpha]$ contains $\mathbb{Z}[\alpha]$ in its image, and the maximal ideals that contain $\mathfrak{p}_f$ are exactly the $p\mathbb{Z}^\times\langle a\rangle$ with p a prime number such that $p^n \in \mathcal{D}_f$ for all n. Also, if p is a prime number such that $\mathfrak{p}_f \subseteq p\mathbb{Z}^\times\langle a\rangle$, then

$$\mathfrak{p}_f = \ker\phi_p = \bigcap_n p^n\mathbb{Z}^\times\langle a\rangle,$$

and then $\mathfrak{p}_f$ is the only non-trivial prime ideal properly contained in $p\mathbb{Z}^\times\langle a\rangle$. (In the example above where $a \equiv 0 \bmod n$ for all $n > 0$, the prime ideal $a\mathbb{Q}[a] = \mathfrak{p}_X$ is contained in $p\mathbb{Z}^\times\langle a\rangle$ for every prime number p, and thus $a\mathbb{Q}[a]$ is the only non-maximal non-trivial prime ideal of $\mathbb{Z}^\times\langle a\rangle$.)

The facts above also imply that $\mathbb{Z}^\times\langle a\rangle$ is a PID (Principal Ideal Domain) iff $\mathcal{D}_f$ is *finite* for each non-constant irreducible polynomial f in $\mathbb{Z}[X]$ with positive leading coefficient. Does this case actually occur? The answer is yes:

Construction of a such that $\mathbb{Z}^\times\langle a\rangle$ is a PID. Fix an injective enumeration (f_n) of the irreducible non-constant polynomials in $\mathbb{Z}[X]$ with positive leading coefficient, such that $f_0 = X$, and put

$$g_n := f_0 \cdots f_n.$$

Let the *size* $s(g)$ of a non-constant polynomial $g \in \mathbb{Z}[X]$ be the maximum of the degree and absolute values of the coefficients of g. Let (q_n) be the sequence of prime powers > 1 in increasing order: $q_0 = 2, q_1 = 3, q_2 = 4, \ldots$.

We shall modify (q_n) to a sequence (q'_n), and construct sequences $(v(n))$ and (a_n) in $\mathbb{N}$ such that for all m and n:

- if $q_n = p^i$, with prime number p and positive integer i, then $q'_n = p^k$ for some integer $k \geq i$, with $k = 1$ if $i = 1$;
- $v(n+1) = v(n)$ or $v(n+1) = v(n) + 1$, and $v(i) \to \infty$ as $i \to \infty$;
- if $p_1 < p_2$ are successive prime numbers, and $q_m = p_1$ and $q_n = p_2$, then $v(m) = v(i)$ for $m \leq i < n$;
- $q_n > s(g_{v(n)})$ if q_n is prime;
- $0 \leq a_n < q'_n$ and $g_{v(n)}(a_n) \not\equiv 0 \bmod q'_n$;
- if $m < n$ and $q'_m = p^k$ and $q'_n = p^l$ with prime number p and positive integers k and l, then $k < l$, and $a_n \equiv a_m \bmod q'_m$.

Assuming such sequences have been constructed, take $a \in {}^*\mathbb{Z}$ with $a > \mathbb{Z}$ and $a \equiv a_n \bmod q'_n$ for all n. Given any n, choose a natural number N such that $v(N) \geq n$; then the set $\mathcal{D}_{f_n}$ contains no prime numbers $> q_N$, and, for each prime number $p \leq q_N$, contains only finitely many powers of p. Hence each set $\mathcal{D}_{f_n}$ is finite, and thus $\mathbb{Z}^\times \langle a \rangle$ is a PID.

It remains to construct sequences as described. We begin by setting $q_0' = q_0 = 2$, $v(0) = 0$ and $a_0 = 1$, so we do have $g_{v(0)}(a_0) \not\equiv 0 \bmod q_0'$. Having constructed $q_0', \ldots, q_n'$, $v(0), \ldots, v(n)$ and $a_0, \ldots, a_n$ in accordance with the conditions above, let $q_{n+1} = p^i$ with p a prime number and i a positive integer.

If $i = 1$ and $p > s(g_{v(n)+1})$, then we put $v(n+1) := v(n) + 1$. If $i = 1$ and $p \leq s(g_{v(n)+1})$, then we put $v(n+1) := v(n)$; in this case we can assume inductively that the largest prime number $< p$ is $> s(g_{v(n)})$, hence $p > s(g_{v(n+1)})$. In both cases we choose the integer a_{n+1} such that $0 \leq a_{n+1} < p$ and $g_{v(n+1)}(a_{n+1}) \not\equiv 0 \bmod p$.

Suppose $i > 1$. Then $p^{i-1} = q_m$ with $m \leq n$, and $q_m' = p^\kappa$ with $\kappa \geq i - 1$ (inductive hypothesis). We put $v(n+1) := v(n)$, and

$$ g(X) := g_{v(n+1)}(p^\kappa X + a_m), $$

and choose an integer $b \geq 0$ such that $g(b) \neq 0$. Next choose an integer $\lambda > 0$ such that $b < p^\lambda$ and $|g(b)| < p^{\kappa+\lambda}$. Then $g(b) \not\equiv 0 \bmod p^{\kappa+\lambda}$. Setting $q_{n+1}' := p^{\kappa+\lambda}$ and $a_{n+1} := p^\kappa b + a_m$ we have $g_{v(n+1)}(a_{n+1}) \not\equiv 0 \bmod q_{n+1}'$, $0 \leq a_{n+1} < q_{n+1}'$ and $a_{n+1} \equiv a_m \bmod q_m'$.

The opposite extreme where no $\mathfrak{p}_f$ is maximal also occurs:

Construction of a such that $\mathbb{Z}^\times \langle a \rangle$ has no maximal ideal of the form $\mathfrak{p}_f$. Let (f_n) be as in the previous construction. Choose inductively for each n a prime number $p(n)$ and an $a_n \in \mathbb{Z}$ such that $f_n(a_n) \equiv 0 \bmod p(n)$ and $f_n'(a_n) \not\equiv 0 \bmod p(n)$, and $p(n) \notin \{p(0), \ldots, p(n-1)\}$. Then there is for all m, n an $a_{m,n} \in \mathbb{Z}$ such that $f_n(a_{m,n}) \equiv 0 \bmod p(n)^m$ and $a_{m+1,n} \equiv a_{m,n} \bmod p(n)^m$. It follows that we can take $a \in {}^*\mathbb{Z}$ with $a > \mathbb{Z}$ such that $a \equiv a_{m,n} \bmod p(n)^m$ for all m, n. Then for each n the set $\mathcal{D}_{a,f_n}$ will contain all powers $p(n)^m$, so $\mathfrak{p}_{f_n}$ is not a maximal ideal.

We finish this section with a question: Does there exists an a such that some maximal ideal of $\mathbb{Z}^\times \langle a \rangle$ is not finitely generated?

§7. Model-theoretic appendix.

The elementary model-theoretic facts used in Section 2 are routine for experts. Here we give short proofs using only the Stone representation theorem for boolean algebras in addition to the basics of first-order logic.

Below $\mathcal{M} = (M, \ldots)$ is an infinite L-structure with $\operatorname{card}(M) \geq \operatorname{card}(L)$, and ${}^*\mathcal{M} = ({}^*M, \ldots)$ is a $\operatorname{card}(M)^+$-saturated elementary extension of $\mathcal{M}$. For $a \in {}^*M$, let $\mathcal{M}\langle a \rangle$ be the L-substructure of ${}^*\mathcal{M}$ generated by a over $\mathcal{M}$, so the underlying set of $\mathcal{M}\langle a \rangle$ is

$$ M\langle a \rangle := \{\tau(a) : \tau(x) \text{ is an } L_M\text{-term}\}. $$

Here and below x denotes a (syntactic) variable.

LEMMA 7.1. *Suppose $T(x)$ is a set of L_M-terms $t(x)$ such that $M\langle a \rangle = \{t(a) : t(x) \in T(x)\}$ for each $a \in {}^*M$. Let $\tau(x)$ be any L_M-term. Then there are terms $t_1(x), \ldots, t_k(x)$ in $T(x)$ such that for each $b \in M$ we have $\tau(b) \in \{t_1(b), \ldots, t_k(b)\}$.*

PROOF. Otherwise, for any terms $t_1(x), \ldots, t_k(x)$ in $T(x)$ there is $b \in M$ such that $\tau(b) \notin \{t_1(b), \ldots, t_k(b)\}$. Saturation then yields an $a \in {}^*M$ such that $\tau(a) \notin \{t(a) : t(x) \in T(x)\} = M\langle a \rangle$, a contradiction. ⊣

Each set $\{b \in M : \tau(b) = t_i(b)\}$ is quantifier-free L_M-definable in $\mathcal{M}$, but for use in Section 2 we need to characterize such sets in a more specific way. This is accomplished by the following result.

PROPOSITION 7.2. *Let $\Psi(x)$ be a collection of quantifier-free L_M-formulas $\psi(x)$ with the following property: whenever $a, a' \in {}^*M$ and*

$$ {}^*\mathcal{M} \models \psi(a) \iff {}^*\mathcal{M} \models \psi(a') $$

for all $\psi(x) \in \Psi(x)$, then there is an isomorphism $\mathcal{M}\langle a \rangle \cong \mathcal{M}\langle a' \rangle$ of L-structures that is the identity on M and sends a to a'.

Then every quantifier-free L_M-formula $\theta(x)$ is equivalent in $\mathcal{M}$ to a boolean combination of formulas in $\Psi(x)$.

The proposition will easily follow from a fact on boolean algebras:

LEMMA 7.3. *Let B be a boolean algebra and $S(B)$ be its Stone space of ultrafilters. Let Ψ be a subset of B and suppose the map $F \mapsto F \cap \Psi : S(B) \to \mathcal{P}(B)$ is injective. Then Ψ generates B: every element of B is a boolean combination of elements of Ψ.*

PROOF. Let B_Ψ be the boolean subalgebra of B generated by Ψ. The inclusion $B_\Psi \hookrightarrow B$ induces the *surjective* map $F \mapsto F \cap B_\Psi : S(B) \twoheadrightarrow S(B_\Psi)$. This map is also *injective*: if $F_1, F_2 \in S(B)$ and $F_1 \cap B_\Psi = F_2 \cap B_\Psi$, then $F_1 \cap \Psi = F_2 \cap \Psi$, hence $F_1 = F_2$ by the hypothesis of the lemma. The bijectivity of $F \mapsto F \cap B_\Psi : S(B) \to S(B_\Psi)$ yields $B = B_\Psi$ by the Stone representation theorem. ⊣

The *quantifier-free type of $a \in {}^*M$ over M* is defined by

$$ \mathrm{qftp}(a) := \{\theta(x) : \theta(x) \text{ is a quantifier-free } L_M\text{-formula and } {}^*\mathcal{M} \models \theta(a)\}. $$

PROOF OF THE PROPOSITION. We apply the last lemma to the boolean algebra B of quantifier-free L_M-formulas $\theta(x)$ modulo equivalence in $\mathcal{M}$, and with Ψ the image of $\Psi(x)$ in B. The saturation assumption on ${}^*\mathcal{M}$ yields that the ultrafilters on B are exactly the images in B of the quantifier-free types of elements of ${}^*M$ over M. The hypothesis on $\Psi(x)$ in the proposition means that the quantifier-free type $\mathrm{qftp}(a)$ of any element $a \in {}^*M$ over M is uniquely determined by its intersection with $\Psi(x)$. It follows that the hypothesis of the last lemma is satisfied. This concludes the proof. ⊣

REFERENCES

[1] C. Chang and J. Keisler, *Model Theory*, third ed., North-Holland, Amsterdam, 1990.

[2] P. Cijsouw, *Transcendence measures of exponentials and logarithms of algebraic numbers*, *Compositio Mathematica*, vol. 28 (1974), pp. 163–178.

[3] Y. Mansour, B. Schieber, and P. Tiwari, *A lower bound for integer greatest common divisor computations*, *Journal of the Association for Computing Machinery*, vol. 38 (1991), no. 2, pp. 453–471.

[4] A. Prestel, *Lectures on Formally Real Fields*, Lecture Notes in Mathematics, vol. 1093, Springer, Berlin, 1984.

[5] C. Smoryński, *Logical Number Theory. I*, Springer, Berlin, 1991.

[6] L. van den Dries, *Generating the greatest common divisor, and limitations of primitive recursive algorithms*, *Foundations of Computational Mathematics*, vol. 3 (2003), pp. 297–324.

[7] L. van den Dries and Y. Moschovakis, *Is the euclidean algorithm optimal among its peers?*, *The Bulletin of Symbolic Logic*, vol. 10 (2004), no. 3, pp. 390–418.

[8] ———, *Arithmetic complexity*, in preparation.

UNIVERSITY OF ILLINOIS
DEPARTMENT OF MATHEMATICS
1409 W. GREEN STREET
URBANA, IL 61801, USA
E-mail: vddries@math.uiuc.edu

FROM BOUNDED ARITHMETIC TO SECOND ORDER ARITHMETIC VIA AUTOMORPHISMS

ALI ENAYAT

Abstract. In this paper we examine the relationship between automorphisms of models of $I\Delta_0$ (bounded arithmetic) and strong systems of arithmetic, such as PA, ACA_0 (arithmetical comprehension schema with restricted induction), and Z_2 (second order arithmetic). For example, we establish the following characterization of PA by proving a "reversal" of a theorem of Gaifman:

THEOREM. *The following are equivalent for completions T of $I\Delta_0$:*

(a) $T \vdash PA$;

(b) *Some model $\mathfrak{M} = (M, \dots)$ of T has a proper end extension $\mathfrak{N}$ which satisfies $I\Delta_0$ and for some automorphism j of $\mathfrak{N}$, M is precisely the fixed point set of j.*

Our results also shed light on the metamathematics of the Quine-Jensen system NFU of set theory with a universal set.

§1. Introduction. The classical work of Ehrenfeucht and Mostowski introduced the powerful method of indiscernibles to show that any first order theory with an infinite model has a proper class of models with rich automorphism groups [CK, Section 3.3]. In the context of models of arithmetic, the first substantial results concerning automorphisms that extend the work of Ehrenfeucht and Mostowski are to be found in Gaifman's seminal work [G] on the model theory of *Peano arithmetic PA*. Gaifman refined the MacDowell-Specker method [MS] of building elementary end extensions by introducing the machinery of *minimal types*, which can be used to produce a variety of models of *PA* with special properties. For example, they can be used to establish the striking result below. Here Aut($\mathfrak{N}$) is the group of automorphisms of $\mathfrak{N}$, and Aut($\mathfrak{N}, M$) is the pointwise stabilizer of M (i.e., the subgroup of Aut($\mathfrak{N}$) consisting of automorphisms of $\mathfrak{N}$ that fix every element of M).

THEOREM 1.1 (Gaifman). *Suppose $\mathfrak{M} = (M, \dots)$ is a model of PA, and $\mathbb{L}$ is a linear order.*

(a) *There is an elementary end extension $\mathfrak{N}$ of $\mathfrak{M}$ such that* Aut($\mathfrak{N}, M$) $\cong$ Aut($\mathbb{L}$) *[G, Theorem 4.11].*

(b) *There is an elementary end extension $\mathfrak{N}$ of $\mathfrak{M}$ such that for some $j \in$* Aut($\mathfrak{N}$), *M is the fixed point set of j [G, Theorems 4.9–4.11].*

Logic in Tehran
Edited by A. Enayat, I. Kalantari, and M. Moniri
Lecture Notes in Logic, 26

Schmerl [Sc] has recently established a strong generalization of part (a) of Theorem 1.1 by showing that $Aut(\mathbb{L})$ can be replaced by any *closed subgroup* of $Aut(\mathbb{L})$. This shows that the class of left-orderable groups coincides with the class of groups that can occur as $Aut(\mathfrak{M})$ for models $\mathfrak{M}$ of PA. A major trend in the study of automorphism groups of models of PA was initiated in the early 1980's with the work of Smorynski and Kotlarski (independently) on automorphisms of *countable recursively saturated models*. This has proved to be a fertile area of research, and has resulted in a number of striking results by Kaye, Kossak, Kotlarski, Lascar, and Schmerl, to name a few. The reader interested in becoming familiar with the rudiments of the subject is referred to the volume [KM].

This paper provides model theoretic characterizations of the strong systems of arithmetic PA, ACA_0, and Z_2 in terms of automorphisms of models of the weak system of arithmetic $I\Delta_0$ (commonly known as *bounded arithmetic*). Previously, Ressayre [Re] provided elegant characterizations of PA and the fragment $I\Sigma_1$ of PA in terms of *endomorphisms*, but there is no overlap between Ressayre's results and ours. For other model theoretic characterizations of PA, see [Kay].

The plan of the paper is as follows. After dealing with preliminaries in Section 2, we concentrate on the relationship between automorphisms of models of bounded arithmetic and the axiomatic systems PA and ACA_0 in Section 3. The principal results of Section 3 are Theorems A and B. Theorem A (Section 3.1) establishes a strong reversal of Theorem 1.1(b), while Theorem B (Section 3.2) is a refined form of Theorem 1.1(b) for models of ACA_0 (Theorem B is implicit in Gaifman [G], but the proof here is new). Theorems A and B together yield a model theoretic characterization of ACA_0 in terms of automorphisms. Section 4 focuses on the relationship between automorphisms of models of bounded arithmetic and models of second order arithmetic. The key notion in Section 4 is that of an "M-amenable automorphism", shown in Theorems C and D to be closely tied to models of full second order arithmetic. Section 5 includes a brief discussion of the consequences of the results in Sections 3 and 4 for the metamathematics of NFU set theory, and a discussion of further work and open questions.

The results of this paper were discovered in the context of the study of Jensen's modification NFU [Jen] of Quine's *New Foundations* system NF of set theory [Q] with a universal set. They have been used by Robert Solovay and the author to pinpoint the "arithmetical content" of certain natural extensions of NFU, such as the theory $NFUA^{-\infty}$ obtained by strengthening NFU with the axioms "every set is finite" and "every Cantorian set is strongly Cantorian". This topic will be fully treated in a forthcoming paper and we have therefore provided only a brief summary of our results for the metamathematics of NFU in Section 5.1. We should mention that there is also a *set theoretical* counterpart to the theme of this paper. This is partly explained in [E-1], in which automorphisms of models of weak systems of set theory are shown to be intimately connected to ZF-set theory with Mahlo cardinals. Roughly speaking, the results in [E-1] are the set theoretical analogues of Theorems A and B of this paper. The set theoretical analogues of Theorems D and E will appear in [E-3].

Brief history. In the early 1990's Holmes [Ho-1] made a breakthrough by using a large cardinal hypothesis (measurability) to establish the consistency of certain natural extensions ($NFUA$ and $NFUB$) of the Quine-Jensen system NFU. Holmes' work prompted Solovay[1] to work out the precise consistency strengths of $NFUA$ and $NFUB$, by showing that (a) $NFUA$ is equiconsistent with

$$ZFC + \{\text{"there is an } n\text{-Mahlo cardinal"} : n \in \omega\},$$

and that (b) $NFUB$ is equiconsistent with

$$ZFC \setminus \{\text{Power Set}\} + \text{"there is a weakly compact cardinal"}.$$

The work of Holmes and Solovay unearthed a deep, unexpected relationship between strong set theoretical hypotheses and models of $NFUA/B$ in which the axiom of infinity holds. This inspired the author to seek a parallel relationship between strong *arithmetical* hypotheses and models of $NFUA/B$ in which the axiom of infinity *fails*. My initial result in this direction (a slightly weaker form of Theorem C) was an arithmetical analogue of a key result in [Sol]. The communication of this result to Solovay in January 2002 led to an extensive (e-mail) correspondence during the following year. It was during the course of this intense and inspiring period that I managed to obtain the results of this paper in their current form. Solovay has also established a number of results concerning the metamathematics of NFU that remain unpublished, which will hopefully appear in the near future.

Acknowledgments. I am indebted to Robert Solovay for his patience and insights offered through meticulously crafted e-mail communiqués. I also wish to thank Randall Holmes for helpful discussions about NFU; Roman Kossak and Joel Hamkins for inviting me to present my results at the CUNY Logic Workshop; Albert Visser for formulating probing questions which led me to the results in Section 3.3; Steve Simpson for alerting me to the crucial role of the dependent choice scheme in Mostowski's forcing construction [Mo-1, 2]; and Iraj Kalantari and Mojtaba Moniri for unfailing camaraderie. I am also grateful to Andreas Blass and the anonymous referees for detailed constructive comments on earlier drafts of this paper.

---

[1]Solovay's work on $NFUB$ appears in [Sol], but his work on $NFUA$ is unpublished. Holmes [Ho-1] contains an extension of one direction of Solovay's equiconsistency result on $NFUA$, and [E-1] contains a generalization of both directions of Solovay's equiconsistency result on $NFUA$.

§2. Preliminaries.

2.1. Bounded arithmetic.

- The language of first order arithmetic, $\mathcal{L}_A$, is $\{+, \cdot, Succ(x), <, 0\}$.
- Models of $\mathcal{L}_A$ are of the form $\mathfrak{M} = (M, +^{\mathfrak{M}}, \ldots)$, $\mathfrak{N} = (N, +^{\mathfrak{N}}, \ldots)$, etc. For models $\mathfrak{M}$ and $\mathfrak{N}$ of $\mathcal{L}_A$, we say that $\mathfrak{N}$ *end extends* $\mathfrak{M}$ (equivalently: $\mathfrak{M}$ is an *initial* submodel of $\mathfrak{N}$), written $\mathfrak{M} \subseteq_e \mathfrak{N}$, if $\mathfrak{M}$ is a submodel of $\mathfrak{N}$ and $a < b$ for every $a \in M$, and $b \in N \setminus M$. We abbreviate the phrase "elementary end extension" by "e.e.e.".
- I is a *cut* of $\mathfrak{M}$, where $\mathfrak{M}$ is a model of Robinson's Q, if I is a *proper* initial segment of $\mathfrak{M}$ with no last element.
- A first order $\mathcal{L}_A$-formula φ is said to be a Δ_0-*formula* if all the quantifiers of φ are bounded, i.e., they are of the form $\exists x \leq y$, or of the form $\forall x \leq y$, where x and y are (meta)variables. Δ_0-formulae are also known as *bounded* formulae.
- *Bounded arithmetic*, or $I\Delta_0$, is the fragment of Peano arithmetic with the induction scheme limited to Δ_0-formulae. More specifically, it is a theory formulated in the language $\mathcal{L}_A$, and is obtained by adding the scheme of induction for Δ_0-formulae to Robinson's arithmetic Q. The metamathematical study of bounded arithmetic has close ties with the subject of computational complexity. See [HP] or [Kr] for thorough introductions.
- Bennett [Be] showed that the graph of the exponential function $y = 2^x$ can be defined by a Δ_0-predicate in the standard model of arithmetic. Later, Paris found another Δ_0-predicate $\varphi(x, y)$ which does the job, and $I\Delta_0$ can prove the familiar algebraic laws about exponentiation for $\varphi(x, y)$ [DG, Appendix][2]. By a classical theorem of Parikh [Pa] however, $I\Delta_0$ can only prove the totality of functions with a polynomial growth rate, hence

$$I\Delta_0 \nvdash \forall x \exists y \varphi(x, y).$$

It is now known that the graphs of many other fast growing recursive functions, such as the superexponential function Superexp[3], the Ackermann function, and indeed all functions $\{F_\alpha : \alpha < \varepsilon_0\}$ in the (fast growing) Wainer hierarchy, can be defined by Δ_0-predicates for which $I\Delta_0$ can prove appropriate recursion schemes. This remarkable discovery is due to Sommer [Som-1], [Som-2], but the reader is also referred to

---

[2]Independently, Pudlák [Pu-1] also provided an $I\Delta_0$-treatment of the exponential function. A detailed exposition is provided in [Bu] and [HP, Chapter V, Section 3(c)]

[3]The superexponential function, $\text{Superexp}(n, x)$, is defined by the recursion scheme: $\text{Superexp}(0, x) = x$, $\text{Superexp}(n + 1, x) = 2^{\text{Superexp}(n,x)}$. Thus for $n > 0$, $\text{Superexp}(n, x)$ is an exponential stack of length $n + 1$, where the top element is x, and the remaining n entries form a tower of 2's.

D'Aquino's paper [D] for a perspicuous Δ_0-treatment of the superexponential function and the Ackermann function.

The following result is well known: a routine proof by contradiction proves (a), with Δ_0 induction applied to $\varphi^*(v) := \forall x \leq v \ \neg\varphi(v)$, (b) follows from (a) since the maximum of S_φ is the least upper bound of S_φ, and (c) follows from (b).

LEMMA 2.1. *Suppose $\mathfrak{M}$ is a model of $I\Delta_0$, and let S_φ be the solution set of some Δ_0-formula $\varphi(x, \overrightarrow{a})$ in $\mathfrak{M}$, where $\overrightarrow{a}$ is a sequence of parameters from M. If $S_\varphi \neq \emptyset$, then:*

(a) $[\Delta_0\text{-}MIN]$ S_φ *has a minimum element;*
(b) $[\Delta_0\text{-}MAX]$ *If S_φ is bounded in $\mathfrak{M}$, then S_φ has a maximum element;*
(c) $[\Delta_0\text{-}OVERSPILL]$ *If S_φ includes a cut I of $\mathfrak{M}$, then for some $b \in M \backslash I$, $[0, b]^{\mathfrak{M}} \subseteq S_\varphi$.*

2.2. The strength of $I\Delta_0 + Exp$. Let $\varphi(x, y)$ be a reasonable Δ_0-formula expressing "$2^x = y$". $I\Delta_0 + \text{Exp}$ is the extension of $I\Delta_0$ obtained by adding the axiom

$$\text{Exp} := \forall x \exists y \varphi(x, y).$$

At first sight $I\Delta_0 + \text{Exp}$ is a rather weak theory since it cannot even prove the totality of the superexponential function or any faster growing function. But, experience has shown that it is a remarkably robust theory that is able to prove a large variety of theorems of number theory and finite combinatorics[4]. One explanation for this phenomenon is offered by the fact that one can use *Ackermann coding* to simulate a workable set theory within $I\Delta_0 + \text{Exp}$. Let $E(x, y)$ be a Δ_0-predicate that expresses "the x-th digit in the binary expansion of y is a 1". We shall henceforth refer to E as "*Ackermann's $\in$*". It is well known that $\mathfrak{M}$ is a model of PA iff (M, E) is a model of $ZF\backslash\{\text{Infinity}\} \cup \{\neg\text{Infinity}\}$, but if $\mathfrak{M}$ is a model of $I\Delta_0 + \text{Exp}$, then (M, E) is still a model of a decent fragment of $ZF\backslash\{\text{Infinity}\} \cup \{\neg\text{Infinity}\}$. More specifically:

THEOREM 2.2 (Dimitracopoulos-Gaifman ([DG], [HP, Ch. I, Sec. 1(b)])). *If $\mathfrak{M} \models I\Delta_0 + \text{Exp}$, and E is Ackermann's $\in$, then (M, E) satisfies the following axioms:*

(1) *Extensionality;*
(2) *Pairs;*
(3) *Union;*
(4) *Powerset;*
(5) Δ_0-*Comprehension Scheme; and*
(6) *the negation of Infinity.*

---

[4]Indeed, Harvey Friedman has conjectured that all "arithmetical theorems" proved in the journal *Annals of Mathematics* (such as Wiles' proof of Fermat's Last Theorem), can be implemented within $I\Delta_0 + \text{Exp}$. We refer the reader to Avigad's paper [Av] for an excellent discussion of the foundational role of $I\Delta_0 + \text{Exp}$.

- Suppose $\mathfrak{M} \vDash I\Delta_0$ and E is Ackermann's $\in$ in the sense of $\mathfrak{M}$.
  - (a) For $c \in M$, $c_E := \{m \in M : mEc\}$.
  - (b) $X \subseteq M$ is *coded* in $\mathfrak{M}$ if there is some $c \in M$ such that $X = c_E$.
  - (c) Suppose I is a cut of $\mathfrak{M}$,

$$SSy_I(\mathfrak{M}) := \{c_E \cap I : c \in N\}.$$

In particular, if I is the standard cut of $\mathfrak{M}$, then $SSy_I(\mathfrak{M})$ is what is known in the literature as the *standard system* of $\mathfrak{M}$.

2.3. Second order arithmetic.

- The systems Z_2 and ACA_0 are fully discussed in Simpson's encyclopedic reference [Si-2]. Z_2 is often referred to as *second order arithmetic*[5], or as *analysis*. ACA_0 is the subsystem of Z_2 with the comprehension scheme limited to formulas with no second order quantifiers.
- Models of second order arithmetic (and its subsystems) are of the *two-sorted* form $(\mathfrak{M}, \mathcal{A})$, where $\mathfrak{M}$ is a model in the language $\mathcal{L}_A$, and $\mathcal{A}$ is a family of subsets of M. Since coding apparatus is available in the models of arithmetic $\mathfrak{M}$ considered here, we shall use expressions such as "$f \in \mathcal{A}$", where f is a function, as a substitute for the more precise but lengthier expression "the canonical code of f is in $\mathcal{A}$".
- For $\mathcal{L} \supseteq \mathcal{L}_A$, $PA(\mathcal{L})$ is PA augmented by the induction scheme for all $\mathcal{L}$-formulas. Note that if $(\mathfrak{M}, \mathcal{A}) \vDash ACA_0$, then $(\mathfrak{M}, S)_{S \in \mathcal{A}} \vDash PA(\mathcal{L})$, where $\mathcal{L}$ is the extension of $\mathcal{L}_A$ obtained by adding a unary predicate for each $S \in \mathcal{A}$.

§3. Automorphisms and ACA_0. The main results of this section are Theorem A and Theorem B. Theorem A establishes a strong "reversal" of Theorem 1.1(b), and Theorem B is the analogue of Theorem 1.1(b) for models of ACA_0.

3.1. ACA_0 from automorphisms.

THEOREM A. *If $\mathfrak{N} \vDash I\Delta_0$ and j is an automorphism of $\mathfrak{N}$ such that the fixed point set M of j is a proper initial segment of $\mathfrak{N}$, $(\mathfrak{M}, SSy_M(\mathfrak{N})) \vDash ACA_0$.*

Before presenting the proof of Theorem A, let us point out an important corollary obtained by coupling Theorem A with Theorem 1.1(b):

COROLLARY 3.1. *The following are equivalent for completions T of $I\Delta_0$:*

- (a) $T \vdash PA$;
- (b) *Every model $\mathfrak{M}$ of T has a proper e.e.e. $\mathfrak{N}$ such that for some automorphism j of $\mathfrak{N}$, M is the fixed point set of j;*
- (c) *Some model $\mathfrak{M}$ of T has a proper end extension $\mathfrak{N}$ such that $\mathfrak{N} \vDash I\Delta_0$ and for some automorphism j of $\mathfrak{N}$, M is the fixed point set of j.*

---

[5]Some authors, especially those belonging to the Polish school of logic (e.g., [Mo-1]), use A_2^- for the system Z_2 (and A_2 for Z_2 plus the choice scheme).

The proof of Theorem A relies on Lemmas A.0 through A.4 below. Lemmas A.0 and A.1 show the preliminary result that $\mathfrak{M}$ satisfies $I\Delta_0 + \text{Exp} + \text{Superexp}$ (where Superexp is the axiom stating that the function $\text{Superexp}(x, x)$ is total). Indeed, the strategy of the proof of Lemma A.1 can be used to establish that M is closed under all primitive recursive functions, thus showing that $\mathfrak{M}$ is a model of PRA (primitive recursive arithmetic). However, the totality of the Ackermann function does not seem to be obtainable via this strategy. These first two Lemmas are used in Lemma A.2 to show that we can replace the end extension $\mathfrak{N}$ of $\mathfrak{M}$ in Theorem A, if necessary, by a model of $I\Delta_0 + \text{Exp}$. Lemma A.2 and Theorem 2.2 together allow us the luxury of accessing a decent amount of set theory within an initial segment of $\mathfrak{N}$ containing M via Ackermann coding, thereby providing streamlined proofs of the central Lemmas A.3 and A.4 without having to go through laborious calculations dealing with Ackermann coding.

- For the rest of this section we make the blanket assumption that $\mathfrak{M}, \mathfrak{N}$, and j are as in the statement of Theorem A. In particular, M is the fixed point set of j, and $\mathfrak{N}$ is a proper end extension of $\mathfrak{M}$.

LEMMA A.0. $\mathfrak{M} \vDash I\Delta_0$.

PROOF. Clearly M is closed under the operations of $\mathfrak{N}$. Since Δ_0-predicates are absolute for end extensions, this shows that $\mathfrak{M}$ inherits $I\Delta_0$ from $\mathfrak{N}$. $\dashv$

LEMMA A.1. Exp *and* Superexp *both hold in* $\mathfrak{M}$.

PROOF. We only verify Exp in $\mathfrak{M}$ since the verification of the totality of the superexponential function uses an identical strategy and is left to the reader. Recall that there is a Δ_0-predicate that reasonably expresses "$2^x = y$". Let

$$I := \{x \in N : \mathfrak{N} \vDash \exists y \, (2^x = y)\}.$$

Note that I is closed downward in $\mathfrak{M}$ and $I \cap M$ has no last element since $\mathfrak{M} \vDash I\Delta_0$ and $I\Delta_0$ is able to prove that the set of numbers x on which 2^x is defined is closed under both predecessors and immediate successors. To show that Exp holds in $\mathfrak{M}$, it suffices to show that $M \subseteq I$ since if x is fixed by j, and 2^x exists in $\mathfrak{N}$, then 2^x is definable from x within $\mathfrak{N}$ and must therefore also be fixed by j. Next, let

$$J := \{y \in N : \mathfrak{N} \vDash \exists x \, (2^x = y)\}.$$

It is easy to see that if J is unbounded in N then $M \subseteq I$, so our proof would be complete once we establish that J is unbounded in $\mathfrak{N}$. Suppose, on the contrary, that some $a \in N$ is an upper bound of J. Then the set

$$\{y < a : \mathfrak{N} \vDash \exists x < a \, (2^x = y)\}$$

has a maximum element by Lemma A.0 and Δ_0-MAX (Lemma 2.1(b)) since it is the solution set of a Δ_0-predicate, thus leading to the absurd conclusion that I has a maximum element. $\dashv$

LEMMA A.2. *There is an initial segment N^* of $\mathfrak{N}$ that properly contains M such that $\mathfrak{N}^* := (N^*, \ldots)$ is a model of $I\Delta_0 + \mathrm{Exp}$, and $j \upharpoonright \mathfrak{N}^*$ is an automorphism.*

PROOF. Let $\psi(x, y)$ be a Δ_0 predicate for $y = \mathrm{SuperExp}(x, x)$ and let $b \in N \backslash M$. By Lemma A.1 for every $m \in M$,

$$\mathfrak{N} \vDash \exists y < b \ \psi(m, y).$$

Therefore, by Δ_0-OVERSPILL (Lemma 2.1(c)) there is an element $a \in N \backslash M$ for which $\mathrm{SuperExp}(a, a)$ is well-defined in $\mathfrak{N}$. This implies that the elements

$$2^a, 2^{2^a}, \ldots, \mathrm{SuperExp}(n, a), \ldots \quad (n \in \omega)$$

are all well-defined within $\mathfrak{N}$. To define N^*, assume without loss of generality that $a < j(a)$ (else replace j by j^{-1}), and let

$$N^* := \bigcup_{k \in \omega} \bigcup_{n \in \omega} \left[0, \mathrm{SuperExp}\left(n, j^k(a)\right) \right]^{\mathfrak{N}}.$$

It is easy to verify that $\mathfrak{N}^* \vDash I\Delta_0 + \mathrm{Exp}$, *and $j \upharpoonright \mathfrak{N}^*$ is an automorphism.* ⊣

Before establishing the next lemma[6], we need to recall the key notion of strong cuts, first introduced by Kirby and Paris [KP]:

- Suppose $\mathfrak{N}$ is a model of $I\Delta_0$ and M is a cut of $\mathfrak{N}$. M is a *strong cut* of $\mathfrak{N}$, if for each function f whose graph is coded in $\mathfrak{N}$ (via Ackermann's $\in$) and whose domain includes M, there is some s in N, such that for all $m \in M$,

$$f(m) \notin M \quad \text{iff} \quad s < f(m).$$

LEMMA A.3. *M is a strong cut of $\mathfrak{N}$.*

PROOF. We first observe that it suffices to show that $\mathfrak{M}$ is a strong cut of the model $\mathfrak{N}^*$ of Lemma A.3. Recall that by Theorem 2.2, we have access to "bounded" set theoretic reasoning within $\mathfrak{N}^*$. Suppose $\overline{f} \in N^*$ codes the graph of a function f whose domain includes M. It is easy to see that $\overline{f} \notin M$. So if $\overline{g} := j(\overline{f})$, then $\overline{g} \notin M$, and $\overline{f} \neq \overline{g}$. Therefore, if g is the function that is coded by $\overline{g}$, then:

$$\forall m \in M \ \left[f(m) = g(m) \Longleftrightarrow f(m) \in M \right].$$

We wish to find $s \in N^*$ such that for all $m \in M$, $f(m) \notin M$ iff $s < f(m)$. Without loss of generality there is some $m_0 \in M$ with $f(m_0) \notin M$. Fix $c \in N^*$ such that c_E contains $\overline{f}, \overline{g}$, and every $m \in M$ (recall: c_E is $\{x \in N : xEc\}$, where E is Ackermann's $\in$ in the sense of $\mathfrak{N}$). Consider the function $h(x)$ defined within $\mathfrak{M}$ on the interval $[m_0, c]$ by

$$h(x) := \mu y \le c \ \left[\exists z \le x(y = f(z) \neq g(z)) \right],$$

---

[6]This lemma was inspired by the results of Kaye, Kossak, and Kotlarski [KKK].

where $\mu y \leq c$ is the (truncated) least number operator, defined via the equation

$$[z := \mu y \leq c\, \varphi(y)] \quad \text{iff} \quad [z \text{ is the first solution } y \text{ of } \varphi, \text{ if } y \leq c; \text{ else } z = c].$$

Note that if $m \in M$ with $m_0 \leq m$ then $h(m) \notin M$, and if $m_0 \leq m \leq m'$, then $h(m') \leq h(m)$. Moreover,

(1) the graph of h is defined by a Δ_0-formula $\varphi(x, y)$ with parameters $\overline{f}$ and $\overline{g}$; and

(2) $m < h(m)$ for all $m \in M$ with $m \geq m_0$.

Therefore, (1), (2), and Δ_0-OVERSPILL (Lemma 2.1(c)) within $\mathfrak{N}^*$ together imply that there is some $s \in N^* \backslash M$ such that $s < h(s)$ holds in $\mathfrak{N}^*$. This shows that s is the desired lower bound for elements of the form $f(m)$, where $m \in M$ and $f(m) \notin M$.   ⊣

Kirby and Paris proved that strong cuts of models of PA are themselves models of PA [KP, Proposition 8]. An analysis of their proof reveals the stronger result below[7].

LEMMA A.4. *Let* $\mathcal{A} := SSy_M(\mathfrak{N})$ *and* $\mathcal{L} := \mathcal{L}_A \cup \{S : S \in \mathcal{A}\}$. *For every* $\mathcal{L}$-*formula*

$$\varphi(x_1, \ldots, x_m),$$

with free variables $x_1 \ldots, x_m$, *there is some* Δ_0-*formula*

$$\theta_\varphi(x_1, \ldots, x_m, b_1, \ldots, b_n),$$

where $b_1, \ldots, b_n$ *is a sequence of parameters from* N, *such that for all sequences* $a_1, \ldots, a_m$ *of elements of* M,

$$(\mathfrak{M}, S)_{S \in \mathcal{A}} \vDash \varphi(a_1, \ldots, a_m) \quad \text{iff} \quad \mathfrak{N} \vDash \theta_\varphi(a_1, \ldots, a_m, b_1, \ldots, b_n).$$

PROOF. In what follows $\mathfrak{N}^*$ is as in Lemma A.2. θ_φ is built by recursion on the complexity of φ:

- If φ is an atomic formula of the form $S_i(v)$, where v is a term, then choose $b \in N$ such that $b_F \cap M = S_i$, and define $\theta_\varphi := (v \in b)$. For other atomic formulas φ, $\theta_\varphi := \varphi$.
- $\theta_{\neg \delta} := \neg \theta_\delta$;
- $\theta_{\delta_1 \vee \delta_2} := \theta_{\delta_1} \vee \theta_{\delta_2}$;
- If $\varphi = \exists v \delta(v, x_1, \ldots, x_t)$, then fix some $c \in N \backslash M$ and consider the function $f(x_1, \ldots, x_t)$ defined in $\mathfrak{N}^*$ on $[0, c]^t$ by:

$$f(x_1, \ldots, x_t) := \begin{cases} \mu v \leq c \text{ such that } \theta_\delta(v, x_1, \ldots, x_t), \\ \qquad\qquad\qquad\qquad \text{if } \exists v \in c\ \theta_\delta(v, x_1, \ldots, x_t); \\ 0, \qquad\qquad\qquad\quad\ \text{otherwise.} \end{cases}$$

---

[7]As noted by one of the referees, this result also appears in Kirby's dissertation [Ki-1].

Note that the graph of f is defined by a $\Delta_0(\mathcal{L})$-formula within $\mathfrak{N}^*$ and so by Theorem 2.2 f is coded in $\mathfrak{N}^*$ and therefore in $\mathfrak{N}$. Hence, we can use Lemma A.3 to invoke the strength of M in $\mathfrak{N}$ to find some $s \in N$, such that for all $m \in M$, $f(m) \in M$ iff $f(m) \leq s$. Now define:

$$\theta_\varphi := \exists v \leq s \; \theta_\delta(v, x_1, \ldots, x_t). \qquad \dashv$$

PROOF OF THEOREM A. Let $\mathcal{A}$ and $\mathcal{L}$ be as in Lemma A.4. It is easy to see that every nonempty member of $\mathcal{A}$ has a first element in $\mathfrak{M}$ (since $\mathfrak{N}$ satisfies $I\Delta_0$). To establish the arithmetical comprehension scheme in $(\mathfrak{M}, \mathcal{A})$, consider any $\mathcal{L}$-formula $\varphi(x)$ with precisely one free variable x. We wish to show that

$$\{m \in M : (\mathfrak{M}, \mathcal{A}) \vDash \varphi(m)\} \in \mathcal{A}.$$

Let θ_φ be as in Lemma A.4 and fix some $c \in N \backslash M$. By Theorem 2.2 (part 5), there is an element $d \in N$ that codes $\{x < c : N \vDash \theta_\varphi(x)\}$. Therefore, by Lemma A.4

$$\{m \in M : (\mathfrak{M}, \mathcal{A}) \vDash \varphi(m)\} = d_E \cap M \in \mathcal{A}. \qquad \dashv$$

3.2. Automorphisms from ACA_0.
The principal result of this section is Theorem B. We should emphasize that Theorem B follows from Gaifman's work in [G], but we have decided to present a detailed proof here for two reasons. Firstly, this theorem is only implicit in Gaifman's paper, and therefore a detailed presentation of this significant result is of some value. Secondly, the method of *iterated ultrapowers modulo generic ultrafilters* developed here for the proof of Theorem B is also employed in the proof of Theorem C (Section 4.1) and a detailed development in this section allows us to later skip some details in the proof of Theorem C.

THEOREM B. *Suppose* $(\mathfrak{M}, \mathcal{A})$ *is a countable model of* ACA_0. *There is a proper elementary end extension* $\mathfrak{N}$ *of* $\mathfrak{M}$ *which satisfies the following two properties*:

(a) $\mathfrak{N}$ *possesses an automorphism* j *whose fixed point set is precisely* M;
(b) $SSy_M(\mathfrak{N}) = \mathcal{A}$.

The proof of Theorem B is presented at the end of this section once the machinery of generic ultrafilters and iterated ultrapowers have been put into place. However, we can easily describe the high-level strategy of the proof: $\mathfrak{N}$ is obtained by an iterated $\mathfrak{M}$-ultrapower along the linearly ordered set of integers $\mathbb{Z}$ modulo a "generic ultrafilter", and the desired automorphism j of $\mathfrak{N}$ is induced by the automorphism $n \mapsto n + 1$ of $\mathbb{Z}$.

3.2.1. Generic ultrafilters.
Suppose $(\mathfrak{M}, \mathcal{A})$ is a countable model of ACA_0. Clearly $\mathcal{A}$ is a Boolean algebra. Our goal is to construct ultrafilters $\mathcal{U}$ over $\mathcal{A}$ with certain desirable combinatorial properties. We shall employ the conceptual framework of forcing in order to efficiently present the necessary

bookkeeping arguments in our construction[8]. Let $\mathbb{P}$ be the poset

$$\{S \in \mathcal{A} : S \text{ is unbounded in } (M, <)\},$$

ordered under inclusion.

- A subset $\mathcal{D}$ of $\mathbb{P}$ is *dense* if for every $X \in \mathbb{P}$ there is some $Y \in \mathcal{D}$ with $Y \subseteq X$.
- $\mathcal{U} \subseteq \mathbb{P}$ is a *filter* if it is (1) closed under intersections and (2) is upward closed.
- A filter $\mathcal{U} \subseteq \mathbb{P}$ is $\mathcal{A}$-*generic over* $(\mathfrak{M}, \mathcal{A})$ if $\mathcal{U}$ meets every dense subset $\mathcal{D}$ of $\mathbb{P}$ which is parametrically definable in $(\mathfrak{M}, \mathcal{A})$.
- A filter $\mathcal{U} \subseteq \mathbb{P}$ is $(\mathfrak{M}, \mathcal{A})$-*complete* if for every $f : M \to [0, a]^{\mathfrak{M}}$, where $a \in M$ and $f \in \mathcal{A}$, there is some $X \in \mathcal{U}$ such that f is constant on X.

Note that if $\mathcal{U}$ is $(\mathfrak{M}, \mathcal{A})$-complete, then $\mathcal{U}$ is a nonprincipal ultrafilter on $\mathcal{A}$ since for each $Y \in \mathcal{A}$, the characteristic function of Y is constant on some member of $\mathcal{U}$. We therefore refer to $(\mathfrak{M}, \mathcal{A})$-complete filters as *ultrafilters*. Generic ultrafilters have some special combinatorial properties. To discuss them we need the following definitions and theorems.

- Let Γ be a canonical bijection between $M \times M$ and M. Every $g : M \to \{0, 1\}$ codes a sequence $\langle S_a^g : a \in M \rangle$ of subsets of M, where

$$S_a^g := \{b \in M : g(\Gamma(a, b)) = 1\}.$$

- A filter $\mathcal{U} \subseteq \mathbb{P}$ is $(\mathfrak{M}, \mathcal{A})$-*iterable*[9] if $\mathcal{U}$ is $(\mathfrak{M}, \mathcal{A})$-complete, and for every $g \in \mathcal{A}$ and $g : M \to \{0, 1\}$,

$$\{a \in M : S_a^g \in \mathcal{U}\} \in \mathcal{A}.$$

- Given a linearly order set $(M, <)$, $[M]^n$ is the set of *increasing n*-tuples from M.
- Suppose $(M, <)$ is a linear order and $f : [M]^n \to M$. A subset X of M is f-*canonical* if there is some $S \subseteq \{1, \ldots, n\}$ such that for all sequences $s_1 < \cdots < s_n$, and $t_1 < \cdots < t_n$ of elements of X,

$$f(s_1, \ldots, s_n) = f(t_1, \ldots, t_n) \iff \forall i \in S \ (s_i = t_i).$$

Note that if $S = \emptyset$, then f is constant on $[X]^n$, and if $S = \{1, \ldots, n\}$, then f is injective on $[X]^n$.

- A filter $\mathcal{U} \subseteq \mathbb{P}$ is $(\mathfrak{M}, \mathcal{A})$-*canonically Ramsey* if for every $f : [M]^n \to M$, where n is a standard natural number, with $f \in \mathcal{A}$, there is some $X \in \mathcal{U}$ on which f is canonical.

---

[8]A closely related notion of forcing, formulated by Gaifman, was employed in [AH, Section 1].

[9]This terminology is motivated by the fact (discussed in Section 3.2.2) that the formation of ultrapowers modulo iterable ultrafilters is amenable to iteration. Iterable ultrafilters are also referred to as *definable* ultrafilters, e.g., as in [Ki-2], motivated by their intimate link with the model theoretic notion of *definable type*.

- $\omega \rightarrow *(\omega)^n$ is the statement in the language of second order arithmetic which asserts that for every $f : [\omega]^n \rightarrow \omega$ there is an unbounded $X \subseteq \omega$ such that X is f-canonical.

Erdös and Rado [ER] proved that $\omega \rightarrow *(\omega)^n$ holds for all $n < \omega$. Their proof derives $\omega \rightarrow *(\omega)^n$ from $\omega \rightarrow (\omega)^{2n}$ and is readily formalizable[10] in ACA_0 for each fixed standard n, i.e.,

THEOREM 3.2. $\forall n \in \omega$, $ACA_0 \vdash \omega \rightarrow *(\omega)^n$.

REMARK 3.2.1. If ACA_0 is replaced by Z_2 (or just ACA_0 plus the full schema of induction) then "$\forall n \in \omega$" can be moved to the right hand side of the provability symbol $\vdash$. It is known that $ACA_0 \nvdash \forall n \in \omega$ $\omega \rightarrow (\omega)^n$. This follows from a theorem of Jockusch [Jo], which states that for each natural number $n \geq 2$ there is a recursive partition P_n of $[\omega]^n$ into two parts such that P_n has no infinite Σ_n-homogeneous subset[11].

The usual proof establishing the existence of filters meeting countably many dense sets shows:

PROPOSITION 3.3. There is a generic filter $\mathcal{U}$ over every countable model $(\mathfrak{M}, \mathcal{A})$.

The following result reveals the key properties of generic ultrafilters.

THEOREM 3.4. If $(\mathfrak{M}, \mathcal{A})$ is a model of ACA_0 and $\mathcal{U}$ is $(\mathfrak{M}, \mathcal{A})$-generic, then

(a) $\mathcal{U}$ is $(\mathfrak{M}, \mathcal{A})$-complete;
(b) $\mathcal{U}$ is $(\mathfrak{M}, \mathcal{A})$-iterable;
(c) $\mathcal{U}$ is $(\mathfrak{M}, \mathcal{A})$-canonically Ramsey.

PROOF. (a) Given $f \in \mathcal{A}$ with $f : M \rightarrow [0, a]^{\mathfrak{M}}$, let

$$\mathcal{D}_1^f := \{ Y \in \mathbb{P} : f \upharpoonright Y \text{ is constant} \}.$$

$\mathcal{D}_1^f$ is dense since for each $X \in \mathbb{P}$, $(\mathfrak{M}, X, f) \vDash PA(X, f)$.

(b) For X and Y in $\mathbb{P}$, let us write $X \subseteq_* Y$ (read: "X is almost contained in Y") if $X \backslash Y$ is bounded in $(M, <)$. Also, let "X decides Y" abbreviate

$$\text{"} X \subseteq_* Y \text{ or } X \subseteq_* M \backslash Y \text{"}.$$

Observe that to establish (b) it suffices to show that if $g : M \rightarrow \{0, 1\}$, with $g \in \mathcal{A}$, then

$$\mathcal{D}_2^g = \{ Y \in \mathbb{P} : \forall a \in M, \ Y \text{ decides } S_a^g \} \text{ is dense.}$$

To show the density of $\mathcal{D}_2^g$ suppose $X \in \mathbb{P}$. We first claim that there is an $\mathcal{A}$-coded sequence $F = \langle F_a : a \in M \rangle$ satisfying the following two properties:

(*) $\forall a \in M, F_a = S_a^g \cap X$ or $F_a = X \backslash S_a^g$;

---

[10]The text [GRS] includes a detailed proof of a special case of Theorem 3.2. See also [Ra] and [Mile] for more perspicuous proofs of the full result.

[11]See [W, p. 25] for more detail on this matter. Note that ACA_0 is referred to as PPA (predicative Peano arithmetic) in [W].

($**$) $\forall a \in M \bigcap_{b \leq a} F_b$ is unbounded in X.

Argue within $(\mathfrak{M}, \mathcal{A})$. For each $s : [0, a] \to \{0, 1\}$, define $\langle F_b^s : b \leq a \rangle$ by:

$$F_b^s := \begin{cases} S_b^g \cap X, & \text{if } s(b) = 1; \\ X \backslash S_b^g, & \text{if } s(b) = 0. \end{cases}$$

Consider the subtree τ of $(2^{<\omega})^{\mathfrak{M}}$ consisting of functions $s : [0, a] \to \{0, 1\}$ such that $\bigcap_{b \leq a} F_b^s$ is unbounded in X. It is easy to see that τ has nodes of every rank $b \in M$, because each level of τ gives rise to a partition of X into 2^b pieces, so one of the pieces must be unbounded since X itself is unbounded. By König's lemma, τ has a branch, which yields the desired sequence $\langle F_a : a \in M \rangle$.

We can now define $Y = \{ y_a : a \in M \} \in \mathbb{P}$ by induction within $(\mathfrak{M}, \mathcal{A})$ such that Y is almost contained in every F_a as follows:

- y_0 is the first element of F_0;
- y_{a+1} is the least member of $\bigcap_{b \leq a} F_b$ above $\{ y_b : b \leq a \}$.

It is clear that Y decides each S_a^g. Therefore $\mathcal{D}_2^g$ is dense.

(c) Suppose $f : [M]^n \to M$, where n is a standard natural number, and $f \in \mathcal{A}$. Let

$$\mathcal{D}_3^f := \{ Y \in \mathbb{P} : f \text{ is canonical on } Y \}.$$

By Theorem 3.2, $\mathcal{D}_3^f$ is dense. $\dashv$

REMARK 3.4.1. By a theorem of Kunen, a Rudin-Keisler minimal ultrafilter on $\mathcal{P}(\omega)$ is already a Ramsey ultrafilter [Jec, Lemma 38.1]. Moreover, the proof of the Erdös-Rado canonical partition theorem can be used to show that a Ramsey ultrafilter on $\mathcal{P}(\omega)$ is also canonically Ramsey. In the context of models of ACA_0, it is known that if $\mathcal{U}$ is 3-Ramsey[12] over $(\mathfrak{M}, \mathcal{A})$, then $\mathcal{U}$ is $(\mathfrak{M}, \mathcal{A})$-iterable and n-Ramsey for all $n \in \omega$ [Ki-2, Theorem 2.4]. Coupled with [Ki-2, Theorem 6.5] and the aforementioned Erdös-Rado proof, this shows that the following are equivalent for an ultrafilter $\mathcal{U}$ over a model $(\mathfrak{M}, \mathcal{A})$ of ACA_0:

(i) $\mathcal{U}$ is 3-Ramsey over $(\mathfrak{M}, \mathcal{A})$;

(ii) $\mathcal{U}$ is both iterable and canonically Ramsey over $(\mathfrak{M}, \mathcal{A})$;

(iii) $\mathcal{U}$ is a minimal end extension type over $(\mathfrak{M}, S)_{S \in \mathcal{A}}$ in the sense of Gaifman [G] (i.e., $\mathcal{U}$ is an iterable ultrafilter over $(\mathfrak{M}, \mathcal{A})$ and for every function $f \in \mathcal{A}$ with $f : M \to M$, f is one-to-one or constant on a member of $\mathcal{U}$).

It is also worth pointing out that the converse of Theorem 3.4 is false, i.e., "$\mathcal{U}$ is generic over $(\mathfrak{M}, \mathcal{A})$" is stronger than the above three conditions. This is a

---

[12]Here $\mathcal{U}$ is n-Ramsey over $(\mathfrak{M}, \mathcal{A})$ if for every $f : [M]^n \to \{0, 1\}$ with $f \in \mathcal{A}$, there is some $X \in \mathcal{U}$ on which f is homogeneous.

consequence of the fact that (a) generic ultrafilters are not first order definable in $(\mathfrak{M}, \mathcal{A})$, and (b) there is a Ramsey ultrafilter on $\mathcal{P}^L(\omega)$ (the powerset of ω in the sense of Gödel's constructible universe) that is first order definable within the model $(\omega, +, \cdot, \mathcal{P}^L(\omega))$. (a) follows from a standard forcing argument, and (b) can be established by coupling the fact that there is a well-ordering of $\mathcal{P}^L(\omega)$ that is definable in $(\omega, +, \cdot, \mathcal{P}^L(\omega))$ [Jec, Theorem 97] with the proof of the existence of a Ramsey ultrafilter assuming the continuum hypothesis [Jec, p. 478].

3.2.2. *Ultrapowers and iterations.* Gaifman [G] refined the MacDowell-Specker Theorem by showing that if $\mathcal{L}$ is a *countable*[13] language extending $\mathcal{L}_A$, $\mathfrak{M}$ is a model of $PA(\mathcal{L})$ of *any cardinality*, and $\mathcal{A}$ is the family of definable subsets of $\mathfrak{M}$, then there is an e.e.e. $\mathfrak{N}$ of $\mathfrak{M}$ such that $\mathcal{A} = SSy_M(\mathfrak{N})$. In the jargon of model theorists of arithmetic, this is rephrased as: if $\mathcal{L}$ is countable, then every model of $PA(\mathcal{L})$ has a *conservative* e.e.e. The first result of this section is an adaptation of Gaifman's result tailormade for our purposes.

LEMMA 3.5. *Suppose* $(\mathfrak{M}, \mathcal{A})$ *is a model of* ACA_0. *The following two conditions are equivalent*:

(a) *There exists a nonprincipal* $(\mathfrak{M}, \mathcal{A})$-*iterable ultrafilter* $\mathcal{U}$ *over* $(\mathfrak{M}, \mathcal{A})$.
(b) $(\mathfrak{M}, S)_{S \in \mathcal{A}}$ *has a proper e.e.e.* $(\mathfrak{N}, S^*)_{S \in \mathcal{A}}$ *such that* $\mathcal{A} = SSy_M(\mathfrak{N})$.

PROOF. To show $(a \Rightarrow b)$, let $(\mathfrak{N}, S^*)_{S \in \mathcal{A}}$ be the ultrapower of $(\mathfrak{M}, S)_{S \in \mathcal{A}}$ modulo $\mathcal{U}$, i.e., the universe N of $\mathfrak{N}$ consists of the $\mathcal{U}$-equivalence classes $[f]$ of functions f from M into M such that f is coded by some element of $\mathcal{A}$, and the operations on N are defined as in the classical theory of ultrapowers, e.g., $+^{\mathfrak{N}}$ is defined by

$$[f] +^{\mathfrak{N}} [g] = [h] \quad \text{iff} \quad \{m \in M : f(m) +^{\mathfrak{M}} g(m) = h(m)\} \in \mathcal{U}.$$

Similarly, for each $S \in \mathcal{A}$,

$$[f] \in S^* \quad \text{iff} \quad \{m \in M : f(m) \in S\} \in \mathcal{U}.$$

The Łoś Theorem for ultrapowers goes through in this limited context, thanks to the fact that every parametrically definable subset of $(\mathfrak{M}, S)_{S \in \mathcal{A}}$ has a $<^{\mathfrak{M}}$-least element (and therefore the model $(\mathfrak{M}, S)_{S \in \mathcal{A}}$ has definable Skolem functions). Consequently, if $\mathcal{U}$ is a non-principal ultrafilter, then $\mathfrak{N}$ is a proper elementary extension of $\mathfrak{M}$ (with the obvious identification of the $\mathcal{U}$-equivalence classes of constant maps with elements of M). It remains to verify (i) and (ii) below:

(i) $\mathfrak{M} \subseteq_e \mathfrak{N}$, and
(ii) $\mathcal{A} = SSy_M(\mathfrak{N})$.

---

[13] Mills [Mill] used a forcing construction to show that the countability assumption cannot be dropped from Gaifman's result.

To verify (i), suppose $\mathfrak{N} \vDash [f] \leq m$ for some $m \in M$. Then for some $X \in \mathcal{U}$, $\forall x \in X \; f(x) < m$. Let f^* be the function in $\mathcal{A}$ defined by $f^*(x) = f(x)$ if $x \in X$, and 0 otherwise. By $(\mathfrak{M}, \mathcal{A})$-completeness of $\mathcal{U}$, there is some $m_0 \leq m$ and some $Y \in \mathcal{U}$ such that $\forall x \in Y \; f^*(x) = m_0$. It is now easy to verify that $\mathfrak{N} \vDash [f] = [f^*] = m_0$. To establish (ii), first, note that for each $X \in \mathcal{A}$,

$$(\mathfrak{M}, X) \prec_e (\mathfrak{N}, X^*) \vDash PA(X^*).$$

This shows that $\mathcal{A} \subseteq SSy_M(\mathfrak{N})$ since if $d \in N \backslash M$, there is some $c \in N$ such that c precisely codes those elements of X^* which are less than d. Therefore, $X = c_E \cap M$. To see that $SSy_M(\mathfrak{N}) \subseteq \mathcal{A}$ we need to invoke the assumption of iterability of $\mathcal{U}$. Given an element $[f] \in N$, we wish to show

(1) $\{m \in M : \mathfrak{N} \vDash mE[f]\} \in \mathcal{A}$.

Observe that (1) is equivalent to

(2) $\{m \in M : \{n \in M : \mathfrak{M} \vDash mEf(n)\} \in \mathcal{U}\} \in \mathcal{A}$.

Let $X_m = \{n \in M : \mathfrak{M} \vDash mEf(n)\}$. By the iterability assumption,

(3) $\{m \in M : X_m \in \mathcal{U}\} \in \mathcal{A}$.

Therefore (1) holds. This completes the proof of (ii).

To show $(b \Rightarrow a)$, assume (b) holds and fix $c \in N \backslash M$. Consider $\mathcal{U}$ defined by

$$\mathcal{U} := \{S \in \mathcal{U} : c \in S^*\}.$$

The assumption that $(\mathfrak{M}, S)_{S \in \mathcal{A}}$ is elementarily end extended by $(\mathfrak{N}, S^*)_{S \in \mathcal{A}}$ can now be invoked to verify that $\mathcal{U}$ is $(\mathfrak{M}, \mathcal{A})$ complete, for if $f \in \mathcal{A}$, $a \in M$, and $(\mathfrak{M}, f) \vDash$ "$f : M \to [0, a]$", then $(\mathfrak{N}, f^*) \vDash$ "$f^* : N \to [0, a]$". Note that since $\mathfrak{N}$ is end extended by $\mathfrak{M}$, $f^*(c) \in M$. It is now easy to verify that

$$\{m \in M : f(m) = f^*(c)\}$$

is the desired member of $\mathcal{U}$ on which f is constant. Similarly, by invoking the assumption $\mathcal{A} = SSy_M(\mathfrak{N})$ we can show that $\mathcal{U}$ is also $(\mathfrak{M}, \mathcal{A})$-iterable, since if $(\mathfrak{M}, g) \vDash$ "$g : M \to \{0, 1\}$", where $g \in \mathcal{A}$, then $X_g := \{m \in M : c \in (S_m^g)^*\}$ is a member of $\mathcal{U}$, and therefore

$$\{m \in M : S_a^g \in \mathcal{U}\} = X_g \in \mathcal{A}. \qquad \dashv$$

For an $(\mathfrak{M}, \mathcal{A})$-iterable ultrafilter $\mathcal{U}$, the fact that the $\mathcal{U}$-based ultrapower does not introduce new subsets of $\mathfrak{M}$ allows one to *iterate the ultrapower formation any finite number of times* to obtain the n-fold iterations $Ult_{\mathcal{U},n}(\mathfrak{M}, S)_{S \in \mathcal{A}}$ for each positive natural number n. Indeed, a finite iteration of length n can be obtained in *one step* by defining an ultrafilter $\mathcal{U}^n$ on M^n. To do so, suppose $X \subseteq M^{n+1}$ is coded in $\mathcal{A}$. By definition[14],

(♣) $X \in \mathcal{U}^{n+1}$ iff $\{\alpha_1 : \{\langle \alpha_2, \ldots, \alpha_{n+1} \rangle : \langle \alpha_1, \alpha_2, \ldots, \alpha_{n+1} \rangle \in X\} \in \mathcal{U}^n\} \in \mathcal{U}$.

---

[14]The iterability condition is invoked to ensure that $\mathcal{U}^{n+1}$ is well-defined via (♣).

REMARK 3.6. It is easy to see that $\mathcal{U}^n$ concentrates on $[M]^n$. Moreover, if $\mathcal{U}$ is n-Ramsey over $(\mathfrak{M}, \mathcal{A})$ for some $n \in \omega$, then

$$\mathcal{U}^n = \{ Y \subseteq M^n : \exists X \in \mathcal{U} \, [X]^n \subseteq Y \}.$$

The process of ultrapower formation modulo $\mathcal{U}$ can be iterated *along any linear order* $\mathbb{L}$ to yield the iterated ultrapower $Ult_{\mathcal{U},\mathbb{L}}(\mathfrak{M}, S)_{S\in\mathcal{A}}$. To describe the isomorphism type of $Ult_{\mathcal{U},\mathbb{L}}(\mathfrak{M}, S)_{S\in\mathcal{A}}\,\mathcal{U}$ one can either use a direct limit construction (as originally formulated by Kunen [Ku], and often used in set theoretic literature) or, equivalently, one can take the following model theoretic route (as in Gaifman [G]). Given an iterable ultrafilter $\mathcal{U}$ we can define, for each positive natural number n, a complete n-type Γ_n over the model $(\mathfrak{M}, S)_{S\in\mathcal{A}}$ by defining $\Gamma_n(x_1, \ldots, x_n)$ as the set of formulas $\varphi(x_1, \ldots, x_n)$ such that

$$\{ \langle \alpha_1, \ldots, \alpha_n \rangle : (\mathfrak{M}, S)_{S\in\mathcal{A}} \vDash \varphi(\alpha_1, \ldots, \alpha_n) \} \in \mathcal{U}^n.$$

Here φ is a formula in the language $\mathcal{L} = \mathcal{L}_A \cup \{ S : S \in \mathcal{A} \}$ (since for each $m \in M$, $\{m\} \in \mathcal{A}$, for all intents and purposes $\mathcal{L}$ has constant symbols for elements of M as well). Then we augment the language $\mathcal{L}$ with a set of new constant symbols $\{ \bar{l} : l \in \mathbb{L} \}$, and define $T_{\mathcal{U},\mathbb{L}}$ to consist of formulas of the form $\varphi(\bar{l}_1, \bar{l}_2, \ldots, \bar{l}_n)$, where $\varphi(x_1, \ldots, x_n) \in \Gamma_n(x_1, \ldots, x_n)$ and $l_1 <_{\mathbb{L}} \cdots <_{\mathbb{L}} l_n$. Since $T_{\mathcal{U},\mathbb{L}}$ is a *complete Skolemized theory*, $Ult_{\mathcal{U},\mathbb{L}}(\mathfrak{M}, S)_{S\in\mathcal{A}}$ can be meaningfully defined as the *prime model* of $T_{\mathcal{U},\mathbb{L}}$.

The following theorem, due to Gaifman [G], summarizes the key properties of iterated ultrapowers[15].

THEOREM 3.7. *Suppose* $\mathcal{U}$ *is an* $(\mathfrak{M}, \mathcal{A})$-*iterable ultrafilter over a model* $(\mathfrak{M}, \mathcal{A})$ *of* ACA_0, *and* $\mathbb{L}$ *is a linearly ordered set. Let* $(\mathfrak{N}, S^*)_{S\in\mathcal{A}} := Ult_{\mathcal{U},\mathbb{L}}(\mathfrak{M}, S)_{S\in\mathcal{A}}$, *and* $c_l := (\bar{l})^{\mathfrak{N}}$.

(a) *Elements of* N *are of the form* $f^*(c_{l_1}, \ldots, c_{l_n})$, *where* $f \in \mathcal{A}$, *and* $l_1 <_{\mathbb{L}} \cdots <_{\mathbb{L}} l_n$;

(b) *For every* $\mathcal{L}$-*formula* $\varphi(x_1, \ldots, x_n)$, *and every increasing sequence* $l_1 <_{\mathbb{L}} \cdots <_{\mathbb{L}} l_n$

$$\mathfrak{N} \vDash \varphi(c_{l_1}, \ldots, c_{l_n}) \quad iff \quad \{ \langle \alpha_1, \ldots, \alpha_n \rangle \in M^n : \mathfrak{M} \vDash \varphi(\alpha_1, \ldots, \alpha_n) \} \in \mathcal{U}^n;$$

(c) $\{ c_l : l \in \mathbb{L} \}$ *is a set of order indiscernibles in* $(\mathfrak{N}, S^*)_{S\in\mathcal{A}}$;

(d) *Every automorphism* h *of* $\mathbb{L}$ *induces an automorphism*

$$j_h : (\mathfrak{N}, S^*)_{S\in\mathcal{A}} \longrightarrow (\mathfrak{N}, S^*)_{S\in\mathcal{A}}$$

defined by

$$j_h(f^*(c_{l_1}, \ldots, c_{l_n})) = f^*(c_{h(l_1)}, \ldots, c_{h(l_n)}).$$

---

[15]The analogue of this result for models of set theory with a weakly compact cardinal is due to Kunen [Ku], and fully developed in [Jec] and [Kan].

- If U is also canonically Ramsey, then Theorem 3.7(d) can be strengthened as follows:

THEOREM 3.8. *Suppose* $(\mathfrak{M}, A) \models ACA_0$, *and let* h *is an automorphism of a linearly ordered set* $\mathbb{L}$ *with no fixed points. If* U *is iterable and canonically Ramsey over* $(\mathfrak{M}, A)$, *then the fixed point set of the automorphism* j_h *of* $Ult_{U,\mathbb{L}}(\mathfrak{M}, S)_{S \in A}$ *is precisely* M.

PROOF. Clearly j_h fixes each $a \in M$ since the constant map $f_a(x) = a$ is in A. To see that j_h fixes no member of $N \backslash M$, suppose that

(1) $f^*(c_{h(l_1)}, \ldots, c_{h(l_n)}) = f^*(c_{l_1}, \ldots, c_{l_n})$

for some $f^*(c_{l_1}, \ldots, c_{l_n}) \in N$. Since $f \in A$, by Theorem 3.4(c) there is some $X \in U$, and some $S \subseteq \{1, \ldots, n\}$ such that for all sequences $a_1 < \cdots < a_n$, and $b_1 < \cdots < b_n$ of elements of X,

(2) $f(a_1, \ldots, a_n) = f(b_1, \ldots, b_n) \Leftrightarrow \forall i \in S \ (a_i = b_i)$.

Moreover, since $X^n \in U^n$,

(3) $Ult_{U,\mathbb{L}}(\mathfrak{M}, S)_{S \in A} \models \langle c_{l_1}, \ldots, c_{l_n} \rangle \in X^n$.

(1), (2), and (3) together imply that $S = \emptyset$, which in turn implies that f must be *constant* on X. Therefore, $f^*(c_{l_1}, \ldots, c_{l_n}) \in M$. $\dashv$

PROOF OF THEOREM B. Let $(\mathfrak{M}, A)$ be a model of ACA_0. Fix some $(\mathfrak{M}, A)$-generic ultrafilter U and let

$$(\mathfrak{N}, S^*)_{S \in A} := Ult_{U,\mathbb{L}}(\mathfrak{M}, S)_{S \in A},$$

where $\mathbb{Z}$ is the ordered set of integers. Consider the automorphism

$$n \longmapsto_h n + 1$$

of $\mathbb{Z}$. By Theorems 3.4 and 3.8 j_h is an automorphism of $(\mathfrak{N}, S^*)_{S \in A}$ whose fixed point set is precisely M. $\dashv$

3.3. An arithmetical theory with a built-in automorphism. Consider the theory VA formulated in $\mathcal{L}_A \cup \{j\}$, where j is a unary function symbol, obtained by augmenting the axioms of $I\Delta_0$ with a single axiom expressing

   "j is a nontrivial $\{+, \cdot\}$-automorphism whose fixed-point set is closed downwards".

This theory was formulated by Albert Visser who noted that Corollary 3.1 implies that PA can be interpreted in VA. This led Visser to ask:

- **Visser's Question:** what is the *interpretability*[16] *relationship* between ACA_0 and VA?

---

[16]See Sections 1 and 2 of Visser's paper [V] in this volume for the precise definition of interpretability. The intuitive idea can be explained as follows: a theory T_1 formulated in a language $\mathcal{L}_1$, is interpretable in a theory T_2 formulated in a language $\mathcal{L}_2$, if there is a "well-behaved" function δ, which translates formulae ψ from $\mathcal{L}_1$ into formulae δ_ψ in $\mathcal{L}_2$ such that for all sentences ψ of $\mathcal{L}_1$,

$$T_1 \vdash \psi \text{ implies } T_2 \vdash \delta_\psi.$$

In this section we partially answer Visser's question by establishing that ACA_0 can be faithfully interpreted within VA. Since the proofs of Theorem 1.1(b) and Theorem A are both formalizable within ACA_0, and ACA_0 is a conservative extension of PA for arithmetical sentences, the statement "VA is equiconsistent with PA" is provable within PA (see Remark 3.9.3 for a refinement). As we shall see, an analysis of the proof of Theorem A yields an interpretation δ of ACA_0 within VA, and Theorem B will show that δ is indeed a faithful interpretation.

THEOREM 3.9. *There is a faithful interpretation δ of ACA_0 within VA.*

PROOF. Suppose $(\mathfrak{N}, j)$ is a model of VA. Let M be the fixed point set of j, and $\mathcal{A} := SSy_M(\mathfrak{N})$. By Theorem A, all axioms of PA are true in $\mathfrak{M}$. This can be syntactically reformulated by saying that if for each formula φ of $\mathcal{L}_A$, φ^M is the formula in $\mathcal{L}_A \cup \{j\}$ obtained by restricting all the quantifiers of φ to the $(\mathfrak{N}, j)$-definable cut M, then by Theorem 1.1(b), Theorem A, and the completeness theorem for first order logic:

$$\text{For all sentences } \varphi \text{ of } L_A, \ PA \vdash \varphi \text{ iff } VA \vdash \varphi^M.$$

This shows that the map $\varphi \mapsto \varphi^M$ describes a *faithful* interpretation of PA within VA. In order to interpret ACA_0 within $(\mathfrak{N}, j)$ define an equivalence relation $\equiv$ by

$$a \equiv b \quad \text{iff} \quad \forall x \forall y (M(x) \wedge M(y) \rightarrow (E(x, a) \leftrightarrow E(x, b))),$$

where $M(x)$ is the formula "$x = j(x)$" and $E(x, y)$ is Ackermann's $\in$. Note that

$$\left[(\mathfrak{N}, j) \models a \equiv b \right] \quad \text{iff} \quad \left[a_E \cap M = b_E \cap M \right],$$

which shows that $\equiv$ interprets the equality relation among sets. Therefore, we can interpret the two-sorted model $(\mathfrak{M}, \mathcal{A}, \in, =_A)$ within $(\mathfrak{N}, j)$ by interpreting $\mathfrak{M}$ via $I(x)$, $\mathcal{A}$ via $N/\equiv$, and the membership relation $\in$ (between members of M, and members of $\mathcal{A}$), via $E(x, y)$. So, by Theorem A, ACA_0 is uniformly interpretable in every model of VA. In syntactical terms, this idea can be used to show:

PROPOSITION 3.9.1. *For every formula $\psi(v_1, \ldots, v_s, X_1, \ldots, X_t)$ in the language of second order arithmetic, whose first order free variables are $v_1, \ldots, v_s$, and whose second order free variables are $X_1, \ldots, X_t$, there is a formula*

$$\delta_\psi(x_1, \ldots, x_s, x_{s+1}, \ldots, x_{s+t})$$

in the language $\mathcal{L}_A \cup \{j\}$ such that the following are equivalent for all models $(\mathfrak{N}, j)$ of VA, all sequences $a_1, \ldots, a_s$ from M, $b_1, \ldots, b_t$ from N, and $S_1, \ldots, S_t$ from $SSy_M(\mathfrak{N})$ such that $S_i = (b_i)_E \cap M$ for $1 \le i \le t$, where M is the fixed

---

If, in additon, the converse of the above implication holds for all sentences ψ of $\mathcal{L}_1$, δ is said to be a *faithful* interpretation.

point set of j:

(i) $(\mathfrak{M}, SSy_M(\mathfrak{N})) \models \psi(a_1/v_1, \ldots, a_s/v_s, S_1/X_1, \ldots, S_t/X_t)$.

(ii) $(\mathfrak{N}, j) \models \delta_\psi(a_1/x_1, \ldots, a_s/x_s, b_1/x_{s+1}, \ldots, b_t/x_{s+t})$.

We can now use Theorem A, Theorem B, Proposition 3.9.1, and the completeness theorem of first order logic together to conclude that for all sentences ψ of second order arithmetic, $ACA_0 \vdash \psi$ iff $VA \vdash \delta_\psi$. Therefore, ACA_0 is *faithfully* interpretable in VA via the interpretation δ. ⊣

COROLLARY 3.9.2. VA *has superexponential speed-up over* PA (*assuming the consistency of* PA). *More specifically, for every natural number k there is a theorem φ_k of PA whose interpretation has a proof of length d_k within VA such that the shortest proof of φ_k within PA is longer than* $\mathrm{Superexp}(k, d_k)$.

PROOF. This is a direct consequence of interpretability of ACA_0 within VA and the independently obtained results of Friedman and Pudlák on the speed-up of ACA_0 over PA. More specifically, let us write $T \vdash_{\leq k} \psi$ for "there is a proof of φ from T of length k", and $T \vdash_{>k} \psi$ for "$T \vdash \psi$ and all proofs of φ from T are longer than k". Given a sentence φ in the language of Peano arithmetic, let $\overline{\varphi}$ be the canonical interpretation of φ within ACA_0. As shown by Friedman ([Fr], [Sm]) and Pudlák[17] [Pu-2, Corollary 4.5]:

(1) There is a sequence $\langle \varphi_k : k \in \omega \rangle$ of theorems of PA and an increasing sequence $\langle d_k : k \in \omega \rangle$ of natural numbers such that for all $k \in \omega$:

$$ACA_0 \vdash_{\leq d_k} \overline{\varphi_k}, \text{ but } PA \vdash_{>\mathrm{Superexp}(k, d_k)} \varphi_k.$$

On the other hand, ACA_0 is finitely axiomatizable[18] and therefore there is a single theorem τ of ACA_0 with the same set of consequences as ACA_0 itself. Since VA interprets ACA_0 via δ of Theorem 3.9, $VA \vdash_{\leq c} \delta_\tau$ for some c. Therefore, for all sentences ψ in the language of second order arithmetic,

(2) $\tau \vdash_{\leq k} \psi \to VA \vdash_{\leq c+k} \delta_\psi$.

This is easy to see: if $\langle \varphi_n : 1 \leq n \leq k \rangle$ is a Hilbert-style proof of ψ from τ (so $\varphi_k = \psi$), then we can obtain a proof of δ_ψ from VA of length $k + c$ by first proving δ_τ in c-steps from VA, and then following the resulting proof with $\langle \delta_{\varphi_n} : 1 \leq n \leq k \rangle$. The result now easily follows from coupling (1) and (2). ⊣

REMARK 3.9.3. The proof of Theorem 3.9 can be used to show that $I\Delta_0 + \mathrm{Exp}$ proves $\mathrm{Con}(VA) \to \mathrm{Con}(ACA_0)$, and therefore

$$I\Delta_0 + \mathrm{Exp} \vdash \mathrm{Con}(VA) \longrightarrow \mathrm{Con}(PA).$$

---

[17]The exposition in [Pu-2] is geared toward the speed-up of GB (Gödel-Bernays theory of classes) over ZF. It is well-known that the same machinery can be used to show the speed-up of ACA_0 over PA.

[18]See [HP, Chapter III, Section 1(b)] or [Si-2, Lemma VIII.1.5].

Coupled with $I\Delta_0 + \mathrm{Exp} \nvdash \mathrm{Con}(PA) \to \mathrm{Con}(ACA_0)$ ([Pu-2], [Fr]), this shows that

$$I\Delta_0 + \mathrm{Exp} \nvdash \mathrm{Con}(PA) \longrightarrow \mathrm{Con}(VA).$$

§4. **Automorphisms and second order arithmetic.** In the previous section we saw that there is a close relationship between models of PA and ACA_0 and fixed point sets of automorphisms of models $\mathfrak{N}$ of $I\Delta_0$. In this section we pursue this theme by investigating a minimal condition (M-amenability) under which the fixed point sets of automorphisms of bounded arithmetic give rise to models of *full second order arithmetic* Z_2.

4.1. Amenable automorphisms from Z_2. The following definition is suggested by the work of Solovay on automorphisms of models of set theory with a weakly compact cardinal [Sol, Section 3.5, Criteria 1 and 2].

- Suppose $\mathfrak{N}$ is a model of $I\Delta_0$, and M is a cut of $\mathfrak{N}$. An automorphism j of $\mathfrak{N}$ is M-*amenable* if the fixed point set of j is precisely M, and for every formula $\varphi(x, j)$ in the language $\mathcal{L}_A \cup \{j\}$, possibly with suppressed parameters from N,

$$\{m \in M : (\mathfrak{N}, j) \vDash \varphi(m, j)\} \in SSy_M(\mathfrak{N}).$$

THEOREM C. *Suppose $(\mathfrak{M}, \mathcal{A})$ is a countable model of $Z_2 + \Pi^1_\infty\text{-}DC$. There exists an e.e.e. $\mathfrak{N}$ of $\mathfrak{M}$ that has an M-amenable automorphism j such that $SSy_M(\mathfrak{N}) = \mathcal{A}$.*

PROOF. Before beginning the proof, recall that $\Pi^1_\infty\text{-}DC$ is the scheme in the language of second order arithmetic consisting of formulas of the form

$$\forall n \forall X \exists Y \theta(n, X, Y) \longrightarrow \left[\forall X \exists Z \left(X = (Z)_0 \text{ and } \forall n\, \theta\left(n, (Z)_n, (Z)_{n+1}\right) \right) \right],$$

where φ is allowed to have number or set parameters, and $(Z)_n = \{i : \Gamma(i, n) \in Z\}$, where Γ is a canonical pairing function. See [Si-2, Section VII.6] for more on choice schemes in second order arithmetic[19].

The proof of Theorem C has two distinct stages. In the first stage, a well-behaved Ramsey ultrafilter $\mathcal{U}$ is constructed by forcing, while in the second stage, an internal iterated ultrapower modulo $\mathcal{U}$ is used to exhibit the desired model $\mathfrak{N}$ and the M-amenable automorphisms j of $\mathfrak{N}$.

Stage 1: Forcing a Ramsey ultrafilter. Forcing was used only as an efficient bookkeeping tool in Section 3.2. In contrast, here it is invoked in an essential manner to adjoin a generic ultrafilter to a model of second order arithmetic[20].

---

[19]N.B. the formulation of DC in [Si-2] is slightly different from the above, but equivalent.

[20]I am indebted to one of the referees for suggesting the self-contained approach for this stage. In the original proof of Theorem C, I used a forcing construction of Mostowski [Mo-1] to adjoin a global well-ordering $\vartriangleleft$ of $\mathcal{A}$ so that the comprehension scheme of Z_2 continues to hold even for formulas mentioning $\vartriangleleft$. It is then routine to define a Ramsey ultrafilter within $(\mathfrak{M}, \mathcal{A}, \vartriangleleft)$ by implementing the classical proof of the existence of a Ramsey ultrafilter using CH. Note that in his original paper [Mo-1], Mostowski claimed that his forcing construction works for countable

Our notion of forcing $\mathbb{P}$ (and therefore our notion of genericity) is the same as the one used already in Section 3.2, but in this section we shall invoke substantive properties of forcing to show that $\mathbb{P}$-forcing over a countable model $(\mathfrak{M}, \mathcal{A})$ of second order arithmetic with dependent choice produces a generic ultrafilter $\mathcal{U}$ such that the expansion $(\mathfrak{M}, \mathcal{A}, \mathcal{U})$ continues to satisfy the comprehension schema in the language of second order arithmetic for formulae that refer to $\mathcal{U}$. To verify this, we begin with some definitions.

- Let $\mathcal{L}_2(U)$ be the result of augmenting the language of second order arithmetic $\mathcal{L}_2$ with a new predicate U with the understanding that U is a *predicate of sets*, i.e., models of $\mathcal{L}_2(U)$ are of the form $(\mathfrak{M}, \mathcal{A}, \mathcal{U})$ where $(\mathfrak{M}, \mathcal{A})$ is an $\mathcal{L}_2$-structure, and $\mathcal{U} \subseteq \mathcal{A}$.
- The *forcing language* Φ is obtained by augmenting $\mathcal{L}_2(U)$ with constant symbols for each element of $M \cup \mathcal{A}$.
- Recall from Section 3.2 that $\mathbb{P}$ is $\{X \in \mathcal{A} : X \text{ is unbounded in } M\}$, ordered under inclusion. The forcing relation is inductively defined as follows:
  (1) $X \Vdash (Y \in U)$ iff $X \subseteq Y$ (where $Y \in \mathcal{A}$); for all other atomic formulae φ, $X \Vdash \varphi$ iff φ holds in $(\mathfrak{M}, \mathcal{A})$.
  (2) $X \Vdash (\varphi_1 \vee \varphi_2)$ iff $X \Vdash \varphi_1$ or $X \Vdash \varphi_2$.
  (3) $X \Vdash (\neg\varphi)$ iff $\forall Y \subseteq X(Y \nVdash \varphi)$.
  (4) $X \Vdash (\exists x\varphi(x))$ iff for some $m \in M$ such that $X \Vdash \varphi(m)$.

The following lemma is standard and is stated without proof. Note that it holds for all $\mathcal{L}_2$-structures $(\mathfrak{M}, \mathcal{A})$.

LEMMA C.1. (1) (Monotonicity) *If* $X \Vdash \varphi$ *and* $Y \subseteq X$, *then* $Y \Vdash \varphi$.

(2) (Definability) *For every formula* $\varphi(v_1, \ldots, v_s, X_1, \ldots, X_t)$ *of* $\mathcal{L}_2(U)$, *there is a formula* $\text{Force}_\varphi(X, v_1, \ldots, v_s, X_1, \ldots, X_t)$ *of* $\mathcal{L}_2(U)$ *such that for every model* $(\mathfrak{M}, \mathcal{A})$ *of* $\mathcal{L}_2$, *every* $X \in \mathbb{P}$, *every* $m_1, \ldots, m_s \in M$, *and every*

$$S_1, \ldots, S_t \in \mathcal{A}, \ X \Vdash \varphi(m_1, \ldots, m_s, S_1, \ldots, S_t)$$
$$\text{iff} \ (\mathfrak{M}, \mathcal{A}) \vDash \text{Force}_\varphi(X, m_1, \ldots, m_s, S_1, \ldots, S_t).$$

(3) (Truth-and-Forcing) *If* $\mathcal{U}$ *is* $\mathbb{P}$-*generic over* $(\mathfrak{M}, \mathcal{A})$, *then for every* Φ-*sentence* φ, $(\mathfrak{M}, \mathcal{A}, \mathcal{U}) \vDash \varphi$ *iff* $X \Vdash \varphi$ *for some* $X \in \mathcal{U}$.

LEMMA C.2. *Suppose* X *and* Y *are elements of* $\mathbb{P}$ *whose symmetric difference* $X \Delta Y$ *is finite in the sense of* $\mathfrak{M}$. *For any sentence* φ *of* Φ, $X \Vdash \varphi$ *iff* $Y \Vdash \varphi$.

PROOF. Recall from Theorem 3.4(a) that generic filters are $(\mathfrak{M}, \mathcal{A})$-complete. Also note that for any $(\mathfrak{M}, \mathcal{A})$-complete ultrafilter $\mathcal{U}$, $X \in \mathcal{U}$ iff $Y \in \mathcal{U}$. The result now easily follows from Truth-and-Forcing. ⊣

The next two results unveil the key properties of generic ultrafilters. From here on, we use the abbreviation $X \parallel \varphi$ for "$X \Vdash \varphi$ or $X \Vdash \neg\varphi$".

models of Z_2 with the *choice scheme*. However, as observed by Simpson [Si-1], Mostowski's proof relies on the stronger scheme of *dependent choice*. This is acknowledged in [Mo-2].

LEMMA C.3. *If $(\mathfrak{M}, \mathcal{A})$ is a model of $Z_2 + \Pi^1_\infty$-DC, then for any unary formula $\varphi(x)$ of Φ, the following set D_φ is dense in $\mathbb{P}$*

$$D_\varphi := \{Y \in \mathbb{P} : \forall m \in M \ (Y \parallel \varphi(m))\}.$$

PROOF. Let $\theta(X, Y, n)$ be the formula "$X \supseteq Y$ and $Y \parallel \varphi(n)$". It is easy to see that

$$(\mathfrak{M}, \mathcal{A}) \vDash \forall n \forall X \exists Y \theta(n, X, Y).$$

Given any $X \in \mathbb{P}$, by the dependent choice scheme there is some element of $\mathcal{A}$ that codes a sequence $\langle X_0, X_1, X_2, \ldots, X_m, \ldots \rangle_{m \in M}$ of elements of $\mathbb{P}$ such that (1) and (2) below hold in $(\mathfrak{M}, \mathcal{A})$.

  (1) $X_0 := X$ and $\forall m \in M \ X_{m+1} \subseteq X_m$;
  (2) $\forall n \theta(n, X_n, X_{n+1})$.

Next, construct Y by setting $Y := \{y_m : m \in M\} \in \mathbb{P}$, where y_m is defined within $(\mathfrak{M}, \mathcal{A})$ via the recursion:

  • y_0 is the first element of X_0;
  • y_{m+1} is the least member of X_m above $\{y_i : i \leq m\}$.

Clearly $Y \subseteq X$ and $Y \backslash X_m$ is $\mathfrak{M}$-finite for all $m \in M$. Therefore, since by Monotonicity, $Y \cap X_m \parallel \varphi(m)$ for all $m \in M$, by Lemma C.2, $Y \in D_\varphi$.   ⊣

LEMMA C.4. *If $(\mathfrak{M}, \mathcal{A})$ is a model of $Z_2 + \Pi^1_\infty$-DC and $\mathcal{U}$ is $\mathbb{P}$-generic over $(\mathfrak{M}, \mathcal{A})$, then for any unary Φ-formula $\varphi(x)$,*

$$S_\varphi := \{m \in M : (\mathfrak{M}, \mathcal{A}, \mathcal{U}) \vDash \varphi(m)\} \in \mathcal{A}.$$

PROOF. By Lemma C.3 there is a condition $Y_0 \in \mathcal{U}$ such that for all $m \in M$, $Y_0 \parallel \varphi(m)$. It is routine to verify (using Truth-and-Forcing) that

$$\{m \in M : (\mathfrak{M}, \mathcal{A}, \mathcal{U}) \vDash \varphi(m)\} = \{m \in M : (\mathfrak{M}, \mathcal{A}, \mathcal{U}) \vDash \text{``}Y_0 \Vdash \varphi(m)\text{''}\}.$$

Therefore S_φ is the solution set of a unary formula $\mathcal{L}_2$-formula (by definability of the forcing relation), and therefore by the comprehension scheme, $S_\varphi \in \mathcal{A}$.   ⊣

Stage 2: Internally building an iterated ultrapower. In this stage of the proof, we employ the machinery of iterated ultrapowers discussed in Section 3.2.2, except that the entire construction is carried out internally within $(\mathfrak{M}, \mathcal{A}, \mathcal{U})$. To see how this works, consider a generic ultrafilter $\mathcal{U}$ over $(\mathfrak{M}, \mathcal{A})$. By Theorem 3.4, $\mathcal{U}$ is $(M, \mathcal{A})$-iterable. Moreover, in light of Remark 3.2.1 it is easy to see that $\mathcal{U}$ is also m-canonically Ramsey[21] over $(M, \mathcal{A})$ for all $m \in M$. Since the construction of the m-type Γ_m uses the ultrafilter $\mathcal{U}^m$, and m might be nonstandard, we need to overcome the following obstacle: $\mathcal{U}^n$ was defined by an *external* induction in Section 3.2.2 via equation (♣) for *standard natural numbers n*. Therefore, to define $\mathcal{U}^m$ for nonstandard m, we seem need to work

---

[21]$\mathcal{U}$ is m-canonically Ramsey over $(M, \mathcal{A})$, where $m \in M$, if for every $f : [M]^m \to M$ with $f \in \mathcal{A}$, there is some $X \in \mathcal{U}$ that is f-canonical.

within *third order* arithmetic in order to carry out the necessary recursion. However, in light of Remark 3.6, there is a way out: since $\mathcal{U}$ is m-Ramsey over $(\mathfrak{M}, \mathcal{A})$, we can use the following recursion-free definition of $\mathcal{U}^m$ within $(\mathfrak{M}, \mathcal{A}, \mathcal{U})$:

$$\mathcal{U}^m := \left\{ Y \subseteq M^m : \exists X \in \mathcal{U} \, [X]^m \subseteq Y \right\}.$$

Therefore for any linear order $\mathbb{L} \in \mathcal{A}$ we can define the *internally* iterated ultrapower $Ult^*_{\mathcal{U}, \mathbb{L}}(\mathfrak{M}, S)_{S \in \mathcal{A}}$ by carrying out the construction of Section 3.3 entirely within $(\mathfrak{M}, \mathcal{A}, \mathcal{U})$. Note that the key difference between the internal and the external iterated ultrapower is that the external iterated ultrapower can be viewed as a direct limit of models that result from iterating the ultrapower formation process finitely many times, while the internal iteration can be viewed as a direct limit of models that result from iterating the ultrapower formation process $\mathfrak{M}$-*finitely* many times. We can therefore choose $\mathbb{L} \in \mathcal{A}$ such that $\mathbb{L}$ has an automorphism $h \in \mathcal{A}$ with no fixed points (e.g., $\mathbb{L} =$ the ordered set of integers in the sense of $\mathfrak{M}$, and $h(n) = n+1$). By minor variants of Theorems 3.7 and 3.8, there is an automorphism

$$j_h^* : Ult^*_{\mathcal{U}, \mathbb{L}}(\mathfrak{M}, S)_{S \in \mathcal{A}} \longrightarrow Ult^*_{\mathcal{U}, \mathbb{L}}(\mathfrak{M}, S)_{S \in \mathcal{A}}$$

that is definable within $(\mathfrak{M}, \mathcal{A}, \mathcal{U})$, and whose fixed point set is precisely M. Since j_h^* is outright definable in $(\mathfrak{M}, \mathcal{A}, \mathcal{U})$, by Lemma C.4 j is M-amenable. This concludes the proof of Theorem C.   ⊣

4.2. Z_2 from amenable automorphisms. We now show that the full strength of second order arithmetic is needed in the proof of Theorem C.

THEOREM D. *If $\mathfrak{M} \vDash I\Delta_0$ and $\mathfrak{N}$ is an end extension of $\mathfrak{M}$ satisfying $I\Delta_0$ such that $\mathfrak{N}$ has an M-amenable automorphism, then $(\mathfrak{M}, SSy_M(\mathfrak{N})) \vDash Z_2$.*

PROOF. By Theorem A, $(\mathfrak{M}, SSy_M(\mathfrak{N})) \vDash ACA_0$. Therefore, we only need to verify the comprehension scheme of Z_2. Recall the mapping $\psi \mapsto \delta_\psi$ of formulas of second order arithmetic to formulas of $\mathcal{L}_A \cup \{j\}$ of Proposition 3.9.1 in the proof of Theorem 3.9. If $\psi(x)$ is a unary formula of second order arithmetic (possibly with suppressed set or number parameters), then by Proposition 3.9.1

$$\{a \in M : (\mathfrak{M}, SSy_M(\mathfrak{N})) \vDash \psi(a)\} = \{a \in M : (\mathfrak{N}, j) \vDash \delta_\psi(a)\}.$$

Coupling this with the M-amenability of j, it now becomes evident that $(\mathfrak{M}, SSy_M(\mathfrak{N}))$ satisfies the comprehension scheme.   ⊣

Let T^* be the extension of the theory T of Section 3.3 obtained by adding a *scheme* asserting that j is an M-amenable automorphism (where M as usual is the fixed point set of j). The proof of Theorem C, coupled with the well-known fact that the theory $Z_2 + \Pi^1_\infty\text{-}DC$ can be interpreted within Z_2 via the "ramified analytical hierarchy" [Si-2] shows that T^* can be interpreted within Z_2. Furthermore, Proposition 3.9.1 and Theorem D together show that Z_2 is interpretable within T^*. Hence:

THEOREM 4.1. *The theories Z_2 and T^* are equiconsistent.*

§5. Further results and open questions.

5.1. Consequences for *NFU*. As mentioned in the introduction, the main results of this paper were obtained by the author in the context of the meta-mathematical study of certain extensions of the theory *NFU*, where *NFU* is Jensen's variant [Jen] of Quine's system of set theory *New Foundations NF* [Q]. *NFU* is obtained from *NF* by relaxing the extensionality axiom in order to allow urelements. The consistency of *NF* relative to any *ZF*-style set theory remains an open problem, but Jensen showed the consistency of *NFU* relative to a fragment of *ZF*-set theory. Theorems A, B, C, and D have been used in the joint work of Robert Solovay and the author to establish the results reported in this section. Here we only briefly define the concepts needed to state our results, and refer the reader to [Fo] or [Ho-1] for detailed background information and references.

- X is *Cantorian* if there is a one-to-one correspondence between X and the set of its singletons $\{\{v\} : v \in X\}$;
- X is *strongly Cantorian* if the map sending v to $\{v\}$ (as v varies in X) exists;
- $NFU^{-\infty}$ is *NFU* plus the axiom "every set is finite";
- $NFUA^{-\infty}$ is $NFU^{-\infty}$ plus the axiom "every Cantorian set is strongly Cantorian"; and
- $NFUB^{-\infty}$ is the extension of $NFUA^{-\infty}$ obtained by adding a scheme asserting that the intersection of any parametrically definable class with the class of Cantorian sets is the result of the intersection of the extension of some element with the class of Cantorian sets.

Of course, in *ZF*-style set theories every set is strongly Cantorian, but in *NF* and *NFU* this is no longer true, e.g., the universal set of a model of *NF* or *NFU* is not even Cantorian, and there are models of *NFU* + "there is an infinite set" + the axiom of choice, in which the set of finite cardinals is Cantorian, but not strongly Cantorian. We are now ready to state the ramifications of Theorems A and B for *NFU*:

THEOREM 5.1. *The following are equivalent for complete theories T in the language $\mathcal{L}_A$ of arithmetic:*

(a) *There is a model of $NFUA^{-\infty}$ whose class of Cantorian cardinals satisfies T.*

(b) *T is an extension of PA.*

COROLLARY 5.1.1. *$NFUA^{-\infty}$ is equiconsistent with PA.*

Furthermore, Theorem 4.1 can be used to show:

THEOREM 5.2. $NFUB^{-\infty}$ is equiconsistent with Z_2.

5.2. A Characterization of $I\Delta_0 + B\Sigma_1 + \text{Exp}$. In recent work [E-2], the author has established the following characterization of the fragment $I\Delta_0 + B\Sigma_1 + \text{Exp}$ of PA in terms of automorphisms. Here $B\Sigma_1$ is the scheme consisting of the universal closure of formulae of the form

$$\big[\forall x < a \; \exists y \; \varphi(x, y)\big] \longrightarrow \big[\exists z \forall x < a \; \exists y < z \; \varphi(x, y)\big].$$

In what follows $I_{fix}(j)$ denotes the largest initial segment of a model $\mathfrak{N}$ of $I\Delta_0$ that is pointwise fixed under an automorphism j of $\mathfrak{N}$.

THEOREM 5.3. (a) *Suppose $\mathfrak{M}$ is a countable model of $I\Delta_0 + B\Sigma_1 + \text{Exp}$. $\mathfrak{M}$ has a proper end extension to a model $\mathfrak{N}$ of $I\Delta_0$ such that for some automorphism j of $\mathfrak{N}$, $I_{fix}(j) = M$.*

(b) *If j is a nontrivial automorphism of some model $\mathfrak{N}$ of $I\Delta_0$, then $I_{fix}(j)$ is a model of $I\Delta_0 + B\Sigma_1 + \text{Exp}$.*

COROLLARY 5.3.1. *$I\Delta_0 + B\Sigma_1 + \text{Exp}$ is the theory of the class of models whose universes are of the form $I_{fix}(j)$ for some nontrivial automorphism j of a model of $I\Delta_0$.*

5.3. Open questions.

- QUESTION 1. Let VA be the theory discussed in Section 3.3. Can VA be interpreted in ACA_0?
- QUESTION 2. Can Theorem D be strengthened by including the clause "$(\mathfrak{M}, SSy_M(\mathfrak{N}))$ satisfies $\Pi^1_\infty\text{-}DC$" in the conclusion?
- QUESTION 3. Besides PA, Z_2, and $I\Delta_0 + B\Sigma_1 + \text{Exp}$, are there other arithmetical theories that can be naturally characterized in terms of automorphisms?

REFERENCES

[AH] F. ABRAMSON and L. HARRINGTON, *Models without indiscernibles*, **The Journal of Symbolic Logic**, vol. 43 (1978), no. 3, pp. 572–600.

[Av] J. AVIGAD, *Number theory and elementary arithmetic*, **Philosophia Mathematica**, vol. 11 (2003), no. 3, pp. 257–284.

[Be] J. H. BENNETT, *On Spectra*, Ph.D. dissertation, Princeton University, 1962.

[Bu] S. BUSS, *First-order proof theory of arithmetic*, **Handbook of Proof Theory** (S. Buss, editor), North-Holland, Amsterdam, 1998, pp. 79–147.

[CK] C. C. CHANG and H. J. KEISLER, *Model Theory*, North-Holland, Amsterdam, 1973.

[D] P. D'AQUINO, *A sharpened version of McAloon's theorem on initial segments of models of $I\Delta_0$*, **Annals of Pure and Applied Logic**, vol. 61 (1993), no. 1-2, pp. 49–62.

[DG] C. DIMITRACOPOULOS and H. GAIFMAN, *Fragments of Peano's arithmetic and the MRDP theorem*, **Logic and Algorithmic**, Monograph. Enseign. Math., vol. 30, Univ. Genève, Geneva, 1982, pp. 187–206.

[E-1] A. ENAYAT, *Automorphisms, Mahlo cardinals, and NFU*, **Nonstandard Models of Arithmetic and Set Theory** (A. Enayat and R. Kossak, editors), Contemporary Mathematics Series, vol. 361, American Mathematical Society, Providence, RI, 2004, pp. 37–59.

[E-2] ——, *Automorphisms of models of bounded arithmetic*, to appear.

[E-3] ——, *Weakly compact cardinals and automorphisms*, to appear.

[ER] P. ERDÖS and R. RADO, *A combinatorial theorem*, **Journal of the London Mathematical Society**, vol. 25 (1950), pp. 249–255.

[Fe] U. FELGNER, *Comparison of the axioms of local and universal choice*, **Fundamenta Mathematicae**, vol. 71 (1971), no. 1, pp. 43–62. (errata insert).

[Fo] T. E. FORSTER, *Set Theory with a Universal Set*, Oxford Logic Guides, vol. 31, Oxford University Press, New York, 1995.

[Fr] H. FRIEDMAN, *Translatability and Relative Consistency, II*, Ohio State University, unpublished notes, 1979.

[G] H. GAIFMAN, *Models and types of Peano's arithmetic*, **Annals of Pure and Applied Logic**, vol. 9 (1976), no. 3, pp. 223–306.

[GRS] R. GRAHAM, B. ROTHSCHILD, and J. SPENCER, *Ramsey Theory*, John Wiley & Sons, New York, 1980.

[HP] P. HÁJEK and P. PUDLÁK, *Metamathematics of First-Order Arithmetic*, Springer, Berlin, 1993.

[Ho-1] R. HOLMES, *Strong axioms of infinity in NFU*, **The Journal of Symbolic Logic**, vol. 66 (2001), no. 1, pp. 87–116.

[Jec] T. JECH, *Set Theory*, Academic Press, New York, 1978.

[Jen] R. B. JENSEN, *On the consistency of a slight (?) modification of quine's new foundations*, **Synthese**, vol. 19 (1969), pp. 250–263.

[Jo] C. JOCKUSCH, *Ramsey's theorem and recursion theory*, **The Journal of Symbolic Logic**, vol. 37 (1972), pp. 268–280.

[Kan] A. KANAMORI, *The Higher Infinite*, Springer-Verlag, Berlin, 1994.

[Kay] R. KAYE, *Model-theoretic properties characterizing Peano arithmetic*, **The Journal of Symbolic Logic**, vol. 56 (1991), no. 3, pp. 949–963.

[KKK] R. KAYE, R. KOSSAK, and H. KOTLARSKI, *Automorphisms of recursively saturated models of arithmetic*, **Annals of Pure and Applied Logic**, vol. 55 (1991), no. 1, pp. 67–99.

[KM] R. KAYE and D. MACPHERSON, *Automorphisms of First-Order Structures*, Oxford University Press, New York, 1994.

[Ki-1] L. KIRBY, *Initial Segments of Models of Arithmetic*, Ph.D. thesis, University of Manchester, 1977.

[Ki-2] ——, *Ultrafilters and types on models of arithmetic*, **Annals of Pure and Applied Logic**, vol. 27 (1984), no. 3, pp. 215–252.

[KP] L. KIRBY and J. PARIS, *Initial segments of models of Peano's axioms*, **Set Theory and Hierarchy Theory, V**, Lecture Notes in Mathematics, vol. 619, Springer, Berlin, 1977, pp. 211–226.

[Kr] J. KRAJÍČEK, *Bounded Arithmetic, Propositional Logic, and Complexity Theory*, Cambridge University Press, Cambridge, 1995.

[Ku] K. KUNEN, *Some applications of iterated ultrapowers in set theory*, **Annals of Pure and Applied Logic**, vol. 1 (1970), pp. 179–227.

[MS] R. MAC DOWELL and E. SPECKER, *Modelle der Arithmetik*, **Infinitistic Methods (Proc. Sympos. Foundations of Math., Warsaw, 1959)**, Pergamon, Oxford, 1961, pp. 257–263.

[Mile] J. MILETI, *The canonical ramsey theorem and computability theory*, to appear.

[Mill] G. MILLS, *A model of Peano arithmetic with no elementary end extension*, **The Journal of Symbolic Logic**, vol. 43 (1978), no. 3, pp. 563–567.

[Mo-1] A. MOSTOWSKI, *Models of second order arithmetic with definable Skolem functions*, **Fundamenta Mathematicae**, vol. 75 (1972), no. 3, pp. 223–234.

[Mo-2] ——, *Errata to "Models of second order arithmetic with definable Skolem functions"*, **Fundamenta Mathematicae**, vol. 84 (1974), no. 2, p. 173.

[Pa] R. PARIKH, *Existence and feasibility in arithmetic*, **The Journal of Symbolic Logic**, vol. 36 (1971), pp. 494–508.

[Pu-1] P. PUDLÁK, *A definition of exponentiation by a bounded arithmetical formula*, **Commentationes Mathematicae Universitatis Carolinae**, vol. 24 (1983), no. 4, pp. 667–671.

[Pu-2] ——, *Cuts, consistency statements and interpretations*, **The Journal of Symbolic Logic**, vol. 50 (1985), no. 2, pp. 423–441.

[Q] W. V. O QUINE, *New Foundations for Mathematical Logic*, **The American Mathematical Monthly**, vol. 44 (1937), no. 2, pp. 70–80.

[Ra] R. RADO, *Note on canonical partitions*, **The Bulletin of the London Mathematical Society**, vol. 18 (1986), no. 2, pp. 123–126.

[Re] J.-P. RESSAYRE, *Nonstandard universes with strong embeddings, and their finite approximations*, **Logic and Combinatorics** (S. Simpson, editor), Contemporary Mathematics, vol. 65, Amer. Math. Soc., Providence, RI, 1987, pp. 333–358.

[Sc] J. SCHMERL, *Automorphism groups of models of Peano arithmetic*, **The Journal of Symbolic Logic**, vol. 67 (2002), no. 4, pp. 1249–1264.

[Si-1] S. SIMPSON, *Review of* [Fe] *and* [Mo-1], **The Journal of Symbolic Logic**, vol. 38, pp. 652–653.

[Si-2] ——, *Subsystems of Second Order Arithmetic*, Springer, Berlin, 1999.

[Sm] C. SMORYŃSKI, *Nonstandard models and related developments*, **Harvey Friedman's Research on the Foundations of Mathematics** (L. A. Harrington et al., editors), North-Holland, Amsterdam, 1985, pp. 179–229.

[Sol] R. SOLOVAY, *The consistency strength of NFUB*, preprint available at Front for the Mathematics ArXiv, http://front.math.ucdavis.edu.

[Som-1] R. SOMMER, *Transfinite Induction and Hierarchies Generated by Transfinite Recursion within Peano Arithmetic*, Ph.D. thesis, University of California, Berkeley, 1990.

[Som-2] ——, *Transfinite induction within Peano arithmetic*, **Annals of Pure and Applied Logic**, vol. 76 (1995), no. 3, pp. 231–289.

[V] A. VISSER, *Categories of theories and interpretations*, this volume.

[W] H. WANG, *Popular Lectures on Mathematical Logic*, Dover Publications, New York, 1993.

DEPARTMENT OF MATHEMATICS AND STATISTICS
AMERICAN UNIVERSITY
4400 MASS. AVE., N.W.
WASHINGTON, D.C. 20016-8050, USA
E-mail: enayat@american.edu

LOCAL-GLOBAL PRINCIPLES AND APPROXIMATION THEOREMS

YURI L. ERSHOV

Abstract. It is known for a long time that local-global principles in field theory imply the independence of localities (linear orders or valuation rings) and, moreover, various approximation theorems. Some theorems of this kind are established in the paper.

§1. Introduction. Each local-global principle reads as follows: if some property (usually, existence of a simple point for a variety of special form) holds for all extensions of a field in a distinguished class then it holds for the field.

Local-global principles were discovered firstly in algebraic number theory. As the distinguished class of extensions there was used the class of all completions of the field of algebraic numbers relative to all its field topologies. The celebrated Minkowski–Hasse theorem on representation of zero by a quadratic form, Hasse's theorem on a norm for cyclic extensions provide examples of these principles. Searching for such "Hasse principles" is so far an actual direction of research in number theory (for example, see the survey [1]).

Model-theoretic studies of the real closed fields (A. Tarski) and the fields of p-adic numbers (J. Ax–S. Kochen; Yu. Ershov) show that instead of completions it is possible to use real closure (for the topology defined by an order) and Henselization (for the topology defined by a valuation ring) correspondingly.

It happens that the fields satisfying some local-global principle (usually for arbitrary absolutely irreducible affine varieties) relative to the local closures for localities (the terminology of [12]; see below) (which are linear orders or valuation rings) from a distinguished class of localities often enjoy "a satisfactory model theory." This can be illustrated by the examples of this sort from the studies of a field satisfying a local-global principle for finitely many orders (L. van den Dries [18]), a finite family of valuation rings (Yu. Ershov [3]), the family of all orders (PRC-fields) (A. Prestel [15]), a finite family of orders and valuation rings (B. Heinemann–A. Prestel [12]), the family of all p-adically closed valuation rings ($P_p C$-fields) (C. Grob [11]), and a Boolean family of

Logic in Tehran
Edited by A. Enayat, I. Kalantari, and M. Moniri
Lecture Notes in Logic, 26

valuation rings (Yu. Ershov [5]). Other forms of local-global principles were considered in [13], [17].

Another direction of research is connected with the remarkable discovery by R. Rumely [16] of a local-global principle for existence of *integer* algebraic points of a variety. This article gave rise to some interesting extensions of the principle ([14], [10]) and new model-theoretic studies (for some surveys of this direction see [7], [2]).

These studies lead the present author to considering the interesting (wonderful, nice) extensions of fields of algebraic numbers [6], [8]. In [9] the author suggested how to use these extensions for obtaining a new presentation of global class field theory (in characteristic 0). One of the main ingredients in that was a very strong approximation theorem ([9, Theorem 2]) which appeared there with no indication of proof. Here this theorem is demonstrated under much less restrictive assumptions.

In fact, it is known for a long time (see [12], [3], [15], [5], [17], [4], etc.) that the local-global principles imply the independence of localities and, moreover, various approximation theorems. Some theorems of this kind are established below.

§2. Block approximation property for linear orders.
We start with a series of definitions. Let F be a field. A *locality* (of F) is either a (linear) order $L \subseteq F$ on F or a nontrivial ($\neq F$) valuation ring of F. The set $\Lambda(F)$ of all localities of F is the disjoint union of the family $X(F)$ of all orders on F and the family $S(F)$, the abstract Riemann surface of F, of all nontrivial valuation rings of F. Both families $X(F)$ and $S(F)$ possess natural topologies which will be discussed below.

Let us call an *equipped field* a pair $\mathbb{F} = \langle F, \Lambda_{\mathbb{F}} \rangle$, where F is a field and $\Lambda_{\mathbb{F}} \subseteq \Lambda(F)$ is a family of localities of F.

Recall the definition of the Harrison topology on $X(F)$: the family of sets

$$V_a \rightleftharpoons \{L \mid L \in X(F), \ a \in L\}$$

for all $a \in F^\times \rightleftharpoons F \setminus \{0\}$ forms a subbase for the topology. The topological space $X(F)$ is *Boolean*, for $X(F)$ is compact Hausdorff, and the base sets V_a are clopen.

If the subbase V_a, $a \in F^\times$, is in fact a multiplicative base, which means that for all $a, b \in F^\times$ there exists $c \in F^\times$ such that $V_a \cap V_b = V_c$, then the field F satisfies the *Strong Approximation Property* or SAP for short. (Actually, it might be more correct to call it the "order independence property.")

REMARK. *If F satisfies SAP then every clopen subset of $X(F)$ is V_a for some $a \in F^\times$.*

A field F satisfies the *Block Approximation Property for (linear) orders* or BAP$_L$ for short (see [4]) if for every finite partition $X(F) = X_0 \dot{\cup} \cdots \dot{\cup} X_{n-1}$

of $X(F)$ into compact subsets, and every choice of elements $a_0, \ldots, a_{n-1} \in F$ and $\varepsilon_0, \ldots, \varepsilon_{n-1} \in P(F)^\times = P(F) \setminus \{0\}$, where $P(F) \rightleftharpoons \cap \{L \mid L \in X(F)\}$ $(= \Sigma F^2)$, there exists an element $a \in F$ such that the inequality

$$|a - a_i|_L <_L \varepsilon_i \ \left(\iff \varepsilon_i \pm (a - a_i) \in L^\times \rightleftharpoons L \setminus \{0\} \right)$$

holds for all $i < n$ and $L \in X_i$.

It is known [4] that every RRC- or PRC-field satisfies BAP_L. We will now establish that BAP_L follows from a rather general local-global principle LG for equipped fields under the condition that the family $W_\mathbb{F} \rightleftharpoons \Lambda_\mathbb{F} \cap S(F)$ is uniformly nonreal.

If $\lambda \in \Lambda(F)$, then the *local closure* F_λ of F (relative to λ) is the real closure $R_L(F)$ of the linearly ordered field $\langle F, L \rangle$ when $\lambda = L \in X(F)$, and the Henselization $H_R(F)$ of F relative to the valuation ring R when $\lambda = R \in S(F)$ $(H_R(F) = q(H(R))$, where $H(R)$ is a Henselization of R).

An equipped field $\mathbb{F} = \langle F, \Lambda_\mathbb{F} \rangle$ satisfies the *local-global principle LG* (written $\mathbb{F} \models LG$), if each absolutely irreducible affine variety V over F has a simple F-rational point whenever V has simple F_λ-rational points for all localities $\lambda \in \Lambda_\mathbb{F}$.

The family $W_\mathbb{F}$ of valuation rings is *uniformly nonreal*, if there exists a natural number $k > 0$ such that for all $R \in W_\mathbb{F}$ every element of the field $H_R(F)$ can be written as a sum of k squares:

$$H_R(F) \models \forall x \exists y_0, \ldots, y_{k-1} \left(x = \sum_{i<k} y_i^2 \right).$$

PROPOSITION 1. *Let* $\mathbb{F} = \langle F, \Lambda_\mathbb{F} \rangle$ *satisfy the LG principle, and* $W_\mathbb{F}$ *is uniformly nonreal. Then* F *satisfies both SAP and* BAP_L.

PROOF. In order to prove this, we have to slightly modify the proof of Lemma 5 in [4], and the remarks on it.

Let $k > 0$ be a natural number such that for all $R \in W_\mathbb{F}$ we have

$$H_R(F) \models \forall x \exists y_0, \ldots, y_{k-1} \left(x = \sum_{i<k} y_i^2 \right).$$

Pick $a, b \in F^\times$ so that $|a|_L > 1$ and $|b|_L > 1$ for all $L \in X(F)$. Note that for all $d \in F^\times$ the equality $V_d = V_{d^*}$ holds, where $d^* \rightleftharpoons (1 + d^2)d^{-1}$ and $|d^*|_L > 1$ for all $L \in X(F)$. Consider the absolutely irreducible polynomial

$$f(x, y_0, \ldots, y_{k-1}) \rightleftharpoons \sum_{i<k} y_i^2 - \left(\frac{1}{4} + a(x - a)^2 \right) \left(\frac{1}{4} - a(x - b)^2 \right).$$

It is not hard to see that for every locality $\lambda \in \Lambda_\mathbb{F}$ the polynomial $f(x, \bar{y})$ has a simple root in F_λ. By LG there exist $c, d_0, \ldots, d_{k-1} \in F$ such that $f(c, \bar{d}) = 0$. An easy check shows that $V_c = V_a \cap V_b$.

Take a compact partition $X(F) = X_0 \dot{\cup} \cdots \dot{\cup} X_{n-1}$ and elements $a_0, \dots,$ $a_{n-1} \in F$ and $\varepsilon_0, \dots, \varepsilon_{n-1} \in P(F)^\times$. The proof above and the remarks imply that there exist such elements $b_0, \dots, b_{n-1} \in F^\times$ that $X_i = V_{b_i}$ and $|b_i|_L > 1$ for all $i < n$ and $L \in X(F)$. Considering, as in [4], the absolutely irreducible polynomial

$$g(x, \bar{y}) \rightleftharpoons \sum_{i<k} y_i^2 - \prod_{i<n} \left(\varepsilon_i^2 - b_i(x - a_i)^2 \right),$$

we find that for $a, d_0, \dots, d_{k-1} \in F$ satisfying $g(a, \bar{d}) = 0$, which exist by LG, the element a enjoys the inequality $|a - a_i| <_L \varepsilon_i$ for all $i < n$ and $L \in X_i$. $\dashv$

§3. Approximations in multi-valued fields.

In this section we treat the case of multi-valued fields. Most of the known facts about multi-valued fields can be found in the monograph [7].

Let F be a field. The *Zariski topology* on $S(F)$ is given by the base

$$V_A \rightleftharpoons \{R \mid R \in S(F), \ A \subseteq R\},$$

where A is a finite subset of F, see [19].

A family $W \subseteq S(F)$ of valuation rings of F is *affine* if the following conditions hold:

1. the *holomorphy ring* $H(W) \rightleftharpoons \bigcap\{R \mid R \in W\}$ of W is a Prüfer ring (recall [7], [19] that a ring (domain) R is a Prüfer ring iff each localization R_m of R at a maximal ideal m is a valuation ring);
2. the field of fractions $q(H(W))$ of $H(W)$ coincides with F;
3. $W = \{H(W)_m \mid m \in \mathrm{mSpec}\, H(W)\}$, where $\mathrm{mSpec}\, H(W)$ is the family of all maximal ideals of $H(W)$, and $H(W)_m$ is the localization of $H(W)$ at m.

REMARK. *If H is a Prüfer subring of F such that $F = q(H)$ then the family* $W(H) \rightleftharpoons \{H_m \mid m \in \mathrm{mSpec}\, H\}$ *is affine, and $H(W(H)) = H$, see* [7].

If W is affine then the Zariski topology on W can also be defined by the base

$$U_a \rightleftharpoons \{R \mid R \in W, \ a \notin m(R)\},$$

where $a \in H(W)^\times$. Here $m(R)$ is the unique maximal ideal of the valuation ring R. Put $W_a \rightleftharpoons W \setminus U_a$ for all $a \in H(W)^\times$. The affine family W is compact in the Zariski topology; thus so are its closed subfamilies W_a, $a \in H(W)^\times$.

The following remark is useful:

REMARK. *A subfamily $W_0 \subseteq W$ of an affine family W is also affine if and only if W_0 is compact in the Zariski topology* ([7, Proposition 2.3.7]).

A necessary condition for the suitable approximation theorems, generalizing the Chinese Remainder Theorem to hold for an affine family W is the following *independence* property of W: given every two distinct rings R_0, R_1 from W, we have $R_0 R_1 = F$.

An affine family W satisfies the *Block Approximation Property for Valuations* (BAP_V) if for every compact partition $W = W_0 \dot{\cup} \cdots \dot{\cup} W_{n-1}$, and every choice of elements $a_0, \ldots, a_{n-1} \in F$ and $\varepsilon_0, \ldots, \varepsilon_{n-1} \in H(W)^\times$, there exists $a \in F$ such that $v_R(a - a_i) \geq v_R(\varepsilon_i)$ for all $i < n$ and $R \in W_i$. Here v_R is the valuation of F corresponding to the valuation ring R, see [7].

Recall the following fact [7, Proposition 2.6.2]:

PROPOSITION 2. *Every independent affine family of valuation rings satisfies* BAP_V.

Strengthenings of BAP_V arise from the possibility of replacing nonstrict inequalities with the corresponding strict inequalities (equalities).

The next statement, analogous to Proposition 2.6.6 in [7], yields yet another type of approximation properties.

PROPOSITION 3. *Suppose that W is an independent affine family and $H(W)$ is a Bezout ring (i.e., every finitely generated ideal is principal). Suppose also that $W = W_0 \dot{\cup} \cdots \dot{\cup} W_{n-1} \dot{\cup} W_n$ is a compact partition for which W_i for $i < n$ are closed, and $a_0, \ldots, a_{n-1} \in F$. There exists $a \in F$ such that $v_R(a) = v_R(a_i)$ for all $i < n$ and $R \in W_i$, while $v_R(a) = 0$ for $R \in W_n$.*

We may assume that all families W_i for $i < n$ differ from $\varnothing$ and W. Let us show that in this case there exist $d_0, \ldots, d_n \in H(W)^\times$ such that $W_i = W_{d_i}$ for $i < n$, and $W_n = U_{d_n}$ when $W_n \neq \varnothing$.

LEMMA 1. *For every nonempty open compact subfamily U of W there exists $d \in H(W)^\times$ such that $U = U_d$.*

PROOF. Since U is open, there exists a family $d_\lambda \in H(W)^\times$, $\lambda \in \Lambda$, such that $U = \bigcup_{\lambda \in \Lambda} U_{d_\lambda}$. Since U is compact, this family can be chosen finite. Hence, if $d \in H(W)^\times$ is such that $(d)_{H(W)} = (d_\lambda \mid \lambda \in \Lambda)_{H(W)}$, where $(S)_{H(W)}$ denotes the ideal in $H(W)$ generated by $S \subseteq H(W)$, then $U = \bigcup_{\lambda \in \Lambda} U_{d_\lambda}$ and $U = U_d$. Such an element d exists because $H(W)$ is a Bezout ring while Λ is finite. $\dashv$

COROLLARY 1. *If $W' \neq W$ is a closed subfamily of W such that $W \setminus W'$ is compact then there exists $d \in H(W)^\times$ such that $W' = W_d$.*

Let us now establish, under the assumptions of Proposition 3, another auxiliary statement which is also of interest in its own right.

LEMMA 2. *Suppose that $d \in H(W)^\times$ and U_d is compact. For every $\varepsilon \in H(W)^\times$ there exists $\delta \in H(W)^\times$ such that*

$$v_R(\delta) > v_R(\varepsilon) \quad \text{for } R \in W_d,$$
$$v_R(\delta) = 0 \quad \text{for } R \in U_d.$$

PROOF. By Proposition 2 there exists $\delta_0 \in H(W)^\times$ such that

$$v_R(\delta_0) \geq v_R(d\varepsilon) > v_R(\varepsilon) \quad \text{for } R \in W_d,$$
$$v_R(\delta_0 - 1) \geq v_R(d\varepsilon) \geq 0 \quad \text{for } R \in U_d.$$

Since U_d is compact and W_{δ_0} is closed, $U_d \cap W_{\delta_0}$ is compact; moreover, $W_d \cap (U_d \cap W_{\delta_0}) = \varnothing$. Applying Lemma 2.6.2 of [7], we find $\delta_1 \in H(W)^\times$ such that

$$v_R(\delta_1) \geq v_R(\delta_0) \quad \text{for } R \in W_d,$$
$$v_R(\delta_1 - 1) \geq v_R(\delta_0) \quad \text{for } R \in U_d \cap W_{\delta_0}.$$

Note that $v_R(\delta_1 - 1) \geq v_R(\delta_0) > 0$ implies that $v_R(\delta_1) = 0$ for $R \in U_d \cap W_{\delta_0}$. Pick $\delta \in H(W)^\times$ such that $(\delta)_{H(W)} = (\delta_0, \delta_1)_{H(W)}$, and check that δ is the required element. If $R \in W_d$ then by Lemma 2.3.5 of [7]

$$v_R(\delta) = \min\{v_R(\delta_0), v_R(\delta_1)\} = v_R(\delta_0) > v_R(\varepsilon).$$

If $R \in U_d \cap W_{\delta_0}$, then $v_R(\delta) = 0$, for $v_R(\delta_1) = 0$ and $v_R(\delta_0) \geq 0$ since $\delta_0 \in H(W)^\times$. If $R \in U_d \cap U_{\delta_0}$, then again $v_R(\delta) = 0$, for $v_R(\delta_0) = 0$ and $v_R(\delta_1) \geq 0$ since $\delta_1 \in H(W)^\times$. Therefore, $v_R(\delta) = 0$ for all $R \in U_d$. ⊣

PROOF OF PROPOSITION 3. We settle the case of $W_n \neq \varnothing$, leaving the simpler case of $W_n = \varnothing$ to the reader. Suppose that $d_0, \ldots, d_n \in H(W)^\times$ are such that $W_i = W_{d_i}$ for $i < n$, and $W_n = U_{d_n}$.

Assume first that $a_0, \ldots, a_{n-1} \in H(W)^\times$ and put $\varepsilon \rightleftharpoons (\prod_{i<n} a_i d_i) \cdot d_n$; then $v_R(a_i) < v_R(\varepsilon)$ for all $i < n$ and $R \in W_i = W_{d_i}$, since $v_R(d_i) > 0$. By Proposition 2 there exists $b \in H(W)^\times$ such that

$$v_R(b - a_i) \geq v_R(\varepsilon) \quad \text{for } R \in W_i, \text{ where } i < n,$$
$$v_R(b - 1) \geq v_R(\varepsilon) \quad \text{for } R \in W_n.$$

Then we have

$$v_R(b) = v_R(a_i) \quad \text{for } R \in W_i = W_{d_i}, \text{ where } i < n,$$
$$v_R(b) \geq 0 \quad \text{for } R \in W_n = U_{d_n}.$$

By Lemma 2 there exists $\delta \in H(W)^\times$ such that $v_R(\delta) > v_R(\varepsilon)$ for $R \in W_b \cap U_{d_n}$. Pick $a \in H(W)^\times$ so that $(a)_{H(W)} = (b, \delta)_{H(W)}$. By Lemma 2.3.5 of [7] the last equality implies that

$$v_R(a) = \min\{v_R(b), v_R(\delta)\}$$

for all $R \in W$. Now if $i < n$ and $R \in W_i = W_{d_i} \subseteq W_{d_n}$ then

$$v_R(\delta) > v_R(\varepsilon) > v_R(a_i) = v_R(b),$$

thus $v_R(a) = v_R(a_i)$. If $R \in W_b \cap W_n = W_b \cap U_{d_n}$, then $v_R(\delta) = 0$, forcing $v_R(a) = 0$. If $R \in U_b \cap W_n = U_b \cap U_{d_n}$, then $v_R(b) = 0$, and so again $v_R(a) = 0$. We thus have $v_R(a) = 0$ for all $R \in W_n = U_{d_n}$, which proves the Proposition in the case $a_0, \ldots, a_{n-1} \in H(W)^\times$.

Now we can deal with the general case $a_0, \ldots, a_{n-1} \in F^\times$. Pick $c \in H(W)^\times$ such that $a_i' \rightleftharpoons a_i c \in H(W)^\times$ for $i < n$. By above, there exists $a' \in H(W)^\times$ such that

$$v_R(a' - a_i') > v_R(a_i') \quad \text{for } R \in W_i, \text{ where } i < n,$$
$$v_R(a') = 0 \quad \text{for } R \in W_n.$$

Also, there exists $\bar{c} \in H(W)^\times$ such that

$$v_R(c - \bar{c}) > v_R(c) \quad \text{for } R \in W_{d_n} = \bigcup_{i<n} W_{d_i},$$
$$v_R(\bar{c}) = 0 \quad \text{for } R \in W_n = U_{d_n}.$$

Put $a \rightleftharpoons a' \bar{c}^{-1}$. Then for all $R \in W$ we have $v_R(a) = v_R(a') - v_R(\bar{c})$. If $i < n$ and $R \in W_i$ then $v_R(a' - a_i') > v_R(a_i')$ implies that

$$v_R(a') = v_R(a_i') = v_R(a_i c) = v_R(a_i) + v_R(c),$$

while $v_R(\bar{c} - c) > v_R(c)$ implies that $v_R(\bar{c}) = v_R(c)$; thus

$$v_R(a_i) = v_R(a_i') - v_R(c) = v_R(a') - v_R(\bar{c}) = v_R(a).$$

If $R \in W_n$ then $v_R(a') = 0$ and $v_R(\bar{c}) = 0$, thus $v_R(a) = v_R(a' \bar{c}^{-1}) = 0$. This completes the proof of Proposition 3. ⊣

In order to prove stronger approximation theorems, we consider a new *linear* local-global principle $LLG_\le$.

Let F be a field, and let $H \le F$ be a subring of F such that $F = q(H)$. Define on F a preorder $\le_H$ by putting $a \le_H b \rightleftharpoons \exists h \in H(ah = b)$ for $a, b \in F$.

REMARKS. 1. $a \le_H 0$ *for all $a \in F$;*
2. $H = \{a \mid 1 \le_H a\}$;
3. *if H is a valuation ring then $a \le_H b \Longleftrightarrow v_H(a) \le v_H(b)$;*
4. *if H is a Prüfer subring then $a \le_H b \Longleftrightarrow v_R(a) \le v_R(b)$ for all $R \in W(H)$.*

A multi-valued field $\langle F, H \rangle$ satisfies the *principle $LLG_\le$* if for every system

$$f_0^0, f_0^1; \ldots; f_{k-1}^0, f_{k-1}^1$$

of such linear polynomials in $x_0, \ldots, x_{n-1}$ that for every $R \in W(H)$ there exists an n-tuple $\bar{a}_R = a_R^0, \ldots, a_R^{n-1}$ of elements of $H_R(F)$ such that

$$f_j^0(\bar{a}_R) \le_{H(R)} f_j^1(\bar{a}_R) \quad \text{for } j < k,$$

then there exists an n-tuple $\bar{a} = a_0, \ldots, a_{n-1}$ of elements of F such that

$$f_j^0(\bar{a}) \le_H f_j^1(\bar{a}) \quad \text{for } j < k.$$

REMARK. *If $\langle F, H \rangle \models LLG_\leq$ then H is a Bezout ring: if $a, b \in H$ and $\bar{e} = e_0, e_1, e_2$ solves a system of the inequalities*

$$x_0 \leq_H a, \qquad x_0 \leq_H b, \qquad ax_1 + bx_2 \leq_H x_0, \qquad 1 \leq_H x_1, \qquad 1 \leq_H x_2,$$

then $(e_0)_H = (a, b)_H$.

THEOREM 1 (Strong Approximation Theorem). *Suppose that a multi-valued field $\langle F, H \rangle$ satisfies $LLG_\leq$ and $W(H)$ is independent. For every compact partition $W(H) = W_0 \dot{\cup} \cdots \dot{\cup} W_{n-1} \dot{\cup} W_n$ with closed subfamilies $W_i \neq W(H)$, $i < n$, and arbitrary $a_0, \ldots, a_{n-1} \in F$ and $\varepsilon_0, \ldots, \varepsilon_{n-1} \in H^\times$, there exists b such that*

$$v_R(b - a_i) > v_R(\varepsilon_i) \quad \text{for } R \in W_i, \text{ where } i < n,$$
$$v_R(b) = 0 \quad \text{for } R \in W_n.$$

First, we establish a statement that our proof of the theorem will reduce to.

PROPOSITION 4. *Suppose that $\langle F, H \rangle \models LLG_\leq$ and $W(H)$ is independent. Then for all $a \in F^\times$ and all $\varepsilon, d \in H^\times$ there exists $b \in F$ such that*

$$v_R(b - a) > v_R(\varepsilon) \quad \text{for } R \in W_d;$$
$$v_R(b) = 0 \quad \text{for } R \in U_d.$$

PROOF. Pick $\delta \in H^\times$ such that

$$v_R(\delta) > \max \{0, v_R(a), v_R(\varepsilon)\} \quad \text{for } R \in W_d;$$
$$v_R(\delta) = 0 \quad \text{for } R \in U_d.$$

Since $F = q(H)$, there exist $c, e \in H^\times$ such that $a = ce^{-1}$, so for all $R \in W(H)$ we have

$$v_R(a) = v_R(c) - v_R(e) \leq v_R(c)$$

for $v_R(e) \geq 0$. Put $\delta_0 \rightleftharpoons c\varepsilon d (\neq 0)$. Then $v_R(\delta_0) > \max\{0, v_R(c), v_R(\varepsilon)\}$ for all $R \in W_d$. Since $W(H) = W_d \dot{\cup} U_d$ is a compact partition, by Proposition 3 there is $\delta \in F^\times$ such that

$$v_R(\delta) = v_R(\delta_0) \quad \text{for } R \in W_d;$$
$$v_R(\delta) = 0 = v_R(1) \quad \text{for } R \in U_d.$$

By Proposition 2 there is $c \in F^\times$ such that

$$v_R(c - a) \geq v_R(\delta) \quad \text{for } R \in W_d;$$
$$v_R(c) = v_R(c - 0) \geq 0 = v_R(1) \quad \text{for } R \in U_d.$$

Consider now the system of linear polynomials $\delta, x - c; x, \delta \in F[x]$, and check that it is "locally solvable."

Suppose that $R \in W_d$. If we put $a_R \rightleftharpoons c$, then $\delta \leq_R 0 \; (= a_R - c = c - c)$ and $a_R = c \leq_R \delta$, since $v_R(c - a) \geq v_R(\delta) > v_R(a)$ implies $v_R(c) = v_R(a) < v_R(\delta)$. Suppose that $R \in U_d$. If we put $a_R = \delta$ then

$$v_R(a_R - c) = v_R(\delta - c) \geq \min\{v_R(\delta), v_R(c)\} \geq 0 = v_R(\delta), \quad \delta \leq_R a_R - c;$$
$$v_R(a_R) = v_R(\delta) \leq v_R(\delta), \quad a_R \leq_R \delta.$$

This "local solvability" and $LLG_\leq$ imply the existence of $b \in F$ such that $\delta \leq_H b - c$ and $b \leq_H \delta$. Let us check that this very element b satisfies the claim of the Proposition.

If $R \in W_d$ then

$$v_R(a - b) = v_R(a - c + c - b)$$
$$\geq \min\{v_R(a - c), v_R(b - c)\} \geq v_R(\delta) > v_R(\varepsilon).$$

If $R \in U_d$ then $b \leq_H \delta$ implies that $v_R(b) \leq v_R(\delta) = 0$. If $v_R(b) < 0$, then $v_R(b - c) = v_R(b) < 0$, since $v_R(c) \geq 0$. This contradicts $0 = v_R(\delta) \leq v_R(b - c)$. Thus, $v_R(b) = 0$, proving the Proposition.   ⊣

PROOF OF THEOREM 1. The hypotheses imply that there exist $d_0, \ldots, d_{n-1}$, $d_n \in H^\times$ such that $W_i = W_{d_i}$ for $i < n$, and $W_n = U_{d_n}$.

Since $d_i \in J(H(W_i))^\times$, we may assume that $\varepsilon_i \in J(H(W_i))^\times$. Using Proposition 2 with ε_i^2 in place of ε_i for $i < n$, we find $a \in F$ and $\varepsilon \in H^\times$ such that $v_R(a - a_i) \geq v_R(\varepsilon_i^2) > v_R(\varepsilon_i)$ and $v_R(\varepsilon) \geq v_R(\varepsilon_i^2) > v_R(\varepsilon_i)$ for all $i < n$ and $R \in W_i$. Applying Proposition 4 to a, ε, and d_n, we find the required element b.   ⊣

§4. A very strong approximation theorem.

In this section we deal with the equipped fields defined as preordered multi-valued fields, or PMV-fields [8]. These are systems $\mathbb{F} = \langle F, H, P \rangle$, where F is a field, H is a Prüfer subring of F such that $F = q(H)$, and P is a preorder on F. The equipment on F is given as follows: $W_{\mathbb{F}} \rightleftharpoons W(H)$, $X_{\mathbb{F}} \rightleftharpoons X(P)$, and $\Lambda_{\mathbb{F}} = W_{\mathbb{F}} \cup X_{\mathbb{F}}$, where

$$X(P) \rightleftharpoons \{L \mid L \text{ is an order on } F \text{ such that } L \supseteq P\}.$$

The article [8] defines the local-global principle $LG_\leq^+$ for PMV-fields. Now we give its (more "homogeneous") modification.

We say that a PMV-field $\mathbb{F} = \langle F, H, P \rangle$ satisfies the *local-global principle* $LG_\leq$, and write $\mathbb{F} \models LG_\leq$, if the following holds: for every absolutely irreducible affine variety $V \subseteq U^n$, where $U > F$ is the universal extension, and every choice of polynomials

$$f_0^0, f_0^1; \ldots; f_{k-1}^0, f_{k-1}^1; g_0^0, g_0^1; \ldots; g_{l-1}^0, g_{l-1}^1 \in F[x_0, \ldots, x_{n-1}]$$

such that

1. for every valuation ring $R \in W_{\mathbb{F}} = W(H)$ the variety V has a simple $H_R(F)$-rational point $\bar{a}_R$ satisfying $f_i^0(\bar{a}_R) \leq_{H(R)} f_i^1(\bar{a}_R)$ for $i < k$;

2. for every order $L \in X_{\mathbb{F}} = X(P)$ the variety V has a simple $R_L(F)$-rational point $\bar{a}_L$ satisfying $g_j^0(\bar{a}_L) \leq_{R(L)} g_j^1(\bar{a}_L)$ for $j < l$,

the variety V has a simple F-rational point $\bar{a}$ satisfying

$$f_i^0(\bar{a}) \leq_H f_i^1(\bar{a}) \quad \text{for } i < k;$$
$$g_i^0(\bar{a}) \leq_P g_j^1(\bar{a}) \quad \text{for } j < l.$$

THEOREM 2 (Very Strong Approximation Theorem). *Suppose that a PMV-field* $\mathbb{F}$ *is such that* $\mathbb{F} \models LG_{\leq}$, *and that* $W_{\mathbb{F}}$ *is uniformly nonreal. For every compact partition* $W_{\mathbb{F}} = W_0 \dot{\cup} \cdots \dot{\cup} W_{n-1} \dot{\cup} W_n$, *where* W_i *are closed for* $i < n$, *arbitrary* $a_0, \ldots, a_{n-1} \in F$ *and* $\varepsilon_0, \ldots, \varepsilon_{n-1} \in H^{\times}$, *every compact partition* $X_{\mathbb{F}} = X_0 \dot{\cup} \cdots \dot{\cup} X_{l-1}$, *and arbitrary* $b_0, \ldots, b_{l-1} \in F$ *and* $\eta_0, \ldots, \eta_{l-1} \in P^{\times}$, *there exists* $c \in F$ *such that*

$$v_R(c - a_i) > v_R(\varepsilon_i) \quad \text{for } R \in W_i, \text{ where } i < n;$$
$$v_R(c) = 0 \quad \text{for } R \in W_n;$$
$$|c - b_j|_L <_L \eta_j \quad \text{for } L \in X_j, \text{ where } j < l.$$

PROOF. Note that the validity of $LG_{\leq}$ implies the validity of $LLG_{\leq}$ for $\langle F, H \rangle$. Furthermore, $W_{\mathbb{F}} = W(H)$ is independent, and H is a Bezout ring (the proof of Proposition 3.2.1 in [7] applies in this case as well). Then Theorem 1 and Proposition 1 show that the conclusion of the theorem can be fulfilled "separately" for $W_{\mathbb{F}}$ and $X_{\mathbb{F}}$.

Pick $c_0 \in F$ so that

$$v_R(c_0 - a_i) > v_R(\varepsilon_i) \quad \text{for } R \in W_i, \text{ where } i < n;$$
$$v_R(c_0) = 0 \quad \text{for } R \in W_n.$$

Pick $c_1 \in F$ so that

$$|c_1 - b_j|_L <_L \frac{1}{2}\eta_j \quad \text{for } L \in X_j, \text{ where } j < l.$$

Let $\varepsilon \rightleftharpoons \prod_{i<n} \varepsilon_i$; then $v_R(\varepsilon) \geq v_R(\varepsilon_i)$ for all $i < n$ and $R \in \bigcup_{j<n} W_j$. Take $d \in H^{\times}$ so that $W_n = U_d$, and $\delta \in H^{\times}$ so that

$$v_R(\delta) > \max \{0, v_R(c_0), v_R(\varepsilon)\} \quad \text{for } R \in W_d \left(= \bigcup_{j<n} W_j \right),$$
$$v_R(\delta) = 0 \quad \text{for } R \in U_d.$$

The proof of Proposition 4 explains how to find such δ.

Let $\eta \in P^{\times}$ be such that $\eta <_P \frac{1}{2}\eta_j$ for all $j < l$.

Consider the absolutely irreducible affine variety V defined by the two polynomials

$$\sum_{i<k} y_i^2 - (\eta + x - c_1) = 0,$$

$$\sum_{i<k} z_i^2 - (\eta - x + c_1) = 0.$$

Here $k > 0$ is a natural number from the definition of uniform nonreality of W_F.

Put $f_0^0 \rightleftharpoons \delta$, $f_0^1 \rightleftharpoons x - c_0$; $f_1^0 \rightleftharpoons x$, $f_1^1 \rightleftharpoons \delta$.

It is easy to check that the local conditions of $LG_{\leq}$ are satisfied. Hence there exist elements $u_0, \ldots, u_{k-1}, v_0, \ldots, v_{k-1}, c \in F$ such that

$$\eta + x - c_1 = \sum_{i<k} u_i^2, \qquad \eta - c + c_1 = \sum_{i<k} v_i^2,$$

and also $\delta \leq_H c - c_0$ and $c \leq_H \delta$.

The first two equalities imply that $|c - c_1|_L <_L \eta$ for all $L \in X(F)$, but then

$$|c - b_j|_L = |c - c_1 + c_1 - b_j|_L <_L \frac{1}{2}\eta_j + \frac{1}{2}\eta_j = \eta_j$$

for all $L \in X_j$.

The inequalities $\delta \leq_H c - c_0$, and $c \leq_H \delta$, together with the choice of δ, imply that

$$v_R(c - a_i) = v_R(c - c_0 + c_0 - a_i)$$

$$\geq \min\{v_R(c - c_0), v_R(c_0 - a_i)\} > v_R(\varepsilon_i) \quad \text{for } R \in W_i, \text{ where } i < n.$$

If $R \in W_n = U_d$ then

$$v_R(c) = v_R(c_0 + c - c_0) \geq \min\{v_R(c_0), v_R(c - c_0)\} \geq 0,$$

since $v_R(c_0) = 0$, and $\delta \leq_H c - c_0$ implies $v_R(c - c_0) \geq v_R(\delta) = 0$. Besides, $c \leq_H \delta$ implies $v_R(c) \leq v_R(\delta) = 0$. Consequently, $v_R(c) = 0$ for $R \in W_n$. This proves the theorem. ⊣

REFERENCES

[1] J.-L. COLLIOT-THÉLÈNE, *The Hasse principle in a pencil of algebraic varieties*, **Number Theory (Tiruchirapalli, 1996)**, Contemporary Mathematics, vol. 210, AMS, Providence, RI, 1998, pp. 19–39.

[2] L. DARNIÈRE, *Decidability and local-global principles*, **Hilbert's Tenth Problem: Relations With Arithmetic and Algebraic Geometry (Ghent, 1999)**, Contemporary Mathematics, vol. 270, AMS, Providence, RI, 2000, pp. 145–167.

[3] YU. L. ERSHOV, *Multiple valued fields*, **Russian Mathematical Surveys**, vol. 37 (1982), no. 3, pp. 63–107.

[4] ———, *Two theorems on regularly r-closed fields*, **Journal für die Reine und Angewandte Mathematik**, vol. 347 (1984), pp. 154–167.

[5] ———, *Relative regular closeness and π-valuations*, **Algebra and Logic**, vol. 31 (1992), no. 6, pp. 342–360.

[6] ———, *On surprising (wonderful) extensions of the field of rationals*, **Rossiĭskaya Akademiya Nauk. Doklady Akademii Nauk**, vol. 62 (2000), no. 1, pp. 8–9.

[7] ———, *Multi-Valued Fields*, Kluwer Academic/Plenum Publishers, 2001.

[8] ———, *Preordered multivalued fields*, **Rossiĭskaya Akademiya Nauk. Doklady Akademii Nauk**, vol. 65 (2002), no. 1, pp. 75–79.

[9] ———, *Nice extensions and global class field theory*, **Rossiĭskaya Akademiya Nauk. Doklady Akademii Nauk**, vol. 67 (2003), no. 1, pp. 21–23.

[10] B. GREEN, F. POP, and P. ROQUETTE, *On Rumely's local-global principle*, **Jahresbericht der Deutschen Mathematiker-Vereinigung**, vol. 97 (1995), no. 2, pp. 43–74.

[11] C. GROB, *Die Entscheidbarkeit der Theorie der maximalen pseudo-p-adisch abgeschlossenen Körper*, Dissertation, Universität Konstanz, 1987.

[12] B. HEINEMANN and A. PRESTEL, *Fields regularly closed with respect to finitely many valuations and orderings*, **Quadratic and Hermitian Forms (Hamilton, Ont., 1983)**, CMS Conf. Proc., vol. 4, AMS, Providence, RI, 1984, pp. 297–336.

[13] M. JARDEN, *Algebraic realization of p-adically projective groups*, **Compositio Mathematica**, vol. 79 (1991), no. 1, pp. 21–62.

[14] L. MORET-BAILLY, *Groupes de Picard et problèmes de Skolem. I, II*, **Annales Scientifiques de l'École Normale Supérieure. Quatrième Série**, vol. 22 (1989), no. 2, pp. 161–179, 181–194.

[15] A. PRESTEL, *Pseudo real closed fields*, **Set Theory and Model Theory (Bonn, 1979)**, Lecture Notes in Mathematics, vol. 872, Springer, Berlin, 1981, pp. 127–156.

[16] R. RUMELY, *Arithmetic over the ring of all algebraic integers*, **Journal für die Reine und Angewandte Mathematik**, vol. 368 (1986), pp. 127–133.

[17] J. SCHMID, *Regularly T-closed fields*, **Hilbert's Tenth Problem: Relations with Arithmetic and Algebraic Geometry (Ghent, 1999)**, Contemporary Mathematics, vol. 270, AMS, Providence, RI, 2000, pp. 187–212.

[18] L. VAN DEN DRIES, *Model Theory of Fields*, Thesis, Utrecht, 1978.

[19] O. ZARISKI and P. SAMUEL, *Commutative Algebra. Vol. II*, D. Van Nostrand Co., Inc., Princeton, N. J.-Toronto-London-New York, 1960.

INSTITUTE OF MATHEMATICS SB RAS
AVENU KOPTUGA 4
NOVOSIBIRSK 630090, RUSSIA
E-mail: ershov@math.nsc.ru

BEATTY SEQUENCES AND THE ARITHMETICAL HIERARCHY

MEHDI GHASEMI AND MOJTABA MONIRI

Abstract. Thanks to an observation by R. Robinson, for any positive real α the function $\lambda n. \lfloor n\alpha \rfloor$ mapping positive integers n to the integer part $\lfloor n\alpha \rfloor$ of $n\alpha$ (called the Beatty sequence corresponding to α) is computable if and only if α is (Cauchy) computable. For any $\alpha \in \mathbb{R}^{\geq 0}$, oracle A, and $r \geq 1$, we show that $\lambda n. \lfloor n\alpha \rfloor$ has a Δ_r^A graph if and only if α is of class $\Delta_r^A(\mathbb{R})$ in the recently introduced Zheng-Weihrauch (Z-W) hierarchy. If $\alpha \in \Sigma_r^A(\mathbb{R}) \cup \Pi_r^A(\mathbb{R})$, then the above function is of class Δ_{r+1}^A. We consider $\lambda n. \lfloor n\alpha + \gamma \rfloor$ and show that the *if* part corresponding to the next to the last statement still holds (when now both coefficients are in $\Delta_r^A(\mathbb{R})$). We prove the converse when α is irrational. If α is rational, then the function is always primitive recursive (even for γ's beyond the Z-W hierarchy). Finally, we show that the set of (indices of) computable Beatty sequences is Π_2.

§1. Introduction and preliminaries.

1.1. Metric number theoretic issues exposed to computability.
Various properties of integer sequences which depend on one or more real parameters are of interest in Diophantine approximation and metric number theory. For example, by Weyl's theorem (proved independently by Bohl and also Sierpinski, see [N, p. 44] for this historical remark), for any irrational number α, the sequence $(n\alpha)_{n \in \mathbb{N}}$ is uniformly distributed modulo 1 (i.e. the sequence $(n\alpha - \lfloor n\alpha \rfloor)_{n \in \mathbb{N}}$ is uniformly distributed in the unit interval). Also the sequence $(\lfloor n\alpha \rfloor)_{n \in \mathbb{N}}$ is uniformly distributed modulo any other positive integer. Here, as usual, $\lfloor \cdot \rfloor$ denotes the integer part function, see, e.g., [N, Theorem 3.2 and Theorem 3.3] for proofs of the uniform distribution facts. Weyl's theorem of course implies that the set of fractional parts of multiples of any irrational is dense in the unit interval and this plays a crucial role in one of the results in this paper. Roughly speaking that result says if α is a positive irrational and $\gamma \in (0, 1)$, then γ can be recovered from the values of the function $\lambda m. \lfloor m\alpha + \gamma \rfloor$. The same does not hold if α is rational. Also let us quote the following about the sequence $a_n = n\theta + \gamma$: *If $\frac{\theta}{\gamma}$ is very well approximable by rationals, then we are dealing*

2000 *Mathematics Subject Classification.* Primary: 03F60, Secondary: 03D55, 11B25, 68Q25.

Key words and phrases. (Computable / Recursive / Arithmetical Hierarchy of) Reals (Functions), Real Arithmetic Progressions, Integer Part Function, Beatty Sequences.

Logic in Tehran
Edited by A. Enayat, I. Kalantari, and M. Moniri
Lecture Notes in Logic, 26

with a sequence which for much of the time resembles na + b with a, b ∈ Z, and so . . . . On the other hand, if $\frac{\theta}{\gamma}$ is not very well approximable, then . . . should be more 'random' in behavior . . . ; see [H, p. 188].

In this paper, we study computability and complexity of sequences $(\lfloor n\alpha + \gamma \rfloor)_{n \in \mathbb{N}}$ of integer parts of terms in real arithmetic progressions in terms of the real parameters involved and vice versa. When $\gamma = 0$, these are called Beatty sequences. By a result due to R. Robinson [R], it is known that an arbitrary Beatty sequence $(\lfloor n\alpha \rfloor)_{n \in \mathbb{N}}$ is computable if and only if its parameter α is computable.

We show that the Zheng-Weihrauch arithmetical hierarchy of reals, introduced recently, enjoys a nice correspondence with the arithmetical hierarchy of functions on $\mathbb{N}$ via the map $\lambda n.\lfloor n\alpha \rfloor$. This can also be relativized to any oracle. For real arithmetic progressions starting with non-zero reals γ, most of the correspondence still goes through, but rational numbers α always give rise to primitive recursive functions no matter where γ is situated in relation to the Z-W hierarchy. We show that indices of computable functions of the form $\lambda n.\lfloor n\alpha \rfloor$, form a Π_2 set.

In the remaining subsections of this section, we present the necessary preliminaries for our findings.

1.2. Some basic notation in computability theory and a fact. We adopt the usual notation $(\varphi_e)_{e \in \mathbb{N}}$ for a two-way effective enumeration of all partial computable functions $\mathbb{N} \to \mathbb{N}$. If $A \subset \mathbb{N}$ is taken as oracle, we have a similar enumeration $(\varphi_e^A)_{e \in \mathbb{N}}$ of the class $\mathcal{C}^A$ consisting of functions partial computable in A. Total functions in this class form Γ^A. The jump of A is $A' = \{n \in \mathbb{N} \mid \varphi_n^A(n) \downarrow\}$. Higher jumps are defined inductively by $A^{(n)} = (A^{(n-1)})'$. We take a *pairing* function which effectively codes all triples of non-negative integers, each one into a single such number, and whose 'inverses' effectively decode such numbers into unique triples, see [K, p. 31]. We use the notation $\langle x \rangle_n$ for the n-th component of x in this decoding, $n = 1, 2$, or 3. By a c.e. set, we mean as usual a computably enumerable set of (possibly tuples of) non-negative integers.

We will use the following, see [S, Chapter IV, 4.1 (ii) and (iii)]:

FACT 1.1. (Relativized Post's Theorem)

(i) $B \in \Sigma_{n+1}^A \Leftrightarrow B \in \Sigma_1^{A^{(n)}}$.

(ii) For all $n \geq 0$, we have $B \leq_T A^{(n)} \Leftrightarrow B \in \Delta_{n+1}^A$.

1.3. Computable real numbers. The function $\nu_{\mathbb{Q}} : \mathbb{N} \to \mathbb{Q}$ defined by $\nu_{\mathbb{Q}}(n) = \frac{\langle n \rangle_1 - \langle n \rangle_2}{\langle n \rangle_3 + 1}$ provides us with an effective enumeration of $\mathbb{Q}$. A function $f : \mathbb{N}^n \to \mathbb{Q}$ into rationals is said to be A-computable when there exists an A-computable $g : \mathbb{N}^n \to \mathbb{N}$ such that $f = \nu_{\mathbb{Q}} \circ g$. We denote the set of all total unary A-computable functions into rationals by $\Gamma_{\mathbb{Q}}^A$. It is clear that $f \in \Gamma_{\mathbb{Q}}^A$

if and only if there are $a, b, s \in \Gamma^A$ such that $f(n) = (-1)^{s(n)} \frac{a(n)}{b(n)}$. We will invoke whichever of the two characterizations that applies more naturally to a given occasion.

A real number α is A-computable (in the sense of Cauchy), if there is $f \in \Gamma_{\mathbb{Q}}^A$ such that $(\forall n \in \mathbb{N})(|\alpha - f(n)| < 2^{-n})$. This turns out to be equivalent to other well known formulations such as the one in the sense of Dedekind, see the last paragraph in [R] for a sketch of the proof of their equivalence. Robinson's proof is based on the important fact that if a real is (Cauchy) computable, then the corresponding Beatty sequence is computable. The set of all A-computable reals is denoted $\Sigma_0^A(\mathbb{R})$.

1.4. The Z-W arithmetical hierarchy of reals. A generalization of computable reals, which gives rise to a hierarchy of arithmetical reals, was presented in [ZW]. This interesting hierarchy is kind of similar to the arithmetical hierarchy of definable functions from $\mathbb{N}$ to $\mathbb{N}$ in plus and times. As mentioned already, in this paper we provide a natural correspondence between the two. Zheng and Weihrauch proved another equivalence as well, see [ZW, Theorem 7.8].

DEFINITION 1.2. [ZW, Definition 7.1] $\Sigma_0^A(\mathbb{R}) = \Pi_0^A(\mathbb{R}) = \Delta_0^A(\mathbb{R}) = \{x \in \mathbb{R} \mid x \text{ is } A\text{-computable}\}$,

$\Sigma_n^A(\mathbb{R}) = \{x \in \mathbb{R} \mid (\exists f \in \Gamma_{\mathbb{Q}}^A)(x = \sup_{i_1} \inf_{i_2} \sup_{i_3} \dots \Theta_{i_n} f(i_1, \dots, i_n))\}$, $n \geq 1$,

$\Pi_n^A(\mathbb{R}) = \{x \in \mathbb{R} \mid (\exists f \in \Gamma_{\mathbb{Q}}^A)(x = \inf_{i_1} \sup_{i_2} \inf_{i_3} \dots \overline{\Theta}_{i_n} f(i_1, \dots, i_n))\}$, $n \geq 1$,

where Θ_{i_n} is $\sup_{i_n}$ if n is odd, and $\inf_{i_n}$ if n is even; $\overline{\Theta}_{i_n}$ is the other way around,

$\Delta_n^A(\mathbb{R}) = \Sigma_n^A(\mathbb{R}) \cap \Pi_n^A(\mathbb{R}), n \geq 1$.

Shoenfield's limit lemma (see [ZW, Lemma 4.1]) implies the following:

FACT 1.3. [ZW, Lemma 7.2] For all $n \geq 1$ and all $x \in \mathbb{R}$, we have:

(i) $(x \in \Sigma_{n+1}^A(\mathbb{R})) \Leftrightarrow (\exists f \in \Gamma_{\mathbb{Q}}^{A^{(n)}})(x = \sup_{i \in \mathbb{N}} f(i))$,

(ii) $(x \in \Pi_{n+1}^A(\mathbb{R})) \Leftrightarrow (\exists f \in \Gamma_{\mathbb{Q}}^{A^{(n)}})(x = \inf_{i \in \mathbb{N}} f(i))$,

(iii) $(x \in \Delta_{n+1}^A(\mathbb{R})) \Leftrightarrow (\exists f \in \Gamma_{\mathbb{Q}}^{A^{(n)}})(x = \lim_{i \to \infty} f(i) \text{ effectively})$.

This hierarchy does not collapse:

FACT 1.4. [ZW, Theorem 7.8] For all $n \geq 1$, we have $\Delta_n^A(\mathbb{R}) \subsetneq \Sigma_n^A(\mathbb{R})$ and $\Delta_n^A(\mathbb{R}) \subsetneq \Pi_n^A(\mathbb{R})$.

§2. Complexity of Beatty sequences. In [ZW, Theorem 7.8] it was shown that the well known map $A \mapsto x[A] = \sum_{a \in A} 2^{-a}$ from arithmetical sets to arithmetical reals respects the levels Δ_n in the two hierarchies (i.e. maps Δ_n sets to numbers in $\Delta_n(\mathbb{R})$ and vice versa). Part (ii) of the result below is another level-by-level equivalence, this time from arithmetical reals to arithmetical functions.

THEOREM 2.1. *Fix $r \in \mathbb{N}$ and oracle A, and let α be an arbitrary positive real.*

(i) *If $\alpha \in \Sigma_r^A(\mathbb{R})$, then the function $\lambda n.\lfloor n\alpha \rfloor$ is of class Δ_{r+1}^A. Similarly, if $\alpha \in \Pi_r^A(\mathbb{R})$, then the function $\lambda n.\lfloor n\alpha \rfloor$ is of class Δ_{r+1}^A.*

(ii) *For $r \geq 1$, we have $\alpha \in \Delta_r^A(\mathbb{R})$ if and only if the function $\lambda n.\lfloor n\alpha \rfloor$ is of class Δ_r^A.*

PROOF. (i) We prove the first assertion; the second one will be similar (essentially by replacing the least upper bounds in the argument by greatest lower bounds). Since we are dealing with a total function, it suffices to show that it is of class Σ_{r+1}^A. If α is rational, then the conclusion is trivial, so suppose α is irrational. Take functions a and b in Δ_r^A such that the sequence $\frac{a(n)}{b(n)}$ is increasing and $\sup \frac{a(n)}{b(n)} = \alpha$. By the definition of α as a least upper bound, for all N we have $(\exists n)\lfloor N\alpha \rfloor = \lfloor N\frac{a(n)}{b(n)} \rfloor$, or equivalently, $(\exists n)(|N(\frac{a(n)}{b(n)} - \alpha)| < \|N\frac{a(n)}{b(n)}\|)$. Recall that norm denotes distance to nearest integer. The reason for this equivalence is that for any two reals ε and δ, if $|\delta - \varepsilon| < \max(\|\varepsilon\|, \|\delta\|)$, then $\lfloor \varepsilon \rfloor = \lfloor \delta \rfloor$; under the condition $|\varepsilon - \delta| < \frac{1}{3}$, the converse holds too. Also for all N and n, we have $|N(\frac{a(n)}{b(n)} - \alpha)| < \|N\frac{a(n)}{b(n)}\|$ if and only if $(\forall m > n)(N\frac{a(m)}{b(m)} - N\frac{a(n)}{b(n)} < \|N\frac{a(n)}{b(n)}\|)$. Let $l(n)$ be the nearest integer to $N\frac{a(n)}{b(n)}$ (a Δ_0-definable function of n, N being fixed). Then $\lfloor N\alpha \rfloor = k$ if and only if $(\exists n)(\forall m > n)(N(a(m)b(n) - a(n)b(m)) < b(m)|b(n)l(n) - Na(n)| \wedge \lfloor N\frac{a(n)}{b(n)} \rfloor = k)$.

(ii) *only if:* Using Fact 1.1 and Fact 1.3, one can give a proof similar to the one sketched in [R].

if: Suppose that $\lambda n.\lfloor n\alpha \rfloor \in \Delta_r^A$. Let $B(0) = \lfloor \alpha \rfloor$ and, for $k \geq 1$, $B(k) = \lfloor 10^k\alpha \rfloor - 10\lfloor 10^{k-1}\alpha \rfloor$. Note that the function B is of class Δ_r^A. Now, for every k, let $\alpha_k = \Sigma_{i=0}^k \frac{B(i)}{10^i}$. These form a Δ_r^A sequence. This sequence effectively tends to its limit α (since for the corresponding infinite series, the tails are dominated by the same-index tails of a convergent geometric sequence). Therefore, by Fact 1.1 and Fact 1.3, we have $\alpha \in \Delta_r^A(\mathbb{R})$. ⊣

§3. The more general Sturmian case.

We now consider sequences of the form $\lfloor n\alpha + \gamma \rfloor$ (the first half of the section is presented for integer parts of polynomials of arbitrary degree though). The related sequences $\lfloor (n + 1)\alpha + \gamma \rfloor - \lfloor n\alpha + \gamma \rfloor$ for irrational $0 < \alpha < 1$ are called Sturmian sequences. The result in the last section generalizes to the following:

PROPOSITION 3.1. *Fix $r \geq 1$ and suppose the positive reals $\alpha_0, \alpha_1, \ldots, \alpha_n$ are in $\Sigma_r^A(\mathbb{R})$ (respectively $\Delta_r^A(\mathbb{R})$). Then the function $f(t) = \lfloor \alpha_n t^n + \alpha_{n-1} t^{n-1} + \cdots + \alpha_1 t + \alpha_0 \rfloor$ is of class Δ_{r+1}^A (respectively $\Delta_r^A(\mathbb{R})$).*

PROOF. This is similar to what we did for the function $\lambda n.\lfloor n\alpha \rfloor$ in the last section (for the "respectively" part, a la Robinson again). ⊣

Regarding the converse of the "respectively" part, the situation is as follows:

PROPOSITION 3.2. *For any $r \geq 1$ and positive reals $\alpha_0, \alpha_1, \ldots, \alpha_n$, if the function $f(t) = \lfloor \alpha_n t^n + \alpha_{n-1} t^{n-1} + \cdots + \alpha_1 t + \alpha_0 \rfloor$ is of class Δ_r^A, then all the coefficients, possibly except the constant α_0, are of class $\Delta_r^A(\mathbb{R})$.*

PROOF. We will have, effectively,

$$\alpha_n = \lim_{i \to \infty} \left(\lfloor \alpha_n (10^i)^n + \alpha_{n-1}(10^i)^{n-1} + \cdots + \alpha_1 10^i + \alpha_0 \rfloor / 10^{ni} \right).$$

This shows that $\alpha_n \in \Delta_r^A(\mathbb{R})$. Therefore the function $\lfloor t\alpha_n \rfloor$ is of class Δ_r^A. Next we have, effectively,

$$\alpha_{n-1} = \lim_{i \to \infty} \left((\lfloor \alpha_n (10^i)^n + \alpha_{n-1}(10^i)^{n-1} + \cdots + \alpha_1 10^i + \alpha_0 \rfloor \right.$$
$$\left. - \lfloor 10^{ni} \alpha_n \rfloor) / 10^{(n-1)i} \right).$$

This implies that $\alpha_{n-1} \in \Delta_r^A(\mathbb{R})$. Continuing this way, we get that the coefficients $\alpha_{n-2}, \ldots, \alpha_1$ are also of the same class. ⊣

REMARK 3.3. There can be no such conclusions for the constant α_0 in general. Indeed, with $\alpha = 1$ and $\gamma \in (0,1)_\mathbb{R}$, the function $\lambda n.\lfloor n + \gamma \rfloor$ is nothing but the identity. But it could be that $\gamma \notin \cup_{j=1}^\infty \Delta_j^A(\mathbb{R})$. If α is *rational*, then for all γ (even those beyond the Z-W hierarchy), the function $\lambda n.\lfloor n\alpha + \gamma \rfloor$, is primitive recursive.

In the above Remark, for rational numbers α, *bad* constants γ could give rise to *good* functions. This can be compared with the following result which says that for all *irrational* numbers α, bad constants γ necessarily give rise to bad functions $\lambda m.\lfloor m\alpha + \gamma \rfloor$.

THEOREM 3.4. *Let α be a positive irrational and $\gamma \in (0,1)$, and suppose $\lambda n.\lfloor n\alpha + \gamma \rfloor \in \Delta_r^A$ for some $r \geq 1$. Then we have $\alpha, \gamma \in \Delta_r^A(\mathbb{R})$.*

PROOF. Assume the hypothesis and note that by an application of Proposition 3.2, as we have $\lambda n.\lfloor n\alpha + \gamma \rfloor \in \Delta_r^A$, we conclude $\alpha \in \Delta_r^A(\mathbb{R})$. Thus $\alpha \in \Pi_r^A(\mathbb{R})$ and $\alpha \in \Sigma_r^A(\mathbb{R})$. Below we show that (i) $\alpha \in \Pi_r^A(\mathbb{R})$ implies $\gamma \in \Sigma_r^A(\mathbb{R})$ and (ii) $\alpha \in \Sigma_r^A(\mathbb{R})$ implies $\gamma \in \Pi_r^A(\mathbb{R})$. The fact that α is irrational will be crucial (as it has to be by the above remark).

(i) Consider a Δ_r^A sequence $(b_m)_{m \in \mathbb{N}}$ of rationals which *decreases* to α. Then for all m and n we have $\lfloor n\alpha + \gamma \rfloor \leq n\alpha + \gamma < nb_m + \gamma$. We claim that $\gamma = \sup_{m,n}(\lfloor n\alpha + \gamma \rfloor - nb_m)$ which will show $\gamma \in \Sigma_r^A(\mathbb{R})$. Here is the argument. Given $0 < \varepsilon < 2\gamma$, there exists $n \in \mathbb{N}$ such that $1 - \gamma < n\alpha - \lfloor n\alpha \rfloor < 1 - \gamma + \frac{\varepsilon}{2}$. The reason is that as mentioned before, multiples of α are uniformly distributed modulo 1 and in particular form a dense set in the unit interval. We will have $\lfloor n\alpha \rfloor + 1 < n\alpha + \gamma < \lfloor n\alpha \rfloor + 1 + \frac{\varepsilon}{2} < \lfloor n\alpha \rfloor + 1 + \gamma < \lfloor n\alpha \rfloor + 2$. Therefore $\lfloor n\alpha + \gamma \rfloor = \lfloor n\alpha \rfloor + 1$. Let m be so that $b_m - \alpha < \frac{\varepsilon}{2n}$. Then we have

$nb_m + \gamma - \lfloor n\alpha + \gamma \rfloor = n(b_m - \alpha) + n\alpha + \gamma - \lfloor n\alpha + \gamma \rfloor < n(\frac{\varepsilon}{2n}) + n\alpha + \gamma - (\lfloor n\alpha \rfloor + 1) = \frac{\varepsilon}{2} + (n\alpha - \lfloor n\alpha \rfloor) + \gamma - 1 < \frac{\varepsilon}{2} + (1 - \gamma + \frac{\varepsilon}{2}) + \gamma - 1 = \varepsilon.$

(ii) Consider a Δ_r^A sequence $(r_m)_{m\in\mathbb{N}}$ of rationals which *increases* to α. Then for all m and n we have $nr_m + \gamma < n\alpha + \gamma < \lfloor n\alpha + \gamma \rfloor + 1$. We claim that $\gamma = \inf_{m,n}(\lfloor n\alpha + \gamma \rfloor + 1 - nr_m)$ which will show $\gamma \in \Pi_r^A(\mathbb{R})$. Given $0 < \varepsilon < 2(1 - \gamma)$, there exists $n \in \mathbb{N}$ such that $1 - \gamma - \frac{\varepsilon}{2} < n\alpha - \lfloor n\alpha \rfloor < 1 - \gamma$ (by multiples of α being dense mod 1 again). By the right inequality we have, $n\alpha + \gamma < \lfloor n\alpha \rfloor + 1$ and so, as γ is positive, $\lfloor n\alpha + \gamma \rfloor = \lfloor n\alpha \rfloor$. Let m be so that $\alpha - r_m < \frac{\varepsilon}{2n}$. Then we have $\lfloor n\alpha + \gamma \rfloor + 1 - nr_m - \gamma = n(\alpha - r_m) + 1 - \gamma - (n\alpha - \lfloor n\alpha + \gamma \rfloor) < \frac{\varepsilon}{2} + 1 - \gamma - (n\alpha - \lfloor n\alpha \rfloor) < \frac{\varepsilon}{2} + 1 - \gamma - (1 - \gamma - \frac{\varepsilon}{2}) = \varepsilon.$ ⊣

REMARK 3.5. One can try to devise relative algorithms for specific cases, algorithms using, among other things, primitive recursion with respect to $\lambda n. \lfloor n\alpha + \gamma \rfloor$ to *compute* γ (along with α). But computation of γ in what sense? It might be that if this computation is performed in terms of computable nested intervals with rational endpoints rather than decimal representation, the task is easier.

§4. Indices of computable Beatty sequences.

In this section we prove Π_2-definability of $\{\ulcorner \lambda n. \lfloor n\alpha \rfloor \urcorner \mid \alpha$ is a computable positive real$\}$. As a prelude, notice that by Robinson's result, and using the Π_2 (complete) set Tot (indices of total computable functions), we have the definition $e \in$ Tot $\wedge$ $(\exists j)(\forall n)(\exists k)(\forall m \geq k)(\varphi_e(n) = \lfloor n \cdot$ (the rational output of the machine j on $m)\rfloor)$. Saving some complexity, we have the definition $e \in$ Tot $\wedge$ $(\exists j)(\forall n)(\varphi_e(n) = \lfloor n \cdot$ (the rational output of the machine j on $n)\rfloor)$ which is Σ_3. In this section we provide two Π_2 definitions for the same set. The lemma below is not concerned with computability.

LEMMA 4.1. *Let $f : \mathbb{N} \to \mathbb{N}$ be an arbitrary function. Then $f = \lambda n. \lfloor n\alpha \rfloor$ for some $\alpha \in \mathbb{R}^{>0}$ if and only if $(\forall x, y)(f(x) + f(y) \leq f(x + y) \leq f(x) + f(y) + 1)$.*

PROOF. The *only if* part is clear. To prove the *if* part, suppose f satisfies the inequalities for all x, y. We proceed through the following claims:

CLAIM 1. $(\forall k \geq 1)(\forall x)(kf(x) \leq f(kx) \leq kf(x) + k - 1)$.

PROOF OF CLAIM 1. By induction on k; the induction basis is trivial. Suppose $kf(x) \leq f(kx) \leq kf(x) + k - 1$. In the assumed inequalities $f(x) + f(y) \leq f(x + y) \leq f(x) + f(y) + 1$, let $y = kx$. Then $f(kx) + f(x) \leq f((k+1)x) \leq f(kx) + f(x) + 1$. This shows $(k + 1)f(x) \leq f((k+1)x) \leq (k + 1)f(x) + k.$ ⊣

CLAIM 2. The sequence $(\frac{f(n)}{n})_{n\in\mathbb{N}}$ is convergent to a real number α.

PROOF OF CLAIM 2. Presume, for the sake of arriving at a contradiction, that the sequence above is not Cauchy. Then $(\exists \varepsilon > 0)(\forall n)(\exists m > n)|\frac{f(n)}{n} - \frac{f(m)}{m}| \geq \varepsilon.$

Fix such an ε. Let $n \in \mathbb{N}$ be such that $\frac{1}{n} < \varepsilon$. Choose $m > n$ such that $\left|\frac{mf(n)-nf(m)}{mn}\right| \geq \varepsilon$. By Claim 1 we have

$$f(mn) - m + 1 \leq mf(n) \leq f(mn) \leq nf(m) + n - 1 \leq f(mn) + n - 1.$$

Therefore $-n < nf(m) - mf(n) < m$. Hence $|nf(m) - mf(n)| < m$ and so $\frac{|nf(m)-mf(n)|}{mn} < \frac{1}{n} < \varepsilon$. This is a contradiction.   ⊣

CLAIM 3. $(\forall k \geq 1)(\forall x)\lfloor \frac{xf(k)}{k} \rfloor \leq f(x) = \lfloor \frac{f(kx)}{k} \rfloor \leq \lfloor \frac{xf(k)+x}{k} \rfloor$.

PROOF OF CLAIM 3. By Claim 1, if $k \geq 1$ then $kf(x) \leq f(kx) \leq kf(x) + k - 1$, so $f(x) \leq \frac{f(kx)}{k} \leq f(x) + 1 - \frac{1}{k}$, whence $f(x) = \lfloor \frac{f(kx)}{k} \rfloor$.

By Claim 1 we have $2f(0) \leq f(2 \cdot 0)$. Hence if $x = 0$ it is clear that for any $k \geq 1$, $\lfloor \frac{xf(k)}{k} \rfloor \leq f(x) \leq \lfloor \frac{xf(k)+x}{k} \rfloor$. So suppose $x \geq 1$. Then by Claim 1, if $k \geq 1$ we have $xf(k) \leq f(kx) \leq xf(k)+x-1$, so $\frac{xf(k)}{k} \leq \frac{f(kx)}{k} \leq \frac{xf(k)+x-1}{k}$, and thus $\lfloor \frac{xf(k)}{k} \rfloor \leq f(x) \leq \lfloor \frac{xf(k)+x-1}{k} \rfloor \leq \lfloor \frac{xf(k)+x}{k} \rfloor$.   ⊣

Returning to the proof of the lemma, Claim 3 implies that for every $x \geq 1$, $\lim_{k\to\infty}\lfloor \frac{xf(k)}{k} \rfloor \leq f(x) \leq \lim_{k\to\infty}\lfloor \frac{xf(k)+x}{k} \rfloor$. Therefore, by Claim 2, $f(x) = \lfloor x\alpha \rfloor$.   ⊣

THEOREM 4.2. *For any oracle A, the set of all nonnegative integers e for which the function φ_e^A is of the form $\lambda k.\lfloor k\alpha \rfloor$, for some $\alpha \in \mathbb{R}^{>0}$ (necessarily A-computable), is of class Π_2^A.*

PROOF. The preceding lemma, shows that the set above equals $\{e \mid (\forall x, y) (\varphi_e^A(x) + \varphi_e^A(y) \leq \varphi_e^A(x + y) \leq \varphi_e^A(x) + \varphi_e^A(y) + 1)\}$. The result follows since the predicate $\varphi_e^A(z) \leq q$ with respect to e, z, and q is in Σ_1^A. In more detail, using Kleene's $T^A(e, x, r)$ relativized predicate, which means r codes computation of A-program number e on the input x, and a primitive recursive function f reading off the output u from such a computation code r, we see that a number e has the property in question if and only if $(\forall x, y)(\exists u, v, w)(\exists r, s, t)(T^A(e, x, r) \wedge f(r) = u \wedge T^A(e, y, s) \wedge f(s) = v \wedge T^A(e, x + y, t) \wedge f(t) = w \wedge u + v \leq w \leq u + v + 1)$.   ⊣

PROPOSITION 4.3. *Let $f : \mathbb{N} \to \mathbb{N}$ be an arbitrary function. Then $f = \lambda n.\lfloor n\alpha \rfloor$ for some $\alpha \in \mathbb{R}^{>0}$ if and only if $(\forall x)(\exists y)(yf(x) \leq xf(y) < yf(x) + y)$.*

PROOF. We use the fact that for f to be a Beatty sequence, it is necessary and sufficient that $(\forall x)(\exists y)(f(x) = \lfloor x\frac{f(y)}{y} \rfloor)$ (the similar idea appeared a couple of times earlier in the paper).   ⊣

COROLLARY 4.4. *The set of indices e of computable Beatty sequences is defined by $(\forall x)(\exists y)(y\varphi_e(x) \leq x\varphi_e(y) < y\varphi_e(x) + y)$ too. The matrix of the formula is Σ_1 and the whole formula is Π_2.*

We leave it open whether the set $\{ \lceil \lambda n. \lfloor n\alpha \rfloor \rceil \mid \alpha \in \mathbb{R}^{>0} \text{ is computable}\}$ is Π_2-complete, that is whether it is not Σ_2. The combinatorial properties of the Sturmian and Beatty sequences might have a role in this regard, see [A].

Acknowledgments. This work was in part supported by a grant (No. 83030213) from Institute for Studies in Theoretical Physics and Mathematics (IPM). The results here formed part of the first author's MS thesis under second author's supervision at Tarbiat Modarres University. We express our gratitude to A. Enayat and I. Kalantari for encouraging us to submit the paper here. Later on the comments by IK, especially those on Thm. 3.4, greatly improved the paper. We are thankful to the two referees, especially for detailed comments by one who also suggested the present simplification of part of the Proof of Lemma 4.1. Last but not least, we thank J.-P. Allouche for his comments on an earlier version and introducing us to Sturmian and Beatty sequences.

REFERENCES

[A] P. ARNOUX, *Sturmian sequences*, **Substitutions in Dynamics, Arithmetics and Combinatorics** (V. Berthé et al., editors), Lecture Notes in Mathematics, vol. 1794, Springer, Berlin, 2002, pp. 143–198.

[H] G. HARMAN, *Metric Number Theory*, The Clarendon Press Oxford University Press, New York, 1998.

[K] R. KAYE, *Models of Peano Arithmetic*, The Clarendon Press Oxford University Press, New York, 1991.

[N] I. NIVEN, *Diophantine Approximations*, Interscience Publishers, a division of John Wiley & Sons, New York, 1963.

[R] R. M. ROBINSON, *Review of rekursive funktionen by R. Péter*, **The Journal of Symbolic Logic**, vol. 16 (1951), pp. 280–282.

[S] R. I. SOARE, *Recursively Enumerable Sets and Degrees*, Perspectives in Mathematical Logic, Springer-Verlag, Berlin, 1987.

[ZW] X. ZHENG and K. WEIHRAUCH, *The arithmetical hierarchy of real numbers*, **Mathematical Logic Quarterly**, vol. 47 (2001), no. 1, pp. 51–65.

INSTITUTE FOR STUDIES IN THEORETICAL
    PHYSICS AND MATHEMATICS
        TEHRAN, IRAN
and
    DEPARTMENT OF MATHEMATICS
        TARBIAT MODARRES UNIVERSITY
            TEHRAN, IRAN
E-mail: mehdigh@ipm.ir
E-mail: mojmon@ipm.ir

SPECKER'S THEOREM, CLUSTER POINTS, AND COMPUTABLE QUANTUM FUNCTIONS

IRAJ KALANTARI AND LARRY WELCH

Abstract. The present paper continues the work we began in [KalWel95, KalWel96, KalWel03, KalWel05].

Specker [Spe49] proved existence of a computable sequence of computable reals whose limit is not a computable real by using a noncomputable c.e. (computably enumerable) set. He did this by requiring his sequence to evade every computable point as a limit.

We study similar and generalized results in our filter-based approach to computable analysis and computable topology. We strategically construct *evading sequences* of basic open sets to generalize Specker's work. An evading sequence is a sequence of basic open sets such that any sequence of points with one point chosen to lie in each basic open set has the same cluster points as any other such sequence. This set of cluster points is the set of *Specker cluster points* of the evading sequence. We devise two methods for obtaining evading sequences such that all the Specker cluster points lie in the spectrum of a given avoidance function. We then use these methods to construct evading sequences where their cluster points can be of one or more types (computable, avoidable, or shadow) and of any possible cardinality for each type. Finally, we use the acquired machinery to reveal two facts of the fine structure of the lower semi-lattice of domains of computable quantum functions under the relation of 'subset'. Specifically, we show that we cannot always interpolate a computable quantum function between two nested computable quantum functions by constructing a pair of nested computable quantum functions whose domains differ in exactly one point. In contrast, we show that any time two nested computable quantum functions exist whose domains differ by at least two points, we can interpolate another computable quantum function between them.

§1. Introduction. There are three classical approaches to general topology: Fréchet's abstract spaces, Hausdorff's neighborhood classes and Kuratowski's

---

Key words and phrases. Recursion, Computability, Topology, Analysis; Π^0_1 trees.

We learned recursion theory in the 1970s, and in those days anything 'effective' was defined as 'recursive'. That experience and inertia induced us to stay with that nomenclature throughout our mathematical work despite a general migration of researchers to the more accurate terminology and Soare's explicit call (see [Soa96]) for adoption of the word 'computable'. Because our approach to the study of effectiveness in analysis and topology is quite intertwined with present research subscribing to the more accurate terminology, we believe it is time for us, too, to adopt the nomenclature that is more appropriately descriptive and harmonious. Thus, in contrast to our previous work, we use 'computable' (and its derivatives) throughout this paper and will do so in the future.

Logic In Tehran
Edited by A. Enayat, I. Kalantari, and M. Moniri
Lecture Notes in Logic, 26
© 2006, ASSOCIATION FOR SYMBOLIC LOGIC

closure classes. The fundamental objects with which one works in all these approaches are points. But a point can be viewed as a limit of a sequence of other points or the sole member of an intersection of a nested sequence of open intervals. These approaches to points are often useful. In computable/recursive mathematics we investigate the 'constructive' aspects of mathematical theorems, interpreting 'constructive' via Turing computability, while maintaining the classical laws of logic.

One way to 'construct' the real numbers from the rationals is by means of Cauchy sequences. A sequence $\{a_n\}$ is Cauchy if for every n there is a number $f(n)$ such that whenever $i, j > f(n)$ then $|a_i - a_j| < 2^{-n}$. If both the sequence and f are computable, then the sequence converges to a computable real number. In this sense, all algebraic real numbers and well-known constants such as π and e are computable. Turing [Tur36] gave mathematical precision to this definition of computable reals in his paper '*On computable numbers, ...* '. In fact, the need for an approach to computable reals was a basic motivating factor for his invention of the Turing machine. Besides the Cauchy sequence representation, there are several other representations of the real numbers (continued fraction representation, representation by nested intervals) which, when effectivized, all induce the same notion of computability of real numbers. At first Turing used the decimal expansion representation, but in a correction to his paper [Tur37], he noted that the decimal representation is not appropriate for the study of real-valued functions. For instance, even an elementary function such as $\lambda x f(x) = 3x$ is not computable with respect to the decimal representation. Mostowski [Mos57] found that even the notion of a *computable sequence* of real numbers depends on the chosen representation for the reals. The computable reals and computable analysis have been extensively researched (see the bibliography by Brattka & Kalantari [BraKal98]), and have many interesting properties: for example, Rice [Ric54] proved that computable real numbers form a real algebraically closed field.

Grzegorczyk [Grz55, Grz57] and Lacombe [Lac55, Lac55a, Lac55b] defined *computable real functions* as follows: a real-valued function f is *computable* if there is a Turing machine M which transforms each Cauchy sequence of rational numbers rapidly converging to x into a Cauchy sequence of rational numbers rapidly converging to $f(x)$. All well-known continuous functions, such as polynomials with computable real coefficients, the trigonometric functions and the exponential function, are computable in this sense. Computable real functions are *necessarily* continuous, and thus some familiar phenomena hold for them. But there are also some interesting pathologies. For instance, Myhill [Myh71] shows that the derivative of a computable function need not be computable, and a result of Specker's [Spe59] (also see Lacombe [Lac57a]) says that a computable real function need not attain its maximum (a computable value) at a computable real. In the Grzegorczyk/Lacombe definition, computable real functions include all real numbers in their domains; but some authors in the 'Russian school' have studied computable functions whose domains are restricted to the computable reals.

In a series of papers, Lacombe [Lac57a, Lac58a] and Kreisel & Lacombe [KL57] defined a *recursively enumerable* open subset of the real numbers. An open subset A of the real numbers is *recursively enumerable* if there is a computable sequence of pairs of rational numbers $(r_n, s_n)_{n \in \omega}$ such that the union of the open intervals (r_n, s_n) is equal to A. This notion yields interesting results in topology and analysis. (See Kalantari [Kal82], and Brattka & Kalantari [BraKal98].)

1.1. Specker's and Russians' results. Both Specker and various Russian researchers such as Markov, Ceĭtin, Šanin, Zaslavskiĭ, and Orevkov have considered the behavior of computable functions only at computable points. Kushner [Kus83] also studies computable functions in this way, but at times shows an interest in the behavior of noncomputable reals. (We elaborate on this below, in Section 5.) These researchers asked the natural question, 'In the context of an effective platform for the reals, which theorems of classical analysis carry over and which don't, and how badly do some fail?' For instance, Specker [Spe49] proved the existence of a computable function on $I = [0, 1]$ such that $f(I) \subseteq I$, and although f achieves a maximum computable value, that value is not achieved at a computable real. Ceĭtin [Cei64a] gave a new proof of a theorem of Mučnik showing that there exists a computable function f on $[-1, 1]$ which is defined at 0 but not defined on all points of any neighborhood of 0. Zaslavskiĭ [Zas62], *Some Properties of constructive Real Numbers and Constructive Functions*, Theorem 5.1, constructs a 'recursive function' from $[0, 1]$ to $\mathbb{R}$ which, though continuous on the computable points of a closed set, has *unbounded* range.

Similarly, Orevkov [Ore63a] constructs two 'computable functions' from the unit square to itself that violate Brouwer's Fixed Point Theorem. He produces the first one by continuously retracting the square to its boundary and then rotating it by 90 degrees to prevent fixed points on the boundary. In order to accomplish the retraction, Orevkov takes this function to be classically continuous, but restricts its domain to a proper subset of the unit square that contains all computable points.

Orevkov's second 'computable function' is in fact continuous on the entire unit square. He constructs it in such a way that none of the points retracted to the boundary by the first function is a fixed point of the second one. Hence all of the fixed points occur at noncomputable points. So while Brouwer's Fixed Point Theorem holds for the second function in the classical sense, it fails computably.

On the unit interval in $\mathbb{R}$, the situation is different. Of course the Fixed Point Theorem classically holds of the closed interval $[0, 1]$, but unlike the unit square, it holds computably too, in the sense that a computable function defined on all computable points of $[0, 1]$ and mapping these points into $[0, 1]$ has a computable fixed point, as shown, for instance, in [KalWel96, Theorem 8.2]. The failure of Brouwer's Fixed Point Theorem on the unit square has been studied by J. Miller in his thesis [Mil02], where he characterizes the set of fixed points of a computable map from $[0, 1]^n$ to itself as a Π_1^0 class which contains a nonempty, connected Π_1^0 subclass, generalizing Orevkov's approach.

1.2. The sequence approach. In contrast to the Russian approach, Goodstein [Goo61], Pour-El & Richards [PR89], and others have investigated computable functions on entire spaces such as an interval in $\mathbb{R}^n$ or an open set in $\mathbb{C}^n$. In [PR89] Pour-El & Richards introduced an axiomatic approach to computability in Banach spaces involving a *computable sequence*. A subset of sequences in a Banach space is called a *computability structure* on that space if it fulfills three axioms which state, in effect, that computable sequences are closed under linear combinations, under norm, and under limit (in a specific sense). They were motivated to pursue this approach by the fact that a real function (defined on the unit interval, say) is computable in the sense of Grzegorczyk/Lacombe if and only if it maps computable sequences to computable sequences and admits a computable modulus of continuity. Pour-El & Richards applied their theory to L^p and l^p spaces, investigated computability of eigenvalues and formulated a criterion, for closed linear operators $f : X \to Y$ and Banach spaces X, Y, which characterizes the pathology that 'there is a computable point x such that $f(x)$ is non-computable'. If X is separable, and if there exists a computable sequence $(e_n)_{n \in \omega}$ whose linear span is dense in X, and furthermore, if $(f(e_n))_{n \in \omega}$ is a computable sequence, then they conclude that $f(x)$ is computable for each computable point x if and only if f is bounded.

The sequence approach of Pour-El & Richards has been generalized to Fréchet spaces by Washihara and Yasugi [WY96] and to metric spaces by Mori, Tsujii & Yasugi [MTY97].

1.3. The representation approach. A more recent approach to computable analysis by Kreitz & Weihrauch [KW84] and Weihrauch [Wei00], which has become well known, and which often now serves as the basis for investigations in computable analysis, is the 'Type 2 Theory of Effectivity'. This theory is based on the idea of a *representation* of a set X, which is a mapping from Cantor space (the space of all infinite sequences over a finite set) onto X. Essentially, a representation is a way to code arbitrary points of X by infinite sequences of symbols. Examples of representations are the Cauchy sequence representation and the binary representation of the real numbers. This approach allows the investigator to single out well-behaved (admissible) representations and is a natural way to find appropriate definitions of effectivity for many spaces. It is therefore easy to define computability of points, sequences, and functions for a large class of spaces (see M. Schröder's [Sch02] for an interesting extension of the representation approach to a vast class of topological spaces). Extensively investigated spaces include the real numbers, the space of continuous functions, the hyperspace of closed subsets, L^p spaces, and measure spaces [Wei93, Wei97b, BW97]. Hertling [Her97b] has shown that the real number structure is *effectively categorical*: up to equivalence, there is only *one* representation which makes the basic operations of the structure computable.

Using the Type 2 Theory of Effectivity, effective versions of several classical theorems have been produced. For example, Kreitz and Weihrauch [Kre87] show that an effective version of the Heine-Borel Theorem holds and Hertling [Her97a] shows that an effective version of the Riemann Mapping Theorem holds.

Brattka [Bra96, Bra97] has shown that representation-free means can be used to accomplish the same results in the case of the space of reals, and more generally for complete separable metric spaces. He does this by applying certain basic operations and closure schemes to such spaces, thereby generating exactly the computable operations of Type 2 Theory.

1.4. The filter approach. In Kalantari & Welch [KalWel95, KalWel96] we have developed an intrinsically point-free approach to computable topology which uses *filters* to define points, computable points, functions and computable functions for certain topological spaces. Our primitive objects are therefore not points, but neighborhoods in the sense of Hausdorff. These neighborhoods form a subbasis of the space. Though the theory is point-free at the level of our object language, in the metalanguage we do not hesitate to refer to points, and many of our results are theorems about the properties of points. In recognition of the Russian results on functions whose domains do not classically contain intervals, we use this point-free approach to investigate the connections between classical topology and computable topology. Such an approach is particularly useful here since computable points are defined as limits of sequences of approximants.

In analogy with $\mathbb{R}^n$ we consider connected, second countable, regular spaces with at least two points. Because each member of our subbasis must be named with a nonnegative integer, second countability is necessary to study a space from a point-free, computable perspective. Because of the regularity and second countability, each of our spaces is homeomorphic to a nontrivial connected subspace of the Hilbert cube $[0, 1]^\omega$. Thus our spaces are metrizable, and some, such as $\mathbb{R}$ topologized by open intervals with rational endpoints, are also computably metrizable.

Our approach to addressing points can be viewed from the perspective of Type 2 Theory of Effectivity in the following way: Each basic open set of the subbasis of a space X is represented by a nonnegative integer, and our indexing of the subbasis is one-to-one. Each point in the space is 'named' with a closure-nested sequence of open sets of the subbasis that converges to it. Of course each point has infinitely many such names. We call such a name for a point a *sharp filter*. In $\mathbb{R}$, where the basic open sets are intervals with rational endpoints, this is similar to the naming system Weihrauch calls ρ^a (see [Wei00, p. 88, Lemma 4.1.6]). But in distinction to Weihrauch's system, we do not require the lengths of the intervals to decrease rapidly, but instead impose a different sort of requirement, that the sharp filter *resolve* each target (see 2.1, below). The recursion theoretic complexity is the same in both cases: the determination of whether a sequence A is a sharp filter is $\Pi_2^0(A)$, just as the determination of whether an infinite binary string p is a name of type ρ^a is $\Pi_2^0(p)$.

In order to allow ourselves to study both the approach of the Russians and that of Pour-El & Richards, et al., we consider two distinct effectivizations for functions, as described in [KalWel96], below, which we call *computable functions* and *computable quantum functions*. When the space is a closed interval I in $\mathbb{R}$, a *computable function* in our sense is identical to a computable function in the ordinary sense that has domain I. On the other hand, a *computable quantum function* on I is a computable function that has all computable points of I in its domain, though certain noncomputable points may be excluded from that domain. Points that can be excluded from domains of computable quantum functions we call *avoidable* points, and those that cannot we call *shadow* points. A computable quantum function on I displays some properties of a classically continuous function with domain I. For example, it satisfies an effective version of the Intermediate Value Theorem. However, a computable quantum function can also be significantly pathological, since it is essentially the same sort of function studied by the Russian school.

§2. **Background, definitions, and basic results.** In this section, we recall some key definitions and results from our previous work [KalWel95] and [KalWel96], and fix the basic topological and recursion theoretic properties of the spaces of our study at the end.

2.1. **Points, sharp filters, and enumerations.** In this subsection, we summarize our machinery for capturing (computable) points through (computable) *sharp filters*, specify our topological and recursion theoretic settings, and describe an *acceptable enumeration capturing all computable sharp filters*.

DEFINITION 2.1. Let X be a topological space with basis Δ. A sequence $A = \{\alpha_i : i \in \omega\}$ of basic open sets is a *sharp filter* in Δ if (1) $(\forall i)(\overline{\alpha_{i+1}} \subseteq \alpha_i)$, and (2) $(\forall \beta, \gamma)[(\overline{\beta} \subseteq \gamma) \Rightarrow \exists i[(\alpha_i \cap \beta = \emptyset) \vee (\alpha_i \subseteq \gamma)]]$.

We say α *resolves* $\langle \beta, \gamma \rangle$ if $(\overline{\beta} \subseteq \gamma) \Rightarrow [(\alpha \cap \beta = \emptyset) \vee (\alpha \subseteq \gamma)]$. (Note that if $\overline{\beta}$ is not a subset γ, then α resolves $\langle \beta, \gamma \rangle$ trivially.) When a sequence $\{\alpha_i : i \in \omega\}$ satisfies clause (2) for fixed β and γ, we say that it *resolves* $\langle \beta, \gamma \rangle$. We will refer to property (2) as the *resolution property*. We say A *converges* to x, or x is the *limit* of A, and write $A \searrow x$, if $\bigcap \alpha_i = \{x\}$.

For $A = \{\alpha_i : i \in \omega\}$ and $B = \{\beta_i : i \in \omega\}$ sharp filters, we say A is *equivalent* to B, and write $A \equiv B$, if $(\forall i)[\alpha_i \cap \beta_i \neq \emptyset]$.

DEFINITION 2.2. Let X be a topological space with basis Δ. Δ is *compactible* if every member of Δ has compact closure. $\langle X, \Delta \rangle$ is *resolvable* if X is regular and $1°$ countable and Δ is a compactible basis, and $\langle X, \Delta \rangle$ is *completely connected* if X a connected space and Δ is a basis of connected sets.

DEFINITION 2.3. $\langle X, \Delta \rangle$ is said to be *semi-computably presentable* if $\Delta = \{\delta_n : n \in \omega\}$ is countable and for $\alpha, \beta \in \Delta$, the predicates '$\alpha \subseteq \beta$', '$\overline{\alpha} \subseteq \beta$', and '$\alpha \cap \beta = \emptyset$' are computable.

DEFINITION 2.4. Let $\langle X, \Delta \rangle$ be a semi-computably presentable space, $\Delta = \{\delta_i : i \in \omega\}$, and let $A = \{\alpha_i : i \in \omega\}$ be a sharp filter in Δ. Then A is *computable* if there is a computable function $f : \omega \to \omega$ such that for every i, $\alpha_i = \delta_{f(i)}$. For $x \in X$, we say x is Δ-*computable* if there is a computable sharp filter A with $A \searrow x$. When Δ is understood we simply say x is *computable*. We also define $\text{Comp}(X) = \{x : x \in X \text{ and } x \text{ is } \Delta\text{-computable}\}$.

DEFINITION 2.5. A sequence of partial computable functions $\{\psi_n : n \in \omega\}$ is called *an acceptable enumeration capturing all computable sharp filters in* X if for each $n \in \omega$, $\psi_n : \omega \to \Delta_X$:

1. $(\forall n, s, k)[\psi_n^s(k+1)\downarrow \Rightarrow \psi_n^s(0)\downarrow \wedge \cdots \wedge \psi_n^s(k)\downarrow]$ (here s is a stage in the computation and $\psi_n^s(k)$ denotes the outcome of $\psi_n(k)$ at stage s);
2. $(\forall n, k)[\psi_n(k+1)\downarrow \Rightarrow \overline{\psi_n(k+1)} \subseteq \psi_n(k)]$; and
3. for each computable sharp filter A, there is n such that $\psi_n = A$.

Clearly, acceptable enumerations capturing all computable sharp filters exist.

2.2. Correspondences. Here we recall the basic definitions for *correspondences*, devices which yield functions in the classical sense.

We study two different effectivizations for functions. First are those that are defined at all points of the space, which we call *computable functions*. If the space is a closed interval in $\mathbb{R}$, our computable functions behave, on computable points, exactly as the classically continuous computable functions studied by Aberth [Abe80], Goodstein [Goo61], Lacombe [Lac55], Mazur [Maz63], Myhill [Myh71], Pour-El & Richards [PR89], Rice [Ric54], Šanin [Sha68], and many others. In this approach a key and interesting fact is the Ceïtin-Kreisel-Lacombe-Schoenfield theorem, that every total computable function on I (an interval in $\mathbb{R}$) is continuous, with a rather technical proof. In our approach, the same theorem (see Theorem 2.16), is true in a more general setting and it has an immediate and transparent proof (see [KalWel95]).

Second are those functions that are maneuvered not to be defined at some noncomputable points, which we call *computable quantum functions*. Since we are interested in analyzing the Russian work, we require, at a minimum, that these functions be defined on all computable points.

NOTATION 2.6. For a partial function $F : \Delta_X \to \Delta_Y$, we write $F(\alpha) \downarrow$ to indicate F converges on input α. If $A = \{\alpha_i : i \in \omega\}$ is a sharp filter, we write $F(A) \downarrow$ to mean $(\forall i)[F(\alpha_i) \downarrow]$.

DEFINITION 2.7. Let $\langle X, \Delta_X \rangle$ and $\langle Y, \Delta_Y \rangle$ be semi-computably presented, resolvable, topological spaces. A partial function $F : \Delta_X \to \Delta_Y$ is a *correspondence*, or a *full correspondence* if

1. $(\forall \alpha, \beta)[[F(\alpha) \downarrow \wedge F(\beta) \downarrow \wedge \alpha \subseteq \beta] \Rightarrow [F(\alpha) \subseteq F(\beta)]]$,
2. $(\forall \alpha, \beta)[[F(\alpha) \downarrow \wedge F(\beta) \downarrow \wedge \overline{\alpha} \subseteq \beta] \Rightarrow [\overline{F(\alpha)} \subseteq F(\beta)]]$ (if F has properties (1) and (2), we say F is *monotone*;), and
3. $(\forall B$ a sharp filter in $\Delta_X)(\exists A$ a sharp filter in $\Delta_X)$ $[(A \equiv B) \wedge (F(A) \downarrow) \wedge (F(A)$ is a sharp filter in $\Delta_Y)]$.

   If F is also partial computable, we say it is a *computable correspondence*.

DEFINITION 2.8. A partial function $F : \Delta_X \to \Delta_Y$ is a *quantum correspondence* if

1. F is monotone, and
2. $(\forall B$ a computable sharp filter in $\Delta_X)(\exists A$ a computable sharp filter in $\Delta_X)$ $[(A \equiv B) \wedge (F(A) \downarrow) \wedge (F(A)$ is a sharp filter in $\Delta_Y)]$.

   If F is also partial computable, we say it is a *computable quantum correspondence*.

DEFINITION 2.9. Let $F, G : \Delta_X \to \Delta_Y$ be correspondences. We say F is *equivalent* to G, and write $F \equiv G$, if, given any sharp filters A and B in X such that $A \equiv B$, $F(A) \downarrow$, $G(B) \downarrow$, and $F(A)$ and $G(B)$ are sharp filters in Y, then $F(A) \equiv G(B)$ (or equivalently, $(\forall i)[F(\alpha_i) \cap G(\beta_i) \neq \emptyset]$).

PROPOSITION 2.10. *Let $F : \Delta_X \to \Delta_Y$ be a correspondence, and suppose $B = \{\beta_i : i \in \omega\}$ is a sharp filter and $\{i : F(\beta_i) \downarrow\}$ is infinite. Then F has the following two properties:*

1. *There exists a sharp filter $C = \{\gamma_i : i \in \omega\} \subseteq B$ such that $(\forall i)[F(\gamma_i) \downarrow]$ and $F(C)$ is a sharp filter in Δ_Y.*
2. *For all sharp filters $D = \{\delta_i : i \in \omega\}$ in X, if $D \equiv B$, $(\forall i)[F(\delta_i) \downarrow]$, and $F(D)$ is a sharp filter in Δ_Y, then $F(D) \equiv F(C)$.*

DEFINITION 2.11. Let $F : \Delta_X \to \Delta_Y$ be a correspondence. Define f_F to be the unique function generated by F as in the above Proposition: namely, $f_F : X \to Y$ is defined by $f_F(x) = $ the unique point in $\bigcap F(A)$, where $x \in X$, and A is a sharp filter in Δ_X such that $A \searrow x$, $F(A) \downarrow$ and $F(A)$ is a sharp filter in Δ_Y. (See (2) of the Proposition above.)

Note that because F is not defined on all sharp filters, f_F is not total and thus the domain of f, $\text{dom}(f)$, could be a proper subset of X. We define the *exdomain* of f, $\text{exdom}(f)$, to be the complement of $\text{dom}(f)$.

PROPOSITION 2.12. *Let $F, G : \Delta_X \to \Delta_Y$ be correspondences. Then $F \equiv G$ iff $f_F = f_G$.*

THEOREM 2.13. *Let $F : \Delta_X \to \Delta_Y$ be a correspondence. Then f_F is a continuous function from X to Y.*

THEOREM 2.14. *Let $\langle X, \Delta_X \rangle$ and $\langle Y, \Delta_Y \rangle$ be resolvable with Δ_X and Δ_Y countable. Suppose $f : X \to Y$ is a continuous function. Then there exists a correspondence $F : \Delta_X \to \Delta_Y$ such that $f_F = f$.*

2.3. Functions and quantum functions. In this subsection, we list some results that demonstrate our approach is unifying, revealing and thus successful.

DEFINITION 2.15. For a computable correspondence F we refer to the function it generates on the space, f_F, as a *computable function*. If F is a computable quantum correspondence, we refer to the function f_F as a *computable quantum function*.

THEOREM 2.16. *Let $F : \Delta_X \to \Delta_Y$ be a computable quantum correspondence. Then f_F is continuous.*

Although a computable quantum function on a closed interval in $\mathbb{R}$ may be undefined at certain points of that interval, and may behave pathologically in some regards, it also displays some of the properties of classically continuous functions on the interval. For example, a computable quantum function satisfies an effective version of the Intermediate Value Theorem. Such functions were studied by Ceĭtin [Cei64a], Ceĭtin & Zaslavskiĭ [Cei62b], Šanin [Sha62], Zaslavskiĭ [Zas62], *et al.*

A natural question that can be raised in this context, similar to the original question raised by the researchers of the 1960's, is which of the results of classical computable analysis about total computable functions carry over to all computable quantum functions, and which can in some way be refuted when totality is not considered.

It turns out that computable quantum functions have some surprisingly well-behaved structure as well as expected pathologies. Among well-behaved properties are the facts that a computable quantum function is continuous on its domain, and that computable quantum functions on $\mathbb{R}$ classically satisfy the Intermediate Value Theorem just as if they were total continuous functions (see [KalWel96]). Furthermore, in [KalWel96] we prove that any function computable in the sense of Goodstein [Goo61] or Pour-El & Richards [PR89] can be generated by a computable correspondence. Among pathologies, we prove given any computable quantum function $f : \mathbb{R} \to \mathbb{R}$, we can find a computable quantum function $g \subseteq f$ such that the Lebesgue measure of the domain of g is 0. For another example, in [KalWel96], we construct a computable quantum function on [0, 1], that is *nonextendible* to a classically continuous function of larger domain. This function is of unbounded variation and its domain can be made of positive Lebesgue measure less than ε for any $\varepsilon > 0$. While a resolvable, semi-computably presentable space need not have a measure on it, we can still build a computable quantum function with dense exdomain.

Another well-behaved and fundamentally needed result is that the intersection of a uniformly computably enumerable sequence of computable quantum functions is also a computable quantum function.

THEOREM 2.17 (*Uniform Function Intersection Theorem*). *Suppose* $\{F_k : k \in \omega\}$ *is a uniformly computable sequence of computable quantum correspondences from* Δ_X *to* Δ_Y *such that for all* $x \in \text{Comp}(X)$ *and all* $j, k \in \omega$, $f_{F_j}(x) = f_{F_k}(x)$. *Then there is a computable quantum correspondence* $H \subseteq F_0$ *such that* $f_H = \bigcap_k f_{F_k}$.

As we build our computable functions through point-free techniques, we must be careful that the mappings from open sets to open sets that define them deliver as expected; namely, that they indeed generate mappings from points to points. This we call the concept of *honesty*. It turns out that despite existence of not honest computable quantum functions, any such computable quantum function, even if *dishonest* on some noncomputable input, has an honest equivalent.

DEFINITION 2.18. A computable quantum correspondence $G : \Delta_X \to \Delta_Y$ is *honest* if for every sharp filter A (computable or not), with $A \subseteq \text{dom}(G)$, there is a sharp filter $B \subseteq A$ where $G(B)$ is a sharp filter in Δ_Y.

THEOREM 2.19. *There is a computable quantum correspondence from* $\Delta_\mathbb{R}$ *to* $\Delta_\mathbb{R}$ *which is not honest.*

THEOREM 2.20. *Every computable quantum correspondence has an honest equivalent. That is, if* $F : \Delta_X \to \Delta_Y$ *is a computable quantum correspondence, then there is an honest computable quantum correspondence* $G \subseteq F$ *such that* $f_G = f_F$.

THEOREM 2.21. *Let* $F, G : \Delta_X \to \Delta_Y$ *be quantum correspondences with* $G \subseteq F$. *Let* H *be a set such that* $G \subseteq H \subseteq F$. *Then* H *is a quantum correspondence. Furthermore, if* F *is honest, so is* H.

2.4. Computable, avoidable, & shadow points, avoidance functions, and spectra. In this subsection we collect definitions of various *types* of points starting with those of partial computable functions that allow us to classify and include some fundamental results about them.

DEFINITION 2.22. Let $\langle X, \Delta_X \rangle$ be a resolvable space. Suppose there is a partial computable function $\phi : \omega \to \omega$ and $x \in X$, such that for any n

1. if ψ_n is a computable sharp filter, then $\phi(n) \downarrow$; and
2. if $\phi(n) \downarrow$ and $\psi_n(\phi(n)) \downarrow$ (which is true if ψ_n is a sharp filter), then $x \notin \psi_n(\phi(n))$.

We then say ϕ is an *avoidance function* for x, or x is *avoidable via ϕ*. If x is avoidable via some ϕ, we say x is *avoidable*. ϕ is an *avoidance function* if it is an avoidance function for some x.

For an avoidance function ϕ, let

$$S_\phi = \{x \in X : \phi \text{ is an avoidance function for } x\},$$

and refer to S_ϕ as the *spectrum* of ϕ.

PROPOSITION 2.23. *Let ϕ be an avoidance function. Then S_ϕ contains no isolated points and is a perfect set.*

DEFINITION 2.24. If a point x is noncomputable and not avoidable, we say x is a *shadow* point.

Similar to $\text{Comp}(X)$ denoting the set of computable points of X, we use $\text{Av}(X)$ and $\text{Shad}(X)$ to denote the sets of avoidable and shadow points of X respectively.

It turns out that the set of all avoidable points is a set of first Baire category which, in $\mathbb{R}^n$, contains the full measure of the space. Hence, the set of all shadow points is of second category, and in $\mathbb{R}^n$ has measure 0. Each set is dense, regardless of the space. In $\mathbb{R}^n$, $n \geq 2$, the set of avoidable points is connected, while the set of shadow points is not. The natural method of accessing a set of avoidable points is using an avoidance function, and its spectrum. Every avoidable point lies in many such spectra.

As we noted, a quantum computable function is continuous on its domain; the complement of its domain, if nonempty, is a countable union of perfect sets of avoidable points. The domain contains all the computable points, all the shadow points, and at least some of the avoidable points.

THEOREM 2.25. *Let $F : \Delta_X \to \Delta_Y$ be a computable quantum correspondence. Then $\text{exdom}(f_F) \subseteq \text{Av}(X)$, and thus $\text{dom}(f_F) \supseteq \text{Shad}(X)$.*

COROLLARY 2.26. *If $F : \Delta_X \to \Delta_Y$ is a computable quantum correspondence, and $x \in \text{exdom}(f_F)$, then for some avoidance function ϕ for x, $\text{dom}(f_F)$ does not include any point in the spectrum of ϕ. That is, we have $S_\phi \cap \text{dom}(f_F) = \emptyset$.*

THEOREM 2.27 (*Spectrum Subtraction Theorem*). *Let $\phi : \omega \to \omega$ be a computable avoidance function. Then there is a computable quantum correspondence H such that $\text{dom}(f_H) = X - S_\phi$.*

COROLLARY 2.28. *If $F : \Delta_X \to \Delta_Y$ is a computable quantum correspondence, and ϕ is a computable avoidance function, then there is a computable quantum correspondence $G \subseteq F$ such that $\text{dom}(f_G) = \text{dom}(f_F) - S_\phi$.*

THEOREM 2.29 (*Restriction Theorem for Avoidance Functions*). *Let ϕ be an avoidance function and let $\alpha, \beta \in \Delta$ be such that $\overline{\alpha} \subseteq \beta$ and, for each n, if $\psi_n(\phi(n)) \downarrow$ then $\psi_n(\phi(n))$ resolves $\langle \alpha, \beta \rangle$. Then there is an avoidance function ϕ' such that for each n, if $\psi_n(\phi'(n)) \downarrow$ then $\psi_n(\phi'(n))$ resolves $\langle \alpha, \beta \rangle$, and $S_{\phi'} = S_\phi \cap \overline{\alpha}$.*

2.5. Trees and points. In this subsection we define what a *tree* is in our setting. It is important to point out that for our study of trees the 'addresses' of nodes (strings of 0s and 1s) as well as the content of the nodes (basic open sets) play crucial and *independent* roles (see [KalWel05]). We end with two theorems on how designated points can be captured through trees of sharp filters.

DEFINITION 2.30. Let Σ be the set of all finite binary strings on the set $\{0, 1\}$. We usually denote members of Σ by σ or τ. Length of σ is denoted by $\mathrm{lh}(\sigma)$. If σ is a substring of τ, we write $\sigma \subseteq_{\mathrm{str}} \tau$. If σ is lexicographically before τ, we write $\sigma \leq_{\mathrm{lex}} \tau$. A *tree of sharp filters* T is the range of a partial function $\Theta : \Sigma \to \Delta$ satisfying the following conditions:

1. $\Theta(\emptyset)\downarrow$;
2. for all $\sigma, \tau \in \Sigma$, if $\Theta(\tau)\downarrow$ and $\sigma \subseteq_{\mathrm{str}} \tau$, then $\Theta(\sigma)\downarrow$ and $\overline{\Theta(\tau)} \subseteq \Theta(\sigma)$;
3. if $\sigma \not\subseteq_{\mathrm{str}} \tau, \tau \not\subseteq_{\mathrm{str}} \sigma$, $\Theta(\sigma)\downarrow$ and $\Theta(\tau)\downarrow$, then $\Theta(\sigma) \cap \Theta(\tau) = \emptyset$; and
4. if $b : \omega \to \Sigma$ is a total function such that for every n, $\mathrm{lh}(b(n)) = n$, $\Theta(b(n))\downarrow$, and $\overline{\Theta(b(n+1))} \subseteq \Theta(b(n))$ (i.e., if $\Theta \circ b$ is a *branch through* T), then $\Theta \circ b$ is a sharp filter. (Note that this condition requires that $b(n)$ is a substring of $b(n+1)$.)

T is a *complete* tree of sharp filters if Θ is a total function. For an infinite branch $b = \{\beta_i : i \in \omega\}$ through T, which is a sharp filter and converges to a point, we denote that point by x_b and say that the point *lies on* T.

For notational convenience, for $\sigma \in \Sigma$ we denote $\Theta(\sigma)$ by θ_σ; thus we have

$$T = \{\Theta(\sigma) : \sigma \in \Sigma'\} = \{\theta_\sigma : \sigma \in \Sigma'\},$$

where $\Sigma' = \mathrm{dom}(\Theta)$ is a subset of Σ and a tree under the ordering $\subseteq_{\mathrm{str}}$.
For a tree T, let

$$\mathcal{T} = \{x_b : b \text{ is a branch in } T\}.$$

In order to refer to the collection of the basic open sets at a fixed height on a tree T, define the *n-nodes* of a tree T as follows:

$$n\text{-Nodes}(T) = \{\theta_\sigma \in T : \mathrm{lh}(\sigma) = n\}.$$

DEFINITION 2.31. T is *computable as a tree*, or a Π_1^0 *tree*, if there is a computable procedure to determine, for each $n \in \omega$ and each finite sequence $\langle \delta_0, \ldots, \delta_n \rangle$ of members of Δ, whether there is $\sigma \in \mathrm{dom}(\Theta)$ with $\mathrm{lh}(\sigma) = n$, such that

$$(\forall \tau \subseteq_{\mathrm{str}} \sigma)(\forall k \leq n)\big[(\mathrm{lh}(\tau) = k) \implies (\theta_\tau = \delta_k)\big].$$

It should be pointed out that a Π_1^0 tree of sharp filters corresponds to a strong Π_2^0 class. (See [KalWel05].)

DEFINITION 2.32. A tree T is *computably bounded* if there is a computable function $f : \omega \to \Delta^{<\omega}$ such that

$$(\forall \sigma \in \mathrm{dom}(\Theta))(\forall n \in \omega)\big[(\mathrm{lh}(\sigma) = n) \implies (\theta_\sigma \in f(n))\big].$$

Some sets of avoidable points can also be described through the use of Π_1^0 trees with no computable branches, and in some cases those trees can be made to be computably bounded. In $\mathbb{R}$ every avoidable point lies on a complete binary Π_1^0 tree, which of course is computably bounded. In general, computably bounded Π_1^0 trees with $2^{\aleph_0}$ avoidable points can be constructed. Some sets of shadow points can also be described via Π_1^0 trees, but in this case the trees cannot be computably bounded. These results are captured below. (See [KalWel03, KalWel05].)

THEOREM 2.33. *Let $x \in \mathrm{Av}(\mathbb{R})$. Then there is a complete computable tree T of sharp filters in $\mathbb{R}$ such that $x \in T$.*

THEOREM 2.34 (*Excision Theorem*). *Let $\alpha \in \Delta$, and let f be a computable quantum function. Then there is a computable quantum function $g \subseteq f$ such that $\|(\mathrm{dom}(f) - \mathrm{dom}(g)) \cap \alpha\| = 2^{\aleph_0}$. In fact, we can additionally have $\mathrm{dom}(f) - \mathrm{dom}(g) = T = \{x_b : b \text{ is a branch of } T\} \subseteq \alpha \cap \mathrm{Av}(X)$, for some computably bounded Π_1^0 tree of sharp filters T with $2^{\aleph_0}$ infinite branches.*

THEOREM 2.35. *Given any $\delta \in \Delta$, there is a Π_1^0, not computably bounded, tree T_∞ of sharp filters having $2^{\aleph_0}$ infinite branches, such that $T_\infty = \{x_b : b \text{ is a branch of } T_\infty\} \subseteq \delta \cap \mathrm{Shad}(X)$.*

2.6. Basic properties. Throughout this paper, we take $\langle X, \Delta \rangle$ to be resolvable, completely connected, and semi-computably presentable. We also assume that Δ is compactible, and that X contains at least two points. These assumptions imply that the space X is regular, second countable, and of second category. We let $\{\psi_n : n \in \omega\}$ denote a fixed acceptable enumeration capturing all computable sharp filters. Fundamental examples of these spaces are $\mathbb{R}^n$, for any $n \geq 1$, with appropriate bases. Familiar examples are $\mathbb{R}$ with the basis of open intervals with rational endpoints, $\mathbb{R}^2$ with the basis of open rectangles whose corners have rational coordinates and whose sides are parallel to the axes, and $\mathbb{R}^2$ with the basis of open balls with rational radii whose centers have rational coordinates.

§3. **Trees of avoidable points and spectra.** While in [KalWel03] we constructed a tree of avoidable points inside a given $\alpha \in \Delta$, in this section we build a tree of avoidable points for a fixed avoidance function inside a fixed α.

Specifically, while in [KalWel03] we prove Theorem 2.34 (see Section 2), we now prove the following.

THEOREM 3.1. *Let $\alpha \in \Delta$ and let T_0 be a complete computable tree of sharp filters with $T_0 \subseteq \alpha$ and let ϕ be an avoidance function such that $S_\phi \cap T_0 \neq \emptyset$.*

Then there is a computably bounded Π_1^0 *tree* $T \subseteq T_0$ *of sharp filters such that* $T \subseteq S_\phi \cap \alpha$.

PROOF. Let $T_0 = \{\theta_\sigma : \sigma \in \Sigma\}$. We prune T_0 to get T as follows. For each $\theta_\sigma \in n$-Nodes(T_0), we let $\theta_\sigma \in n$-Nodes(T) iff the following condition is satisfied:

$$(\dagger) \qquad (\forall m, s \leq n)\big[\psi_m^s(\phi^s(m)) \downarrow \Longrightarrow \theta_\sigma \not\subseteq \psi_m^s(\phi^s(m))\big].$$

(Here, $\phi^s(m)$ denotes the outcome of $\phi(m)$ at stage s.) First we note that T is a tree of sharp filters, for if $\theta_\sigma \not\subseteq T$ then there is some pair m, s with $\psi_m^s(\phi^s(m)) \downarrow$, $\theta_\sigma \subseteq \psi_m^s(\phi^s(m))$, and lh$(\sigma) \geq \max\{m, s\}$. Thus for each $i \in \{0, 1\}$, lh$(\sigma^\frown i) > \max\{m, s\}$ and $\theta_{\sigma^\frown i} \subseteq \theta_\sigma \subseteq \psi_m^s(\phi^s(m))$, so $\theta_{\sigma^\frown i} \not\subseteq T$.

Next we note that T is computable and computably bounded, because T_0 has both these properties and $(\dagger)$ is a computable condition.

Finally we note that $T \subseteq S_\phi$, for if $y \not\subseteq S_\phi$ then there is some m such that $y \in \psi_m(\phi(m))$. If $y \not\subseteq T_0$ then since $T \subseteq T_0$, of course $y \not\subseteq T$. But if $y \in T_0$ then there is a branch b of T_0 converging to y, and there is some $\theta \in b$ such that $\theta \subseteq \psi_m(\phi(m))$. Now take s such that $\psi_m^s(\phi^s(m)) \downarrow$ and take $\theta_\sigma \in b$ such that $\theta_\sigma \subseteq \theta$ and lh$(\sigma) \geq \max\{m, s\}$. Then $\theta_\sigma \not\subseteq T$, so $y \not\subseteq T$.

Of course $T \subseteq T_0 \subseteq \alpha$, so $T \subseteq S_\phi \cap \alpha$. ⊣

COROLLARY 3.2. *Let* $x \in$ Av$(\mathbb{R})$, *let* $\alpha \in \Delta_{\mathbb{R}}$ *be such that* $x \in \alpha$, *and let* ϕ *be an avoidance function for* x. *Then there is a computably bounded* Π_1^0 *tree of sharp filters* T *such that* $T \subseteq S_\phi \cap \alpha$.

PROOF. Form a complete computable tree of sharp filters T_0 with $\theta_\emptyset \subseteq \alpha$ such that $x \in T_0$, as per Theorem 2.33 (proved in [KalWel03]). Then $x \in S_\phi \cap T_0$ so we may form a computably bounded Π_1^0 tree T as in the theorem above, and thus $T \subseteq S_\phi \cap \alpha$. ⊣

§4. Left branches of trees.

In this section we introduce some needed definitions and notation to identify and work with the *left branch* of a given tree. Recalling the construction of a noncomputably bounded Π_1^0 tree of sharp filters T_∞ such that $T_\infty \subseteq$ Shad(X) (see [KalWel05]), we make new observations about the left branch of T_∞.

DEFINITION 4.1. Let $T = \{\theta_\sigma : \sigma \in \Sigma'\}$ be a tree of sharp filters. For each n, let λ_n denote the *lexicographically least member* of $\{\sigma : \sigma \in \Sigma' \wedge$ lh$(\sigma) = n\}$. T is *left-bounded* if $\{\lambda_n : n \in \omega\}$ is a computable set.

PROPOSITION 4.2. *Every computably bounded* Π_1^0 *tree of sharp filters is left-bounded*.

PROOF. Obvious. ⊣

DEFINITION 4.3. For a given tree of sharp filters $T = \{\theta_\sigma : \sigma \in \Sigma'\}$ with at least one infinite branch, let b_l be the 'leftmost' infinite branch of T in the sense that if $\theta_\tau \in b_l$ and $\sigma \leq_{\text{lex}} \tau$, then either $\sigma \subseteq \tau$ or $\{v : v \in \Sigma \wedge v \supseteq \sigma\}$ is a finite set. Let λ_n^* denote the string such that $\mathrm{lh}(\lambda_n^*) = n$ and $\theta_{\lambda_n^*} \in b_l$, so that $\theta_{\lambda_n^*} \in n\text{-Nodes}(T)$.

In general, for the tree T, we will use the notations λ_n and λ_n^* with these meanings throughout this paper. If a tree is called T_m, we will use $\lambda_{m,n}$ and $\lambda_{m,n}^*$.

PROPOSITION 4.4. *Let T be a fixed infinite tree. For each n, $\lambda_n \leq_{\mathrm{lex}} \lambda_n^*$. Also for each n there is $N \geq n$ such that for all $m \geq N$, $\lambda_n^* \subseteq_{\mathrm{str}} \lambda_m$.*

PROOF. Obvious.   ⊣

In Theorem 2.35 of Section 2, proved in [KalWel03], we have a noncomputably bounded Π_1^0 tree of sharp filters T_∞ such that $T_\infty \subseteq \mathrm{Shad}(X)$. This was done by forming a c.e. sequence of finitely branching computable trees $T_n = \{\theta_{n,\sigma} : \sigma \in \Sigma_n\}$ and letting $T_\infty = \bigcup_n \{\theta_{n,\sigma} : \mathrm{lh}(\sigma) \leq n\}$. We draw attention to two properties of $\{T_n : n \in \omega\}$ that derive expressly from the construction. First, each T_n is computably bounded, hence left-bounded. As a consequence, $\{\lambda_{n,n} : n \in \omega\}$ is computable. Second, if there is some $m > n$ and some $\theta_{m,\sigma} \in (n\text{-Nodes}(T_m)) - (n\text{-Nodes}(T_n))$ then $\sigma \geq_{\mathrm{lex}} \lambda_{n,n}$; this is due to the fact that $\theta_{n,\lambda_{n,n}}$, the leftmost node of T_n at level n, is never dormant at any stage $s \geq n$ of the construction. Consequently, $\lambda_{n,n}$ is the 'leftmost' node of T_∞ at level n; that is, $\lambda_{\infty,n} = \lambda_{n,n}$. We conclude that $\{\lambda_{\infty,n} : n \in \omega\}$ is computable, so T_∞ is left-bounded.

It is worth our while to state this as a theorem. We incorporate into the theorem the fact that the nodes of T_∞ are taken from a complete binary tree of sharp filters T. It should be noted that T_∞ is *not* actually a subtree of T in the ordinary sense of 'subtree'. In fact, Θ_{T_∞} is *not* a subfunction of Θ_T, while $\mathrm{rng}(\Theta_{T_\infty}) \subseteq \mathrm{rng}(\Theta_T)$. We refer the reader to [KalWel05] for a detailed discussion of this subtlety.

THEOREM 4.5. *Let T be a complete binary tree of sharp filters. Then there is a left-bounded tree T_∞ such that $T_\infty \subseteq \mathrm{Shad}(X)$ and $\{\theta : \theta \in T_\infty\} \subseteq \{\theta : \theta \in T\}$. Furthermore, there is a uniformly computable process to find such a tree T_∞ from any such tree T.*

§5. Cluster points.

Specker [Spe49] (also see Rice [Ric54]) proved the existence of a monotonic computable sequence of computable reals in $[0, 1]$ whose limit is not a computable real. While the limit of Specker's sequence exists classically, some researcher are interested in 'rapidness' of convergence. They consider only *rapidly* Cauchy and *rapidly* convergent sequences to be Cauchy and convergent, respectively, in a computable sense. For instance, Kushner [Kus83] presents Specker's theorem and related results in this light.

As per Kushner [Kus83], a sequence of real numbers $\{x_n : n \in \omega\}$ is *fundamental*, or *rapidly Cauchy*, if there is a computable function f such that for each l, m, and n, if $l, m > f(n)$, then $|x_l - x_m| < 2^{-n}$. A sequence of real numbers $\{x_n : n \in \omega\}$ is *(rapidly) convergent* to a real number y if there is a computable function f such that for each m and n, if $m > f(n)$, then $|y - x_m| < 2^{-n}$. He also defines *pseudofundamental* and *pseudoconvergent*, which essentially mean Cauchy and convergent in the classical sense, respectively, whether rapid or not. He further defines a *pseudonumber* to be an ordinary, classical real number.

Kushner [Kus83] gives several theorems about pseudoconvergence: It is possible, for example, to construct a computable sequence of computable numbers that pseudoconverges to 0, even though it does not (rapidly) converge to 0. On the other hand, if a *monotonic* computable sequence of real numbers pseudoconverges to a computable real number, then it rapidly converges to that number. Also, any pseudofundamental computable sequence of computable real numbers pseudoconverges to a pseudonumber.

In this section and the next, we build the machinery necessary to discuss cluster points of sequences in our topological spaces in a general way. Consequently, we determine the classical (that is, *pseudo-*) behavior of the cluster points of computable sequences of computable points. Accordingly, we begin by defining the concepts of *external cluster points* and *Specker cluster points* and show they exist for sequences of sets under certain conditions.

DEFINITION 5.1. Let $S \subseteq X$. The set of *external cluster points* of S is $\overline{S} - S$, which we denote by

$$\dot{S} = \overline{S} - S.$$

Let $S = \{S_i : i \in \omega\}$ be a sequence of closed sets in X. The set of *external cluster points* of S is the set $\dot{S} = \overline{(\cup_i S_i)} - (\cup_i S_i)$.

Let $M = \{\alpha_i : i \in \omega\}$ be a sequence of basic open sets in Δ. A sequence $S = \{S_i : i \in \omega\}$ of sets in X *persists through* the sequence M if for each $i \in \omega$, $S_i \neq \emptyset$ and $S_i \subseteq \alpha_i$.

We will be most interested in the case where each S_i is closed (hence compact, since each α_i has compact closure).

DEFINITION 5.2. Suppose $M = \{\alpha_i : i \in \omega\} \subseteq \Delta$, and suppose that whenever U_1, U_2 are two sequences of closed sets each persisting through M, then $\dot{U}_1 = \dot{U}_2$. Then we define the set of *Specker cluster points* of M to be $\dot{M} = \dot{S}$ for any sequence S of closed sets persisting through M.

Note that if $\dot{M}$ exists and each $S_i \in S$ is a singleton, say $S_i = \{y_i\}$, then $\dot{M}$ is exactly the set of cluster points of $\{y_i : i \in \omega\}$.

DEFINITION 5.3. Given a left-bounded tree T, we define the tree

$$L_T = \{\theta_\sigma : \theta_\sigma \in T \wedge \exists n(\sigma \leq_{\text{lex}} \lambda_n)\},$$

and let $\mathcal{L}_T$ be its set of points (i.e. leaves).

Note that $L_T = \{\theta_\sigma : \theta_\sigma \in T \wedge (\exists n)(\sigma \leq_{\text{lex}} \lambda_n^*)\}$.

We state the following observation in the form of a proposition because it is critically used in the forthcoming sections.

PROPOSITION 5.4. *If* $T \neq \emptyset$, *then* $\mathcal{L}_T$ *is a singleton.*

PROOF. Obvious.   ⊣

DEFINITION 5.5. A sequence $M = \{\alpha_i : i \in \omega\}$ of pairwise disjoint basic open sets is an *evading sequence* if for any pair $\langle \beta, \gamma \rangle$, there is $N \in \omega$ such that α_n resolves $\langle \beta, \gamma \rangle$ for all $n \geq N$. (Recall α *resolves* $\langle \beta, \gamma \rangle$ if $(\overline{\beta} \subseteq \gamma) \Rightarrow [(\alpha \cap \beta = \emptyset) \vee (\alpha \subseteq \gamma)]$. Note that if $\overline{\beta}$ is not a subset γ, then α resolves $\langle \beta, \gamma \rangle$ trivially.)

THEOREM 5.6. *If* $M = \{\alpha_i : i \in \omega\}$ *is an evading sequence, then* $\dot{M}$ *exists.*

PROOF. Let $S_1 = \{\mathcal{S}_{1,i} : i \in \omega\}$, and $S_2 = \{\mathcal{S}_{2,i} : i \in \omega\}$ be sequences of closed sets persisting through M and presume $\dot{S}_1 \neq \dot{S}_2$. Without loss of generality, let $x \in \dot{S}_2 - \dot{S}_1$, and let $\{y_i : i \in \omega\}$ be a sequence of points in $\bigcup \mathcal{S}_{2,i}$ such that $\lim_i y_i = x$. First we note that there cannot be any single $\alpha_j \in M$ containing infinitely many of the points y_k, because if $y_k \in \alpha_j$ then we must have $y_k \in \mathcal{S}_{2,j}$, which is closed. If infinitely many of the points y_k lay in α_j we would therefore have $x \in \mathcal{S}_{2,j}$, contradicting $x \in \overline{\bigcup \mathcal{S}_{2,i}} - \bigcup \mathcal{S}_{2,i}$. So each α_j contains finitely many of the points y_k. Without loss of generality, we may assume $y_i \in \alpha_{k_i}$ where for all $i, j, i < j \Rightarrow k_i < k_j$.

Now, $x \notin \bigcup \mathcal{S}_{1,i}$ because, first, each $\mathcal{S}_{1,j} \subseteq \alpha_j$, and at most one of the points y_i lies in α_j; and second, since $\mathcal{S}_{1,j}$ is closed, α_j is open with compact closure, and X is a regular topological space, $x = \lim_i y_i \notin \mathcal{S}_{1,j}$ for each j.

But also $x \notin \overline{\bigcup \mathcal{S}_{1,i}} - \bigcup \mathcal{S}_{1,i}$ by assumption, so $x \notin \overline{\bigcup \mathcal{S}_{1,i}}$. Use the regularity of X to find two basic open sets β and γ with $x \in \beta$, $\overline{\beta} \subseteq \gamma$, $\gamma \cap \overline{\bigcup \mathcal{S}_{1,i}} = \emptyset$. Take $N \in \omega$ such that for all $n \geq N$, α_n resolves $\langle \beta, \gamma \rangle$ and take $i \geq N$ such that $y_i \in \beta$. Then $k_i \geq N$ too, so $\alpha_{k_i} \cap \beta \neq \emptyset$. Because α_{k_i} resolves $\langle \beta, \gamma \rangle$ we have $\alpha_{k_i} \subseteq \gamma$, so $\mathcal{S}_{1,k_i} \subseteq \gamma$. This contradicts $\beta \cap \overline{\bigcup \mathcal{S}_{1,i}} = \emptyset$.   ⊣

DEFINITION 5.7. For T a tree of sharp filters, we denote the set of its nodes by

$$\text{Nodes}(T) = \bigcup_n (n\text{-Nodes}(T)).$$

THEOREM 5.8. *Let* T *be a tree of sharp filters such that* T *is nonempty. Then* T *is compact.*

DEFINITION 5.9. Let T be a tree of sharp filters and let $f : \text{Nodes}(T) \to \Delta$ be such that:

1. for each $\theta \in T$, $\overline{f(\theta)} \subseteq \theta$, and
2. if $\theta_1 \neq \theta_2$, then $f(\theta_1) \cap f(\theta_2) = \emptyset$.

Define the set $M = \{f(\theta) : \theta \in \text{Nodes}(T)\}$. We say such a set M *persists through* the tree T.

For convenience we write $\alpha_\theta = f(\theta)$, so that $M = \{\alpha_\theta : \theta \in \text{Nodes}(T)\}$, and assume an enumeration of M (not necessarily computable): say $M = \{\alpha_n : n \in \omega\}$, and consider it as a sequence.

It should be noted that if M persists through T, then M is an evading sequence, and the set of its Specker cluster points exists and is exactly the same as T; we prove that next.

THEOREM 5.10. *Let T be a tree of sharp filters and let M be a sequence persisting through T. Then M is an evading sequence and $\dot{M} = T$.*

PROOF. Assume T is a tree of sharp filters with M a sequence persisting through it. In light of (2) of the definition above, M consists of pairwise disjoint sets; so we need only prove that M resolves every pair of sets in Δ. Let $\langle \alpha, \beta \rangle$ be a pair of sets in Δ. If b is a branch of T then b is a sharp filter, so b resolves $\langle \alpha, \beta \rangle$. By König's Lemma, since T is finitely branching, there is some $n_0 \in \omega$ such that each $\theta \in n_0$-Nodes(T) resolves $\langle \alpha, \beta \rangle$. Now for each $m \geq n_0$, if $\theta \in m$-Nodes(T) then there is some $\theta' \in n_0$-Nodes(T) with $\theta \subseteq \theta'$, so θ resolves $\langle \alpha, \beta \rangle$. Thus there are only finitely many members of Nodes(T) that fail to resolve $\langle \alpha, \beta \rangle$. By (2) of the definition above, f is one-to-one, so there is some n, such that if $m \geq n$, then α_m resolves $\langle \alpha, \beta \rangle$. Thus M is an evading sequence.

Next, by Theorem 5.6, it follows that $\dot{M}$ exists; so we need to show that $\dot{M} = T$. In preparation, as per Theorem 5.6, for each $\alpha \in M$ select a point y_α, so that for $\mathcal{Y} = \{y_\alpha : \alpha \in M\}$, $\dot{\mathcal{Y}} = \dot{M}$.

(To see $T \subseteq \dot{M}$.) Let $x \in T$ and let b be the branch of T such that $b \searrow x$. Pick $\gamma \in \Delta$ such that $x \in \gamma$, and pick $\theta_0 \in b$ such that $\theta_0 \subseteq \gamma$. Then infinitely many members of b lie in γ. Let $\mathcal{Y}_1 = \{y_\theta : \theta \in b \wedge \theta \subseteq \theta_0\}$. Then $\|\mathcal{Y}_1\| = \aleph_0$ and $\mathcal{Y}_1 \subseteq \theta_0 \subseteq \gamma$. Hence $x \in \dot{M}$.

(To see $\dot{M} \subseteq T$.) Suppose $x \in \dot{M}$. We will show that for each $n \in \omega$, there is a unique member θ_n of n-Nodes(T) such that $x \in \theta_n$. Let $\{y_k : k \in \omega\}$ be a subsequence of $\{y_\alpha : \alpha \in M\}$ such that $\lim_k y_k = x$. Choose $n \in \omega$. Since T is finitely branching, $(n+1)$-Nodes(T) is a finite subset of Δ, so there is some $\theta \in (n+1)$-Nodes(T) such that an infinite subset of $\{y_k : k \in \omega\}$ lies in θ. Thus $x \in \overline{\theta}$. Now let $\theta_n \in n$-Nodes(T) be such that $\overline{\theta} \subseteq \theta_n$; such θ_n exists by the definition of a tree of sharp filters, and of course $x \in \theta_n$. Since the members of n-Nodes(T) are mutually disjoint, θ_n is the unique member of n-Nodes(T) containing x. It follows that $b = \{\theta_n : n \in \omega\}$ is a branch of T, and that $b \searrow x$, so that $x \in T$.   ⊣

§6. **Sequences with controlled sorts of Specker cluster points.** By applying the results 2.33, 2.34, 4.2 and 4.5, given any $\alpha \in \Delta$, we can form left-bounded computable trees $T_{\alpha,C}, T_{\alpha,A}, T_{\alpha,S} \subseteq \alpha$, where $T_{\alpha,C}$ is a complete tree, $T_{\alpha,A}$ a tree of avoidable points, and $T_{\alpha,S}$ a tree of shadow points.

We can also form, as per Definition 5.3, left-bounded computably enumerable trees $L_{\alpha,C}$, $L_{\alpha,A}$, $L_{\alpha,S}$, each containing exactly *one* infinite branch, such that $\mathcal{L}_{\alpha,C} = \{x_{\alpha,C}\} \subseteq \text{Comp}(X)$, $\mathcal{L}_{\alpha,A} = \{x_{\alpha,A}\} \subseteq \text{Av}(X)$, and $\mathcal{L}_{\alpha,S} = \{x_{\alpha,S}\} \subseteq \text{Shad}(X)$. Finally, we can form sequences $M_{T_{\alpha,C}}$, $M_{T_{\alpha,A}}$, $M_{T_{\alpha,S}}$, $M_{L_{\alpha,C}}$, $M_{L_{\alpha,A}}$, and $M_{L_{\alpha,S}}$ each persisting through the corresponding tree (suggested by the subscript). Furthermore, there is a way to accomplish all this uniformly computably in α. Also, if $L = L_{\alpha,C}$, $L_{\alpha,A}$, or $L_{\alpha,S}$, then for each $\theta \in L$ we can computably find $n \in \omega$ such that $\theta \in n$-Nodes(L).

THEOREM 6.1 (*Specker Sequence Engine*). *Let* $\alpha \in \Delta$. *Given cardinal numbers* κ, λ, μ, *not all zero, such that*

1. $\kappa \leq \aleph_0$,
2. $\lambda \leq \aleph_0$ *or* $\lambda = 2^{\aleph_0}$, *and*
3. $\mu \leq \aleph_0$ *or* $\mu = 2^{\aleph_0}$,

we can form an evading sequence $M = \{\rho_n : n \in \omega\} \subseteq \Delta$ *such that each* $\rho_n \subseteq \alpha$, $\dot{M}$ *is defined, and* $\dot{M}$ *contains* κ *computable points,* λ *avoidable points, and* μ *shadow points, all lying in* α.

PROOF. Suppose we start with a computably enumerable tree $L = L_{\alpha,C}$, $L_{\alpha,A}$, or $L_{\alpha,S}$, and take a specific computable enumeration of it. Let $M_L = \{\alpha_\theta : \theta \in L\}$ be a sequence persisting through L. For each $n \in \omega$ let θ_n be the first member of n-Nodes(L) to be enumerated in the computable enumeration of L, and let $\alpha_n = \alpha_{\theta_n}$. Let $M_\omega = \{\alpha_n : n \in \omega\}$, so that $\dot{M}_\omega$ is defined and $\dot{M}_\omega = \mathcal{L}$.

Now for each $n \in \omega$ build a tree T_n in θ_n as follows, provided n meets the condition stipulated.

1. If $n = 3m$ and $m < \kappa$, let $T_n = L_{\theta_n,C}$.
2. a. If $n = 3m + 1$ and $m < \lambda \leq \aleph_0$, let $T_n = L_{\theta_n,A}$.
   b. If $n = 1$ and $\lambda = 2^{\aleph_0}$, let $T_n = T_{\theta_1,A}$.
3. a. If $n = 3m + 2$ and $m < \mu \leq \aleph_0$, let $T_n = L_{\theta_n,S}$.
   b. If $n = 2$ and $\mu = 2^{\aleph_0}$, let $T_n = T_{\theta_2,S}$.

If n does not meet any of these conditions, do not build a tree T_n. Follow this procedure uniformly computably in n.

Let $I = \{n : T_n \text{ is defined.}\}$. For each $n \in I$ form a sequence M_n persisting through T_n; note that each M_n is an evading sequence. Let $M = \bigcup_{n \in I} M_n$; then M is also an evading sequence and computably enumerable. It is easily checked that if $\kappa + \lambda + \mu < \aleph_0$ then

$$\dot{M} = \left(\bigcup \{\mathcal{L}_{3m} : m < \kappa\}\right) \cup \left(\bigcup \{\mathcal{L}_{3m+1} : m < \lambda\}\right) \cup \left(\bigcup \{\mathcal{L}_{3m+2} : m < \mu\}\right),$$

so that M is as claimed. But if $\kappa + \lambda + \mu \geq \aleph_0$ then

$$\dot{M} = \mathcal{L} \cup \left(\bigcup \{\mathcal{L}_{3m} : m < \kappa\}\right) \cup \left(\bigcup \{\mathcal{L}_{3m+1} : m < \lambda\}\right) \cup \left(\bigcup \{\mathcal{L}_{3m+2} : m < \mu\}\right),$$

so that M is as claimed except that $\dot{M}$ contains the unique point x in $\mathcal{L}$. In order to maintain the proper cardinalities among the types of Specker cluster points of M we therefore stipulate the following about L:

If $\kappa = \aleph_0$, then $L = L_{\alpha,C}$.

If $\kappa < \aleph_0$ and either $\lambda = \aleph_0$ or $\lambda = 2^{\aleph_0}$, then $L = L_{\alpha,A}$.

If $\kappa, \lambda < \aleph_0$ and $\mu = 2^{\aleph_0}$, then $L = L_{\alpha,S}$.

With this further condition, M is as claimed under all circumstances.   ⊣

The following corollary extends this result to sequences with multiple cluster points, and addresses the question of how these cluster points can be configured based on our classification of points as computable, avoidable, or shadow points.

COROLLARY 6.2. *Given $\alpha \in \Delta$ and cardinalities κ, λ, μ as in the Theorem, there is a computable sequence of computable points $\{y_n : n \in \omega\}$ such that the set of cluster points of this sequence contains exactly κ computable points, λ avoidable points, and μ shadow points.*

§7. **An evasive action engine.** In Theorem 6.1 and Corollary 6.2, we build a sequence of points in which some of the points are computable, some are shadow points, and some are avoidable. These points are the Specker cluster points of an evading sequence M of basic open sets. When forming M, not only can we control the cardinalities of the various kinds of Specker cluster points, but to some extent we can also control the specific computable points in $\dot{M}$ and the spectra into which the avoidable points of $\dot{M}$ fall. For every computable point x, it is easy to see that there is a complete computable tree T such that x is the limit of the left branch of T. Thus Theorem 6.1 makes it clear how the computable points in $\dot{M}$ can be controlled.

In regard to the avoidable points, if our space is $\mathbb{R}$ we note that *every* avoidable point in $\mathbb{R}$ lies in T for some complete computable tree T of sharp filters. Thus given any avoidance function ϕ for $\mathbb{R}$ such that $S_\phi \cap \alpha \neq \emptyset$, we can find $x \in S_\phi \cap \alpha$, and using Theorem 3.1 we can form a tree T of avoidable sharp filters in α, where $x \in T \subseteq S_\phi$. Then we can form M such that $\mathrm{Av}(\mathbb{R}) \cap \dot{M} \subseteq S_\phi$, and in fact we can ensure not only this, but also, if $\lambda = 2^{\aleph_0}$, that $\mathrm{Av}(\mathbb{R}) \cap \dot{M}_{L_{\alpha,A}} = T$.

When our space X is not $\mathbb{R}$, we may not be able to put each avoidable point on a complete computable tree, so we may not have such thorough control of Specker cluster points as we have in $\mathbb{R}$. In this section we prove that even in such a space, if ϕ is an avoidance function and $S_\phi \cap \alpha \neq \emptyset$, we can choose the avoidable points of $\dot{M}$ to lie in S_ϕ. The method for achieving what we describe here, however, lacks the control of the cardinality of $\mathrm{Av}(X) \cap \dot{M}$ that the method of Theorem 6.1 gives us.

In the theorem below, we write M to stand for what in Theorem 6.1 is called $M_{L_{\alpha,A}}$.

THEOREM 7.1 (*Evasive Action Engine*). *For every avoidable point x and every $\alpha \in \Delta$ with $x \in \alpha$, there is a partial computable avoidance function ϕ for x and a computable sequence $M = \{\rho_n \subseteq \alpha : n \in \omega\}$ of basic open sets such that:*

1. $\{\rho_n \subseteq \alpha : n \in \omega\}$ *is an evading sequence;*
2. $(\forall i, j)[i \neq j \Rightarrow \overline{\rho_i} \cap \overline{\rho_j} = \emptyset]$;
3. $(\forall m)(\exists N)(\forall i > N)[(\psi_m(\phi(m)) \downarrow) \Rightarrow \overline{(\psi_m(\phi(m))} \cap \overline{\rho_i} = \emptyset)]$; *and*
4. $\dot{M} \subseteq S_\phi \cap \alpha$.

PROOF. The idea behind this proof, firstly, relies on using x as a point of reference, finding a β with $x \in \beta$ and $\overline{\beta} \subseteq \alpha$ and, with the help of an initial and arbitrary avoidance function Φ for it, constructing a subsequent and more restrictive avoidance function ϕ for x. Secondly, we rely on being able to computably enumerate the sets $\psi_m(\phi(m))$ and choosing a preliminary sequence of basic open sets $\{\gamma_s : s \in \omega\}$ by finding, at each stage s, a basic open set $\gamma_s \subseteq \beta$ disjoint from those sets of the form $\psi_m(\phi(m))$ so far enumerated. If the collection of these γ_s's were *mutually* disjoint and in *tempo* with ϕ, they would be the sets ρ_n that we want. But the γ_s's can intersect each other or be out of tempo with ϕ occasionally, so we must do some more work to find our sets ρ_n. The trick is to pick the sets ρ_n to lie inside the sets γ_s, and to do so in such a way that they are disjoint. Since each γ_s is disjoint from each of the sets $\psi_m(\phi(m))$ enumerated by stage s, we choose to make each ρ_n equal to some $\psi_m(\phi(m))$ lying in an appropriate γ_s. Thus the sets ρ_n are built to lie in a subsequence of the sequence of sets γ_s.

We choose to consider a dovetailed computation of the sets $\psi_m(\phi(m))$, and at each stage s to think of those $\psi_m(\phi(m))$ whose computations have halted by that stage as *enumerated* by that stage. Note that $\psi_m^s(\phi^s(m))$ means that portion of the computation of $\psi_m(\phi(m))$ that has been accomplished by stage s.

Let $x \in \alpha \cap \mathrm{Av}(X)$. Choose β such that $x \in \beta$ and $\overline{\beta} \subseteq \alpha$ and let Φ be a partial computable avoidance function for x. For each m let $\phi(m) \doteq \Phi(m) + 1$. Clearly ϕ is a partial computable avoidance function for x. Next, for each m, if $\psi_m(\phi(m)) \downarrow$ then $\overline{\psi_m(\phi(m))} \subseteq \psi_m(\Phi(m))$. So $x \notin \bigcup\{\psi_m(\phi(m)) : (m \in \omega) \wedge (\psi_m(\phi(m)) \downarrow)\}$. For each $s \in \omega$ define

$$C_s = \bigcup \left\{ \overline{\psi_m^s(\phi^s(m))} : (m \leq s) \wedge \left(\psi_m^s(\phi^s(m)) \downarrow \right) \right\}.$$

Then for each s, C_s is a finite union of compact sets, and hence C_s is compact. Furthermore, $x \notin C_s$, so $\beta - C_s$ is a nonempty open set and thus there is a basic open set $\gamma_s \subseteq \beta - C_s$. For any s and $\delta \in \Delta$, we can computably determine whether $\delta \cap C_s \neq \emptyset$ (i.e., whether $(\exists m \leq s)[(\psi_m^s(\phi^s(m)) \downarrow) \wedge (\delta \cap \psi_m^s(\phi^s(m)) \neq \emptyset)]$). Since we can also computably determine whether $\delta \subseteq \beta$, we may assume that $\{\gamma_s : s \in \omega\}$ is a computable sequence.

We now form M in stages, along with sequences $\{k_n : n \in \omega\}$ and $\{s_n : n \in \omega\}$. For uniformity of description, we let $s_{-1} = 0$. We shall ultimately have $0 = s_{-1} < s_0 < s_1 < \ldots$, with $\rho_n = \psi_{k_n}^{s_n}(\phi^{s_n}(k_n))$, and $\overline{\rho_n} \subseteq \gamma_{s_{n-1}}$.

Note that for each s there is a computable sharp filter ψ_n with $\overline{\psi_n(0)} \subseteq \gamma_s$, and so indeed, $\phi(n)\downarrow$, $\psi_n(\phi(n))\downarrow$ and $\overline{\psi_n(\phi(n))} \subseteq \gamma_s$. We will use this fact in each stage of the construction.

Stage 0. Let s_0 be the least $s > 0$ such that for some $k \leq s$, $\psi_k^s(\phi^s(k))\downarrow$ and $\overline{\psi_k^s(\phi^s(k))} \subseteq \gamma_0$, and let k_0 be the least k for which this happens when $s = s_0$. Let

$$\rho_0 = \psi_{k_0}^{s_0}(\phi^{s_0}(k_0)) = \psi_{k_0}(\phi(k_0)).$$

Stage n. Assume $k_0, \ldots, k_{n-1}, 0 < s_0 < \cdots < s_{n-1}$, and $\rho_0, \ldots, \rho_{n-1}$ have been defined so that for each $i \leq n - 1$,

(i) $\overline{\psi_{k_i}^{s_i}(\phi^{s_i}(k_i))}\downarrow$,

(ii) $\overline{\psi_{k_i}(\phi(k_i))} \subseteq \gamma_{s_{i-1}}$, and

(iii) $\rho_i = \psi_{k_i}(\phi(k_i))$.

Let s_n be the least $s \geq s_{n-1}$ such that for some $k \leq s$, $\psi_k^s(\phi^s(k))\downarrow$ and $\overline{\psi_k^s(\phi^s(k))} \subseteq \gamma_{s_{n-1}}$. Let k_n be the least k for which this happens when $s = s_n$. Let

$$\rho_n = \psi_{k_n}^{s_n}(\phi^{s_n}(k_n)) = \psi_{k_n}(\phi(k_n)).$$

Then $M = \{\rho_n : n \in \omega\}$ is clearly a computable sequence of basic open sets each of which is a subset of β and thus of α.

CLAIM 1. $(\forall i, j)[i \neq j \Rightarrow \overline{\rho_i} \cap \overline{\rho_j} = \emptyset]$.

CLAIM 2. $(\forall m)(\exists N)(\forall i \geq N)[(\psi_m(\phi(m))\downarrow) \Rightarrow \overline{(\psi_m(\phi(m))} \cap \overline{\rho_i} = \emptyset)]$, and there is a partial computable function that computes N from m whenever $\psi_m(\phi(m))\downarrow$.

CLAIM 3. $\dot{M} \subseteq S_\phi \cap \alpha$.

PROOF OF CLAIM 1. Without loss of generality, pick i, j with $i < j$. Then $\rho_i = \psi_{k_i}(\phi(k_i)) = \psi_{k_i}^{s_i}(\phi^{s_i}(k_i))$; since $s_{j-1} \geq s_i \geq k_i$, we have $\psi_{k_i}^{s_{j-1}}(\phi^{s_{j-1}}(k_i))\downarrow$ and so $\overline{\rho_i} \cap \gamma_{j-1} = \emptyset$. But $\overline{\rho_j} \subseteq \gamma_{j-1}$, so $\overline{\rho_i} \cap \overline{\rho_j} = \emptyset$. $\dashv$

PROOF OF CLAIM 2. Pick $m \in \omega$ and suppose $\psi_m(\phi(m))\downarrow$. Choose $N \geq m$ so large that $\psi_m^{s_N}(\phi^{s_N}(m))\downarrow$. Then for every $i > N$, we find $\gamma_{s_{i-1}} \subseteq \beta - C_{s_{i-1}}$; but since $s_{i-1} \geq s_N \geq m$, we have $\overline{\psi_m(\phi(m))} \subseteq C_{s_{i-1}}$, and thus $\gamma_{s_{i-1}} \cap \overline{\psi_m(\phi(m))} = \emptyset$. Furthermore, $\overline{\rho_i} \subseteq \gamma_{s_{i-1}}$, and therefore $\overline{\psi_m(\phi(m))} \cap \overline{\rho_i} = \emptyset$. $\dashv$

PROOF OF CLAIM 3. Consider $S = \{y_n : n \in \omega\}$, where $y_n \in \rho_n$ for each n. Let y be either a computable point or a shadow point. Since ϕ is an avoidance function, there is some m such that $\psi_m(\phi(m))\downarrow$ and $y \in \psi_m(\phi(m))$. Let N be so large that $(\forall i \geq N)[\overline{\psi_m(\phi(m))} \cap \overline{\rho_i} = \emptyset]$. $\dashv$

Now $\psi_m(\phi(m))$ is an open set containing y, and for each $i \geq N, y_i \in \rho_i$, so $y_i \notin \psi_m(\phi(m))$. Hence y is not a limit point of S. So $\dot{M} \subseteq S_\phi \cap \bar{\beta} \subseteq S_\phi \cap \alpha$. This completes the proof of the theorem.   ⊣

§8. **Nondensity & interpolation theorems.** In this section we study the difference set of domains of nested computable quantum functions. We first show, by constructing a pair of nested computable quantum functions whose domains differ in *exactly one point*, that we cannot always interpolate a computable quantum function between two nested computable quantum functions. In contrast to this example, we next show that any time two nested computable quantum functions exist whose domains differ by at least two points, we can interpolate another computable quantum function between them.

THEOREM 8.1 (*Nondensity*). *Let $F : \Delta_X \to \Delta_Y$ be a computable quantum correspondence, and let $\alpha \in \delta$. Then there are computable quantum correspondences G and H such that $G \subseteq H \subseteq F$ and $\mathrm{dom}(f_H) - \mathrm{dom}(f_G)$ contains exactly one point, and that point lies in α. (And thus no proper computable quantum correspondence can be interpolated between G and H.)*

PROOF. We may assume F is honest. Form G as per the Excision Theorem, so that there is a Π_1^0 tree of sharp filters T with $\mathcal{T} = \mathrm{dom}(f_F) - \mathrm{dom}(f_G) \subseteq \alpha$. Form L_T as in Definition 5.3.

Define

$$H = G \cup \{(\theta_\sigma, F(\theta_\sigma)) : \theta_\sigma \in L_T\}.$$

Since $G \subseteq H \subseteq F$ and H is computably enumerable, it is immediate by Theorem 2.21 (proved in [KalWel03]) that H is an honest computable quantum correspondence. Let x be the unique point in $\mathcal{L}_T$ as per Proposition 5.4. Since $L_T \subseteq \mathrm{dom}(H)$, we have $x \in \mathrm{dom}(f_H)$ and since $L_T \subseteq T$, we have $x \notin \mathrm{dom}(f_G)$. So $\mathrm{dom}(f_H) - \mathrm{dom}(f_G)$ contains at least the point x.

Now suppose $y \in \mathrm{dom}(f_H) - \mathrm{dom}(f_G)$ and let $A = \{\alpha_i : i \in \omega\}$ be a sharp filter converging to y such that $H(A) \downarrow$. Since $y \notin \mathrm{dom}(f_G)$, there must be some n_0 such that $(\forall n \geq n_0)[\alpha_n \notin \mathrm{dom}(G)]$. Hence by the definition of H, $A' = \{\alpha_n : n \geq n_0\} \subseteq L_T$. It therefore follows that $y = x$.

Hence x is the unique point in $\mathrm{dom}(f_H) - \mathrm{dom}(f_G)$.   ⊣

COROLLARY 8.2. *Let $F : \Delta_X \to \Delta_Y$ be a computable quantum correspondence and let κ be a cardinal number such that $\kappa \leq \aleph_0$. Then there are computable quantum correspondences G and H such that $G \subseteq H \subseteq F$ and $\| \mathrm{dom}(f_H) - \mathrm{dom}(f_G) \| = \kappa$.*

PROOF. Assume F is honest. Use either the Evasive Action Engine or the Specker Sequence Engine to obtain a sequence $M = \{\rho_n : n \in \omega\}$ of mutually disjoint basic open sets. For each n, use Theorem 8.1 to form computable quantum correspondences G_n and H_n such that $\mathrm{dom}(f_{H_n}) \subseteq \mathrm{dom}(f_F)$, and there is a unique point $x_n \in \rho_n$ such that $\mathrm{dom}(f_{G_n}) = \mathrm{dom}(f_{H_n}) - \{x_n\}$.

Now use Theorem 2.17 to obtain H and G as follows: $f_H = \bigcap_{n \in \omega} f_{H_n}$, $f_G = f_H \cap \bigcap_{n < \kappa} f_{G_n}$. Then $\operatorname{dom}(f_G) = \operatorname{dom}(f_H) - \{x_n : n < \kappa\}$.   ⊣

THEOREM 8.3 (*Interpolation*). *Let F and G be computable quantum correspondences from Δ_X to Δ_Y with $G \subseteq F$, and let $\operatorname{dom}(f_F) - \operatorname{dom}(f_G)$ contain at least two points. Then there is a computable quantum correspondence $H : \Delta_X \to \Delta_Y$ such that $H \subseteq F$ and $f_G \subsetneq f_H \subsetneq f_F$.*

PROOF. Let $x, y \in \operatorname{dom}(f_F) - \operatorname{dom}(f_G)$, with $x \neq y$.

As per Corollary 2.26, let ϕ be an avoidance function for x such that $S_\phi \cap \operatorname{dom}(f_G) = \emptyset$. Since $x \neq y$, pick basic open sets $\alpha, \beta \in \Delta_X$ such that $x \in \alpha, \overline{\alpha} \subseteq \beta$ and $y \notin \beta$. Next, form an avoidance function ϕ' as in Theorem 2.29 such that $S_{\phi'} \subseteq S_\phi \cap \overline{\alpha}$. Then $y \notin S_{\phi'}$ and $S_{\phi'} \cap \operatorname{dom}(f_G) = \emptyset$. Also, $S_{\phi'} \cap \alpha = S_\phi \cap \alpha$, so $x \in S_{\phi'}$.

Now, with the help of ϕ', and using Corollary 2.28, form a computable quantum correspondence $H \subseteq F$ such that $\operatorname{dom}(f_H) = \operatorname{dom}(f_F) - S_{\phi'}$. Since $x \in S_{\phi'}$, $f_H \subsetneq f_F$. Since $S_{\phi'} \cap \operatorname{dom}(f_G) = \emptyset$, $f_G \subseteq f_H$. Since $y \notin S_{\phi'}$, $y \in \operatorname{dom}(f_H)$; so $f_G \subsetneq f_H$.   ⊣

Acknowledgements. We are grateful to John Chisholm and Linda Lawton for profitable conversations. We also thank the anonymous referee for a careful reading of the manuscript and for many helpful suggestions leading to improvement of the exposition.

REFERENCES

[Abe80] O. ABERTH, *Computable Analysis*, McGraw-Hill, New York, 1980.

[Bra96] V. BRATTKA, *Recursive characterization of computable real-valued functions and relations*, **Theoretical Computer Science**, vol. 162 (1996), pp. 45–77.

[Bra97] ——, *Order-free recursion on the real numbers*, **Mathematical Logic Quarterly**, vol. 43 (1997), pp. 216–234.

[BraKal98] V. BRATTKA and I. KALANTARI, *A bibliography of recursive analysis and recursive topology*, **Handbook of Recursive Mathematics** (Yu. L. Ershov, S. S. Goncharov, A. Nerode, and J. B. Remmel, editors), Studies in Logic and the Foundations of Mathematics, vol. 138, Elsevier, Amsterdam, 1998, Volume 1, Recursive Model Theory, pp. 583–620.

[BW97] V. BRATTKA and K. WEIHRAUCH, *Computability on subsets of Euclidean space I: Closed and compact subsets*, **Theoretical Computer Science**, vol. 219 (1999), pp. 65–93.

[Cei64a] G. S. CEĬTIN, *Three theorems on constructive functions*, **Trudy Matematicheskogo Instituta Imeni V. A. Steklova**, vol. 72 (1964), pp. 537–543, Russian.

[Sha68] G. S. CEĬTIN, N. A. ŠANIN, and I. D. ZASLAVSKIĬ, *Peculiarities of Constructive Mathematical Analysis*, Izdat. Mir, Moscow, 1968, Proc. Internat. Congr. Math. (Moscow, 1966), Russian.

[Cei62b] G. S. CEĬTIN and I. D. ZASLAVSKIĬ, *Singular coverings and properties of constructive functions connected with them*, **Trudy Matematicheskogo Instituta Imeni V. A. Steklova**, vol. 67 (1962), pp. 458–502, Russian.

[Goo61] R. L. GOODSTEIN, *Recursive Analysis*, Studies in Logic and the Foundations of Mathematics, North-Holland Publishing Co., Amsterdam, 1961.

[Grz55] A. GRZEGORCZYK, *Computable functionals*, **Fundamenta Mathematicae**, vol. 42 (1955), pp. 168–202.

[Grz57] ———, *On the definitions of computable real continuous functions*, **Fundamenta Mathematicae**, vol. 44 (1957), pp. 61–71.

[Her97a] P. HERTLING, *An effective Riemann mapping theorem*, **Theoretical Computer Science**, vol. 219 (1999), pp. 225–265.

[Her97b] ———, *A real number structure that is effectively categorical*, **Mathematical Logic Quarterly**, vol. 45 (1999), no. 2, pp. 147–182.

[Kal82] I. KALANTARI, *Major subsets in effective topology*, **Patras Logic Symposion (Patras, 1980)**, Studies in Logic and the Foundations of Mathematics, 109, North-Holland, Amsterdam-New York, 1982, pp. 77–94.

[KalWel95] I. KALANTARI and L. WELCH, *Point-free topological spaces, functions and recursive points; filter foundation for recursive analysis. I*, **Annals of Pure and Applied Logic**, vol. 93 (1998), no. 1–3, pp. 125–151.

[KalWel96] ———, *Recursive and nonextendible functions over the reals; filter foundation for recursive analysis, II*, **Annals of Pure and Applied Logic**, vol. 98 (1999), no. 1–3, pp. 87–110.

[KalWel03] ———, *A blend of methods of recursion theory and topology*, **Annals of Pure and Applied Logic**, vol. 124 (2003), no. 1–3, pp. 141–178.

[KalWel05] ———, *A blend of methods of recursion theory and topology: a Π_1^0 tree of shadow points*, **Archives for Mathematical Logic**, vol. 43 (2004), pp. 991–1008.

[KL57] G. KREISEL and D. LACOMBE, *Ensembles récursivement mesurables et ensembles récursivement ouverts et fermés*, **Comptes Rendus Académie des Sciences Paris**, vol. 245 (1957), pp. 1106–1109, French.

[KW84] C. KREITZ and K. WEIHRAUCH, *A unified approach to constructive and recursive analysis*, **Computation and Proof Theory** (M. M. Richter, E. Börger, W. Oberschelp, B. Schinzel, and W. Thomas, editors), Lecture Notes in Mathematics, vol. 1104, Springer, Berlin, 1984, pp. 259–278.

[Kre87] ———, *Compactness in constructive analysis revisited*, **Annals of Pure and Applied Logic**, vol. 36 (1987), pp. 29–38.

[Kus83] B. KUSHNER, *A Class of Specker Sequences*, Mathematical Logic, Mathematical Linguistics and Theory of Algorithms, Kalinin. Gos. Univ., Kalinin, 1983, Russian.

[Lac55] D. LACOMBE, *Extension de la notion de fonction récursive aux fonctions d'une ou plusieurs variables réelles I*, **Comptes Rendus Académie des Sciences Paris**, vol. 240 (1955), pp. 2478–2480, French.

[Lac55a] ———, *Extension de la notion de fonction récursive aux fonctions d'une ou plusieurs variables réelles II*, **Comptes Rendus Académie des Sciences Paris**, vol. 241 (1955), pp. 13–14, French.

[Lac55b] ———, *Extension de la notion de fonction récursive aux fonctions d'une ou plusieurs variables réelles III*, **Comptes Rendus Académie des Sciences Paris**, vol. 241 (1955), pp. 151–153, French.

[Lac57a] ———, *Les ensembles récursivement ouverts ou fermés, et leurs applications à l'analyse récursive*, **Comptes Rendus Académie des Sciences Paris**, vol. 245, 246 (1957), pp. 1040–1043, 28–31, French.

[Lac58a] ———, *Les ensembles récursivement ouverts ou fermés, et leurs applications à l'Analyse récursive*, **Comptes Rendus Académie des Sciences Paris**, vol. 246 (1958), pp. 28–31, French.

[Maz63] S. MAZUR, *Computable Analysis*, **Rozprawy Matematyczne**, vol. 33, 1963.

[Mil02] J. MILLER, Π_1^0 *Classes in Computable Analysis and Topology*, Ph.D. thesis, Cornell University, Ithaca NY, USA, 2002.

[Mos57] A. MOSTOWSKI, *On computable sequences*, **Fundamenta Mathematicae**, vol. 44 (1957), pp. 37–51.

[Myh71] J. Myhill, *A recursive function, defined on a compact interval and having a continuous derivative that is not recursive*, **The Michigan Mathematical Journal**, vol. 18 (1971), pp. 97–98.

[Ore63a] V. P. Orevkov, *A constructive map of the square into itself, which moves every constructive point*, **Rossiiskaya Akademiya Nauk. Doklady Akademii Nauk**, vol. 152 (1963), pp. 55–58.

[PR89] M. B. Pour-El and I. Richards, *Computability in Analysis and Physics*, Perspectives in Mathematical Logic, Springer, Berlin, 1989.

[Ric54] H. G. Rice, *Recursive real numbers*, **Proceedings of the American Mathematical Society**, vol. 5 (1954), pp. 784–791.

[Sch02] M. Schröder, *Extended admissibility*, **Theoretical Computer Science**, vol. 284 (2002), no. 2, pp. 519–538.

[Soa96] R. I. Soare, *Computability and Recursion*, **The Bulletin of Symbolic Logic**, vol. 2 (1996), pp. 284–321.

[Spe49] E. Specker, *Nicht konstruktiv beweisbare Sätze der Analysis*, **The Journal of Symbolic Logic**, vol. 14 (1949), pp. 145–158, German.

[Spe59] ———, *Der Satz vom Maximum in der rekursiven Analysis*, **Constructivity in Mathematics** (A. Heyting, editor), North Holland, Amsterdam, 1959, Colloquium at Amsterdam, 1957, pp. 254–265.

[Tur36] A. M. Turing, *On computable numbers, with an application to the "Entscheidungsproblem"*, **Proceedings of the London Mathematical Society**, vol. 42 (1936), no. 2, pp. 230–265.

[Tur37] ———, *On computable numbers, with an application to the "Entscheidungsproblem". A correction*, **Proceedings of the London Mathematical Society**, vol. 43 (1937), no. 2, pp. 544–546.

[Sha62] N. A. Šanin, *Constructive real numbers and constructive functional spaces*, **Trudy Matematicheskogo Instituta Imeni V. A. Steklova**, vol. 67 (1962), pp. 15–294, Russian.

[WY96] M. Washihara and M. Yasugi, *Computability and metrics in a Fréchet space*, **Mathematica Japonica**, vol. 43 (1996), no. 3, pp. 431–443.

[Wei93] K. Weihrauch, *Computability on computable metric spaces*, **Theoretical Computer Science**, vol. 113 (1993), pp. 191–210.

[Wei97b] ———, *Computability on the probability measures on the Borel sets of the unit interval*, **Theoretical Computer Science**, vol. 219 (1999), pp. 421–437.

[Wei00] ———, *Computable Analysis*, Springer, Berlin, 2000.

[MTY97] M. Yasugi, T. Mori, and Y. Tsujii, *Effective properties of sets and functions in metric spaces with computability structure*, **Theoretical Computer Science**, vol. 219 (1999), pp. 467–486.

[Zas62] I. D. Zaslavskiĭ, *Some properties of constructive real numbers and constructive functions*, **Trudy Matematicheskogo Instituta Imeni V. A. Steklova**, vol. 67 (1962), pp. 385–457, Russian.

DEPARTMENT OF MATHEMATICS
WESTERN ILLINOIS UNIVERSITY
MACOMB, IL 61455, USA
E-mail: i-kalantari@wiu.edu
E-mail: l-welch@wiu.edu

ADDITIVE POLYNOMIALS AND THEIR ROLE IN THE MODEL THEORY OF VALUED FIELDS

FRANZ-VIKTOR KUHLMANN

Dedicated to Mahmood Khoshkam († October 13, 2003)

Abstract. We discuss the role of additive polynomials and p-polynomials in the theory of valued fields of positive characteristic and in their model theory. We outline the basic properties of rings of additive polynomials and discuss properties of valued fields of positive characteristic as modules over such rings. We prove the existence of Frobenius-closed bases of algebraic function fields $F \mid K$ in one variable and deduce that F/K is a free module over the ring of additive polynomials with coefficients in K. Finally, we prove that every minimal purely wild extension of a henselian valued field is generated by a p-polynomial.

§1. Introduction. This paper is to some extent a continuation of my introductive and programmatic paper [Ku3]. In that paper I pointed out that the ramification theoretical *defect* of finite extensions of valued fields is responsible for the problems we have when we deal with the model theory of valued fields, or try to prove local uniformization in positive characteristic.

In the present paper I will discuss the connection between the defect and additive polynomials. I will state and prove basic facts about additive polynomials and then treat several instances where they enter the theory of valued fields in an essential way that is particularly interesting for model theorists and algebraic geometers. I will show that non-commutative structures (skew polynomial rings) play an essential role in the structure theory of valued fields

This paper was written while I was a guest of the Equipe Géométrie et Dynamique, Institut Mathématiques de Jussieu, Paris, and of the Equipe Algèbre–Géométrie at the University of Versailles. I gratefully acknowledge their hospitality and support. I was also partially supported by a Canadian NSERC grant and by a sabbatical grant from the University of Saskatchewan. Furthermore I am endebted to the organizers of the conference in Teheran and the members of the IPM and all our friends in Iran for their hospitality and support. I also thank Peter Roquette, Florian Pop and Philip Rothmaler for their helpful suggestions and inspiring discussions, and the two referees for their careful reading of the paper, their corrections and numerous useful suggestions. The final revision of this paper was done during my stay at the Newton Institute at Cambridge; I gratefully acknowledge its support.

Logic in Tehran
Edited by A. Enayat, I. Kalantari, and M. Moniri
Lecture Notes in Logic, 26
© 2006, ASSOCIATION FOR SYMBOLIC LOGIC

in positive characteristic. Further, I will state the main open questions. I will also include some exercises.

In the next section, I will give an introduction to additive polynomials and describe the reasons for their importance in the model theory of valued fields. For the convenience of the reader, I outline the characterizations of additive polynomials in Section 3 and the basic properties of rings of additive polynomials in Section 4. For more information on additive polynomials, the reader may consult [Go], cf. also [O1], [O2], [Wh1], [Wh2], [Ku4], [Dr–Ku]. The remaining sections of this paper will then be devoted to the proofs of some of the main theorems stated in Section 2.

§2. Reasons for the importance of additive polynomials in the model theory of valued fields. A polynomial $f \in K[X]$ is called *additive* if

$$(1) \qquad\qquad f(a + b) = f(a) + f(b)$$

for all elements a, b in every extension field L of K, that is, if the mapping induced by f on L is an endomorphism of the additive group $(L, +)$. If K is infinite, then f is additive already when condition (1) is satisfied for all $a, b \in K$: see part b) of Corollary 23 in Section 3.

It follows from the definition that an additive polynomial cannot have a non-zero constant term. If the characteristic char K is zero, then the only additive polynomials over K are of the form cX with $c \in K$. If char $K = p > 0$, then the mapping $a \mapsto a^p$ is an endomorphism of K, called the *Frobenius*. Therefore, the polynomial X^p is additive over any field of characteristic p. Another famous and important additive polynomial is $\wp(X) := X^p - X$, the additive *Artin-Schreier polynomial*. An extension of a field K of characteristic p generated by a root of a polynomial of the form $X^p - X - c$ with $c \in K$ is called an *Artin-Schreier extension*. We will see later that Artin-Schreier extensions play an important role in the theory of fields in characteristic p.

Note that there are polynomials defined over a finite field which are not additive, but satisfy the condition for all elements coming from that field. For example, we know that $a^p = a$ and thus $a^{p+1} - a^2 = 0$ for all $a \in \mathbb{F}_p$. Hence, the polynomial $g(X) := X^{p+1} - X^2$ satisfies $g(a + b) = 0 = g(a) + g(b)$ for all $a, b \in \mathbb{F}_p$. But it is not an additive polynomial. To show this, let us take an element ϑ in the algebraic closure of $\mathbb{F}_p$ such that $\vartheta^p - \vartheta = 1$. Then $g(\vartheta) = \vartheta(\vartheta^p - \vartheta) = \vartheta$. On the other hand, $g(\vartheta + 1) = (\vartheta + 1)((\vartheta + 1)^p - (\vartheta + 1)) = (\vartheta + 1)(\vartheta^p + 1^p - \vartheta - 1) = \vartheta + 1 \neq \vartheta = g(\vartheta) + g(1)$. Hence, already on the extension field $\mathbb{F}_p(\vartheta)$, the polynomial g does not satisfy the additivity condition.

The following well known theorem gives a very useful characterization of additive polynomials. I will present a proof in Section 3.

THEOREM 1. *Let p be the characteristic exponent of K (i.e., $p = \operatorname{char} K$ if this is positive, and $p = 1$ otherwise). Take $f \in K[X]$. Then f is additive if and only if it is of the form*

$$(2) \qquad f(X) = \sum_{i=0}^{m} c_i X^{p^i} \quad \text{with } c_i \in K.$$

Assume that $\operatorname{char} K = p > 0$. Then as a mapping on K, X^p is equal to the Frobenius endomorphism φ. Similarly, X^{p^2} is equal to the composition of φ with itself, written as φ^2, and by induction, we can replace 2 by every integer n. On the other hand, the monomial X induces the identity mapping, which we may write as φ^0. Note that addition and composition of additive mappings on $(K, +)$ give again additive mappings (in particular, addition of additive polynomials gives additive polynomials). It remains to interpret the coefficients of additive polynomials as mappings. This is easily done by viewing K as a K-vector space: the mapping $c \cdot$ induced by $c \in K$ is given by multiplication $a \mapsto ca$, and it is an automorphism of $(K, +)$ if $c \neq 0$. So cX^{p^n} as a mapping is the composition of φ^n with $c \cdot$. We will write this composition as $c\varphi^n$. Adding these monomials generates new additive mappings of the form $\sum_{i=0}^{m} c_i \varphi^i$, and addition of such mappings gives again additive mappings of this form. Composition of such additive mappings generates again additive mappings, and the reader may compute that they can again be written in the above form. In this way, we are naturally led to considering the ring $K[\varphi]$ of all polynomials in φ over K, where multiplication is given by composition. From the above we see that this ring is a subring of the endomorphism ring of the additive group of K. The correspondence that we have worked out now reads as

$$(3) \qquad \sum_{i=0}^{m} c_i X^{p^i} \longleftrightarrow \sum_{i=0}^{m} c_i \varphi^i \in K[\varphi]$$

which means that both expressions describe the same additive mapping on K. For instance, the additive Artin-Schreier polynomial $X^p - X$ corresponds to $\varphi - 1$. Through the above correspondence, the ring $K[\varphi]$ may be considered as the *ring of additive polynomials over K*. Note that this ring is not commutative; in fact, we have

$$\varphi c = c^p \varphi \quad \text{for all } c \in K.$$

This shows that assigning $\varphi \mapsto z$ induces an isomorphism of $K[\varphi]$ onto the skew polynomial ring $K[z; \varphi]$. But we will keep the notation "$K[\varphi]$" since it is simpler.

Let me state some basic properties of the ring $K[\varphi]$. For more information, I recommend the comprehensive book "Free rings and their relations" by P. M. Cohn ([C1], [C2]). Let R be a ring with $1 \neq 0$. Equipped with a function deg : $R \setminus \{0\} \to \mathbb{N} \cup \{0\}$, the ring R is called *left euclidean* if for all elements $s, s' \in R$, $s \neq 0$, there exist $q, r \in R$ such that

$$s' = qs + r \quad \text{with } r = 0 \text{ or } \deg r < \deg s,$$

and it is called *right euclidean* if the same holds with "$s' = sq + r$" in the place of "$s' = qs + r$". (Usually, the function deg is extended to 0 by setting $\deg 0 = -\infty$.) For example, polynomial rings over fields equipped with the usual degree function are both-sided euclidean rings. Further, an integral domain R is called a *left principal ideal domain* if every left ideal in R is principal (and analogously for "right" in the place of "left"). I leave it to the reader to show that every left (or right) euclidean integral domain is a left (or right) principal ideal domain. Finally, an integral domain R is called a *left Ore domain* if

$$Rr \cap Rs \neq \{0\} \quad \text{for all } r, s \in R \setminus \{0\},$$

and it is called a *right Ore domain* if $rR \cap sR \neq \{0\}$ for all $r, s \in R \setminus \{0\}$. Every left (or right) Ore domain can be embedded into a skew field (cf. [C1, Section 0.8, Corollary 8.7]). The reader may prove that a left (or right) principal ideal domain is a left (or right) Ore domain.

The ring $K[\varphi]$ may be equipped with a degree function which satisfies $\deg 0 = -\infty$ and $\deg \sum_{i=0}^{m} c_i \varphi^i = m$ if $c_m \neq 0$. This degree function is a homomorphism of the multiplicative monoid of $K[\varphi] \setminus \{0\}$ onto $\mathbb{N} \cup \{0\}$ since it satisfies $\deg rs = \deg r + \deg s$. In particular, this shows that $K[\varphi]$ is an integral domain. The following theorem is due to O. Ore [O2]; I will give a proof in Section 4.

THEOREM 2. *The ring $K[\varphi]$ is a left euclidean integral domain and thus also a left principal ideal domain and a left Ore domain. It is right euclidean if and only if K is perfect; if K is not perfect, then $K[\varphi]$ is not even right Ore.*

EXAMPLE 1. Let $\mathbb{F}_p$ denote the field with p elements. The ring $\mathbb{F}_p[\varphi]$ is a both-sided euclidean integral domain, and every field K of characteristic p is a left $\mathbb{F}_p[\varphi]$-module and a left $K[\varphi]$-module, where the action of φ on K is just the application of the Frobenius endomorphism. K is perfect if and only if every element of K is divisible by the ring element φ. But this does not imply that K is a divisible $\mathbb{F}_p[\varphi]$- or $K[\varphi]$-module. For instance, if K admits non-trivial Artin-Schreier extensions, that is, if $K \neq \wp(K) = (\varphi - 1)K$, then there are elements in K which are not divisible by $\varphi - 1$. On the other hand, K is a divisible $\mathbb{F}_p[\varphi]$- and $K[\varphi]$-module if K is algebraically closed.

Observe that K is not torsion free as an $\mathbb{F}_p[\varphi]$- or $K[\varphi]$-module. Indeed, K contains $\mathbb{F}_p$ which satisfies

$$(\varphi - 1)\mathbb{F}_p = \{0\}.$$

EXAMPLE 2. The power series field $K := \mathbb{F}_p((t)) = \{\sum_{i=N}^{\infty} c_i t^i \mid N \in \mathbb{Z}, c_i \in K\}$ (also called "field of formal Laurent series over $\mathbb{F}_p$") is not perfect, since t does not admit a p-th root in K. Hence, the ring $K[\varphi]$ is not right Ore. K is a left $K[\varphi]$-module.

In Section 4, Remark 24, I will collect a few properties of the rings $K[\varphi]$ that follow from Theorem 2, and describe what happens if K is not perfect. We will see that in that case the structure of $K[\varphi]$-modules becomes complicated. It seems that the "bad" properties of $K[\varphi]$, for K non-perfect, are symptomatic for the problems algebraists and model theorists have with non-perfect valued fields in positive characteristic. Let me discuss the most prominent of such non-perfect valued fields.

The field $\mathbb{F}_p((t))$ carries a canonical valuation v_t, called the t-adic valuation. It is given by $v_t \sum_{i=N}^{\infty} c_i t^i = N$ if $c_N \neq 0$ and $v_t 0 = \infty$. $(\mathbb{F}_p((t)), v_t)$ is a complete discretely valued field, with value group $v_t \mathbb{F}_p((t)) = \mathbb{Z}$ (that is what "discretely valued" means) and residue field $\mathbb{F}_p((t))v_t = \mathbb{F}_p$. At the first glance, such fields may appear to be the best known objects in valuation theory. Nevertheless, the following prominent questions about the elementary theory $\mathrm{Th}(\mathbb{F}_p((t)), v_t)$ are still unanswered:

OPEN PROBLEM 1. Is the elementary theory of the valued field $\mathbb{F}_p((t))$ model complete? Does it admit quantifier elimination in some natural language? Is it decidable? If yes, what would be a complete recursive axiomatization?

The corresponding problem for the p-adics was solved in the mid 1960s independently by Ax and Kochen [A–K] and by Ershov [Er]. Since then, the above problem has been well known to model-theoretic algebraists, but resisted all their attacks.

Encouraged by the similarities between $\mathbb{F}_p((t))$ and the field $\mathbb{Q}_p$ of p-adic numbers, one might try to give a complete axiomatization for $\mathrm{Th}(\mathbb{F}_p((t)), v_t)$ by adapting the well known axioms for $\mathrm{Th}(\mathbb{Q}_p, v_p)$. They express that $(\mathbb{Q}_p, v_p)$ has the following elementary properties:

• It is a henselian valued field of characteristic 0. A valued field (K, v) is called henselian if it satisfies Hensel's Lemma: *If f is a polynomial with coefficients in the valuation ring $\mathcal{O}$ of v and if $b \in \mathcal{O}$ such that $vf(b) > 0$ while $vf'(b) = 0$, then there is some $a \in \mathcal{O}$ such that $f(a) = 0$ and $v(a - b) > 0$.* This holds if and only if the extension of v to the algebraic closure of K is unique.

• Its value group a $\mathbb{Z}$-group, i.e., an ordered abelian group elementarily equivalent to $\mathbb{Z}$.

• Its residue field is $\mathbb{F}_p$.

• $v_p p$ is equal to 1 (= the smallest positive element in the value group).

The last condition is not relevant for $\mathbb{F}_p((t))$ since there, $p \cdot 1 = 0$. Nevertheless, we may add a constant name t to $\mathcal{L}$ so that one can express by an elementary sentence that $v_t t = 1$.

A naive adaptation would just replace "characteristic 0" by "characteristic p" and p by t. But there is an elementary property of valued fields that is satisfied by all valued fields of residue characteristic 0 and all formally p-adic fields, but not by all valued fields in general. It is the property of being defectless. A valued field (K, v) is called *defectless* if for every finite extension $L|K$, equality holds in the *fundamental inequality*

$$(4) \qquad\qquad n \geq \sum_{i=1}^{g} e_i f_i,$$

where $n = [L : K]$ is the degree of the extension, $v_1, \ldots, v_g$ are the distinct extensions of v from K to L, $e_i = (v_i L : vK)$ are the respective ramification indices, and $f_i = [Lv_i : Kv]$ are the respective inertia degrees. (Note that $g = 1$ if (K, v) is henselian.) There is a simple example, due to F. K. Schmidt, which shows that there is a henselian discretely valued field of positive characteristic which is not defectless (cf. [Ri, Exemple 1, p. 244]). This field has a finite purely inseparable extension with non-trivial defect. But defect does not only appear in purely inseparable extensions: there is an example, due to A. Ostrowski, of a complete valued field admitting a finite separable extension with non-trivial defect (cf. [Ri, Exemple 2, p. 246]). These and several other examples of extensions with non-trivial defect of various types can also be found in [Ku12] (see also [Ku8]).

However, each power series field with its canonical valuation is henselian and defectless (cf. [Ku12]). In particular, $(\mathbb{F}_p((t)), v_t)$ is defectless. For a less naive adaptation of the axiom system of $\mathbb{Q}_p$, we will thus add "defectless". We obtain the following axiom system in the language $\mathcal{L}(t) = \mathcal{L} \cup \{t\}$:

$$(5) \quad \begin{cases} (K, v) \text{ is a henselian defectless valued field} \\ K \text{ is of characteristic } p \\ vK \text{ is a } \mathbb{Z}\text{-group} \\ Kv = \mathbb{F}_p \\ vt \text{ is the smallest positive element in } vK. \end{cases}$$

Let us note that also $(\mathbb{F}_p(t), v_t)^h$, the henselization of $(\mathbb{F}_p(t), v_t)$, satisfies these axioms. The *henselization* of a valued field (K, v) is a henselian algebraic extension which is minimal in the sense that it admits a (unique!) embedding over K in every henselian extension of (K, v). Henselizations exist for all valued fields, and they are separable extensions (cf. [Ri, Théorème 2, p. 176]). It is well known that $(\mathbb{F}_p(t), v_t)^h$ is a defectless field, being the henselization of a global field (cf. [Ku9]). It is also well known that $(\mathbb{F}_p(t), v_t)^h$ is existentially closed in $(\mathbb{F}_p((t)), v_t)$ (see below for the definition of this notion); this fact follows from work of Greenberg [Gre] and also from Theorem 2 of [Er1] (see also [Ku7] for some related information). But it is not known whether $(\mathbb{F}_p((t)), v_t)$ is an elementary extension of $(\mathbb{F}_p(t), v_t)^h$.

It did not seem unlikely that axiom system (5) could be complete, until I proved in [Ku1] (cf. [Ku4]):

THEOREM 3. *The axiom system* (5) *is not complete.*

I will give an idea of the proof of this theorem in Section 2.3 below. It was inspired by an observation of Lou van den Dries. He had worked with a modified axiom system (with larger residue fields) and had found an elementary sentence which he was not able to deduce from that axiom system (as it turned out, that wasn't van den Dries' fault). This sentence was formulated using only addition, multiplication with the element t and application of the Frobenius, but no general multiplication. This led van den Dries to the question whether one could at least determine the model theory of $\mathbb{F}_p((t))$ as a module which admits multiplication with t and application of the Frobenius, forgetting about general multiplication. But this means that we view $\mathbb{F}_p((t))$ as a left $K[\varphi]$-module, where the field K contains t and should be contained in $\mathbb{F}_p((t))$. But then, K is not perfect, and therefore the structure of $K[\varphi]$-modules may be quite complicated.

There is a common feeling that additive polynomials play a crucial role in the theory of valued fields of positive characteristic. So indeed, van den Dries' question may be the key to the model theory of $\mathbb{F}_p((t))$ (but it could be as hard to solve as the original problem). In this paper, I will give some reasons for this common feeling, but also confront it with our present problem of understanding the notion of extremality.

2.1. Reason #1: Kaplansky's hypothesis A. For a valued field (K, v), we denote by vK its value group and by Kv its residue field. An extension $(K, v) \subset (L, v)$ of valued fields is called *immediate* if the induced embeddings of vK in vL and of Kv in Lv are onto. Henselizations are immediate extensions (cf. [Ri, Corollaire 1, p. 184]). Wolfgang Krull [Kr] (see also [Gra]) proved that every valued field admits a maximal immediate extension. A natural and in fact very important question is whether this is unique up to (valuation preserving) isomorphism. This plays a role when one wishes to embed valued fields in power series fields. In his celebrated paper [Ka1], Irving Kaplansky gave a criterion, called "hypothesis A", which guarantees uniqueness. (We will present it later.) Kaplansky then showed that a valued field (K, v) of positive characteristic having a cross-section and satisfying hypothesis A can be embedded in the power series field $Kv((vK))$ over its residue field Kv with exponents in its value group vK. Kaplansky also gives examples which show that this may fail if hypothesis A is not satisfied. In this case, there are *maximal fields* (= valued fields not admitting any proper immediate extensions) which do not have the form of a power series field (not even if one allows factor systems).

If we are considering an elementary class of valued fields satisfying hypothesis A (which can be expressed by a recursive scheme of elementary sentences in the language of valued rings), then the uniqueness of maximal immediate extensions can be fruitfully used in the proof of model theoretic properties.

Let us give the example of algebraically maximal Kaplansky fields. A valued field is called *algebraically maximal* if it does not admit any proper immediate algebraic extension. It is called a *Kaplansky field* if it satisfies hypothesis A. The following theorem is due to Ershov [Er1] and, independently, Ziegler [Zi].

THEOREM 4. *The elementary theory of an algebraically maximal Kaplansky field is completely determined by the elementary theory of its value group and the elementary theory of its residue field.*

In other words, algebraically maximal Kaplansky fields satisfy the following *Ax–Kochen–Ershov principle*:

$$(6) \qquad vK \equiv vL \wedge Kv \equiv Lv \implies (K, v) \equiv (L, v).$$

Where the first elementary equivalence is in the language of ordered groups, the second in the language of rings and the third in the language of valued rings. In the case of $(K, v) \subseteq (L, v)$ there is also a version of the Ax–Kochen–Ershov principle with "$\equiv$" replaced by "$\prec$" (elementary extension). In the same situation, there is also the more basic version

$$(7) \qquad vK \prec_{\exists} vL \wedge Kv \prec_{\exists} Lv \implies (K, v) \prec_{\exists} (L, v).$$

Where "$\prec_{\exists}$" means "*existentially closed in*", that is, every existential elementary sentence which holds in the upper structure also holds in the lower structure. In fact, it has turned out that proving this version is the essential step in proving Ax–Kochen–Ershov principles and other model theoretic results about valued fields; the further results then often follow by general model theoretic arguments (the reader should think of Robinson's Test).

Hypothesis A implicitly talks about additive polynomials. Following Kaplansky [Ka2], we will call a polynomial $f \in K[X]$ a *p-polynomial* if it is of the form

$$(8) \qquad f(X) = \mathcal{A}(X) + c,$$

where $\mathcal{A} \in K[X]$ is an additive polynomial, and c is a constant in K. A field K of characteristic $p > 0$ will be called *p-closed* if every p-polynomial in $K[X]$ has a root in K. That is,

K is *p-closed* if and only if it is a divisible $K[\varphi]$-module.

In particular, every p-closed field is perfect.

Now hypothesis A for a valued field (K, v) with char $Kv = p > 0$ reads as follows:

(A1) the value group vK is p-divisible, and

(A2) the residue field Kv is p-closed.

For valued fields (K, v) with char $Kv = 0$, hypothesis A is empty. The condition of a field to be p-closed seemed obscure at the time of Kaplansky's paper. But we have learned to understand this condition better. The following theorem was first proved by Whaples in [Wh2], using the cohomology theory of additive polynomials. A more elementary proof was later given in [Del]. Then Kaplansky gave a short and elegant proof in his "Afterthought: Maximal Fields with Valuation" ([Ka2]). We will reproduce this proof in Section 9.

THEOREM 5. *A field K of characteristic $p > 0$ is p-closed if and only if it does not admit any finite extensions of degree divisible by p.*

This theorem lets us understand hypothesis A much better. Based on this insight, Kaplansky's result about uniqueness of maximal immediate extensions is reproved in [Ku–Pa–Ro]. There, it is deduced from the Schur–Zassenhaus Theorem about conjugacy of complements in profinite groups, via Galois correspondence.

As we are shifting our focus to additive polynomials, the original condition "p-closed" regains its independent interest. In Section 9 we will use Theorem 5 to prove:

THEOREM 6. *A henselian valued field of characteristic $p > 0$ is p-closed if and only if it is an algebraically maximal Kaplansky field.*

For a generalization of the notion "p-closed" and of this theorem to fields of characteristic 0 see [V, in particular Corollary 5].

By Theorem 5 we can split condition (A2) into two distinct conditions:
(A2.1) the residue field Kv is perfect, and
(A2.2) the residue field Kv does not admit any finite separable extension of degree divisible by p.

While (A2.1) is a perfectly natural condition about fields, (A2.2) is somewhat unusual. This is the reason for the fact that Kaplansky fields are not often found in applications. Certainly also the other conditions restrict the possible applications (for example, $\mathbb{F}_p((t))$ doesn't satisfy (A1)). But for instance, every perfect valued field of characteristic $p > 0$ satisfies conditions (A1) and (A2.1) (but not necessarily (A2.2)). So we would obtain a more natural condition if we could drop condition (A2.2). To obtain good model theoretic properties for fields satisfying (A1) and (A2.1), one has to require again that they are algebraically maximal. Such fields form a part of an important larger class of valued fields, the tame fields. A *tame field* is a henselian field whose algebraic closure is equal to the ramification field K^r of the normal extension $K^{\text{sep}}|K$, where K^{sep} denotes the separable-algebraic closure of K. The *ramification field* of a normal separable-algebraic extension of valued fields is the fixed field in that extension of a certain subgroup of the Galois group, the *ramification group*. We don't need the definition of this group here; we only need the basic properties of the field K^r which I will put together in Theorem 38 below. By part e) of this theorem, every tame field is defectless, and it follows directly from the definition that every tame field is perfect. In [Ku1] (cf. also [Ku11]) I proved:

THEOREM 7. *The elementary theory of a tame field is completely determined by the elementary theory of its value group and the elementary theory of its residue field.*

All tame fields satisfy conditions (A1) and (A2.1), but not necessarily (A2.2). That means that we have lost the uniqueness of maximal immediate extensions. But the above result shows that uniqueness is not necessary for an elementary class of valued fields to have good model theoretic properties. However, we have to work much harder. This work is again directly related to additive polynomials, and we will describe this connection now.

2.2. Reason #2: the defect and purely wild extensions. Let us assume that (K, v) is henselian. Then for every finite extension L of K, we have $g = 1$ in the fundamental inequality (4). Then the Lemma of Ostrowski (cf. [Ri, Théorème 2, p. 236]) tells us that we have an equality

$$(9) \qquad [L : K] = (vL : vK) \cdot [Lv : Kv] \cdot p^{\delta},$$

where p is the characteristic exponent of the residue field Kv, and δ is a non-negative integer. The factor p^{δ} is called the *defect* of the extension $(L|K, v)$; it is *trivial* if $p^{\delta} = 1$. Consequently, every valued field with residue field of characteristic 0 is defectless.

It follows from Theorem 38 that a valued field is tame if it is henselian and for every finite extension $L|K$,

- the characteristic of Kv does not divide $(vL : vK)$,
- the extension $Lv|Kv$ is separable, and
- the extension $(L|K, v)$ is defectless.

The ramification field K^r is the unique maximal tame extension of every henselian field (K, v).

As I have explained in [Ku3], the presence of non-trivial defect is one of the main obstacles in proving an Ax–Kochen–Ershov principle like (7). Let me quickly sketch this again. Assume that (L, v) is an extension of a henselian field (K, v) such that $vK \prec_\exists vL$ and $Kv \prec_\exists Lv$. Then we take (K^*, v^*) to be an $|L|^+$-saturated elementary extension of (K, v). It follows that v^*K^* is a $|vL|^+$-saturated extension of vK; hence $vK \prec_\exists vL$ yields that vL can be embedded over vK in v^*K^*. It also follows that K^*v^* is an $|Lv|^+$-saturated extension of vL; hence $Kv \prec_\exists Lv$ yields that Lv can be embedded over Kv in K^*v^*. Now we have to lift these embeddings to an embedding of (L, v) in (K^*, v^*) over K. Once this is achieved, we are done, because every existential elementary sentence which holds in (L, v) carries over to its image in (K^*, v^*), from there up to (K^*, v^*), and from there down to the elementary substructure (K, v).

By a general model theoretic argument, the situation can be reduced to the case where L is finitely generated over K. That is, $(L|K, v)$ is a valued function field (by "function field", we will always mean "algebraic function field").

Hence, we need the structure theory of valued function fields to solve our embedding problem (as it is the case for the problem of local uniformization). Let us assume that we can reduce to the case where the transcendence degree of $L|K$ is 1. This can be done for tame fields, but for the model theory of $\mathbb{F}_p((t))$, this is another serious problem, again connected with additive polynomials (see Section 2.3). Assume further that we have lifted the embeddings of vL and Lv to an embedding of some subfield L_0 of L. Then $L|L_0$ is a finite immediate extension, and in general, it will be proper (i.e., $L \neq L_0$). Taking henselizations, we obtain that also $L^h|L_0^h$ is a finite immediate extension. Since we assumed that (K, v) is henselian (which is true for every algebraically maximal and every tame field), its elementary extension (K^*, v^*) is also henselian. Therefore, the embedding of L_0 extends to an embedding of L_0^h in K^* (this is the universal property of the henselization). But if $L^h \neq L_0^h$, we do not know how to lift the extension further (which we would need to get all of L embedded), unless we have uniqueness of maximal immediate extensions. Since $L^h|L_0^h$ is immediate, we have $(vL^h : vL_0^h) = 1$ and $[L^h v : L_0^h v] = 1$; hence if $L^h \neq L_0^h$, then by (9), the extension has non-trivial defect, equal to its degree.

We see that indeed, the presence of non-trivial defect constitutes a serious obstacle for our embedding problem. So we have to avoid the defect. In certain cases, it will not even appear. All tame fields are defectless fields (and so are all other valued fields for which we know good model theoretic results). This does not mean that every valued function field over a tame field is defectless. But for a certain type of valued function fields, this is true. Let $(L|K, v)$ be an extension of valued fields of finite transcendence degree. Then the following inequality holds (cf. [B, Chapter VI, Section 10.3, Theorem 1]):

$$(10) \qquad \operatorname{trdeg} L|K \geq \dim_{\mathbb{Q}}(\mathbb{Q} \otimes (vL/vK)) + \operatorname{trdeg} Lv|Kv.$$

If equality holds then we will say that $(L|K, v)$ is without transcendence defect. For such function fields, we have ([Ku1], [Ku9]):

THEOREM 8 (Generalized Stability Theorem). Let $(L|K, v)$ be a valued function field without transcendence defect. If (K, v) is a defectless field, then also (L, v) is a defectless field.

Using this theorem, one can prove (cf. [Ku9]):

THEOREM 9. Let (K, v) be a henselian defectless field. Then the Ax–Kochen–Ershov principle (7) holds for every extension (L, v) of (K, v) without transcendence defect.

I proved Theorem 8 in [Ku1]. How does this proof work? How can we see whether a given valued field is defectless? First of all, a valued field is defectless if and only if its henselization is (see [Ku9]; a partial proof is also given in [En]). So we can assume that L is the henselization of a valued function field. Second, if (k, v) is any henselian field, then every finite extension of

k inside the ramification field k^r has trivial defect, and if $k_1|k$ is any finite extension, then $k_1|k$ and $k^r.k_1|k^r$ have the same defect (cf. Theorem 38). So in our situation, we have to show that every finite extension of L' has trivial defect. The advantage of working over L' is that general ramification theory tells us that $L^{sep}|L'$ is a p-extension. A normal and separable field extension is called a *p-extension* if its Galois group is a pro-p-group. It follows from the general theory of p-groups (cf. [H], Chapter III, Section 7, Satz 7.2 and the following remark) via Galois correspondence that every finite separable-algebraic extension of L' is a tower of Galois extensions of degree p. Hence we just have to show by induction that each of them has trivial defect. (The complementary case of purely inseparable extensions is much easier and can be disposed of more directly.) Now every Galois extension $k'|k$ of degree p of fields of characteristic p is an Artin-Schreier extension; this well known fact is proved by an application of Hilbert's Theorem 90. We include a proof in Section 7 (Theorem 35), as a special case of a generalization which we will discuss below.

Let ϑ be a root of $X^p - X - a$. Then $k' = k(\vartheta) = k(\vartheta - c)$ for every $c \in k$. As $X^p - X$ is an additive polynomial, we have $(\vartheta - c)^p - (\vartheta - c) = a - c^p + c$, that is, $\vartheta - c$ is a root of the p-polynomial $X^p - X - (a - c^p + c)$. So we may change a by subtracting elements in k of the form $c^p - c$, without changing the extension generated by the polynomial. The idea in our above situation is now to find by this method a suitable normal form for the element a from which we can read off that the extension has trivial defect. The idea of deducing suitable normal forms for Artin-Schreier extensions (and for Kummer extensions in the case of fields of characteristic 0) can already be found in the work of Hasse, Whaples, Epp ([Ep], see also [Ku5]), Matignon and Abhyankar.

Let us quickly discuss two examples. We wish to show that a given Artin-Schreier extension $L'|L'$ has trivial defect. Before we go on, we note that by valuation theoretical arguments, the proof of Theorem 8 can be reduced to the case where K and hence also its residue field Kv is algebraically closed. Assume that the transcendence degree of $L|K$ is 1. Then by (10) with equality, we can have

- $\dim_{\mathbb{Q}}(\mathbb{Q} \otimes (vL/vK)) = 1$ and trdeg $Lv|Kv = 0$, or
- $\dim_{\mathbb{Q}}(\mathbb{Q} \otimes (vL/vK)) = 0$ and trdeg $Lv|Kv = 1$.

In the first case, there is an element $x \in L$ such that vx is rationally independent over vK. Under certain additional conditions, we can then take a to be a polynomial in x with coefficients in K. Since the values of the monomials in this polynomial $a = a(x)$ lie in distinct cosets modulo vK, their values are distinct. By the ultrametric triangle law, this implies that the value of such a monomial is equal to the least value of its monomials. Now we can use the above method to get rid of p-th powers of x in $a(x)$ (we can replace a monomial cx^{kp} by $c^{1/p}x^k$). Therefore, we can assume that all monomials appearing in the polynomial $a(x)$ are of the form $c_i x^i$ with i not divisible by p. Then the value $va(x)$ is not divisible by p in vL'. This value cannot be positive since otherwise, the extension would be trivial by Hensel's Lemma. With the value being negative, the reader may show that if ϑ is a root of $X^p - X - a(x)$, then

$$v\vartheta = \frac{va(x)}{p}.$$

This implies that $(vL' : vL') = p = [L' : L']$, so the extension has trivial defect.

In the second case, we will have an element $x \in L$ of value 0 whose residue xv is transcendental over Kv. Now we will have to deal with finite sums of the form $c_i d_i$ where $d_i \in L$ are representatives of elements in the residue field. We play the same game as before, trying to come up with a residue that has no p-th root, from which it would follow in a similar way as above that $[L'v : L'v] = p = [L' : L']$, showing that the extension has trivial defect. The problem here is that when we replace some monomial $c_i d_i$ by its p-th root $c_i^{1/p} d_i^{1/p}$, then even if the residue $d_i^{1/p} v$ does not have a p-th root in Lv, the element $d_i^{1/p}$ might sum up with some other d_j to an element whose residue has again a p-th root in Lv. We somehow have to see that this process cannot go on infinitely. A good idea would be to take the d_i such that their residues form a basis of $Lv|Kv$. But then we would need that also the residue of $d_i^{1/p}$ is in this basis. It is easily seen that a basis being closed under taking p-th roots (as long as we stay in Lv) is the same as a basis being closed under taking p-th powers (in other words, being closed under the Frobenius). Such a basis will be called a *Frobenius-closed basis*. See Lemma 25 which gives the exact formulation of the property of a Frobenius-closed basis that we need in [Ku9].

The residue field of $K(x)$ is just $Kv(xv)$. Further, L being a function field of transcendence degree 1, $L|K(x)$ is a finite extension. It follows from the fundamental inequality that also $Lv|K(x)v$ is a finite extension. This shows that $Lv|Kv$ is a function field of transcendence degree 1. So in order to prove our theorem in the second case, our task is to find a Frobenius-closed basis for every function field of transcendence degree 1 over an algebraically closed field of positive characteristic. In [Ku1], I proved a more general result:

THEOREM 10. *Let F be an algebraic function field of transcendence degree 1 over a perfect field K of characteristic $p > 0$. If K is relatively algebraically closed in F, then there exists a Frobenius-closed basis for $F|K$.*

The proof and some further background are given in Section 5. There, we will also deduce the following result from Theorem 10, showing the connection between Frobenius-closed bases and additive polynomials:

THEOREM 11. *If F is an algebraic function field of transcendence degree* 1 *over a perfect field K of characteristic $p > 0$ and if K is relatively algebraically closed in F, then F/K is a free $K[\varphi]$-module.*

The second important theorem that I use in the proof of Theorem 7 is needed when the valued function field $(L|K, v)$ has non-trivial transcendence defect, i.e., equality does not hold in (10). In reducing to transcendence degree 1 by induction, one reaches the case where (L, v) is an immediate extension of transcendence degree 1 of the tame field (K, v). The defect is then avoided by means of the following theorem.

THEOREM 12 (*Henselian Rationality*). *Let (K, v) be a tame field and $(L|K, v)$ an immediate function field of transcendence degree* 1. *Then the henselization $(L, v)^h$ of (L, v) is henselian rational, i.e.,*

(11)   there is $x \in L$ such that $L^h = K(x)^h$.

For valued fields of residue characteristic 0, the assertion is a direct consequence of the fact that every such field is defectless (in fact, every $x \in L \backslash K$ will then do the job). In contrast to this, the case of positive residue characteristic requires a much deeper structure theory of immediate algebraic extensions of henselian fields, in order to find suitable elements x. I proved this theorem in [Ku1] (cf. also [Ku10]).

The proof works as follows. Suppose we have chosen the wrong x. Then the extension $L^h|K(x)^h$ is proper and immediate. So by (9) its defect is equal to its degree and thus non-trivial. If $L^h|K(x)^h$ were an Artin-Schreier extension, we could employ the same methods as described above to find a normal form that allows us to find a better x (i.e., one for which the degree $[L^h : K(x)^h]$ is smaller). But in general, even if $L^h|K(x)^h$ is separable, it will not necessarily be a tower of Artin-Schreier extensions. Note that because its degree is a prime, an Artin-Schreier extension does not admit any proper subextensions; such an extension is called *minimal*. This leads us to the following question: what is the structure of minimal subextensions of such extensions $L^h|K(x)^h$?

Since $L^h|K(x)^h$ is immediate, but every finite subextension of $K(x)^r|K(x)^h$ (where $K(x)^r := (K(x)^h)^r$) is defectless by part e) of Theorem 38, it follows that $L^h|K(x)^h$ is linearly disjoint from $K(x)^r|K(x)^h$. Take any henselian field (k, v). Then an algebraic extension k_1 is called *purely wild* if $k_1|k$ is linearly disjoint from $k^r|k$. Hence, our extension $L^h|K(x)^h$ is purely wild. Our question is now answered by the following theorem, which again shows the importance of p-polynomials and hence also of additive polynomials. This theorem is due to Florian Pop ([Pop]).

THEOREM 13. *Let (k, v) be a henselian field of characteristic $p > 0$ and $(k_1|k, v)$ a minimal purely wild extension. Then there exist an additive polynomial $\mathcal{A}(X) \in \mathcal{O}_k[X]$ and an element $\vartheta \in k_1$ such that $k_1 = k(\vartheta)$ and the p-polynomial $\mathcal{A}(X) - \mathcal{A}(\vartheta)$ is the minimal polynomial of ϑ over k.*

It can be shown using Hensel's Lemma that if $k_1 \neq k$, then $\mathcal{A}(\vartheta) \notin \mathcal{O}_k$.

Using the additivity of the polynomial $\mathcal{A}$ like I used the additivity of the Artin-Schreier polynomial $X^p - X$ before, it is indeed possible to deduce a normal form that allows to find a better x. Therefore, Theorem 13 is an important ingredient in the proof of Theorem 12. Three sections of this paper are devoted to the previously unpublished proof of Theorem 13. For the convenience of the reader, G-modules and twisted homomorphisms are introduced in Section 6. In Section 7, a Galois theoretical result of independent interest is proved. It is a generalization of the theorem that I have already used above and that states that every Galois extension of degree p in characteristic p is an Artin-Schreier extension. Then I apply it in Section 8 to the situation of purely wild extensions and derive Theorem 13.

Theorem 13 gains even more importance in conjunction with a result of Matthias Pank (see [Ku–Pa–Ro]):

THEOREM 14. *Let (K, v) be a henselian field. Then K^r admits a field complement W in the algebraic closure $\tilde{K}$, that is, $W.K^r = \tilde{K}$ and $W \cap K^r = K$. Every such complement W is a maximal purely wild extension of K. The quotient group vW/vK is a p-group (where p is the characteristic exponent of Kv), and the extension $Wv|Kv$ is purely inseparable.*

Note that (K, v) is a tame field if and only if $W = K$.

2.3. Reason #3: extremality and elementary properties of additive polynomials. If f is a polynomial in n variables with coefficients in K, then we will say that (K, v) is *extremal with respect to* f if the set

$$(12) \qquad \{vf(a_1, \ldots, a_n) \mid a_1, \ldots, a_n \in K\} \subseteq vK \cup \{\infty\}$$

has a maximum. This means that

$$\exists Y_1, \ldots, Y_n \forall X_1, \ldots, X_n vf(X_1, \ldots, X_n) \leq vf(Y_1, \ldots, Y_n)$$

holds in (K, v). It follows that being extremal with respect to f is an elementary property in the language of valued fields with parameters from K. Note that the maximum is ∞ if and only if f admits a K-rational zero. A valued field (K, v) is called *extremal* if for all $n \in \mathbb{N}$, it is extremal with respect to every polynomial f in n variables with coefficients in K. This notion is due to Ershov. The property of being extremal can be expressed by a countable scheme of elementary sentences (quantifying over the coefficients of all possible polynomials of degree at most n in at most n variables). Hence, it is elementary in the language of valued fields.

The following result was first stated by Delon in [Del], but the proof contained gaps. The gaps were later filled by Ershov in [Er2]. I give an alternative proof in [Ku8].

THEOREM 15. *A valued field is algebraically maximal if and only if it is extremal with respect to every polynomial in one variable.*

The following related results, also proved in [Ku8], illustrate again the importance of additive and p-polynomials. First, using Theorem 13, we can push the result stated in Theorem 15 even further:

THEOREM 16. *A henselian valued field of characteristic $p > 0$ is algebraically maximal if and only if it is extremal with respect to every p-polynomial in one variable.*

A polynomial $A \in K[X_1, \ldots, X_n]$ in n variables is called *additive* if for all elements $a_1, \ldots, a_n, b_1, \ldots, b_n$ in any extension field of K,

$$A(a_1 + b_1, \ldots, a_n + b_n) = A(a_1, \ldots, a_n) + A(b_1, \ldots, b_n).$$

In fact, if A is additive then

$$A(X_1, \ldots, X_n) = \sum_{i=1}^{n} A_i(X_i)$$

where

$$A_i(X_i) := A(0, \ldots, 0, X_i, 0, \ldots, 0)$$

are additive polynomials in one variable. As before, a polynomial $f \in K[X_1, \ldots, X_n]$ in n variables is called *p-polynomial* if it is of the form $A + c$ where $A \in K[X_1, \ldots, X_n]$ is additive, and $c \in K$. From the above we see that also every p-polynomial is a sum of p-polynomials in one variable.

A valued field is called *inseparably defectless* if all purely inseparable extensions have trivial defect. The following is proved in [Ku8]:

THEOREM 17. *A valued field (K, v) of characteristic $p > 0$ is inseparably defectless if and only if it is extremal with respect to every p-polynomial of the form*

$$(13) \qquad b - \sum_{i=1}^{n} b_i X_i^p, \quad n \in \mathbb{N}, \ b, b_1, \ldots, b_n \in K.$$

Observe that again, all of these notions can be axiomatized by recursive elementary axiom schemes.

I will now sketch the basic idea of the proof of Theorem 3. Note that the image of a polynomial f on a valued field K has the *optimal approximation property* in the sense of [Ku4] and [Dr–Ku] if and only if K is extremal with respect to $f - c$ for every $c \in K$. Consequently,

the images of all additive polynomials over (K, v) have the optimal approximation property if and only if K is extremal with respect to all p-polynomials over K.

This holds in one variable as well as in several variables.

In [Ku4], I considered the following additive polynomial over $\mathbb{F}_p((t))$:

$$(14) \qquad X_0^p - X_0 + tX_1^p + \cdots + t^{p-1}X_{p-1}^p.$$

I showed that the image of this polynomial has the optimal approximation property in $\mathbb{F}_p((t))$. Then I constructed an extension (L, v) of $\mathbb{F}_p((t))$ of transcendence degree 1 which is henselian, defectless, has value group a $\mathbb{Z}$-group, with vt the smallest positive element, and residue field $\mathbb{F}_p$, but the image of the above polynomial does not have the optimal approximation property in (L, v). This shows that $\mathbb{F}_p((t))$ with its t-adic valuation is not an elementary substructure of (L, v). This yields Theorem 3.

Further, I proved in [Ku4] that in all maximal fields, the images of all additive polynomials which satisfy a certain elementary condition have the optimal approximation property. Maximal fields are interesting objects in the model theory of valued fields because all maximal immediate extensions of a valued field are maximal. So if we are considering an elementary class of valued fields closed under maximal immediate extensions (so far, this is the case for all classes of valued fields without additional structure that play a role in model theory), and if the Ax–Kochen–Ershov principle (6) holds for this class, then every field in the class should be elementarily equivalent to all of its maximal immediate extensions. Therefore, the following question is very important:

OPEN PROBLEM 2. Is every maximal field of characteristic $p > 0$ extremal with respect to every p-polynomial in several variables? Is every maximal field extremal?

Since every maximal field is algebraically maximal, Theorem 15 shows that it is at least extremal with respect to every polynomial in one variable. To answer the first question to the affirmative, one would have to eliminate the condition in the result mentioned above.

In [Ku4], I also construct an immediate function field (F, v) of transcendence degree 1 over (L, v) such that (L, v) is not existentially closed in (F, v). Any maximal immediate extension (M, v) of (F, v) is also a maximal immediate extension of (L, v). Since (L, v) is not existentially closed in (F, v), it is not existentially closed in (M, v). So it is not an elementary substructure of (M, v), and it cannot lie in an elementary class which has the good properties discussed above.

The function field F is generated over L by two elements x_0, x_1 which satisfy an equation

$$x = x_0^p - x_0 + tx_1^p$$

where x is an element in L which is transcendental over $\mathbb{F}_p((t))$. So the existential sentence

$$\exists X_0 \exists X_1 : x = X_0^p - X_0 + tX_1^p$$

holds in F. On the other hand, L is constructed in such a way that this sentence does not hold in L. This proves that L is not existentially closed in F and, a fortiori, (L, v) is not existentially closed in (F, v).

The function field $F|L$ shows the following interesting symmetry between a generating Artin-Schreier extension and a generating purely inseparable extension of degree p. On the one hand, we have the Artin-Schreier extension

$$L(x_0, x_1)|L(x_1)$$

given by

(15) $$x_0^p - x_0 = x - tx_1^p.$$

On the other hand we have the purely inseparable extension

$$L(x_0, x_1)|L(x_0)$$

given by

$$x_1^p = \frac{1}{t}\left(-x_0^p + x_0 + x\right).$$

From equation (15) it is immediately clear that the function field $L(x_0, x_1)$ becomes rational after a constant field extension by $t^{1/p}$; namely

$$F(t^{1/p}) = L(t^{1/p})(x_0 + t^{1/p}x_1).$$

This shows that the base field L, not being existentially closed in the function field F, becomes existentially closed in the function field after a finite purely inseparable constant extension, although this extension is linearly disjoint from $F|L$.

In our above example there also exists a separable constant field extension $L'|L$ of degree p such that $(F.L')^h$ is henselian rational. To show this, we take a constant $d \in L$ and an element a in the algebraic closure of L satisfying

$$t = a^p - da,$$

and we put $L' = L(a)$. If we choose d with a sufficiently high value, then we will have that $vdax_1^p > 0$. From this we deduce by Hensel's Lemma that there is an element $b \in L'(x_1)^h$ such that $b^p - b = -dax_1^p$. If we put $z = x_0 + ax_1 + b \in L'(x_0, x_1)^h$, we get that

$$z^p - z = x - tx_1^p + a^px_1^p - ax_1 - dax_1^p = x - ax_1 + (a^p - da - t)x_1^p = x - ax_1,$$

which shows that

$$x_1 \in L'(z).$$

This in turn yields that $b \in L'(z)^h$ and consequently,

$$x_0 = z - ax_1 - b \in L'(z)^h.$$

Altogether, we have proved that

$$L'(x_0, x_1)^h = L'(z)^h$$

is henselian rational.

Let us discuss one more problem about the model theory of $\mathbb{F}_p((t))$ that becomes visible through our above example. It can be shown that for every $k \in \mathbb{N}$, the sentence

$$\forall X \exists X_0, \ldots, X_{p^k-1}, Y : X = X_0^{p^k} - X_0 + t X_1^{p^k} + \cdots + t^{p^k-1} X_{p^k-1}^{p^k} + Y \wedge v Y \geq 0$$

holds in $\mathbb{F}_p((t))$ as well as in every maximal field which satisfies axiom system (5). On the other hand, given any $n \in \mathbb{N}$, the construction of (L, v) can be modified in such a way that for some k the above sentence does not hold in (L, v) and that the smallest extension of (L, v) within any maximal immediate extension in which that sentence holds is at least of transcendence degree n over L. This is in drastic contrast to the tame behaviour shown by tame fields:

If $(M \mid K, v)$ is an immediate extension, (M, v) is a tame field and K is relatively algebraically closed in M, then also (K, v) is a tame field.

(For a proof, see [Ku11].) This property of tame fields is used in an essential way in the proof of Theorem 7 in order to reduce to immediate extensions of transcendence degree 1 (so that Theorem 12 can be applied). Apparently, in the case of fields elementarily equivalent to $\mathbb{F}_p((t))$, we have to succeed without this tool.

For the construction of the field L, I needed a handy criterion for defectless fields. The following is proved in [Ku8]:

THEOREM 18. *A valued field of positive characteristic is henselian and defectless if and only if it is separable-algebraically maximal and inseparably defectless.*

The proof uses a classification of Artin-Schreier extensions with non-trivial defect according to whether they can be obtained as a deformation of a purely inseparable extension with non-trivial defect, or not (cf. [Ku8]). This classification is also of independent interest. For instance, S. D. Cutkosky and O. Piltant [Cu–Pi] give an example of a tower of two Artin-Schreier extensions with non-trivial defect of a rational function field in which a certain form of "relative resolution" fails. It would be interesting to know whether such properties depend on the classification. In [Ku6], valued rational function fields are constructed which allow an infinite tower of Artin-Schreier extensions with non-trivial defect, but it is not clear whether one can obtain both sorts of extensions. The classification may also be important for the characterization of all valued fields whose maximal immediate extensions are finite (cf. [V] for the background).

2.4. But what about extremality for all polynomials? Let us come back to the question whether every maximal field is extremal. We know the answer in the case of discrete valued fields:

THEOREM 19. *If (K, v) is a henselian defectless field with value group isomorphic to $\mathbb{Z}$, then (K, v) is extremal.*

In [Del], Delon deduced this from the work of Greenberg [Gre]. An elegant model theoretic proof was given by Ershov. The theorem implies that in particular, $(\mathbb{F}_p((t)), v_t)$ is extremal. It also implies that every henselian defectless field with value group isomorphic to $\mathbb{Z}$ is extremal with respect to all p-polynomials in several variables. An alternative proof for this fact can be found in [Dr–Ku]. It uses the local compactness of $\mathbb{F}_p((t))$. If this could be eliminated in the case of maximal fields, we could at least prove that every maximal field is extremal with respect to all p-polynomials in several variables. This generates the following question:

OPEN PROBLEM 3. If a henselian field of characteristic $p > 0$ is extremal with respect to all p-polynomials in several variables, does this imply that it is extremal?

Are p-polynomials representative for all polynomials when extremality is concerned? Theorem 16 indicates that modulo henselization this is true for polynomials in one variable. But can we associate directly to every polynomial in one variable a p-polynomial in one variable from which we can read off information about extremality? A result of Kaplansky ([Ka1, Lemma 10]), originally proved to be used in the construction of one of the counterexamples to embeddability in power series fields, shows that this can be done over every henselian field with archimedean value group. Using technical machinery developed in [Ku1], this result can be generalized (the proof is implicit in [Ku1]):

PROPOSITION 20. Let (K, v) be a henselian field and $(a_\rho)_{\rho < \lambda}$ a pseudo Cauchy sequence in K without a limit in K. Pick a polynomial f of minimal degree such that the value $vf(a_\rho)$ is not ultimately fixed. Then there is an additive polynomial $\mathcal{A} \in K[X]$ such that for all large enough ρ,

$$v(f(a_\rho) - \mathcal{A}(a_\rho)) > vf(a_\rho)$$

(which in particular implies that $vf(a_\rho) = v\mathcal{A}(a_\rho)$).

OPEN PROBLEM 4. Is Proposition 20 also true for polynomials in several variables?

If this were not the case, then it would destroy our hope to capture the complete theory of $\mathbb{F}_p((t))$ by adjoining axioms about extremality with respect to additive polynomials to axiom system (5). That would mean that additive polynomials are important but do not tell us all the missing information about $\mathbb{F}_p((t))$.

It should be mentioned that the case of several variables is very much different from the case of one variable, and there is not much hope of treating it by induction on the number n of variables starting with $n = 1$. Indeed, if $(L|K, v)$ is an immediate extension generated by a polynomial f, then a pseudo Cauchy sequence of algebraic type can be constructed with respect to which the value of f is not fixed (for these notions, see [Ka1]). This has been done in [Er2] and in [Ku8]. A similar procedure is *not* known for the case of several variables.

2.5. Concluding remarks about valued $K[\varphi]$-modules. Van den Dries' question can be reformulated as: *Determine the model theory of valued $K[\varphi]$-modules.* What do we mean by a "valued module"? There are some notions of "valued module" in the literature, but as far as I know they do not cover the case we are interested in. Basically, one could define a "valued module" to be a module which also has the structure of a valued abelian group. But without any further assumptions on the compatibility between module structure and valuation, this would not lead us far. So we have to choose axioms for the compatibility that cover the case we are interested in. I have done this in [Ku2], but these axioms are not yet in a very satisfactory form. Although the structure of $K[\varphi]$-modules can be nasty when K is not perfect, there is still the valuation on them and it appears that with an appropriate choice of axioms one can tame these modules. Indeed, a first answer to van den Dries' question was given by his student Thomas Rohwer who proved in his thesis [Roh] the following results:

THEOREM 21. *The elementary theory of $\mathbb{F}_p((t))$ as an $\mathbb{F}_p((t))[\varphi]$-module with a predicate for $\mathbb{F}_p[[t]]$ is model complete. The elementary theory of $\mathbb{F}_p((t))$ as an $\mathbb{F}_p(t)[\varphi]$-module with a predicate for $\mathbb{F}_p[[t]]$ is decidable.*

It should be noted that Pheidas and Zahidi [Ph–Za] prove analogous results for $\mathbb{F}_p[t]$ as an $\mathbb{F}_p[t][\varphi]$-module.

Theorem 21 immediately leads to a number of questions:

OPEN PROBLEM 5. What do Rohwer's results tell us about the model theory of the valued field $\mathbb{F}_p((t))$?

OPEN PROBLEM 6. Does the elementary theory of $\mathbb{F}_p((t))$ as an $\mathbb{F}_p((t))[\varphi]$- or $\mathbb{F}_p(t)[\varphi]$-module admit quantifier elimination in some natural language?

Rohwer works with predicates V_i that are interpreted by the sets of all elements of value $\geq i$. This gives less information than a binary predicate $P(x, y)$ interpreted by $vx \leq vy$ ("valuation divisibility").

OPEN PROBLEM 7. What are the model theoretic properties of the elementary theory of $\mathbb{F}_p((t))$ as a valued $\mathbb{F}_p((t))[\varphi]$- or $\mathbb{F}_p(t)[\varphi]$-module in a language which includes a binary predicate for valuation divisibility?

OPEN PROBLEM 8. What is the structure of extensions of valued $K[\varphi]$-modules? Can one prove Ax–Kochen–Ershov principles for valued $K[\varphi]$-modules?

An important tool in the model theory of valued fields is Kaplansky's well known result that a valued field is maximal if and only if every pseudo Cauchy sequence in this field has a limit (cf. [Ka1]). One can ask the same question for other valued structures. In the case of valued modules with value-preserving scalar multiplication, the corresponding result is already in the literature. For the case of valued modules with the above mentioned axioms that cover the case of the valued $K[\varphi]$-module $\mathbb{F}_p((t))$, I proved the corresponding result in [Ku2]. Together with Rohwer's work, this seems to be a good start towards a comprehensive study of valued $K[\varphi]$-modules, including a full answer to van den Dries' question, but quite a bit of work remains to be done.

§3. **Characterization of additive polynomials.** In this section we give the basic characterizations of additive polynomials and prove Theorem 1.

LEMMA 22. *Take $f \in K[X]$ and consider the following polynomial in two variables*:

(16) $$g(X, Y) := f(X + Y) - f(X) - f(Y).$$

If there is a subset A of cardinality at least $\deg f$ in some extension field of K such that g vanishes on $A \times A$, then f is additive and of the form (2).

PROOF. Assume that there is a subset A of cardinality at least $\deg f$ in some extension field of K such that g vanishes on $A \times A$. Take L to be any extension field of K. By field amalgamation, we may assume that A is contained in an extension field L' of L. For all $c \in L'$, the polynomials $g(c, Y)$ and $g(X, c)$ are of lower degree than f. This follows from their Taylor expansion. Assume that there exists $c \in L$ such that $g(c, Y)$ is not identically 0. Since A has more than $\deg g(c, Y)$ many elements, it follows that there must be $a \in A$ such that $g(c, a) \neq 0$. Consequently, $g(X, a)$ is not identically 0. But since A has more than $\deg g(X, a)$ many elements, this contradicts the fact that $g(X, a)$ vanishes on A. This contradiction shows that $g(c, Y)$ is identically 0 for all $c \in L$. That is, g vanishes on $L \times L$. Since this holds for all extension fields L of K, we have proved that f is additive.

By what we have shown, $g(c, Y)$ vanishes identically for every c in any extension field of K. That means that the polynomial $g(X, Y) \in K(Y)[X]$ has infinitely many zeros. Hence, it must be identically 0. Write $f = d_n X^n + \cdots + d_0$. Then $g(X, Y)$ is the sum of the forms $d_j(X + Y)^j - d_j X^j - d_j Y^j$ of degree j, $1 \leq j \leq \deg f$. Since g is identically 0, the same must be true for each of these forms and thus for all $(X + Y)^j - X^j - Y^j$ for which $d_j \neq 0$. But $(X + Y)^j - X^j - Y^j \equiv 0$ can only hold if j is a power of the characteristic exponent of K. Hence, $d_j = 0$ if j is not a power of p. Setting $c_i := d_{p^i}$, we see that f is of the form (2). ⊣

PROOF OF THEOREM 1: Suppose that $f \in K[X]$ is additive. Then the polynomial g defined in (16) vanishes on every extension field L of K. Choosing L to be infinite and taking $A = L$, we obtain from the foregoing lemma that f is of the form (2).

Conversely, for every $i \in \mathbb{N}$, the mapping $x \mapsto x^{p^i}$ is a homomorphism on every field of characteristic exponent p. Hence, every polynomial $c_i X^{p^i}$ is additive, and so is the polynomial $\sum_{i=0}^{m} c_i X^{p^i}$.   ⊣

COROLLARY 23. *Take $f \in K[X]$.*

a) *If f is additive, then the set of its roots in the algebraic closure $\tilde{K}$ of K is a subgroup of the additive group of $\tilde{K}$. Conversely, if the latter holds and f has no multiple roots, then f is additive.*

b) *If f satisfies condition (1) on a field with at least $\deg f$ many elements, then f is additive.*

PROOF. a): If f is additive and a, b are roots of f, then $f(a+b) = f(a) + f(b) = 0$; hence $a+b$ is also a root. Further, $f(0) = f(0+0) = f(0) + f(0)$ shows that $0 = f(0) = f(a-a) = f(a) + f(-a) = f(-a)$, so 0 and $-a$ are also roots. This shows that the set of roots of f form a subgroup of $(\tilde{K}, +)$.

Now assume that the set A of roots of f forms a subgroup of $(\tilde{K}, +)$, and that f has no multiple roots. The latter implies that A has exactly $\deg f$ many elements. Since $A + A = A$, the polynomial $g(X, Y) = f(X + Y) - f(X) - f(Y)$ vanishes on $A \times A$. Hence by Lemma 22, f is additive.

b): This is an immediate application of Lemma 22.   ⊣

EXERCISE 1. a) Let K be any finite field. Give an example of a polynomial $f \in K[X]$ which is not additive but induces an additive mapping on $(K, +)$. b) Show that the second assertion in part a) of Corollary 23 fails if we drop the condition that f has no multiple roots. Replace this condition by a suitable condition on the multiplicity of the roots. c) Deduce Corollary 23 from the theorem of Artin as cited in [L, VIII, Section 11, Theorem 18].

§4. Rings of additive polynomials.

This section is devoted to the structure of rings of additive polynomials. Euclidean division is discussed in the following

PROOF OF THEOREM 2: Take $s = \sum_{i=0}^{m} c_i \varphi^i$ and $s' = \sum_{i=0}^{n} d_i \varphi^i$. If $\deg s' < \deg s$, then we set $q = 0$ and $r = s'$. Now assume that $\deg s' = n \geq m = \deg s$. Then

$$\deg \left(s' - d_n c_m^{-p^{n-m}} \varphi^{n-m} s \right) \leq n - 1 < \deg s'.$$

Now take $q \in K[\varphi]$ such that $\deg(s' - qs)$ is minimal. Then $\deg(s' - qs) < \deg s$. Otherwise, we could apply the above to $s' - qs$ in the place of s', finding some $q' \in R$ such that $\deg(s' - (q+q')s) = \deg(s' - qs - q's) < \deg(s' - qs)$ contradicting the minimality of q. Setting $r = s' - qs$, we obtain $s' = qs + r$ with $\deg r < \deg s$. We have proved that $K[\varphi]$ is left euclidean. If K is perfect, hence $K = K^{p^m}$, then we also have

$$\deg \left(s' - s \left(c_m^{-1} d_n \right)^{1/p^m} \varphi^{n-m} \right) \leq n - 1 < \deg s',$$

and in the same way as above one deduces that $K[\varphi]$ is right euclidean.

Now assume that K is not perfect and choose some element $c \in K$ not admitting a p-th root in K. Then $K^p \cap cK^p = \{0\}$ and

$$\varphi K[\varphi] \cap c\varphi K[\varphi] = \{0\}$$

since every nonzero additive polynomial in the set $\varphi K[\varphi]$ has coefficients in K^p whereas every nonzero additive polynomial in $c\varphi K[\varphi]$ has coefficients in cK^p. ⊣

REMARK 24. Let us state some further properties of the ring $K[\varphi]$ which follow from Theorem 2. More generally, let R be any left principal ideal domain. Then R is a left free ideal ring (fir), and it is thus a semifir, i.e., every finitely generated left or right ideal is free of unique rank (note that this property is left-right symmetrical, cf. [C2, Chapter 1, Theorem 1.1]). Consequently, every finitely generated submodule of a (left or right) free R-module is again free, cf. [C2, Chapter 1, Theorem 1.1]. On the other hand, every finitely generated torsion free (left or right) R-module is embeddable in a (finitely generated) free R-module if and only if R is right Ore, cf. [C2, Chapter 0, Corollary 9.5] and [Ge, Proposition 4.1]. Being a semifir, R is right Ore if and only if it is a right Bezout ring. But if R is not right Ore, then it contains free right ideals of arbitrary finite or countable rank, and R is thus not right noetherian, cf. [C2, Chapter 0, Proposition 8.9 and Corollary 8.10]. Every projective (left or right) R-module is free, cf. [C2, Chapter 1, Theorem 4.1]. A right R-module is flat if and only if it is torsion free, and a left R-module M is flat if and only if every finitely generated submodule of M is free, cf. [C2, Chapter 1, Corollary 4.7 and Proposition 4.5]. In view of the above, the latter is the case if and only if every finitely generated submodule of M is embeddable in a free R-module. Further, a left R-module M is flat if and only if for every $n \in \mathbb{N}$ and all right linearly independent elements $r_1, \ldots, r_n \in R$,

$$\forall x_1, \ldots, x_n \in M : \sum r_i x_i = 0 \Rightarrow \forall i : x_i = 0,$$

cf. [C1, Chapter 1, Lemma 4.3]. As a semifir, R is a coherent ring. Finally, since R is left Ore, it can be embedded in a skew field of left fractions, cf. [C2, Chapter 0, Corollary 8.7].

Note that in particular, the above shows that all finitely generated torsion free (left or right) $K[\varphi]$-modules are free if and only if K is perfect, that is, $K[\varphi]$ is euclidean on both sides.

EXERCISE 2. Describe the relation of the degree functions on $K[X]$ and $K[\varphi]$ via the correspondence (3), giving thereby a proof of $\deg rs = \deg r + \deg s$. Show that it also satisfies $\deg r + s \leq \max\{\deg r, \deg s\}$ with equality holding if $\deg r \neq \deg s$. Can it be transformed into a valuation?

§5. Frobenius-closed bases of function fields. In this section, we prove the existence of Frobenius-closed bases of algebraic function fields $F|K$ of transcendence degree 1, and exhibit the connection between their existence and the structure of F as a $K[\varphi]$-module (for arbitrary transcendence degree).

Take an arbitrary extension $F|K$ of fields of characteristic $p > 0$. Recall that a K-basis B of F is called *Frobenius-closed* if $B^p \subset B$, where $B^p = \{b^p \mid b \in B\} = \varphi B$. In [Ku9] we need Frobenius-closed bases because they have the following property:

LEMMA 25. *Take a Frobenius-closed basis* z_j, $j \in J$, *of* $F|K$. *If the sum*

$$s = \sum_{i \in I} c_i z_i, \quad c_i \in K, \ I \subset J \text{ finite}$$

is a p-th power, then for every $i \in I$ *with* $c_i \neq 0$, *the basis element* z_i *is a p-th power of a basis element.*

PROOF. Assume that

$$s = \left(\sum_{j \in J_0} c_j' z_j \right)^p, \quad c_j' \in K$$

where $J_0 \subset J$ is a finite index set. Then

$$\sum_{i \in I} c_i z_i = s = \sum_{j \in J_0} (c_j')^p z_j^p$$

where the elements z_j^p are also basis elements by hypothesis, which shows that every z_i which appears on the left hand side (i.e., $c_i \neq 0$) equals a p-th power z_j^p appearing on the right hand side. ⊣

We will show the existence of Frobenius-closed bases for algebraic function fields of transcendence degree 1 over a perfect field of characteristic $p > 0$, provided that K is relatively algebraically closed in F. We first prove the following:

LEMMA 26. *If* F *is an algebraic function field of transcendence degree 1 over an algebraically closed field* K *of arbitrary characteristic and* q *is an arbitrary natural number* > 1, *then there exists a basis of* $F|K$ *which is closed under q-th powers.*

If $F = K(x)$ is a rational function field, then our lemma follows from the *Partial Fraction Decomposition*: Every element $f \in F$ has a unique representation

$$f = c + \sum_{n>0} c_n x^n + \sum_{a \in K} \sum_{n>0} c_{a,n} \frac{1}{(x-a)^n}$$

where only finitely many of the coefficients $c, c_n, c_{a,n} \in K$ are nonzero. If we put

$$t_a = \frac{1}{x - a}, \qquad t_\infty = x$$

then it follows that the elements

$$1, t_a^n \text{ with } a \in K \cup \{\infty\}, \ n \in \mathbb{N}$$

form a K-basis of F; this basis has the property that *every* power of a basis element is again a basis element.

For general function fields the Partial Fraction Decomposition remains true in a modified form (according to Helmut Hasse) that we shall describe now. At this point, we need the Riemann-Roch Theorem. In order to apply it, we have to introduce some notation. In what follows, we always assume that K is relatively algebraically closed in F. A *divisor* of $F|K$ is an element of the (multiplicatively written) free abelian group generated by all places of $F|K$. (By a place of $F|K$ we mean a place of F which is trivial on K, i.e., $P|_K = \mathrm{id}$. We identify equivalent places.) The places themselves are called *prime divisors*. A divisor may thus be written in the form

$$A = \prod_P P^{v_P A}$$

where the product is taken over all places of $F|K$ and the $v_P A$ are integers, only finitely many of them nonzero. The *degree of a non-trivial place* P of $F|K$, denoted by $\deg P$, is defined to be the degree $[FP : K]$ (which is finite since $F|K$ is an algebraic function field in one variable). Accordingly, the *degree of a divisor* A, denoted by $\deg A$, is defined to be the integer $\sum_P v_P A \cdot \deg P$. By the symbol "$v_P$" we will also denote the valuation on F which is associated with the place P. Following the notation of [F–Jr], we set

$$\mathcal{L}(A) := \{f \in F \mid v_P f \geq -v_P A \text{ for all places } P \text{ of } F|K\}$$

is a K-vector space. Indeed, $0 \in \mathcal{L}(A)$ since $v_P 0 = \infty > -v_P A$ for all places P of $F|K$. Further, $v_P(K^\times) = \{0\}$, hence $\forall c \in K^\times : v_P(cf) = v_P f$ for all P, so $f \in \mathcal{L}(A)$ implies $cf \in \mathcal{L}(A)$. Finally, if $f, g \in \mathcal{L}(A)$, then $v_P(f - g) \geq \min\{v_P f, v_P g\} \geq -v_P A$ for all P, hence $f - g \in \mathcal{L}(A)$. We write

$$\dim A := \dim_K \mathcal{L}(A).$$

The divisor A determines bounds for the zero and pole orders of the algebraic functions in $\mathcal{L}(A)$. For example, if $A = P^n$ with n a natural number, then $f \in \mathcal{L}(A)$ if and only if f has no pole at all (in which case it is a constant function) or has a pole at P of pole order at most $n = v_P A$.

THEOREM 27 (*Riemann-Roch*). *Let $F|K$ be an algebraic function field in one variable with K relatively algebraically closed in F. There exists a smallest non-negative integer g, called the genus of $F|K$, such that*

$$\dim A \geq \deg A - g + 1$$

for all divisors A of $F|K$. Furthermore,

$$\dim A = \deg A - g + 1$$

whenever $\deg A > 2g - 2$.

For a proof, see [Deu].

Let P_∞ be a fixed place of $F|K$ and R^∞ the ring of all $f \in F$ which satisfy $v_P f \geq 0$ for every $P \neq P_\infty$. The following is an application of the Riemann-Roch Theorem:

COROLLARY 28. *For every $P \neq P_\infty$ there exists an element $t_P \in F$ such that*

$$v_P t_P = -1$$
$$v_Q t_P \geq 0 \quad \text{for } Q \neq P, P_\infty.$$

PROOF. If we choose $n \in \mathbb{N}$ as large as to satisfy $n \deg P_\infty > 2g - 2$, then by the Riemann-Roch Theorem,

$$\dim(PP_\infty^n) = \deg P + n \deg P_\infty - g + 1 > n \deg P_\infty - g + 1 = \dim P_\infty^n.$$

Hence there is an element $t_P \in \mathcal{L}(PP_\infty^n) \setminus \mathcal{L}(P_\infty^n)$. This element has the required properties.   ⊣

We return to the proof of our lemma, assuming that K is algebraically closed. Hence, K is the residue field of every place P of $F|K$ (that is, $\deg P = 1$). Every t_P of the foregoing corollary is the inverse of a uniformizing parameter for P. Every $f \in F$ can be expanded P-adically with respect to such a uniformizing parameter, and the principal part appearing in this expansion has the form

$$h_P(f) = \sum_{n>0} c_{P,n} t_P^n,$$

where only finitely many of the coefficients $c_{P,n} \in K$ are nonzero, namely $n \leq -v_P f$. By construction, t_P has only a single pole $\neq P_\infty$ and this pole is P; the same holds for $h_P(f)$ (if $h_P(f) \neq 0$). Consequently,

$$h = f - \sum_{P \neq P_\infty} h_P(f)$$

has no pole other than P_∞ and is thus an element of R^∞. We have shown that f has a unique representation

$$f = h + \sum_{P \neq P_\infty} \sum_{n>0} c_{P,n} t_P^n$$

with coefficients $c_{P,n} \in K$ and an element $h \in R^\infty$. This shows that the elements

$$t_P^n \text{ with } P \neq P_\infty, \ n \in \mathbb{N}$$

form a K-basis of F modulo R^∞ which has the property that every power of a basis element is again a basis element.

Now it remains to show that R^∞ admits a basis which is closed under q-th powers. An integer $n \in \mathbb{N}$ is called *pole number* of P_∞ if there exists $t_n \in R^\infty$ such that $v_{P_\infty} t_n = -n$. Let $H_\infty \subseteq \mathbb{N}$ be the set of all pole numbers. Fixing a t_n for every $n \in H_\infty$, we get a K-basis

$$1, t_n \text{ with } n \in H_\infty$$

of R^∞. To get a basis which is closed under q-th powers, we have to carry out our choice as follows:

Observe that H_∞ is closed under addition; in particular

$$qH_\infty \subset H_\infty.$$

For every $m \in H_\infty \setminus qH_\infty$ we choose an arbitrary element $t_m \in R^\infty$ with $v_{P_\infty} t_m = -m$. Every $n \in H_\infty$ can uniquely be written as

$$n = q^v m \text{ where } v \geq 0 \text{ and } m \in H_\infty \setminus qH_\infty.$$

Accordingly we put

$$t_n = t_m^{q^v}$$

which implies

$$v_{P_\infty} t_n = q^v \cdot v_{P_\infty} t_m = -q^v m = -n.$$

This construction produces a K-basis

$$1, t_m^{q^v} \text{ with } m \in H_\infty \setminus qH_\infty, \ v \geq 0$$

of R^∞, which is closed under q-th powers. This concludes the proof of our lemma.

For the generalization of this lemma to perfect ground fields of characteristic $p > 0$ we have to choose $q = p$.

PROOF OF THEOREM 10: We have to prove:

Let F be an algebraic function field of transcendence degree 1 over a perfect field K of characteristic $p > 0$. If K is relatively algebraically closed in F, then there exists a Frobenius-closed basis for $F|K$.

If K is not algebraically closed, we have to modify the proof of the previous lemma since not every place P of K has degree 1. (Such a modification is also necessary for the Partial Fraction Decomposition in $K(x)$ if K is not algebraically closed.) The modification reads as follows:

For every place P of $F|K$, let

$$d_P = \deg P = [FP : K]$$

be the degree of P. For every $P \neq P_\infty$ we choose elements $u_{P,i} \in R^\infty$, $1 \leq i \leq d_P$, such that their residues $u_{P,1}P, \ldots, u_{P,d_P}P$ form a K-basis of FP. We note that for every $v \geq 0$, the p^v-th powers $u_{P,i}^{p^v}$ of these elements have the same property: their P-residues also form a K-basis of FP since K is perfect. We write every $n \in \mathbb{N}$ in the form

$$n = p^v m \text{ with } m \in \mathbb{N}, (p, m) = 1, v \geq 0$$

and observe that the elements

$$u_{P,i}^{p^v} t_P^n \text{ with } P \neq P_\infty, n \in \mathbb{N}, 1 \leq i \leq d_P$$

form a Frobenius-closed K-basis of F modulo R^∞.

It remains to construct a Frobenius-closed K-basis of R^∞. This is done as follows: We consider the K-vector spaces

$$\mathcal{L}_n = \mathcal{L}(P_\infty^n) = \{x \in F \mid v_{P_\infty} x \geq -n \text{ and } v_P(x) \geq 0 \text{ for } P \neq P_\infty\}.$$

By our assumption that K is relatively algebraically closed in F, we have $\mathcal{L}_0 = K$. Further,

$$R^\infty = \bigcup_{n \in \mathbb{N}} \mathcal{L}_n.$$

We set

$$d_{\infty,n} := \dim \mathcal{L}_n/\mathcal{L}_{n-1} \geq 0.$$

(Note that by the Riemann-Roch Theorem, $d_{\infty,n} = [FP_\infty : K]$ for large enough n; cf. the proof of the above corollary.) Now for $n = 1, 2, \ldots$ we shall choose successively basis elements $t_{n,i} \in \mathcal{L}_n$ modulo $\mathcal{L}_{n-1}$. Then the elements

$$1, t_{n,i} \text{ with } n \in \mathbb{N}, 1 \leq i \leq d_{\infty,n}$$

form a K-basis of R^∞. To obtain that this basis is Frobenius-closed, we organize our choice as follows:

If $n = pm$, the p-th powers $t_{m,i}^p \in \mathcal{L}_n$ are linearly independent modulo $\mathcal{L}_{p(m-1)}$ and even modulo $\mathcal{L}_{pm-1} = \mathcal{L}_{n-1}$. This fact follows from our hypothesis that K is perfect: the existence of nonzero elements $c_i \in K$ with $\sum c_i t_{m,i}^p \in \mathcal{L}_{pm-1}$, i.e., $v_{P_\infty} \sum c_i t_{m,i}^p > -pm$, would yield $v_{P_\infty} \sum c_i^{1/p} t_{m,i} > -m$, hence $\geq -m + 1$, showing that $\sum c_i^{1/p} t_{m,i} \in \mathcal{L}_{m-1}$, which is a contradiction. In our choice of the elements $t_{n,i}$ we are thus free to take all the elements $t_{m,i}^p$ and to extend this set to a basis of $\mathcal{L}_n$ modulo $\mathcal{L}_{n-1}$ by arbitrary further elements, if necessary (for n large enough, the elements $t_{m,i}^p$ will already form such a basis). This procedure guarantees that the p-th power of every basis element $t_{m,i}$ is again a basis element, namely equal to $t_{pm,j}$ for suitable j. Hence a basis constructed in this way will be Frobenius-closed.   ⊣

Let $F|K$ be an arbitrary extension of fields of characteristic $p > 0$. Both F and K are $K[\varphi]$-modules, and so is the quotient module F/K. Suppose that F/K is a free $K[\varphi]$-module. Then it admits a $K[\varphi]$-basis. Let $B_0 \subset F$ be a set of representatives for such a $K[\varphi]$-basis of F/K. It follows that

$$B = \bigcup_{n=0}^{\infty} B_0^{p^n} \cup \{1\} = \bigcup_{n=0}^{\infty} \varphi^n B_0 \cup \{1\}$$

is a set of generators of the K-vector space F. By our construction of B, every K-linear combination of elements of $B \setminus \{1\}$ may be viewed as a $K[\varphi]$-linear combination of elements of B_0. This shows that the elements of B are K-linearly independent, and B is thus a Frobenius-closed basis of $F|K$. Note that B_0 is the basis of a free $K[\varphi]$-submodule M of F which satisfies $F = M \oplus K$.

The converse to this procedure would mean to extract a $K[\varphi]$-basis B_0 from a Frobenius-closed K-basis B. But B_0 can only be found if for every element $b \in B \setminus \{1\}$ there is some element b_0 which is not a p-th power in F and such that $b = b_0^{p^n}$ for some $n \in \mathbb{N} \cup \{0\}$. This will hold if no element of $F \setminus K$ has a p^n-th root for every $n \in \mathbb{N}$.

LEMMA 29. *If $F|K$ is an algebraic function field (of arbitrary transcendence degree), and if K is relatively algebraically closed in F, then no element of $F \setminus K$ has a p^n-th root for every $n \in \mathbb{N}$.*

PROOF. Let $f \in F \setminus K$. Since K is relatively algebraically closed in F, we know that f is transcendental over K. So we may choose a transcendence basis T of $F|K$ containing f. We may choose a K-rational valuation v on the rational function field $K(T)$ such that the values of all elements in T are rationally independent (cf., e.g., [Ku7]). This yields that $vK(T) = \bigoplus_{t \in T} \mathbb{Z}vt$. In particular, vf is not divisible by p in $vK(T)$. Since $F|K(T)$ is finite, the same is true for $(vF : vK(T))$ by the fundamental inequality (4). This yields that there is some $n \in \mathbb{N}$ such that vf is not divisible by p^n in vF. Hence, f does not admit a p^n-th root in F. ⊣

This lemma shows that if $F|K$ is an algebraic function field with K relatively algebraically closed in F, admitting a Frobenius-closed basis B and if we let B_0 be the set of all elements in B which do not admit a p-th root in F, then we obtain $B = \bigcup_{n=0}^{\infty} B_0^{p^n} \cup \{1\}$. Since the elements of B are K-linearly independent, the elements of B_0 are $K[\varphi]$-linearly independent over K. Moreover, B_0 is a set of generators of the $K[\varphi]$-module F over K. Hence, the set B_0/K is a $K[\varphi]$-basis of F/K. We have thus proved:

PROPOSITION 30. *Let $F|K$ be an algebraic function field (of arbitrary transcendence degree), and K relatively algebraically closed in F. Then F admits a Frobenius-closed K-basis if and only if F/K is a free $K[\varphi]$-module.*

This lemma shows that Theorem 10 implies Theorem 11.

OPEN PROBLEM 9. Do Theorems 10 and 11 also hold for transcendence degree > 1?

OPEN PROBLEM 10. Do Theorems 10 and 11 also hold if the assumption that K be perfect is replaced by the assumption that $F|K$ be separable? Note that if K is not perfect, then there exist places P of $F|K$ such that $FP|K$ is not separable, even if $F|K$ is separable. In this case, the construction of the proof of Theorem 10 breaks down and it cannot be expected that there is a Frobenius-closed K-basis of F which is as "natural" as the ones produced by that construction.

EXERCISE 3. Show that F/K cannot be a free $K[\varphi]$-module if K is not relatively algebraically closed in F. Does there exist an algebraic field extension which admits a Frobenius-closed basis? Prove a suitable version of Proposition 30 which does not use the assumption that K be relatively algebraically closed in F.

§6. **G-modules and group complements.** In this section, we introduce some notions that we will need in the next section. Take any group G. For $\sigma \in G$, *conjugation by σ* means the automorphism

$$G \ni \tau \mapsto \tau^\sigma := \sigma^{-1}\tau\sigma.$$

Note that

(17) $\tau^{\sigma\rho} = \rho^{-1}\sigma^{-1}\tau\sigma\rho = \rho^{-1}(\tau^\sigma)\rho = (\tau^\sigma)^\rho$ for all $\tau, \sigma, \rho \in G$.

Further, we set $\tau^{-\sigma} := (\tau^{-1})^\sigma$ (which indeed is the inverse of τ^σ). As usual, we set $M^\sigma = \{m^\sigma \mid m \in M\}$ for every subset $M \subset G$. A subgroup N is normal in G if and only if $N^\sigma = N$ for all $\sigma \in G$. We always have $G^\sigma = G$. Hence, if H is a *group complement* of the normal subgroup N in G, that is,

(18) $HN = G$ and $H \cap N = \{1\}$,

then so is every conjugate H^σ for $\sigma \in G$. Uniqueness up to conjugation would mean that these are the only group complements of N in G.

We shall now introduce two notions that will play an important role in Section 7. A *right G-module* is an arbitrary group N together with a mapping μ from G into the group of automorphisms of N such that $\mu(\sigma\rho) = \mu(\rho) \circ \mu(\sigma)$. For example, to every $\sigma \in G$ we may associate the conjugation by σ; in view of (17), this turns G into a right G-module. In this setting, a subgroup N of G is normal if and only if it is a G-submodule of G. A mapping ϕ from G into a G-module N is called a *twisted homomorphism* (or *crossed homomorphism*) if it satisfies

(19) $\phi(\sigma\rho) = \phi(\sigma)^\rho\phi(\rho)$ for all $\sigma, \rho \in G$.

As for a usual homomorphism, also the kernel of a twisted homomorphism is a subgroup of G, but it may not be normal in G.

Let us assume that H is a group complement of the normal subgroup N in G. It follows from (18) that every element $\sigma \in G$ admits a unique representation

(20) $\sigma = \sigma_H \sigma_N$ with $\sigma_H \in H, \sigma_N \in N$.

Note that H is a system of representatives for the left cosets of G modulo N. Since $N \lhd G$, we have $HN = NH$, and H is also a system of representatives for the right cosets of G.

Now assume in addition that N is abelian. Then the scalar multiplication of the G-module N given by conjugation reads as

(21) $\sigma^\rho = \rho_N^{-1}(\rho_H^{-1}\sigma\rho_H)\rho_N = \rho_H^{-1}\sigma\rho_H = \sigma^{\rho_H}$ for all $\sigma \in N, \rho \in G$

since ρ_N and $\rho_H^{-1}\sigma\rho_H$ are elements of N. According to (20) and (21) we write

$$\sigma\rho = \sigma_H\sigma_N\rho_H\rho_N = \sigma_H\rho_H\rho_H^{-1}\sigma_N\rho_H\rho_N = \sigma_H\rho_H\sigma_N^\rho\rho_N.$$

Hence, the projection $\sigma \mapsto \sigma_H$ onto the first factor in (20) is the canonical epimorphism from G onto H with kernel N. The other projection $\sigma \mapsto \sigma_N$ is a twisted homomorphism from G onto N, satisfying

(22) $(\sigma\rho)_N = \sigma_N^\rho\rho_N$ for all $\sigma, \rho \in G$;

it induces the identity on N, and its kernel is H.

§7. Field extensions generated by p-polynomials.

In this section, let K be a field of characteristic $p > 0$. By a Galois extension we mean a normal and separable, but not necessarily finite algebraic extension. A field extension $L|K$ is called *p-elementary extension* if it is a finite Galois extension and its Galois group is an elementary-abelian p-group, that is, an abelian p-group in which every nonzero element has order p. In particular, $[L : K]$ is a power of p.

In this section, we will consider the following larger class of all extensions $L|K$ which satisfy the following condition:

(23) $\begin{cases} \text{there exists a Galois extension } K'|K \text{ which is linearly disjoint} \\ \text{from } L|K, \text{ such that } L.K'|K' \text{ is a } p\text{-elementary extension} \\ \text{and also } L.K'|K \text{ is a Galois extension.} \end{cases}$

From the linear disjointness it follows that $\mathrm{Gal}\, L.K'|L \simeq \mathrm{Gal}\, K'|K$ and that $[L : K] = [L.K' : K']$ which yields that $[L : K] = p^n$ for some natural number n. For a further investigation of this situation, we will use the following notation. We set

$$L' := L.K'$$

and define

- $G := \mathrm{Gal}\, L'|K$,
- $N := \mathrm{Gal}\, L'|K' \lhd G$,
- $H := \mathrm{Gal}\, L'|L \simeq \mathrm{Gal}\, K'|K \simeq G/N$.

The group N is abelian of order p^n. Since $K'|K$ is assumed to be a Galois extension, N is a normal subgroup of G. That is, N is a right G-module with scalar multiplication given by conjugation:

$$(\sigma, \tau) \mapsto \sigma^\tau = \tau^{-1}\sigma\tau \quad \text{for all } \sigma \in N, \ \tau \in G.$$

Since $L.K' = L'$ and $L \cap K' = K$, we have that $H \cap N = 1$ and $G = HN$, that is, H is a group complement for N in G. As we have seen in the last section, every element $\sigma \in G$ admits a unique representation (20). Since N is abelian, the scalar multiplication of the G-module N is given by (21). The projection $\sigma \mapsto \sigma_N$ is a twisted homomorphism from G onto N, satisfying (22); it induces the identity on N, and its kernel is H.

LEMMA 31. *If $L|K$ satisfies condition (23) then w.l.o.g., the extension $K'|K$ may assumed to be finite (which yields that also $L'|K$ is finite).*

PROOF. Suppose that $K'|K$ is an arbitrary algebraic extension such that (23) holds. Let $N_0 := \{\tau \in G \mid \forall \sigma_N \in N : \tau^{-1}\sigma_N\tau = \sigma_N\}$ be the subgroup of all automorphisms in G whose action on N is trivial (i.e., N_0 is the centralizer of N in G). Since N is abelian, it is contained in N_0. Consequently, the fixed field K_0 of N_0 in L' is contained in K' (which is the fixed field of N in L' by definition of N). Since N is a normal subgroup of G, also its centralizer N_0 is a normal subgroup of G, showing that $K_0|K$ is a Galois extension. We set $H_0 := G/N_0$. By our choice of N_0, the action of G on N induces an action of H_0 on N which is given by $\rho^{-1}\sigma_N\rho = \tau^{-1}\sigma_N\tau$ for $\rho = \tau N_0 \in H_0$. Consequently, H_0 must be finite, being a group of automorphisms of the finite group N. This proves that $K_0|K$ is a finite Galois extension with Galois group H_0. Recall that it follows from (23) that also $L|K$ is finite.

We claim that $H \cap N_0$ is a normal subgroup of G. Let $\tau \in H \cap N_0$ and $\sigma \in G$; we want to show that $\tau^\sigma \in H \cap N_0$. Write $\sigma = \sigma_H\sigma_N$ according to (20). Then $\tau^\sigma = \sigma_N^{-1}(\sigma_H^{-1}\tau\sigma_H)\sigma_N$; since $\sigma_H \in H$ and $N_0 \triangleleft G$, we find that $\tau' := \sigma_H^{-1}\tau\sigma_H \in H \cap N_0$. In particular, τ' lies in the centralizer of N. In view of $\sigma_N \in N$ we obtain $\tau^\sigma = \sigma_N^{-1}\tau'\sigma_N = \sigma_N^{-1}\sigma_N\tau' = \tau' \in H \cap N_0$. We have proved that $H \cap N_0$ is a normal subgroup of G. With L_0 the fixed field of $H \cap N_0$ in L', we hence obtain a Galois extension $L_0|K$. Since $L_0 = L.K_0$, the extension $L_0|K$ is finite.

Finally, it remains to show that $\text{Gal } L_0|K_0 \simeq \text{Gal } L'|K'$ which also yields that $L_0|K_0$ is p-elementary. Observe that $HN_0 = G$ since it contains $HN = G$. Now we compute: $\text{Gal } L_0|K_0 = \text{Gal } L'|K_0/\text{Gal } L'|L_0 = N_0/(H \cap N_0) \simeq H.N_0/H = G/H \simeq N = \text{Gal } L'|K'$. We have proved that condition (23) also holds with K_0, L_0 in the place of K', L'. $\quad\dashv$

In view of this lemma, we will assume in the sequel that all field extensions are finite.

Like N, also the additive group $(L', +)$ is a right G-module, the scalar multiplication given by

$$(a, \tau) \mapsto a^{\tau} := \tau^{-1} a \quad \text{for all } a \in L', \ \tau \in G.$$

Let us show:

LEMMA 32. *There is an embedding* $\phi : N \longrightarrow (L', +)$ *of right G-modules.*

PROOF. By the Normal Basis Theorem (cf. [L]), the finite Galois extension $K'|K$ admits a normal basis. That is, there exists $b \in K'$ such that b^{ρ}, $\rho \in \text{Gal} \, K'|K$, is a basis of K' over K. Since H is a set of representatives in G for $\text{Gal} \, K'|K$, we may represent these conjugates as b^{ρ}, $\rho \in H$. Let $\psi : N \to (K, +)$ be any homomorphism of groups (there is always at least the trivial one), and set

$$(24) \qquad \phi(\sigma_N) := \sum_{\rho \in H} \psi(\rho \sigma_N \rho^{-1}) b^{\rho} \quad \text{for all } \sigma_N \in N.$$

Since ψ is a group homomorphism from N into $(L', +)$, the same is true for ϕ. Given $\tau \in G$, we write $\tau = \tau_H \tau_N$; then $b^{\rho \tau} = (b^{\rho \tau_H})^{\tau_N} = b^{\rho \tau_H}$ since $b^{\rho \tau_H} \in K'$ and $\tau_N \in N = \text{Gal} \, L'|K'$. Observing also that $H = H\tau_H$ and using (21), we compute:

$$\phi(\sigma_N)^{\tau} = \sum_{\rho \in H} \psi(\rho \sigma_N \rho^{-1}) b^{\rho \tau} = \sum_{\rho \in H} \psi\left(\rho \tau_H^{-1} \sigma_N \left(\rho \tau_H^{-1}\right)^{-1}\right) b^{\rho}$$

$$= \sum_{\rho \in H} \psi(\rho \sigma_N^{\tau_H} \rho^{-1}) b^{\rho} = \phi(\sigma_N^{\tau_H}) = \phi(\sigma_N^{\tau})$$

which shows that ϕ is a homomorphism of right G-modules.

Now we have to choose ψ so well as to guarantee that ϕ becomes injective. If $\phi(\sigma_N) = 0$ then $\psi(\rho \sigma_N \rho^{-1}) = 0$ for all $\rho \in H$ since by our choice of b, the conjugates b^{ρ}, $\rho \in H$, are linearly independent over K. In particular, $\phi(\sigma_N) = 0$ implies $\psi(\sigma_N) = 0$. Hence, ϕ will be injective if we are able to choose ψ to be injective. This is done as follows. The elementary-abelian p-group N may be viewed as a finite-dimensional $\mathbb{F}_p$-vector space. If K is an infinite field (which by our general assumption has characteristic p), then it contains $\mathbb{F}_p$-vector spaces of arbitrary finite dimension; so there exists an embedding ψ of N into $(K, +)$. If K is a finite field, then all finite extensions of K are cyclic, their Galois groups being generated by a suitable power of the Frobenius φ; consequently, N must be cyclic. Since it is also elementary-abelian, N is isomorphic to $\mathbb{Z}/p\mathbb{Z}$ which is the additive group of $\mathbb{F}_p \subset K$. Hence also in this case, N admits an embedding ψ into $(K, +)$. $\dashv$

By composition with ϕ, the twisted homomorphism $\sigma \mapsto \sigma_N$ is turned into a mapping $\sigma \mapsto \phi(\sigma_N)$ from G into $(L', +)$. We shall write $\phi(\sigma)$ instead of $\phi(\sigma_N)$, thereby considering the G-module homomorphism $\phi : N \to (L', +)$ as being extended to $\phi : G \to (L', +)$. By construction, the latter has kernel

H and is injective on N. Further, it satisfies $\phi(\sigma\tau) = \phi((\sigma\tau)_N) = \phi(\sigma_N^\tau \tau_N) = \phi(\sigma_N)^\tau + \phi(\tau_N) = \phi(\sigma)^\tau + \phi(\tau)$ showing that ϕ is a twisted homomorphism in the following sense:

$$(25) \qquad \phi(\sigma\tau) = \phi(\sigma)^\tau + \phi(\tau) \quad \text{for all } \sigma, \tau \in G.$$

We claim that there exists an element $\vartheta \in L'$ such that

$$(26) \qquad \vartheta^\tau = \vartheta + \phi(\tau) \quad \text{for all } \tau \in G.$$

Note that (26) determines ϑ up to addition of elements from K. (Indeed, ϑ' satisfies the same equation if and only if $(\vartheta - \vartheta')^\tau = \vartheta - \vartheta'$, i.e., if and only if $\vartheta - \vartheta' \in K$.)

The element ϑ can be constructed as follows. We choose an element $a \in L'$ such that the trace $s := \mathrm{Tr}_{L'|K}(a) = \sum_{\sigma \in G} \sigma a = \sum_{\sigma \in G} a^\sigma$ is not zero (we have seen in the foregoing proof that such an element exists: we could choose a to be the generator of a normal basis of $L'|K$; the linear independence will then force the trace to be nonzero). We set

$$(27) \qquad \vartheta := -\frac{1}{s} \sum_{\sigma \in G} \phi(\sigma) a^\sigma.$$

Given $\tau \in G$, we have $G\tau = G$ and

$$1 = \frac{1}{s} \sum_{\sigma \in G} a^\sigma = \frac{1}{s} \sum_{\sigma \in G} a^{\sigma\tau}$$

which we use to compute

$$\vartheta^\tau = -\frac{1}{s} \sum_{\sigma \in G} \phi(\sigma)^\tau a^{\sigma\tau} = -\frac{1}{s} \sum_{\sigma \in G} \left((\phi(\sigma)^\tau + \phi(\tau)) a^{\sigma\tau} - \phi(\tau) a^{\sigma\tau} \right)$$

$$= -\frac{1}{s} \sum_{\sigma \in G} \phi(\sigma\tau) a^{\sigma\tau} + \phi(\tau) \frac{1}{s} \sum_{\sigma \in G} a^{\sigma\tau}$$

$$= -\frac{1}{s} \sum_{\sigma \in G} \phi(\sigma) a^\sigma + \phi(\tau) \frac{1}{s} \sum_{\sigma \in G} a^\sigma = \vartheta + \phi(\tau).$$

This proves that ϑ indeed satisfies (26).

REMARK 33. The additive analogue of Hilbert's Satz 90 (cf. [L] or [J, Chapter 1, Section 15]) says that $H^1(G, (L', +)) = 0$. Since the twisted homomorphism $\phi : G \rightarrow (L', +)$ may be interpreted as a 1-cocycle, this implies that ϕ splits, which indicates the existence of ϑ. Replacing the twisted homomorphism ϕ by an arbitrary 1-cocycle in our above computation provides a proof of this additive analogue.

Since H is the kernel of ϕ, (26) yields that H is the group of all automorphisms of $L'|K$ which fix ϑ. Since on the other hand, by definition of $H = \mathrm{Gal}\, L'|L$ the fixed field of H in L' is L, we know from Galois theory that

$L = K(\vartheta)$. Let us now compute the minimal polynomial f of ϑ over K. The group N may be viewed as a system of representatives for the left cosets of G modulo H. Consequently, the elements ϑ^τ, $\tau \in N$, are precisely all conjugates of ϑ over K. So

$$f(X) = \prod_{\tau \in N} (X - \vartheta^\tau) = \prod_{\tau \in N} (X - \vartheta - \phi(\tau)) = \mathcal{A}(X - \vartheta),$$

where

$$\mathcal{A}(X) := \prod_{\tau \in N} (X - \phi(\tau)).$$

The roots of $\mathcal{A}$ form the additive group $\phi(N)$. Since we have chosen ϕ to be injective, we have $|\phi(N)| = |N| = \deg \mathcal{A}$. By part a) of Corollary 23 it follows that $\mathcal{A}$ is an additive polynomial. In particular,

$$f(X) = \mathcal{A}(X - \vartheta) = \mathcal{A}(X) - \mathcal{A}(\vartheta).$$

Since $f(X) \in K[X]$, we have $\mathcal{A}(X) \in K[X]$ and $\mathcal{A}(\vartheta) \in K$. Since $\deg f = \deg \mathcal{A} = |N| = [L : K] = [K(\vartheta) : K]$, f is the minimal polynomial of ϑ over K.

We have proved:

THEOREM 34. *Let $L|K$ be an extension which satisfies condition (23). Then there exist an additive polynomial $\mathcal{A}(X) \in K[X]$ and an element $\vartheta \in L$ such that $L = K(\vartheta)$ and $\mathcal{A}(X) - \mathcal{A}(\vartheta) \in K[X]$ is the minimal polynomial of ϑ over K.*

As an example, let us discuss an important special case. Let us assume that $L|K$ is a Galois extension of degree p. Then its Galois group is just $\mathbb{Z}/p\mathbb{Z}$, and the extension is thus p-elementary. In the above setting, we may then choose $K' = K$ which yields $L' = L$, $G = N = \mathbb{Z}/p\mathbb{Z}$ and $H = 1$. The embedding $\phi : N \longrightarrow (L', +)$ may be chosen "by hand" to be the most natural one: $N = \mathbb{Z}/p\mathbb{Z} = (\mathbb{F}_p, +) \subset (L', +)$. We obtain

$$\mathcal{A}(X) = \prod_{i \in \mathbb{F}_p} (X - i) = X^p - X$$

since the latter is the unique polynomial of degree p which vanishes on all elements of $\mathbb{F}_p$. The extension $L|K$ is thus generated by the root ϑ of the polynomial $f(X) = X^p - X - \mathcal{A}(\vartheta)$ which we call an *Artin-Schreier polynomial*. The extension $L|K$ is an Artin-Schreier extension. So we have shown:

THEOREM 35. *Every Galois extension of degree p of a field of characteristic $p > 0$ is an Artin-Schreier extension.*

Inspired by this special case, we want to investigate whether we can get more information about the additive polynomial $\mathcal{A}$ if we strengthen the hypotheses. For instance, ϕ may be injective even if ψ is not. In our special case, $N = \mathbb{Z}/p\mathbb{Z}$ was an irreducible G-module, that is, it did not admit any proper nonzero

G-submodule. But if N is an irreducible G-module, then every G-module homomorphism ϕ can only have kernel 0 or N, so if it does not vanish, then it is injective. For ϕ as defined in (24), we obtain $\phi \neq 0$ if $\psi \neq 0$. So it will suffice to take $\psi : N \to (\mathbb{F}_p, +)$ as a nonzero (additive) character; it exists since N is a non-trivial p-group. With this choice of ψ, we obtain

$$\phi(N) \subset \sum_{\rho \in H} \mathbb{F}_p b^\rho = \sum_{\rho \in H} \mathbb{F}_p \rho b.$$

Since the coefficients of the polynomial $\mathcal{A}$ are the elementary symmetric polynomials of the elements $\phi(\tau)$, $\tau \in N$, they lie in the ring $\mathbb{F}_p[\rho b \mid \rho \in H]$.

The condition that N be an irreducible G-module has turned out to be of certain importance. It is satisfied in the following special case:

LEMMA 36. *Assume that $L|K$ is minimal with the property* (23), *that is, there is no proper non-trivial subextension with the same property. Then N is an irreducible G-module.*

PROOF. Assume that M is a G-submodule of N, that is, M is a normal subgroup of G. Then HM is a subgroup of G containing H. In view of the unique representation (20), we have $HM = H$ if and only if $M = 1$ and $HM = G$ if and only if $M = N$. Note that the fixed field L'_1 of M in L' is a Galois extension of K containing K'. Further, the fixed field L_1 of HM is contained in L, and it satisfies $L_1.K' = L'_1$ since $HM \cap N = M \cap N = M$. Consequently, also $L_1|K$ has property (23).

Suppose now that $L|K$ is minimal with the property (23). Then $L_1 = L$ or $L_1 = K$. Hence $HM = H$ or $HM = G$, that is, $M = 1$ or $M = N$, showing that the G-module N is irreducible.   ⊣

We summarize our preceding discussion in the following

LEMMA 37. *Assume that $L|K$ is minimal with the property* (23). *If $K'|K$ is infinite, we may replace it by a suitable finite subextension. For every $b \in K'$ generating a normal basis of $K'|K$, and for every nonzero additive character $\psi : N \to (\mathbb{F}_p, +)$, the G-module homomorphism ϕ defined in* (24) *is injective. Moreover, the coefficients of the correponding additive polynomial $\mathcal{A}(X)$ lie in the ring*

$$K \cap \mathbb{F}_p[\rho b \mid \rho \in H].$$

EXERCISE 4. Let char $K = p > 0$ and $L|K$ be an Artin-Schreier extension and ϑ an *Artin-Schreier generator* of $L|K$, that is, $L = K(\vartheta)$ and $\vartheta^p - \vartheta \in K$. Show that all other Artin-Schreier generators of $L|K$ are of the form $i\vartheta + c$ with $i \in \{1, 2, \ldots, p-1\}$ and $c \in K$. Can something similar be said in the setting of Theorem 34? (Hint: use the uniqueness statement following equation (26)).

§8. Minimal purely wild extensions. This section is devoted to the proof of Theorem 13 which shows the important connection between purely wild (and in particular, immediate) extensions of henselian fields of positive characteristic and additive polynomials.

Before we continue, we put together several facts from ramification theory that can be found in [En], [N] and [Ku12] or can be deduced easily from other facts (exercise for the reader). For a field L, we denote by $\tilde{L}$ its algebraic closure and by $\operatorname{Gal} L$ the absolute Galois group $\operatorname{Gal} \tilde{L}|L$ of L. Recall that for a henselian field (K, v), K^r denotes the ramification field of the extension $(K^{\mathrm{sep}}|K, v)$.

THEOREM 38. *Let (K, v) be a henselian field and p the characteristic exponent of Kv. Then the following assertions hold:*

a) *$\operatorname{Gal} K^r$ is a normal subgroup of $\operatorname{Gal} K$ and $K^r|K$ is a Galois extension.*

b) *$\operatorname{Gal} K^r$ is a pro-p-group, so the separable-algebraic closure of K is a p-extension of K^r.*

c) *The value group vK^r consists of all elements in the ordered divisible hull of vK whose order modulo vK is prime to p. The residue field $K^r v$ is the separable-algebraic closure of Kv.*

d) *If $vK^r = vK$ (we say that $K^r|K$ is unramified), then for every Galois subextension $K'|K$ of $K^r|K$, we have $\operatorname{Gal} K'|K \simeq \operatorname{Gal} K'v|Kv$.*

e) *Every finite extension $(K_2|K_1, v)$, where $K \subseteq K_1 \subseteq K_2 \subseteq K^r$, is defectless.*

f) *If L is an algebraic extension of K, then $L^r = L.K^r$, and the extensions $(L^r|K^r, v)$ and $(L|K, v)$ have the same defect.*

g) *If (L, v) is an immediate henselian extension of (K, v), not necessarily algebraic, then $L^r = L.K^r$.*

Let (K, v) be a henselian field which is not tame and thus admits purely wild extensions (see Theorem 14). Theorem 13 will follow from Theorem 34 if we are able to show that every minimal purely wild extension $L|K$ satisfies condition (23) which is the hypothesis the latter theorem. As a natural candidate for an extension $K'|K$ which is Galois and linearly disjoint from $L|K$, we can take the extension $K^r|K$. By part f) of Theorem 38 we know that $L.K^r = L^r$. We set

- $\mathcal{G} := \operatorname{Gal} K$,
- $\mathcal{N} := \operatorname{Gal} K^r$, which is a normal subgroup of $\mathcal{G}$ and a pro-p-group,
- $\mathcal{H} := \operatorname{Gal} L$, which is a maximal proper subgroup of $\mathcal{G}$ since $L|K$ is a minimal non-trivial extension, and which satisfies $\mathcal{N}.\mathcal{H} = \mathcal{G}$ since $K^r|K$ is linearly disjoint from $L|K$
- $\mathcal{D} := \mathcal{N} \cap \mathcal{H} = \operatorname{Gal} L.K^r = \operatorname{Gal} L^r$.

The next lemma examines this group theoretical situation.

LEMMA 39. *Let $\mathcal{G}$ be a profinite group with maximal proper subgroup $\mathcal{H}$. Assume that the non-trivial pro-p-group $\mathcal{N}$ is a normal subgroup of $\mathcal{G}$ not* *contained in* $\mathcal{H}$. *Then* $\mathcal{D} = \mathcal{N} \cap \mathcal{H}$ *is a normal subgroup of* $\mathcal{G}$ *and the finite factor group* $\mathcal{N}/\mathcal{D}$ *is an elementary-abelian p-group. Further,* $\mathcal{N}/\mathcal{D}$ *is an irreducible right* $\mathcal{G}/\mathcal{D}$-*module.*

PROOF. By the maximality of $\mathcal{H}$, we have $\mathcal{H}\mathcal{N} = \mathcal{G}$. Since $\mathcal{N} \not\subset \mathcal{H}$, we have that $\mathcal{D}$ is a proper subgroup of $\mathcal{N}$. Since every maximal proper subgroup of a profinite group is of finite index, we have that $(\mathcal{N} : \mathcal{D}) = (\mathcal{G} : \mathcal{H})$ is finite. Observe that $\mathcal{D}$ is $\mathcal{H}$-*invariant* (which means that $\mathcal{D}^\sigma = \mathcal{D}$ for every $\sigma \in \mathcal{H}$). This is true since $\mathcal{N} \lhd \mathcal{G}$ and $\mathcal{H}$ are $\mathcal{H}$-invariant. Assume that $\mathcal{E}$ is an $\mathcal{H}$-invariant subgroup of $\mathcal{N}$ containing $\mathcal{D}$. Then $\mathcal{H}\mathcal{E}$ is a subgroup of $\mathcal{G}$ containing $\mathcal{H}$. From the maximality of $\mathcal{H}$ it follows that either $\mathcal{H}\mathcal{E} = \mathcal{H}$ or $\mathcal{H}\mathcal{E} = \mathcal{G}$, whence either $\mathcal{E} = \mathcal{D}$ or $\mathcal{E} = \mathcal{N}$ (this argument is as in the proof of Lemma 36). We have proved that $\mathcal{D}$ is a maximal $\mathcal{H}$-invariant subgroup of $\mathcal{N}$.

Now let $\Phi(\mathcal{N})$ be the Frattini subgroup of $\mathcal{N}$, i.e., the intersection of all maximal open subgroups of $\mathcal{N}$. Since $\mathcal{D} \neq \mathcal{N}$, we can pick some maximal proper subgroup of $\mathcal{N}$ containing $\mathcal{D}$, and since it also contains $\Phi(\mathcal{N})$, it follows that $\mathcal{D}\Phi(\mathcal{N}) \neq \mathcal{N}$. Being a characteristic subgroup of $\mathcal{N}$, the Frattini subgroup $\Phi(\mathcal{N})$ is $\mathcal{H}$-invariant like $\mathcal{N}$. Consequently, also the group $\mathcal{D}\Phi(\mathcal{N})$ is $\mathcal{H}$-invariant. From the maximality of $\mathcal{D}$ we deduce that $\mathcal{D}\Phi(\mathcal{N}) = \mathcal{D}$, showing that

$$\Phi(\mathcal{N}) \subset \mathcal{D}.$$

On the other hand, the factor group $\mathcal{N}/\Phi(\mathcal{N})$ is a (possibly infinite dimensional) $\mathbb{F}_p$-vector space (cf. [R–Za, part (b) of Lemma 2.8.7]). In view of $\Phi(\mathcal{N}) \subset \mathcal{D}$, this yields that also $\mathcal{D}$ is a normal subgroup of $\mathcal{N}$ and that also $\mathcal{N}/\mathcal{D}$ is an elementary-abelian p-group. Since $\mathcal{D}$ is $\mathcal{H}$-invariant, $\mathcal{D} \lhd \mathcal{N}$ implies that

$$\mathcal{D} \lhd \mathcal{H}\mathcal{N} = \mathcal{G}.$$

As a normal subgroup of $\mathcal{G}$, $\mathcal{N}$ is a $\mathcal{G}$-module, and in view of $\mathcal{D} \lhd \mathcal{G}$ it follows that $\mathcal{N}/\mathcal{D}$ is a $\mathcal{G}/\mathcal{D}$-module. If it were reducible then there would exist a proper subgroup $\mathcal{E}$ of $\mathcal{N}$ such that $\mathcal{E}/\mathcal{D}$ is a non-trivial $\mathcal{G}/\mathcal{D}$-module. But then, $\mathcal{E}$ must be a normal subgroup of $\mathcal{G}$ properly containing $\mathcal{D}$; in particular, $\mathcal{E}$ would be a proper $\mathcal{H}$-invariant subgroup of $\mathcal{N}$, in contradiction to the maximality of $\mathcal{D}$. $\dashv$

This lemma shows that $L'|K$ is Galois and $L' = L.K'$ is a finite p-elementary extension of K'. Hence $L|K$ satisfies (23) with $K' = K^r$. We apply Theorem 34 to obtain an additive polynomial $\mathcal{A}(X) \in K[X]$ and an element $\vartheta \in L$ such that $L = K(\vartheta)$ and that $\mathcal{A}(X) - \mathcal{A}(\vartheta)$ is the minimal polynomial of ϑ over K. Since $L|K$ is a minimal purely wild extension by our assumption, it is in particular minimal with property (23) and thus satisfies the hypothesis of Lemma 37. Hence, the extension $K^r|K$ can be replaced by a finite subextension $K'|K$, and $\mathcal{A}(X)$ may be chosen such that its coefficients lie in the ring $K \cap \mathbb{F}_p[\rho b \mid \rho \in H]$, where b is the generator of a normal basis of $K'|K$. Since vK is cofinal in $v\tilde{K} = \widetilde{vK}$, we may choose some $c \in K$ such that $vcb \geq 0$. Since (K, v) is henselian by assumption, it follows that $v\sigma(cb) = vcb \geq 0$ for all $\sigma \in \operatorname{Gal} K$. On the other hand, cb is still the generator of a normal basis of $K'|K$. So we may replace b by cb, which yields that $K \cap \mathbb{F}_p[\rho b \mid \rho \in H] \subset \mathcal{O}_K$ and consequently, that $\mathcal{A}(X) \in \mathcal{O}_K[X]$.

Now assume in addition that $(K|k, v)$ is an immediate extension of henselian fields. Then we may infer from part g) of Theorem 38 that $K^r = k^r.K$. So the Galois extension K' of K is the compositum of K with a suitable Galois extension k' of k. In this case, b may be chosen to be already the generator of a normal basis of k' over k; it will then also be the generator of a normal basis of K' over K. With this choice of b, we obtain that the ring $K \cap \mathbb{F}_p[\rho b \mid \rho \in H]$ is contained in $K \cap k' = k$, whence $\mathcal{A}(X) \in \mathcal{O}_k[X]$. Let us summarize what we have proved; the following theorem will imply Theorem 13.

THEOREM 40. *Let (K, v) be a henselian field and $(L|K, v)$ a minimal purely wild extension. Then $L^r|K$ is a Galois extension and $L^r|K^r$ is a p-elementary extension. Hence, $L|K$ satisfies condition (23), and there exist an additive polynomial $\mathcal{A}(X) \in \mathcal{O}_K[X]$ and an element $\vartheta \in L$ such that $L = K(\vartheta)$ and that $\mathcal{A}(X) - \mathcal{A}(\vartheta)$ is the minimal polynomial of ϑ over K. If $(K|k, v)$ is an immediate extension of henselian fields, then $\mathcal{A}(X)$ may already be chosen in $\mathcal{O}_k[X]$.*

Let us conclude this section by discussing the following special case. Assume that the value group vK is divisible by all primes $q \neq p$. Then by part c) of Theorem 38, $K^r|K$ is an unramified extension. Consequently by part d) of Theorem 38, $\operatorname{Gal} K'|K \simeq \operatorname{Gal} \overline{K'}|\overline{K}$ and we may choose the element b such that $\overline{b}$ is the generator of a normal basis of $\overline{K'}|\overline{K}$. It follows that the residue mapping is injective on the ring $\mathbb{F}_p[\rho b \mid \rho \in H]$ and thus, also the mapping $\tau \mapsto \overline{\phi(\tau)}$ is injective. In this case, we obtain that

$$\overline{\mathcal{A}}(X) = \prod_{\tau \in N} (X - \overline{\phi(\tau)})$$

has no multiple roots and is thus separable.

§9. p-closed fields.

This section is devoted to the proofs of the two theorems that deal with p-closed fields of positive characteristic.

PROOF OF THEOREM 5: We have to prove that *A field K is p-closed if and only if it does not admit any finite extensions of degree divisible by p.*

"$\Leftarrow$": Assume that K does not admit any finite extensions of degree divisible by p. Take any p-polynomial $f \in K[X]$. Write $f = \mathcal{A} + c$ where $\mathcal{A} \in K[X]$ is an additive polynomial. Let h be an irreducible factor of f; by hypothesis, it has a degree d not divisible by p. Fix a root b of h in the algebraic closure $\tilde{K}$ of K. All roots of f are of the form $b + a_i$ where the a_is are roots of $\mathcal{A}$. By part a) of Corollary 23 the roots of $\mathcal{A}$ in $\tilde{K}$ form an additive group. The sum of the roots of h lies in K. This gives us $db + s \in K$, where s is a sum of a subset of the a_is and is therefore again a root of $\mathcal{A}$. Likewise, $d^{-1}s$ is a root of $\mathcal{A}$ (as d is not divisible by p, it is invertible in K). Then $b + d^{-1}s = d^{-1}(db + s)$ is a root of f, and it lies in K, as required.

"$\Rightarrow$": (This part of the proof is due to David Leep.) Assume that K is p-closed. Since K is perfect, it suffices to take a Galois extension $L|K$ of degree n and show that p does not divide n. By the normal basis theorem there is a basis $b_1, \ldots, b_n$ of $L|K$ where the b_is are the roots of some irreducible polynomial over K. Since they are linearly independent over K, their trace is non-zero. The elements

$$1, b_1, b_1^p, \ldots, b_1^{p^{n-1}}$$

are linearly dependent over K since $[L : K] = n$. Therefore there exist elements $d_0, \ldots, d_{n-1}, e \in K$ such that the p-polynomial

$$f(X) = d_{n-1}X^{p^{n-1}} + \cdots + d_0X + e$$

has b_1 as a root. It follows that all the b_is are roots of f. Thus the elements $b_2 - b_1, \ldots, b_n - b_1$ are roots of the additive polynomial $f(X) - e$. Since these $n - 1$ roots are linearly independent over K, they are also linearly independent over the prime field $\mathbb{F}_p$. This implies that the additive group G generated by the elements $b_2 - b_1, \ldots, b_n - b_1$ contains p^{n-1} distinct elements, which therefore must be precisely the roots of $f(X) - e$. So $G + b_1$ is the set of roots of f. By hypothesis, one of these roots lies in K; call it ϑ. There exist integers $m_2, \ldots, m_n$ such that

$$\vartheta = m_2(b_2 - b_1) + \cdots + m_n(b_n - b_1) + b_1.$$

In this equation take the trace from L to K. The elements $b_1, \ldots, b_n$ all have the same trace; hence the trace of every $m_i(b_i - b_1)$ is 0. It follows that the trace $n\vartheta$ of ϑ is equal to the trace of b_1; as we have remarked already, this trace is non-zero. Hence $n\vartheta \neq 0$, which shows that n is not divisible by p.   ⊣

PROOF OF THEOREM 6: We have to prove: *A henselian valued field of characteristic $p > 0$ is p-closed if and only if it is an algebraically maximal Kaplansky field.*

We will use Theorem 5 throughout the proof without further mention. Assume first that (K, v) is henselian and that K is p-closed. Since every finite extension of the residue field Kv can be lifted to an extension of K of the same degree, it follows that Kv is p-closed. Likewise, if the value group vK were not p-divisible, then K would admit an extension of degree p; this shows that vK is p-divisible. We have thus proved that (K, v) is a Kaplansky field. Since the degree of every finite extension of K is prime to p, it follows that (K, v) is defectless, hence algebraically maximal.

For the converse, assume that (K, v) is an algebraically maximal Kaplansky field. Since the henselization is an immediate algebraic extension, it follows that (K, v) is henselian. By Theorem 14, there exists a field complement W of K^r in $\tilde{K}$. As vK is p-divisible and Kv is p-closed, hence perfect, the same theorem shows that W is an immediate extension of K. Hence $W = K$, which shows that $K^r = \tilde{K}$. So every finite extension $L|K$ is a subextension of $K^r|K$ and is therefore defectless; that is, $[L : K] = (vL : vK)[Lv : Kv]$. As the right hand side is not divisible by p, (K, v) being a Kaplansky field, we find that p does not divide $[L : K]$. By Theorem 5, this proves that K is p-closed. ⊣

REFERENCES

[A-K] J. Ax and S. Kochen, *Diophantine problems over local fields II. a complete set of axioms for p-adic number theory*, American Journal of Mathematics, vol. 87 (1965), pp. 631–648.

[B] N. Bourbaki, *Elements of Mathematics. Commutative Algebra*, Hermann, Paris, 1972.

[C1] P. M. Cohn, *Free Rings and Their Relations*, LMS Monographs, vol. 2, Academic Press, London, 1971.

[C2] ———, *Free Rings and Their Relations*, London Mathematical Society Monographs, vol. 19, Academic Press Inc. [Harcourt Brace Jovanovich Publishers], London, 1985, Second edition.

[Cu-Pi] S. D. Cutkosky and O. Piltant, *Ramification of valuations*, Advances in Mathematics, vol. 183 (2004), no. 1, pp. 1–79.

[Del] F. Delon, *Quelques Propriétés des Corps Valués en Théories des Modèles*, Thèse Paris VII, 1981.

[Deu] M. Deuring, *Lectures on the Theory of Algebraic Functions of One Variable*, Springer LNM, vol. 314, Springer-Verlag, Berlin, 1973.

[Dr-Ku] L. van den Dries and F.-V. Kuhlmann, *Images of additive polynomials in* $\mathbb{F}_q((t))$ *have the optimal approximation property*, Canadian Mathematical Bulletin. Bulletin Canadien de Mathématiques, vol. 45 (2002), no. 1, pp. 71–79.

[En] O. Endler, *Valuation Theory*, Springer-Verlag, New York, 1972.

[Ep] Helmut P. Epp, *Eliminating wild ramification*, Inventiones Mathematicae, vol. 19 (1973), pp. 235–249.

[Er] Yu. L. Ershov, *On elementary theories of local fields*, Algebra i Logika (Seminar), vol. 6 (1967), no. 2, pp. 5–30, (in Russian).

[Er1] ———, *On the elementary theory of maximal valued fields III*, Algebra i Logika, vol. 6 (1967), no. 3, pp. 31–38, (in Russian).

[Er2] ———, *Multi-Valued Fields*, Kluwer, New York, 2001.

[F-Jr] M. Fried and M. Jarden, *Field Arithmetic*, Springer-Verlag, Berlin, 1986.

[Ge] E. R. Gentile, *On rings with one-sided field of quotients*, Proceedings of the American Mathematical Society, vol. 11 (1960), pp. 380–384.

[Go] D. Goss, *Basic Structures of Function Field Arithmetic*, Springer, Berlin, 1998.

[Gra] K. A. H. Gravett, *Note on a result of Krull*, Mathematical Proceedings of the Cambridge Philosophical Society, vol. 52 (1956), p. 379.

[Gre] M. J. Greenberg, *Rational points in Henselian discrete valuation rings*, Institut des Hautes Études Scientifiques. Publications Mathématiques, vol. 31 (1966), pp. 59–64.

[H] B. Huppert, *Endliche Gruppen. I*, Springer-Verlag, Berlin, 1967.

[J] N. JACOBSON, *Lectures in Abstract Algebra. Vol III: Theory of Fields and Galois Theory*, D. Van Nostrand Co., Inc., Princeton, N.J.-Toronto, Ont.-London-New York, 1964.

[Ka1] I. KAPLANSKY, *Maximal fields with valuations*, Duke Mathematical Journal, vol. 9 (1942), pp. 303–321.

[Ka2] ———, *Selected Papers and Other Writings*, Springer-Verlag, New York, 1995.

[Kr] W. KRULL, *Allgemeine Bewertungstheorie*, Journal für die Reine und Angewandte Mathematik, vol. 167 (1931), pp. 160–196.

[Ku1] F.-V. KUHLMANN, *Henselian Function Fields and Tame Fields*, extended version of Ph.D. thesis, Heidelberg, 1990.

[Ku2] ———, *Valuation Theory of Fields, Abelian Groups and Modules*, Habilitation thesis, Heidelberg, 1995.

[Ku3] ———, *Valuation theoretic and model theoretic aspects of local uniformization*, Resolution of Singularities (Obergurgl, 1997) (Herwig Hauser, Joseph Lipman, Frans Oort, and Adolfo Quiros, editors), Progress in Mathematics, vol. 181, Birkhäuser, Basel, 2000, pp. 381–456.

[Ku4] ———, *Elementary properties of power series fields over finite fields*, The Journal of Symbolic Logic, vol. 66 (2001), no. 2, pp. 771–791.

[Ku5] ———, *A correction to: "Elimination of wild ramification"*, Inventiones Mathematicae, vol. 153 (2003), no. 3, pp. 679–681.

[Ku6] ———, *Value groups, residue fields, and bad places of rational function fields*, Transactions of the American Mathematical Society, vol. 356 (2004), no. 11, pp. 4559–4600.

[Ku7] ———, *Places of algebraic function fields in arbitrary characteristic*, Advances in Mathematics, vol. 188 (2004), no. 2, pp. 399–424.

[Ku8] ———, *A classification of Artin Schreier defect extensions and a characterization of defectless fields*, submitted.

[Ku9] ———, *Elimination of Ramification I: The Generalized Stability Theorem*, submitted.

[Ku10] ———, *Elimination of Ramification II: Henselian Rationality*, in preparation.

[Ku11] ———, *The model theory of tame valued fields*, in preparation.

[Ku12] ———, *Book on Valuation Theory (in preparation)*, Preliminary versions of several chapters available at: http://math.usask.ca/~fvk/Fvkbook.htm.

[Ku-Pa-Ro] F.-V. KUHLMANN, M. PANK, and P. ROQUETTE, *Immediate and purely wild extensions of valued fields*, Manuscripta Mathematica, vol. 55 (1986), no. 1, pp. 39–67.

[L] S. LANG, *Algebra*, Addison-Wesley Publishing Co., Inc., Reading, Mass., 1965.

[N] J. NEUKIRCH, *Algebraic Number Theory*, Springer-Verlag, Berlin, 1999.

[O1] O. ORE, *Theory of non-commutative polynomials*, Annals of Mathematics, vol. 34 (1933), pp. 480–508.

[O2] ———, *On a special class of polynomials*, Transactions of the American Mathematical Society, vol. 35 (1933), pp. 559–584.

[Ph-Za] T. PHEIDAS and K. ZAHIDI, *Elimination theory for addition and the Frobenius map in polynomial rings*, The Journal of Symbolic Logic, vol. 69 (2004), no. 4, pp. 1006–1026.

[Pop] F. POP, *Über die Structur der rein wilden Erweiterungen eines Körpers*, manuscript, Heidelbeg, 1987.

[R-Za] L. RIBES and P. ZALESSKII, *Profinite Groups*, Springer-Verlag, Berlin, 2000.

[Ri] P. RIBENBOIM, *Théorie des Valuations*, Deuxième édition multigraphiée. Séminaire de Mathématiques Supérieures, No. 9, 1964, Les Presses de l'Université de Montréal, Montreal, Que., 1968.

[Roh] T. ROHWER, *Valued difference fields as modules over twisted polynomial rings*, Ph.D. thesis, Urbana, 2003, Available at: http://math.usask.ca/fvk/theses.htm.

[V] P. VÁMOS, *Kaplansky fields and p-algebraically closed fields*, Communications in Algebra, vol. 27 (1999), no. 2, pp. 629–643.

[Wh1] G. WHAPLES, *Additive polynomials*, Duke Mathematical Journal, vol. 21 (1954), pp. 55–65.

[Wh2] ———, *Galois cohomology of additive polynomial and n-th power mappings of fields*, **Duke Mathematical Journal**, vol. 24 (1957), pp. 143–150.

[Zi] M. ZIEGLER, **Die elementare Theorie der Henselschen Körper**, Inaugural Dissertation, Köln, 1972.

MATHEMATICAL SCIENCES GROUP
UNIVERSITY OF SASKATCHEWAN
106 WIGGINS ROAD, SASKATOON
SASKATCHEWAN, S7N 5E6, CANADA
E-mail: fvk@math.usask.ca

DENSE SUBFIELDS OF HENSELIAN FIELDS, AND INTEGER PARTS

FRANZ-VIKTOR KUHLMANN

Abstract. We show that every henselian valued field L of residue characteristic 0 admits a proper subfield K which is dense in L. We present conditions under which this can be taken such that $L|K$ is transcendental and K is henselian. These results are of interest for the investigation of integer parts of ordered fields. We present examples of real closed fields which are larger than the quotient fields of all their integer parts. Finally, we give rather simple examples of ordered fields that do not admit any integer part and of valued fields that do not admit any subring which is an additive complement of the valuation ring.

§1. Introduction. At the "Logic, Algebra and Arithmetic" Conference, Teheran 2003, Mojtaba Moniri asked the following question: *Does every non-archimedean ordered real closed field L admit a proper dense subfield K?* This question is interesting since if such a subfield K admits an integer part I then I is also an integer part for L, but the quotient field of I lies in K and is thus smaller than L. An *integer part* of an ordered field K is a discretely ordered subring I with 1 such that for all $a \in K$ there is $r \in I$ such that $r \leq a < r + 1$. It follows that the element r is uniquely determined, and in particular that 1 is the least positive element in I.

Since the natural valuation of a non-archimedean ordered real closed field L is non-trivial, henselian and has a (real closed) residue field Lv of characteristic 0, the following theorem answers the above question to the affirmative:

THEOREM 1. *Every henselian non-trivially valued field (L, v) with a residue field of characteristic 0 admits a proper subfield K which is dense in (L, v). This subfield K can be chosen such that $L|K$ is algebraic.*

This paper was written while I was a guest of the Equipe Géométrie et Dynamique, Institut Mathématiques de Jussieu, Paris, and of the Equipe Algèbre-Géométrie at the University of Versailles. I gratefully acknowledge their hospitality and support. I was also partially supported by a Canadian NSERC grant and by a sabbatical grant from the University of Saskatchewan. Furthermore I am endebted to the organizers of the conference in Teheran and the members of the IPM and all our friends in Iran for their hospitality and support. I also thank the two referees as well as A. Fornasiero for their careful reading of the paper and their useful suggestions. This paper is dedicated to Salma Kuhlmann who got me interested in the subject and provided the personal contacts that inspired and supported my work.

Logic in Tehran
Edited by A. Enayat, I. Kalantari, and M. Moniri
Lecture Notes in Logic, 26
© 2006, Association for Symbolic Logic

Here, density refers to the topology induced by the valuation; that is, K is dense in (L, v) if for every $a \in L$ and all values α in the value group vL of (L, v) there is $b \in K$ such that $v(a - b) \geq \alpha$. In the case of non-archimedean ordered fields with natural (or non-trivial order compatible) valuation, density in this sense is equivalent to density with respect to the ordering.

In the case where the value group vL has a maximal proper convex subgroup, the proof is quite easy, but does in general not render any subfield K such that $L|K$ is transcendental. In the case of vL having no maximal proper convex subgroup, the proof is much more involved, but leaves us the choice between $L|K$ algebraic or transcendental:

THEOREM 2. *In addition to the assumptions of Theorem* 1, *suppose that* vL *does not have a maximal proper convex subgroup. Then for each integer* $n \geq 1$ *there is a henselian* (*as well as a non-henselian*) *subfield* K *dense in* L *such that* $\operatorname{trdeg} L|K = n$. *It can also be chosen such that* $\operatorname{trdeg} L|K$ *is infinite.*

To see that such valued fields (L, v) exist, take x_i, $i \in \mathbb{N}$, to be a set of algebraically independent elements over an arbitrary field k and define a valuation v on $k(x_i \mid i \in \mathbb{N})$ by setting $0 < vx_1 \ll vx_2 \ll \cdots \ll vx_i \ll \cdots$; then pass to the henselization of $(k(x_i \mid i \in \mathbb{N}), v)$. For a more general approach, see Lemma 26.

REMARK 3. A. Fornasiero [F] has shown that every henselian valued field with a residue field of characteristic 0 admits a truncation closed embedding in a power series field with coefficients in the residue field and exponents in the value group (in general, the power series field has to be endowed with a non-trivial factor system). "Truncation closed" means that every truncation of a power series in the image lies again in the image.

It follows that all of the henselian dense subfields admit such truncation closed embeddings. But also the dense non-henselian subfields can be chosen such that they admit truncation closed embeddings. We will sketch the proof in Section 3 (Remarks 25 and 28).

Our construction developed for the proof of Theorem 2 also gives rise to a counterexample to a quite common erroneous application of Hensel's Lemma. A valuation w is called a *coarsening* of v if its associated valuation ring contains that of v. In this case, v induces a valuation $\overline{w}$ on the residue field Kw whose valuation ring is simply the image of the valuation ring of v under the residue map associated with w. The counterexample proves:

PROPOSITION 4. *There are valued fields* (K, v) *such that* vK *has no maximal proper convex subgroup, the residue field* $(Kw, \overline{w})$ *is henselian for every non-trivial coarsening* $w \neq v$ *of* v, *but* (K, v) *itself is not henselian.*

The proofs of Theorems 1 and 2 and of Proposition 4 are given in Section 3. There, we will also give a more explicit version of Theorem 2.

In general, the quotient fields of integer parts of an ordered field are smaller than the field. The following theorem will show that there are real closed fields for which the quotient field of *every* integer part is a proper subfield. If k is any field, then

$$\mathrm{PSF}(k) := \bigcup_{n \in \mathbb{N}} k\left(\left(t^{\frac{1}{n}}\right)\right)$$

is called the *Puiseux series field over* k; it is a subfield of the power series field $k((t^{\mathbb{Q}}))$ with coefficients in k and exponents in $\mathbb{Q}$, which we also simply denote by $k((\mathbb{Q}))$.

THEOREM 5. *Let $\mathbb{Q}^{\mathrm{rc}}$ denote the field of real algebraic numbers and* $\mathrm{PSF}(\mathbb{Q}^{\mathrm{rc}})$ *the Puiseux series field over* $\mathbb{Q}^{\mathrm{rc}}$. *If I is any integer part of this real closed field, then* $\mathrm{Quot}\, I$ *is a proper countable subfield of* $\mathrm{PSF}(\mathbb{Q}^{\mathrm{rc}})$ *such that the transcendence degree of* $\mathrm{PSF}(\mathbb{Q}^{\mathrm{rc}})$ *over* $\mathrm{Quot}\, I$ *is uncountable. The same holds for the completion of* $\mathrm{PSF}(\mathbb{Q}^{\mathrm{rc}})$.

This answers a question of M. Moniri. An answer was also given, independently, by L. van den Dries at the conference. A larger variety of such fields is presented in Section 4. On the other hand, there are fields that admit integer parts whose quotient field is the whole field:

THEOREM 6. *Let λ be any cardinal number and k any field of characteristic 0. Then there exists a henselian valued field (L, v) with residue field k which has the following properties*:

a) *L contains a k-algebra R which is an additive complement of its valuation ring such that* $\mathrm{Quot}\, R = L$.

b) *At the same time, for each non-zero cardinal number $\kappa \leq \lambda$, L contains a k-algebra R_κ which is an additive complement of its valuation ring such that* $\mathrm{trdeg}\, L | \mathrm{Quot}\, R_\kappa = \kappa$.

If in addition k is an archimedean ordered field and $<$ is any ordering on L compatible with v (see Section 2 for this notion), then $(L, <)$ admits an integer part I such that $\mathrm{Quot}\, I = L$. *At the same time, for each non-zero cardinal number $\kappa \leq \lambda$, L admits an integer part I_κ such that* $\mathrm{trdeg}\, L | \mathrm{Quot}\, I_n = \kappa$.

S. Boughattas [Bg] has given an example of an ordered (and "n-real closed") field which does not admit any integer part. In the last section of our paper, we generalize the approach and consider a notion that comprises integer parts as well as subrings which are additive complements of the valuation ring or of the valuation ideal in a valued field. A subring R of a valued field (K, v) will be called a *weak complement* (in K) if it has the following properties:

- $vr \leq 0$ for all $r \in R$,
- for all $a \in K$ there is $r \in R$ such that $v(a - r) \geq 0$.

Every integer part in a non-archimedean ordered field K is a weak complement with respect to the natural valuation of K (see Lemma 35).

Using a somewhat surprising little observation (Lemma 37) together with a result of [K1] (which is a generalization of a result in [M-S]) we construct examples for valued fields that do not admit any weak complements. From this we obtain ordered fields without integer parts. In particular, we show:

THEOREM 7. *For every prime field k there are valued rational function fields $k(t, x, y)$ of transcendence degree 3 over the trivially valued subfield k which do not admit any weak complements. There are ordered rational function fields of transcendence degree 3 over $\mathbb{Q}$ which do not admit any integer parts.*

There are valued rational function fields of transcendence degree 4 over a trivially valued prime field which do not admit any weak complements, but admit an embedding of their residue field and a cross-section. There are ordered rational function fields of transcendence degree 4 over $\mathbb{Q}$ which do not admit any integer parts, but admit an embedding of their residue field and a cross-section for their natural valuation.

Our example of an n-real closed field without integer parts is the n-real closure of such an ordered rational function field. It is quite similar to the example given by Boughattas, but in contrast to his example, ours is of finite transcendence degree over $\mathbb{Q}$.

OPEN PROBLEM. Are there valued fields of transcendence degree ≤ 2 over a trivially valued ground field that do not admit any weak complements? Are there ordered fields of transcendence degree ≤ 2 over an archimedean ordered field that do not admit any integer parts? Are there examples of transcendence degree ≤ 3 with embedding of their residue field and cross-section?

§2. **Some preliminaries.** For basic facts from general valuation theory we refer the reader to [E], [R], [W], [Z-S], [K2]. For ramification theory, see [N], [E] and [K2]. In the following, we state some well known facts without proofs.

Take any valued field (K, v). If v' is a valuation on the residue field Kv, then $v \circ v'$ will denote the valuation whose valuation ring is the subring of the valuation ring of v consisting of all elements whose v-residue lies in the valuation ring of v'. (Note that we identify equivalent valuations.) While $v \circ v'$ does actually not mean the composition of v and v' as mappings, this notation is used because in fact, up to equivalence the place associated with $v \circ v'$ is indeed the composition of the places associated with v and v'.

Every convex subgroup Γ of vK gives rise to a coarsening v_Γ of v such that $v_\Gamma K$ is isomorphic to vK/Γ. As mentioned in the introduction, v induces a valuation $\bar{v}_\Gamma$ on the residue field Kv_Γ. We then have that $v = v_\Gamma \circ \bar{v}_\Gamma$. The value group $\bar{v}_\Gamma(Kv_\Gamma)$ of $\bar{v}_\Gamma$ is isomorphic to Γ, and its residue field $(Kv_\Gamma)\bar{v}_\Gamma$ is isomorphic to Kv. Every coarsening w of v is of the form v_Γ for some convex subgroup Γ of vK.

If a is an element of the valuation ring $\mathcal{O}_v$ of v on K, then av will denote the image of a under the residue map associated with the valuation v. This map is a ring homomorphism from $\mathcal{O}_v$ onto the residue field Kv. It is only unique up to equivalence, i.e., up to composition with an isomorphism from Kv to another field (and so the residue field Kv is only unique up to isomorphism). If w is a coarsening of v, that is, $\mathcal{O}_v$ contains the valuation ring $\mathcal{O}_w$ of v on L, then the residue map $\mathcal{O}_w \ni a \mapsto aw \in Kw$ can be chosen such that it extends the residue map $\mathcal{O}_v \ni a \mapsto av \in Kv$.

An ordering $<$ on a valued field (K, v) is said to be *compatible with the valuation* v (and v is *compatible with* $<$) if

$$(1) \qquad \forall x, y \in K : 0 < x \le y \Longrightarrow vx \ge vy.$$

This holds if and only if the valuation ring of v is a convex subset of $(K, <)$. This in turn holds if and only if $<$ induces an ordering on the residue field Kv. We will need the following well-known facts (cf. [P]):

LEMMA 8. *Take any valued field (K, v). Every ordering $<_r$ on Kv can be lifted to an ordering $<$ on K which is compatible with v and induces $<_r$ on Kv (that is, if a, b are elements of the valuation ring of v such that $a < b$, then $av = bv$ or $av <_r bv$).*

LEMMA 9. *If an ordering of a field K is compatible with the valuation v of K, then v extends to a valuation of the real closure K^{rc} of $(K, <)$, which is still compatible with the ordering on K^{rc}. This extension is henselian, its value group vK^{rc} is the divisible hull of vK, and its residue field $K^{rc}v$ is the real closure of Kv (with respect to the induced ordering on Kv).*

A compatible valuation of an ordered field $(K, <)$ is called the *natural valuation* of $(K, <)$ if its residue field is archimedean ordered. The natural valuation is uniquely determined, and every compatible valuation is a coarsening of the natural valuation.

Take any valued field (K, v) and a finite extension $L|K$. Then the following *fundamental inequality* holds:

$$(2) \qquad n \ge \sum_{i=1}^{g} e_i f_i,$$

where $n = [L : K]$ is the degree of the extension, $v_1, \ldots, v_g$ are the distinct extensions of v from K to L, $e_i = (v_i L : vK)$ are the respective ramification indices, and $f_i = [Lv_i : Kv]$ are the respective inertia degrees. Note that the extension of v from K to L is unique (i.e., $g = 1$) if and only if (K, v) is henselian (which by definition means that (K, v) satisfies Hensel's Lemma). The following are easy consequences:

LEMMA 10. *If $L|K$ is a finite extension and v is a valuation on L, then $[L : K] \ge (vL : vK)$ and $[L : K] \ge [Lv : Kv]$.*

COROLLARY 11. *Let $L|K$ be an algebraic extension and v a valuation on L. Then vL/vK is a torsion group and the extension $Lv|Kv$ of residue fields is algebraic. If v is trivial on K (i.e., $vK = \{0\}$), then v is trivial on L.*

An extension $(K, v) \subseteq (L, v)$ of valued fields is called *immediate* if the canonical embeddings of vK in vL and of Kv in Lv are onto. We have:

LEMMA 12. *If K is dense in (L, v), then $(K, v) \subseteq (L, v)$ is an immediate extension.*

PROOF. If $a \in L$ and $b \in K$ such that $v(a - b) > va$, then $va = vb \in vK$. If $a \in L$ such that $va = 0$ and $b \in K$ such that $v(a - b) > 0$, then $av = bv \in Kv$. ⊣

The following is a well known consequence of the so-called "Lemma of Ostrowski":

LEMMA 13. *If a valued field (L, v) is an immediate algebraic extension of a henselian field (K, v) of residue characteristic 0, then $L = K$.*

LEMMA 14. *The henselization K^h of a valued field (K, v) (which is unique up to valuation preserving isomorphism over K) is an immediate extension and can be chosen in every henselian valued extension field of (K, v).*

LEMMA 15. *An algebraic extension of a henselian valued field, equipped with the unique extension of the valuation, is again henselian.*

LEMMA 16. *Let (L, v) be any field and $v = w \circ \overline{w}$ where w is non-trivial. Take any subfield L_0 of L. Then L_0 is dense in (L, v) if and only if L_0 is dense in (L, w).*

LEMMA 17. *Let (K, v) be any field and $v = w \circ \overline{w}$. Then (K, v) is henselian if and only if (K, w) and $(Kw, \overline{w})$ are.*

COROLLARY 18. *Let (K, v) be any field and $v = w \circ \overline{w}$. If $(Kw, \overline{w})$ is henselian, then the henselization of (K, v) is equal to the henselization of (K, w) (as fields).*

The value group vK of a valued field (K, v) is *archimedean* if it is embeddable in the ordered additive group of the reals. This holds if and only if every convex subgroup of vK is equal to $\{0\}$ or to vK.

LEMMA 19. *If (K, v) is a valued field such that vK is archimedean, then K is dense in its henselization. In particular, the completion of (K, v) is henselian.*

The following result is an easy application of Hensel's Lemma:

LEMMA 20. *Take (K, v) to be a henselian valued field of residue characteristic char $Kv = 0$. Take any subfield K_0 of K on which v is trivial. Then there is a subfield K' of K containing K_0 and such that v is trivial on K' and the residue map associated with v induces an isomorphism from K' onto Kv. If $Kv|K_0v$ is algebraic, then so is $K'|K_0$.*

A field K' as in this lemma is called a *field of representatives for the residue field Kv.*

PROPOSITION 21.   a) *Take a non-empty set T of elements algebraically independent over K and a finite extension F of $K(T)$. Then no non-trivial valuation on F is henselian. In particular, no non-trivial valuation on an algebraic function field (of transcendence degree at least one) is henselian.*
  b) *Fix $n \in \mathbb{N}$, take $K(T)$ as in a) and take F to be the closure of $K(T)$ under successive adjunction of roots of polynomials of degree $\leq n$. Then no non-trivial valuation on F is henselian.*

PROOF. Choosing any $t \in T$ and replacing K by $K(T \setminus \{t\})$, we may assume in parts a) and b) that T consists of a single element, i.e., $\mathrm{trdeg}\, F|K = 1$.

Take any non-trivial valuation on F. We show that there is some $x \in K(T)$ such that $vx > 0$ and x is transcendental over K. Assume first that v is trivial on K. Since v is non-trivial on F and $F|K(T)$ is algebraic, Corollary 11 shows that v is non-trivial on $K(T)$. Hence there must be some $x \in K(T)$ such that $vx \neq 0$. Replacing x by x^{-1} if necessary, we may assume that $vx > 0$. It follows that $x \notin K$, so x is transcendental over K.

Now assume that v is not trivial on K, and take an arbitrary $x \in K(T)$ transcendental over K. If $vx > 0$, we are done. If $vx < 0$, we replace it by x^{-1} and we are done again. If $vx = 0$, we pick some $c \in K$ such that $vc > 0$. Then $vcx > 0$ and cx is transcendental over K, hence replacing x by cx finishes the proof of our claim.

Pick any positive integer q such that q is not divisible by the characteristic $p := \mathrm{char}\, Kv$ of the residue field Kv. By Hensel's Lemma, any henselian extension of $K(x)$ will contain a q-th root of the 1-unit $y := 1 + x$. We wish to show that any algebraic extension of $K(x)$ containing such a q-th root must be of degree at least q over $K(x)$. A valuation theoretical proof for this fact reads as follows. Take the y-adic valuation v_y on $K(x) = K(y)$. Then $v_y y$ is the least positive element in the value group $v_y K(x) \simeq \mathbb{Z}$, and any q-th root b of y will have v_y-value $\frac{1}{q} v_y y$. This shows that $(v_y K(x)(b) : v_y K(x)) \geq q$. By the fundamental inequality, it follows that $[K(x, b) : K(x)] \geq (v_y K(x, b) : v_y K(x)) \geq q$.

Proof of part a): Since $\mathrm{trdeg}\, F|K = 1$, $x \in K(T)$ is transcendental over K and $F|K$ is finite, also $F|K(x)$ is finite. Pick $q > [F : K(x)]$ not divisible by p. Then it follows that F does not contain a q-th root of y, and so (F, v) cannot be henselian.

Proof of part b): This time, we still have that $K(T)|K(x)$ is finite. Pick a prime $q > \max\{n, [K(T) : K(x)]\}$, $q \neq p$. For every element α in the value group $v_y F$ there is an integer e which is a product of positive integers $\leq n$ such that $e\alpha \in v_y K(T)$. Further, there is a positive integer e' such that $e' e \alpha \in v_y K(x)$. On the other hand, by our choice of q, it does not divide $e' e$. Since the order of the value $\frac{1}{q} v_y y$ modulo $v_y K(x)$ is q, it follows that this value does not lie in $v_y F$. Hence again, (F, v) cannot be henselian.   ⊣

PROPOSITION 22. *Take (L, v) to be a henselian field of residue characteristic char $Lv = 0$, and K a subfield of L such that $L|K$ is algebraic. Then K admits an algebraic extension L_0 inside of L such that the extension of v from K to L_0 is unique, L_0 is linearly disjoint over K from the henselization K^h of K in L, and $L = L_0.K^h = L_0^h$ (where $L_0.K^h$ denotes the field compositum of the fields L_0 and K^h inside of L).*

PROOF. Take any subextension $L_0|K$ of $L|K$ maximal with the property that the extension of v from K to L_0 is unique. By general ramification theory it follows that $L_0|K$ is linearly disjoint from $K^h|K$ and that $L_0^h = L_0.K^h$. We only have to show that $L_0^h = L$. Note that $L|L_0^h$ is algebraic since already $L|K$ is algebraic,

Let us show that $L_0v = Lv$. If this is not the case, then there is be some element $\zeta \in Lv \setminus L_0v$. By Corollary 11, $Lv|L_0v$ is algebraic. Let $g \in L_0v[X]$ be the minimal polynomial of ζ over L_0v. Since char $Kv = 0$, g is separable. We choose some monic polynomial f with integral coefficients in L_0 whose reduction modulo v is g; it follows that $\deg f = \deg g$. Since ζ is a simple root of g, it follows from Hensel's Lemma that the henselian field (L, v) contains a root z of f whose residue is ζ. We have

$$[L_0(z) : L_0] \leq \deg f = \deg g = [L_0v(\zeta) : L_0v]$$
$$\leq [L_0(z)v : L_0v] \leq [L_0(z) : L_0],$$

where the last inequality follows from Lemma 10. We conclude that $[L_0(z) : L_0] = [L_0(z)v : L_0v]$. From the fundamental inequality it follows that the extension of v from L_0 (and hence also from K) to $L_0(z)$ is unique. But this contradicts the maximality of L_0. Hence, $L_0v = Lv$.

Next, let us show that $vL_0 = vL$. If this is not the case, then there is some $\alpha \in vL \setminus vL_0$. By Corollary 11, vL/vL_0 is a torsion group and hence there is some $n > 1$ such that $n\alpha \in vL_0$. We choose n minimal with this property, so that $(vL_0 + \alpha\mathbb{Z} : vL_0) = n$. Further, we pick some $a \in L$ such that $va = \alpha$. Since $n\alpha \in vL_0$, there is some $d \in L_0$ such that $vd = n\alpha = va^n$. It follows that $va^n/d = 0$, and since we have already shown that $Lv = L_0v$, we can choose some $c \in L_0$ such that $(a^n/cd)v = 1$. Consequently, the reduction of $X^n - a^n/cd$ modulo v is the polynomial $X^n - 1$, which admits 1 as a simple root since char $Kv = 0$. Hence by Hensel's Lemma, $X^n - a^n/cd$ admits a root b in the henselian field (L, v). For $z := \frac{a}{b}$ it follows that

$$nvz = v\frac{a^n}{b^n} = vcd = vd = n\alpha,$$

which shows that $\alpha = vz \in vL_0(z)$. We have

$$[L_0(z) : L_0] \leq n = (vL_0 + \alpha\mathbb{Z} : vL_0) \leq (vL_0(z) : vL_0) \leq [L_0(z) : L_0],$$

where again the last inequality follows from Lemma 10. We conclude that $[L_0(z) : L_0] = (vL_0(z) : vL_0)$. From the fundamental inequality it follows that the extension of v from L_0 (and hence also from K) to $L_0(z)$ is unique. But this again contradicts the maximality of L_0. Hence, $vL_0 = vL$.

We have shown that $vL = vL_0$ and $Lv = L_0v$. Hence, $vL = vL_0^h$ and $Lv = L_0^h v$. As $L|L_0$ is algebraic, the same is true for $L|L_0^h$. Since the residue field characteristic of (L, v) is zero, Lemma 13 shows that $L = L_0^h$. This concludes our proof.   ⊣

§3. Dense subfields. In this section we prove the existence of proper dense subfields of henselian fields with residue characteristic 0.

PROPOSITION 23. *Take a henselian valued field* (L, v) *such that* vL *admits a maximal proper convex subgroup* Γ. *Assume that* char $Lv_\Gamma = 0$. *Then L admits a proper dense subfield* L_0 *such that* $L|L_0$ *is algebraic.*

PROOF. By Lemma 16 it suffices to find a subfield L_0 which is dense in L with respect to v_Γ, and such that $L|L_0$ is algebraic. By Lemma 17, (L, v_Γ) is henselian. Since char $Lv_\Gamma = 0$ and hence char $L = 0$, L contains $\mathbb{Q}$ and v_Γ is trivial on $\mathbb{Q}$. Pick a transcendence basis T of $L|\mathbb{Q}$. Since v_Γ is non-trivial on L, T is non-empty. We infer from Lemma 21 that $(\mathbb{Q}(T), v_\Gamma)$ is not henselian. By Proposition 22, there is an algebraic extension L_0 of $\mathbb{Q}(T)$ within L such that L_0 is linearly disjoint over $\mathbb{Q}(T)$ from the v_Γ-henselization $\mathbb{Q}(T)^h$ of $\mathbb{Q}(T)$ in L, and $L = L_0.\mathbb{Q}(T)^h = L_0^h$. Since $(\mathbb{Q}(T), v_\Gamma)$ is not henselian, $\mathbb{Q}(T)^h|\mathbb{Q}(T)$ is a proper extension. By the linear disjointness, the same holds for $L|L_0$. As Γ is the maximal proper convex subgroup of vL, $v_\Gamma L \simeq vL/\Gamma$ must be archimedean. Thus by Lemma 19, (L_0, v_Γ) lies dense in its henselization (L, v_Γ). Hence by Lemma 16, (L_0, v) lies dense in its henselization (L, v). Since $L|\mathbb{Q}(T)$ is algebraic, so is $L|L_0$.   ⊣

In certain cases, even if v has a coarsest non-trivial coarsening, there will also be dense subfields K such that $L|K$ is transcendental. For instance, this is the case for $L = k((t))$ equipped with the t-adic valuation v_t, where a subfield is dense in L as soon as it contains $k(t)$. On the other hand, the henselization $k(t)^h$ of $k(t)$ w.r.t. v_t admits $k(t)$ as a proper dense subfield, and the extension $k(t)^h|K$ is algebraic for every subfield K which is dense in $k(t)^h$. More generally, the following holds:

PROPOSITION 24. *Suppose that* (L, v) *is a valued field and that* v *is trivial on the prime field* k *of L. If*

$$\text{trdeg } L|k = \dim_\mathbb{Q}(\mathbb{Q} \otimes vL) + \text{trdeg } Lv|k < \infty,$$

then $L|K$ *is algebraic for every dense subfield* K.

PROOF. If K is a dense subfield, then by Lemma 12, $(L|K, v)$ is an immediate extension. Hence,

$$\text{trdeg } K|k \geq \dim_{\mathbb{Q}}(\mathbb{Q} \otimes vK) + \text{trdeg } Kv|k$$
$$= \dim_{\mathbb{Q}}(\mathbb{Q} \otimes vL) + \text{trdeg } Lv|k$$
$$= \text{trdeg } L|k,$$

whence $\text{trdeg } K|k = \text{trdeg } L|k$, showing that $L|K$ is algebraic.   ⊣

Note that if (L, v) is a valued field with a subfield L_0 on which v is trivial, and if $\text{trdeg } L|L_0 < \infty$, then in general,

$$(3) \qquad \text{trdeg } L|L_0 \geq \dim_{\mathbb{Q}}(\mathbb{Q} \otimes vL) + \text{trdeg } Lv|L_0.$$

This is a special case of the so-called "Abhyankar inequality". For a proof, see [Br, Chapter VI, §10.3, Theorem 1]. Note that $\mathbb{Q} \otimes vL$ is the divisible hull of vL, and $\dim_{\mathbb{Q}}(\mathbb{Q} \otimes vL)$ is the maximal number of rationally independent elements in vL.

REMARK 25. It can be shown that if $\text{char } Lv = 0$, then the dense subfield L_0 in Proposition 23 can always be constructed in such a way that it admits a truncation closed embedding into a power series field. The idea is as follows. Since (L, v) is henselian, we can use Lemma 20 to find a field k of representatives in L for the residue field Lv. Then we can choose a twisted cross-section as in [F]. The field L_1 generated over k by the image of the cross-section admits a truncation closed embedding in $k((vL))$ with a suitable factor system, and this embedding ι can be extended to a truncation closed embedding of (L, v) in $k((vL))$ (cf. [F]). It is easy to show that $L_\Gamma := \iota^{-1}(\iota L \cap k((\Gamma)))$ is a field of representatives for the residue field Lv_Γ in (L, v_Γ), and that ι induces a truncation closed embedding of L_Γ in $k((\Gamma)) \subset k((vL))$. This can be extended to a truncation closed embedding of $L_2 := L_1.L_\Gamma$ which is obtained from L_Γ by adjoining the image of the cross-section. We note that (L, v_Γ) is an immediate extension of (L_2, v_Γ). If this extension is algebraic, then L is also algebraic over the henselization of L_2 (with respect to v_Γ), and by Lemma 13, the two fields must be equal. That shows that L_2 is dense in (L, v_Γ) and hence in (L, v), and we can take $L_0 = L_2$.

If $L|L_2$ is transcendental, we take a transcendence basis S of $L|L_2$ and pick $s \in S$. Then one shows as before that L is the henselization of $L_2(S)$, and also of the larger field $L_0 := L_2(S \setminus \{s\})^h(s)$. Again, L_0 is dense in (L, v). Following [F], $L_2(S \setminus \{s\})^h$ admits a truncation closed embedding in $k((vL))$. As (L, v) is immediate over (L_2, v), it is also immediate over $L_2(S \setminus \{s\})^h$. Therefore, s is the limit of a pseudo Cauchy sequence in $L_2(S \setminus \{s\})^h$ without a limit in this field. As the field is henselian of residue characteristic 0, this pseudo Cauchy sequence is of transcendental type. Now [F] shows that the truncation closed embedding can be extended to L_0.

Now we turn to the case where vL admits no maximal proper convex subgroup, i.e., v admits no coarsest non-trivial coarsening. Such valued fields exist:

LEMMA 26. *Take any regular cardinal number λ and any field k. Then there is a valued field (L, v) with residue field k and such that λ is the cofinality of the set of all proper convex subgroups of vL, ordered by inclusion.*

PROOF. Take J to be the set of all ordinal numbers $< \lambda$, endowed with the reverse of the usual ordering. Choose any archimedean ordered abelian group Γ. Then take G to be the ordered Hahn product $\mathbf{H}_J\Gamma$ with index set J and components Γ (see [Fu] or [KS] for details on Hahn products). Then the set of all proper convex subgroups of G, ordered by inclusion, has order type λ and hence has cofinality λ. Now take (L, v) to be the power series field $k((G))$ with its canonical valuation.   ⊣

Note that if vL admits no maximal proper convex subgroup, then vL is the union of its proper convex subgroups. Indeed, if $\alpha \in vL$, then the smallest convex subgroup C of vL that contains α ($=$ the intersection of all convex subgroups containing α) admits a largest convex subgroup, namely the largest convex subgroup of vL that does not contain α ($=$ the union of all convex subgroups not containing α). Therefore $C \neq vL$, showing that C is a proper convex subgroup containing α.

PROPOSITION 27. *Take a henselian valued field (L, v) such that vL admits no maximal proper convex subgroup. Assume that $\operatorname{char} L = 0$. Then L admits a proper dense subfield K such that $L|K$ is algebraic. If $\kappa > 0$ is any cardinal number smaller than or equal to the cofinality of the set of convex subgroups of vL ordered by inclusion, then there is also a henselian (as well as a non-henselian) subfield K dense in L such that $\operatorname{trdeg} L|K = \kappa$.*

PROOF. It suffices to prove that there is a subfield K dense in L such that the transcendence degree of $L|K$ is equal to the cofinality λ of the set of convex subgroups of vL. This is seen as follows. Take a transcendence basis T of $L|K$. If κ is a cardinal number $\leq \lambda$, then take a subset T_κ of T of cardinality κ. Then $K_\kappa := K(T \setminus T_\kappa)$ is dense in L because it contains K; furthermore, $\operatorname{trdeg} L|K_\kappa = \kappa$. We may always, even in the case of $\kappa = \lambda$, choose $T_\kappa \neq T$. Then by part a) of Proposition 21, (K_κ, v) is not henselian. In particular, $(K(T), v)$ is not henselian and thus, $K(T)$ is a proper subfield of L such that $L|K(T)$ is algebraic. If $\kappa \neq 0$, then $L|K_\kappa$ will be transcendental. By Lemma 14, the henselian field L contains the henselization K_κ^h of K_κ. Since it is an algebraic extension of K_κ, we have $\operatorname{trdeg} L|K_\kappa^h = \operatorname{trdeg} L|K_\kappa = \kappa$, and it is dense in L, too.

To illustrate the idea of our proof, we first show that there is a dense subfield K such that $\operatorname{trdeg} L|K > 0$. We choose a convex subgroup C_0 of vL as follows. If $\operatorname{char} Lv = 0$, then we set $C_0 = \{0\}$. If $\operatorname{char} Lv = p > 0$, then we observe that $0 \neq p \in L$ since char $L = 0$, so we may take C_0 to be the smallest proper convex subgroup that contains vp. We let $w_0 = v_{C_0}$ be the coarsening of v associated with C_0. We have $w_0 = v$ if char $Lv = 0$. Since C_0 is a proper convex subgroup, w_0 is a non-trivial valuation.

Let λ be the cofinality of the set of all proper convex subgroups of vL, ordered by inclusion. Starting from C_0, we pick a strictly ascending cofinal sequence of convex subgroups C_ν, $\nu < \lambda$, in this set. We denote by w_ν the coarsening of v which corresponds to C_ν.

By Lemma 20 there is a field K_0' of representatives for Lw_0 in L. We pick a transcendence basis $T_0 = \{t_{0,\mu} \mid \mu < \kappa_0\}$ of $K_0'|\mathbb{Q}$, where κ_0 is the transcendence degree of $Lw_0|\mathbb{Q}$. Then we proceed by induction on $\nu < \lambda$. Suppose we have already constructed a field K_ν' of representatives of Lw_ν and a transcendence basis $\bigcup_{\nu' \leq \nu} T_{\nu'}$ for it. By Lemma 20, K_ν' can be extended to a field $K_{\nu+1}'$ of representatives of $Lw_{\nu+1}$, and we choose a transcendence basis $T_{\nu+1} = \{t_{\nu+1,\mu} \mid \mu < \kappa_{\nu+1}\}$ of $K_{\nu+1}'|K_\nu'$. Having constructed K_ν', $\nu < \lambda'$ for some limit ordinal $\lambda' \leq \lambda$, we set $K_{\lambda'}^* = \bigcup_{\nu < \lambda'} K_\nu'$. Again by Lemma 20, $K_{\lambda'}^*$ can be extended to a field of representatives $K_{\lambda'}'$ of $Lw_{\lambda'}$, and we choose a transcendence basis $T_{\lambda'} = \{t_{\lambda',\mu} \mid \mu < \kappa_{\lambda'}\}$ of $K_{\lambda'}'|K_{\lambda'}^*$. Note that $T_{\lambda'}$ may be empty.

We set $K' = \bigcup_{\nu < \lambda} K_\nu'$ and show that K' is dense in L. Take any $a \in L$ and $\alpha \in vL$. Then there is some $\nu < \lambda$ such that $\alpha \in C_\nu$. By construction, K' contains a field of representatives for Lw_ν. Hence there is some $b \in K'$ such that $aw_\nu = bw_\nu$, meaning that $w_\nu(a - b) > 0$ and thus, $v(a - b) > \alpha$. This proves our claim. Hence if $\operatorname{trdeg} L|K' > 0$, we set $K = K'$ and we are done showing the existence of a subfield K with $\operatorname{trdeg} L|K > 0$. But it may well happen that $L|K'$ is algebraic, or even that $L = K'$. In this case, we construct a subfield K of K' as follows.

Note that for all $\nu < \lambda$, (K_ν', v) is henselian. Indeed, it is isomorphic (by the place associated with w_ν) to $(Lw_\nu, \overline{w}_\nu)$, where $\overline{w}_\nu$ is the valuation induced by v on Lw_ν; since (L, v) is henselian, Lemma 17 shows that the same is true for $(Lw_\nu, \overline{w}_\nu)$ and hence for (K_ν', v). Again from Lemma 17 it follows that (K_ν', w_μ) is henselian for all $\mu < \lambda$. Note that w_μ is non-trivial on K_ν' only for $\mu < \nu$, and in this case, $K_\nu' w_\mu = Lw_\mu$ since K_ν' contains the field K_μ' of representatives for Lw_μ.

Note further that for all $\nu < \lambda$ and all $\mu < \kappa_\nu$, $w_\nu t_{\nu,\mu} = 0$. On the other hand, after multiplication with suitable elements in $K_{\nu+1}$ we may assume that $w_\nu t_{\nu+1,\mu} > 0$ for all $\mu < \kappa_{\nu+1}$.

We will now construct inside of K' a chain (ordered by inclusion) of subfields $K_\nu \subset K_\nu'$ ($\nu < \lambda$) such that each K_ν is a field of representatives for Lw_ν and contains the element $t_{0,0} - t_{\nu+1,0}$, but not the element $t_{0,0}$.

Since $T_0 = \{t_{0,\mu} \mid \mu < \kappa_0\}$ is a transcendence basis of $K_0'|\mathbb{Q}$, Lemma 11 shows that the residue field $K_1' w_0 = K_0' w_0$ is algebraic over $\mathbb{Q}(t_{0,\mu} \mid \mu < \kappa_0)w_0$.

Because $(t_{0,0} - t_{1,0})w_0 = t_{0,0}w_0$ by construction, the latter field is equal to $\mathbb{Q}(t_{0,0} - t_{1,0}, t_{0,\mu} \mid 1 \leq \mu < \kappa_0)w_0$. Since char $K_1'w_0 = 0$, we can use Lemma 20 to find inside of the henselian field (K_1', w_0) an algebraic extension K_0 of $\mathbb{Q}(t_{0,0} - t_{1,0}, t_{0,\mu} \mid 1 \leq \mu < \kappa_0)$ which is a field of representatives for $K_1'w_0 = Lw_0$. Note that $t_{0,0}$ is transcendental over $\mathbb{Q}(t_{0,0} - t_{1,0}, t_{0,\mu} \mid 1 \leq \mu < \kappa_0)$ and therefore, $t_{0,0} \notin K_0$, but $t_{0,0} - t_{1,0} \in K_0$.

Suppose we have already constructed all fields K_μ for $\mu \leq \nu$, where ν is some ordinal $< \lambda$. Since $T_{\nu+1} = \{t_{\nu+1,\mu} \mid \mu < \kappa_{\nu+1}\}$ is a transcendence basis of $K_{\nu+1}' \mid K_\nu'$, Lemma 11 shows that the residue field $K_{\nu+2}'w_{\nu+1} = K_{\nu+1}'w_{\nu+1}$ is algebraic over $K_\nu(t_{\nu+1,\mu} \mid \mu < \kappa_{\nu+1})w_{\nu+1}$. Because $(t_{\nu+1,0} - t_{\nu+2,0})w_{\nu+1} = t_{\nu+1,0}w_{\nu+1}$ by construction, the latter field is equal to $K_\nu(t_{\nu+1,0} - t_{\nu+2,0}, t_{\nu+1,\mu} \mid 1 \leq \mu < \kappa_{\nu+1})w_{\nu+1}$. Since char $K_{\nu+2}'w_{\nu+1} = 0$, we can use Lemma 20 to find inside of the henselian field $(K_{\nu+2}', w_{\nu+1})$ an algebraic extension $K_{\nu+1}$ of $K_\nu(t_{\nu+1,0} - t_{\nu+2,0}, t_{\nu+1,\mu} \mid 1 \leq \mu < \kappa_{\nu+1})$ which is a field of representatives for $K_{\nu+2}'w_{\nu+1} = Lw_{\nu+1}$. Since $t_{0,0} - t_{\nu+1,0}, t_{\nu+1,0} - t_{\nu+2,0} \in K_{\nu+1}$ we have that $t_{0,0} - t_{\nu+2,0} \in K_{\nu+1}$. Again, $t_{0,0} \notin K_{\nu+1}$ as $t_{0,0}$ is transcendental over $K_\nu(t_{\nu+1,0} - t_{\nu+2,0}, t_{\nu+1,\mu} \mid 1 \leq \mu < \kappa_{\nu+1})$.

Suppose we have already constructed all fields K_ν for $\nu < \lambda'$, where λ' is some limit ordinal $\leq \lambda$. We note that $t_{0,0} \notin \bigcup_{\nu < \lambda'} K_\nu =: K_{\lambda'}^{**}$. But $K_{\lambda'}^{**}(t_{0,0})$ contains the entire transcendence basis of $K_{\lambda'}^* \mid \mathbb{Q}$ because $t_{0,0} - t_{\nu+1,0} \in K_\nu$ for every $\nu < \lambda'$ (recall that $K_{\lambda'}^*$ is the field we constructed above before constructing $K_{\lambda'}'$). It follows that $T_{\lambda'} \cup \{t_{0,0}\}$ is a transcendence basis of $K_{\lambda'}' \mid K_{\lambda'}^{**}$, and therefore the residue field $K_{\lambda'+1}'w_{\lambda'} = Lw_{\lambda'}$ is an algebraic extension of $K_{\lambda'}^{**}(T_{\lambda'} \cup \{t_{0,0}\})w_{\lambda'}$. Because $(t_{0,0} - t_{\lambda'+1,0})w_{\lambda'} = t_{0,0}w_{\lambda'}$, the latter field is equal to $K_{\lambda'}^{**}(T_{\lambda'} \cup \{t_{0,0} - t_{\lambda'+1,0}\})w_{\lambda'}$. Again by Lemma 20, there is an algebraic extension $K_{\lambda'}$ of $K_{\lambda'}^{**}(T_{\lambda'} \cup \{t_{0,0} - t_{\lambda'+1,0}\})$ inside of the henselian field $(K_{\lambda'+1}', w_{\lambda'})$ which is a field of representatives for $K_{\lambda'+1}'w_{\lambda'} = Lw_{\lambda'}$. By construction, $t_{0,0} - t_{\lambda'+1,0} \in K_{\lambda'}$. As before, $t_{0,0} \notin K_{\lambda'}$ as $t_{0,0}$ is transcendental over $K_{\lambda'}^{**}(T_{\lambda'} \cup \{t_{0,0} - t_{\lambda'+1,0}\})$.

We set

$$(4) \qquad\qquad K := \bigcup_{\nu < \lambda} K_\nu.$$

By construction, $t_{0,0} \notin K$, but $K(t_{0,0})$ contains $t_{\nu,\mu}$ for all $\nu < \lambda$ and $\mu < \kappa_\nu$. Hence, $K' \mid K(t_{0,0})$ is algebraic and therefore, trdeg $K' \mid K = 1$. With the same argument as for K', one shows that K is dense in L. (This also follows from the fact that $t_{0,0}$ is limit of the Cauchy sequence $(t_{0,0} - t_{\nu+1,0})_{\nu < \lambda}$ in K and K' is dense in L.)

Now we indicate how to achieve trdeg $K' \mid K = \lambda$. By passing to a cofinal subsequence of $(C_\nu)_{\nu < \lambda}$ if necessary, we can assume that every T_ν contains at least $|\nu|$ many elements, where $|\nu|$ denotes the cardinality of the ordinal number ν. Then it is possible to re-order the elements of T_ν in such a way that $T_\nu = \{t_{\nu,\mu} \mid \mu < \mu_\nu\}$ where μ_ν is some ordinal number $\geq \nu$. Now we modify the above construction of K as follows: at every step v where $v = 0$ or v is a successor ordinal, we replace $t_{v,\mu}$ by $t_{v,\mu} - t_{v+1,\mu}$ for all $\mu \leq v$. In the limit case for $\lambda' < \lambda$, we then have that $T_{\lambda'} \cup \{t_{v,v} \mid v < \lambda'\}$ is a transcendence basis of $K'_{\lambda'} | K^{**}_{\lambda'}$. Here, we replace every $t_{v,v}$ for $v < \lambda'$ by $t_{v,v} - t_{\lambda'+1,v}$. In this way we achieve that the elements $t_{v,v}$, $v < \lambda$ will be algebraically independent over K, but K will still be dense in L.   ⊣

REMARK 28. We can replace the field $K_\kappa = K(T \setminus T_\kappa)$ mentioned in the first paragraph of the proof by the larger field $K(T \setminus T_\kappa \setminus \{t\})^h(t)$ where $t \in T \setminus T_\kappa$. By the same argument as given at the end of Remark 25, this field admits a truncation closed embedding into the corresponding power series field.

Propositions 23 and 27 together prove Theorem 1. Theorem 2 follows immediately from Proposition 27 since if vL admits no maximal proper convex subgroup, then the cofinality of the set of convex subgroups of vL is an infinite cardinal number. It remains to give the

PROOF OF PROPOSITION 4. By Lemma 26 we may take a henselian valued field (L, v) of residue characteristic 0 such that vL admits no maximal proper convex subgroup. Using Proposition 27 we pick a non-henselian proper subfield K which is dense in L. Lemma 16 shows that for every non-trivial coarsening w of v, (K, w) is dense in (L, w), whence $Kw = Lw$. By Lemma 17, $(Lw, \overline{w})$ is henselian because (L, v) is henselian and $v = w \circ \overline{w}$. Hence, $(Kw, \overline{w})$ is henselian, which finishes our proof.   ⊣

EXAMPLE 1. A more direct construction of a counterexample works as follows: Take an ascending chain of convex subgroups C_i, $i \in \mathbb{N}$, in some ordered abelian group. Take k to be any field and set

$$K := \bigcup_{i \in \mathbb{N}} k((C_i)).$$

As a union of an ascending chain of henselian valued fields, K is itself a henselian valued field. But K is not complete. For instance, if $0 < \alpha_i \in C_i \setminus C_{i-1}$, then the element

$$x := \sum_{i \in \mathbb{N}} t^{\alpha_i} \in k\left(\left(\bigcup_{i \in \mathbb{N}} C_i\right)\right)$$

lies in the completion of K, but not in K. Since every henselian field is separable-algebraically closed in its henselization (cf. [W], Theorem 32.19), x is either transcendental or purely inseparable over K. But it cannot be purely inseparable over K because if $p = \operatorname{char} K > 0$, then $x^{p^v} = \sum_{i \in \mathbb{N}} t^{p^v \alpha_i} \notin K$ for all $v \geq 0$. Hence by part a) of Proposition 21, $K(x)$ (endowed with the restriction v of the valuation of the completion of K) is not henselian. But for every non-trivial coarsening w of v, $K(x)w = Kw$ since K is dense in $(K(x), v)$, and we leave it to the reader to prove that $(Kw, \overline{w})$ is henselian.

§4. **Small integer parts.** We will use a cardinality argument to show that there are real closed fields that are larger than the quotient fields of all its integer parts.

LEMMA 29.   a) *Take any valued field (L, v). Then all additive complements of the valuation ring of L, if there are any, have the same cardinality.*

b) *All integer parts in an ordered field, if there are any, are isomorphic as ordered sets and thus have the same cardinality.*

PROOF. a) As an additive group, any additive complement of the valuation ring $\mathcal{O}$ of L is isomorphic to $L/\mathcal{O}$. .

b) Take two integer parts I_1 and I_2 of a given ordered field $(L, <)$. Since I_2 is an integer part, for every $a \in I_1$ there is a unique element $a' \in I_2$ such that $a' \le a < a' + 1$. Hence, we have a mapping $I_1 \ni a \mapsto a' \in I_2$. Conversely, since I_1 is an integer part, there is a unique $a'' \in I_1$ such that $a'' < a' \le a''+1$. Consequently, $a = a'' + 1$ and a is the only element that is sent to a', showing that the map is injective and even order preserving. On the other hand, since $a'' + 1$ is sent to a', the mapping is also proved to be onto.   ⊣

We also need the following facts, which are well known (note that a similar statement holds for weak complements):

LEMMA 30.   a) *If K is dense in (L, v), then every additive complement of the valuation ring of (K, v) is also an additive complement of the valuation ring of (L, v).*

b) *If K is dense in $(L, <)$, then every integer part of $(K, <)$ is also an integer part of $(L, <)$.*

PROOF. We only prove a) and leave the proof of b) to the reader. Let A be an additive complement of the valuation ring $\mathcal{O}_K$ of (K, v), that is, $A \cap \mathcal{O}_K = \{0\}$ and $A + \mathcal{O}_K = K$. Denote the valuation ring of (L, v) by $\mathcal{O}_L$. Since the valuation on L is an extension of the valuation on K, we have that $K \cap \mathcal{O}_L = \mathcal{O}_K$ and thus, $A \cap \mathcal{O}_L = A \cap \mathcal{O}_K = \{0\}$. Now take any $a \in L$. Since K is dense in (L, v), there is $b \in K$ such that $v(a - b) \ge 0$, that is, $a - b \in \mathcal{O}_L$. Consequently, $a = b + (a - b) \in K + \mathcal{O}_L = A + \mathcal{O}_K + \mathcal{O}_L = A + \mathcal{O}_L$. This proves that $A + \mathcal{O}_L = L$.   ⊣

We cite the following fact; for a proof, see for instance [B-K-K].

LEMMA 31. *If K is an ordered field and R is a subring which is an additive complement of the valuation ring for the natural valuation of K, then $R + \mathbb{Z}$ is an integer part of K.*

For every ordered abelian group G, written additively, we set

$$G^{<0} := \{g \in G \mid g < 0\}.$$

PROPOSITION 32. *Suppose that k is a countable field. Then the countable ring $k[\mathbb{Q}^{<0}] := k[t^g \mid 0 > g \in \mathbb{Q}] \subset k((\mathbb{Q}))$ is an additive complement of* *the valuation ring of the uncountable henselian valued field* $\mathrm{PSF}(k)$. *The same remains true if* $\mathrm{PSF}(k)$ *is replaced by its completion.*

If in addition k *is an ordered* (*respectively, real closed*) *field, then* $k[\mathbb{Q}^{<0}] + \mathbb{Z}$ *is an integer part of the ordered* (*respectively, real closed*) *field* $\mathrm{PSF}(k)$, *and this also remains true if* $\mathrm{PSF}(k)$ *is replaced by its completion.*

PROOF. It is well known that every field $k((t))$ of formal Laurent series is uncountable. Hence, $\mathrm{PSF}(k)$ is uncountable. As the union of an ascending chain of fields $k((t^{\frac{1}{n}}))$ of formal Laurent series, which are henselian, $\mathrm{PSF}(k)$ is itself henselian. Note that the completion of a henselian field is again henselian ([W, Theorem 32.19]).

Every element $a \in \mathrm{PSF}(k)$ lies in $k((t^{\frac{1}{n}}))$ for some $n \in \mathbb{N}$. Hence, it suffices to show that $k[t^{\frac{m}{n}} \mid 0 > m \in \mathbb{Z}]$ is an additive complement of the valuation ring $k[[t^{\frac{1}{n}}]]$ in $k((t^{\frac{1}{n}}))$. Renaming $t^{\frac{1}{n}}$ by t, we thus have to show that $k[t^m \mid 0 > m \in \mathbb{Z}]$ is an additive complement of the valuation ring $k[[t]]$ in $k((t))$. But this is clear since $k((t))$ is the set of formal Laurent series

$$\sum_{i=N}^{\infty} c_i t^i = \sum_{i=N}^{-1} c_i t^i + \sum_{i=0}^{\infty} c_i t^i$$

where $N \in \mathbb{Z}$ and $c_i \in k$. The first sum lies in $k[t^m \mid 0 > m \in \mathbb{Z}]$ and the second sum in $k[[t]]$.

Part a) of Lemma 30 shows that $k[\mathbb{Q}^{<0}]$ is also an additive complement of the valuation ring in the completion of $\mathrm{PSF}(k)$.

The assertions about the ordered case follow from Lemma 31 together with part b) of Lemma 30.   ⊣

From this proposition together with Lemma 29, we obtain the following corollary, which in turn proves Theorem 5.

COROLLARY 33. *Suppose that* k *is a countable field. If* R *is any subring which is an additive complement of the valuation ring of* $\mathrm{PSF}(k)$, *then* $\mathrm{Quot}\, R$ *is countable and* $\mathrm{trdeg}\, \mathrm{PSF}(k) \mid \mathrm{Quot}\, R$ *is uncountable.*

If in addition k *is an ordered field and* I *an integer part of* $\mathrm{PSF}(k)$, *then* $\mathrm{Quot}\, I$ *is countable and* $\mathrm{trdeg}\, \mathrm{PSF}(k) \mid \mathrm{Quot}\, I$ *is uncountable.*

The same remains true if $\mathrm{PSF}(k)$ *is replaced by its completion.*

PROOF. The quotient field of a countable ring is again countable. So it only remains to prove the assertion about the transcendence degree. It follows from the fact that the algebraic closure of a countable field is again countable. So if T would be a countable transcendence basis of $\mathrm{PSF}(k) \mid \mathrm{Quot}\, R$, then $(\mathrm{Quot}\, R)(T)$ and hence also $\mathrm{PSF}(k)$ would be countable, which is not the case.   ⊣

Denote by $k((G)) = k((t^G))$ the power series field with coefficients in k and exponents in G, and by $k(G)$ the smallest subfield of $k((G))$ which contains all monomials ct^g, $c \in k$, $g \in G$. Denote by $k(G)^c$ its completion;

it can be chosen in $k((G))$. Note that the completion of $\mathrm{PSF}(k)$ is equal to $k(\mathbb{Q})^c$. Further, denote by $k[G^{<0}]$ the subring of $k(G)$ generated by k and all monomials ct^g where $c \in k$ and $0 > g \in G$.

PROPOSITION 34. *Suppose that k is a countable field and G is a countable archimedean ordered abelian group. Then the countable ring $k[G^{<0}]$ is an additive complement of the valuation ring of the uncountable henselian valued field $k(G)^c$.*

If R is any subring which is an additive complement of the valuation ring of $k(G)^c$, then $\mathrm{Quot}\, R$ is countable and $\mathrm{trdeg}\, k(G)^c \mid \mathrm{Quot}\, R$ is uncountable.

If in addition k is an ordered field and I is an integer part of the ordered field $k(G)^c$, then $\mathrm{Quot}\, I$ is countable and $\mathrm{trdeg}\, k(G)^c \mid \mathrm{Quot}\, I$ is uncountable.

PROOF. By Lemma 19, $k(G)^c$ is henselian. (Therefore, it is real closed if and only if k is real closed and G is divisible.)

We show that the ring $k[G^{<0}]$ is an additive complement of the valuation ring in $k(G)$. Every element a of the latter is a quotient of the form

$$a = \frac{c_1 t^{g_1} + \cdots + c_m t^{g_m}}{d_1 t^{h_1} + \cdots + d_n t^{h_n}}$$

with $c_1, \ldots, c_m, d_1, \ldots, d_n \in k$ and $g_1, \ldots, g_m, h_1, \ldots, h_n \in G$. Without loss of generality we may assume that h_1 is the unique smallest element among the $h_1, \ldots, h_n$. Then we rewrite a as follows:

$$a = \frac{\frac{c_1}{d_1} t^{g_1 - h_1} + \cdots + \frac{c_m}{d_1} t^{g_m - h_1}}{1 + \frac{d_2}{d_1} t^{h_2 - h_1} + \cdots + \frac{d_n}{d_1} t^{h_n - h_1}}.$$

By our assumption on h_1, all summands in the denominator except for the 1 have positive value. Hence, we can rewrite a as

$$a = \left(\frac{c_1}{d_1} t^{g_1 - h_1} + \cdots + \frac{c_m}{d_1} t^{g_m - h_1} \right)$$
$$\times \left(1 + \sum_{i=1}^{\infty} (-1)^i \left(\frac{d_2}{d_1} t^{h_2 - h_1} + \cdots + \frac{d_n}{d_1} t^{h_n - h_1} \right)^i \right).$$

In the power series determined by this geometric series, only finitely many summands will have negative value; this is true since G is archimedean by hypothesis. Let $b \in k[G^{<0}]$ be the sum of these summands. Then $v(a-b) \geq 0$. This proves that $k[G^{<0}]$ is an additive complement of the valuation ring in $k(G)$. Part a) of Lemma 30 shows that $k[G^{<0}]$ is also an additive complement of the valuation ring in $k(G)^c$.

All other assertions are deduced like the corresponding assertions of Corollary 33.   ⊣

§5. **Proof of Theorem 6.** We take k to be any field of characteristic 0 and

$$L := \bigcup_{v < \lambda} k((C_v))$$

to be the henselian valued field constructed in Example 1. The set $\operatorname{Neg} k((C_v))$ of all power series in $k((C_v))$ with only negative exponents is a k-algebra which is an additive complement of the valuation ring $k[[C_v]]$ of $k((C_v))$. It follows that

$$R := \bigcup_{v < \lambda} \operatorname{Neg} k((C_v))$$

is a k-algebra which is an additive complement of the valuation ring $\bigcup_{v < \lambda} k[[C_v]]$ of L. We wish to show that its quotient field is L. Take any element $a \in L$. Since L is the union of the $k((C_v))$, there is some v such that $a \in k((C_v))$. Pick some negative $\alpha \in C_{v+1} \setminus C_v$. Then $\alpha < C_v$. Denote by t^α the monic monomial of value α in $k((C_{v+1}))$. Then at^α has only negative exponents, so t^α and at^α are both elements of $\operatorname{Neg} k((C_{v+1}))$. Therefore, $a \in \operatorname{Quot} \operatorname{Neg}(k((C_{v+1})) \subseteq \operatorname{Quot} R$.

Now take any non-zero cardinal number $\kappa \leq \lambda$. We modify the construction in the final part of the proof of Proposition 27 in that we start with $K'_v = k((C_v))$, and replace $t_{v,v}$ by $t_{v,v} - t_{v+1,v}$ (or by $t_{v,v} - t_{\lambda'+1,v}$ in the limit case) only as long as $v \leq \kappa$. Then the elements $t_{v,v}$, $v < \kappa$, will be algebraically independent over $K = \bigcup_{v < \lambda} K_v$, we have $\operatorname{trdeg} L|K = \kappa$, and K will be dense in L.

For every $v < \lambda$, w_v induces a valuation preserving isomorphism from $(k((C_v)), v)$ and from (K_v, v) onto $(Lw_v, \overline{w}_v)$. Hence, $\iota_v := (w_v|_{K_v})^{-1} \circ w_v|_{k((C_v))}$ is a valuation preserving isomorphism from $(k((C_v)), v)$ onto (K_v, v). For $v < \mu < \lambda$, ι_μ is an extension of ι_v. Hence, $\iota := \bigcup_{v < \lambda} \iota_v$ is a valuation preserving isomorphism from (L, v) onto (K, v). The image R_κ of R under ι is a k-algebra which is an additive complement of the valuation ring of K and has quotient field K. Consequently, $\operatorname{trdeg} L| \operatorname{Quot} R_\kappa = \kappa$. Since K is dense in L, R_κ is also an additive complement of the valuation ring of L.

If in addition k is an archimedean ordered field and $<$ is any ordering on L compatible with v, then v is the natural valuation of $(L, <)$. Hence by Lemma 31, $R + \mathbb{Z}$ and $R_\kappa + \mathbb{Z}$ are integer parts of $(L, <)$. Since $\operatorname{Quot}(R + \mathbb{Z}) = \operatorname{Quot} R$ and $\operatorname{Quot}(R_\kappa + \mathbb{Z}) = \operatorname{Quot} R_\kappa$, this completes our proof.

§6. **Weak complements.**

LEMMA 35. *Let I be an integer part of the ordered field $(K, <)$. If v denotes the natural valuation of $(K, <)$, then I is a weak complement in (K, v).*

PROOF. Take $0 < x \in K$ and assume that $vx > 0$. Then for all $n \in \mathbb{N}$, also $vnx > 0 = v1$ which by (1) implies that $0 < nx \leq 1$. Consequently, $0 < x < 1$ and thus, $x \notin I$. This proves that $vr \leq 0$ for all $r \in I$.

For every $a \in K$ there is $r \in I$ such that $0 \le a - r < 1$. Again by (1), this implies that $v(a - r) \ge v1 = 0$.   ⊣

In what follows, let R be a weak complement in a valued field (K, v). For every convex subgroup Γ of vK, we define

$$R_\Gamma := \{r \in R \mid vr \in \Gamma \cup \{\infty\}\}.$$

LEMMA 36. *For every convex subgroup Γ of vK, R_Γ is a subring of K. Denote by K_Γ its quotient field. Then $vK_\Gamma = \Gamma$.*

PROOF. Take $r, s \in R_\Gamma$. Then $vr, vs \in \Gamma$. Since $r - s \in R$, we have $0 \ge v(r - s) \ge \min\{vr, vs\}$, showing that $v(r - s) \in \Gamma$ and thus $r - s \in R_\Gamma$. Further, $rs \in R$ and $vrs = vr + vs \in \Gamma$, showing that $rs \in R_\Gamma$. This proves that R_Γ is a subring of K.

Since $vR_\Gamma := \{vr \mid r \in R_\Gamma\} \subseteq \Gamma$, we know that $vK_\Gamma \subseteq \{\alpha - \beta \mid \alpha, \beta \in \Gamma\} = \Gamma$. On the other hand, for every $a \in K$ with $va \in \Gamma^{<0}$ there is some $r \in R$ such that $v(a - r) \ge 0$. It follows that $vr = va \in \Gamma$ and thus $r \in R_\Gamma$ and $va = vr \in vR_\Gamma$. Hence, $\Gamma^{<0} \subseteq vR_\Gamma$, which implies that $vK_\Gamma = \Gamma$.   ⊣

Note that K_Γ is a subfield of the quotient field of R. Since $vK_\Gamma = \Gamma$, we have that $v_\Gamma K_\Gamma = \{0\}$. This means that the residue map associated with v_Γ induces an isomorphism on K_Γ. This is in fact an isomorphism

$$(K_\Gamma, v) \simeq (K_\Gamma v_\Gamma, \bar{v}_\Gamma)$$

of valued fields.

LEMMA 37. *For every non-trivial convex subgroup Γ of vK, the valued residue field $(K_\Gamma v_\Gamma, \bar{v}_\Gamma)$ lies dense in $(Kv_\Gamma, \bar{v}_\Gamma)$.*

PROOF. We have to show: if $a \in K$ such that $va \in \Gamma$, then for every positive $\gamma \in \Gamma$ such that $\gamma > va$ there is some $b \in K_\Gamma$ such that $v(a - b) \ge \gamma$. Since $\Gamma^{<0} \subseteq vR_\Gamma$ by the foregoing lemma, we may pick some $c \in R_\Gamma$ such that $vc = -\gamma$. Then there is some $r \in R$ such that $v(ac - r) \ge 0$. Since $vac = va - \gamma \in \Gamma^{<0}$, we have $vr = vac \in \Gamma^{<0}$ and therefore, $r \in R_\Gamma$. Setting $b = \frac{r}{c} \in K_\Gamma$, we obtain $v(a - b) \ge -vc = \gamma$.   ⊣

Now we give examples for valued fields and ordered fields without weak complements or integer parts.

BASIC CONSTRUCTION. Take an arbitrary field k and t a transcendental element over k. Denote by v_t the t-adic valuation on $k(t)$. Choose some countably generated separable-algebraic extension (k_1, v_t) of $(k(t), v_t)$. Take two algebraically independent elements x, y over $k(t)$. Then by Theorem 1.1 of [K1] there exists a non-trivial valuation w on $K := k(t, x, y)$ whose restriction to $k(t)$ is trivial, whose value group is $\mathbb{Z}$ and whose residue field is k_1; since w is trivial on $k(t)$, we may assume that the residue map associated with w induces the identity on $k(t)$. Now we take the valuation v on the rational function field K to be the composition of w with v_t:

$$v = \dot{w} \circ v_t.$$

EXAMPLE 2. We take k to be one of the prime fields $\mathbb{Q}$ or $\mathbb{F}_p$ for some prime p. We choose k_1 such that $k_1 v_t = k$ and that $v_t k_1 / v_t k(t)$ is infinite. Take Γ to be the convex subgroup of vK such that $v_\Gamma = w$; in fact, Γ is the minimal convex subgroup containing vt.

Suppose K admits a weak complement R. Then by Lemma 37 the isomorphic image $K_\Gamma w$ of the subfield K_Γ of K is dense in the valued residue field (k_1, v_t). It follows from Lemma 12 that $v_t(K_\Gamma w) = v_t k_1$. Note that the isomorphism $K_\Gamma \to K_\Gamma w$ preserves the prime field k of K_Γ. Since $v_t(K_\Gamma w) \neq \{0\}$, it follows from Corollary 11 that $K_\Gamma w$ cannot be algebraic over the trivially valued subfield k. Hence, trdeg $K_\Gamma w | k = 1$, and we take some $t' \in K_\Gamma$ such that $t'w$ is transcendental over k. It follows that $K_\Gamma | k(t')$ is algebraic. As $v_t k_1 / v_t k(t)$ is infinite, Lemma 10 shows that $K_\Gamma w | k(t)$ and hence also $K_\Gamma | k(t')$ must be an infinite extension.

Since trdeg $K | k = 3$, we have that trdeg $K | k(t') = 2$. Let $\{x', y'\}$ be a transcendence basis for this extension. Because the algebraic extension $K_\Gamma | k(t')$ is linearly disjoint from the purely transcendental extension $k(t', x', y') | k(t')$, the extension $K_\Gamma(x', y') | k(t', x', y')$ is infinite. But it is contained in the finite extension $K | k(t', x', y')$. This contradiction shows that K cannot admit weak complements. Note that by construction,

$$Kv = k_1 v_t = k \subset K.$$

EXAMPLE 3. In the foregoing example, take $k = \mathbb{Q}$. By Lemma 8, there is an ordering $<$ on the rational function field $K = k(t, x, y)$ which is compatible with the valuation v. Then $(K, <)$ does not admit an integer part, because any such integer part would be a weak complement for v.

EXAMPLE 4. In Example 2, take $k = \mathbb{Q}$. By Lemma 8 there is an ordering on $k(t)$ compatible with the v_t-adic valuation. The real closure $k(t)^{rc}$ of $k(t)$ with respect to this ordering is a countably generated infinite algebraic extension of $k(t)$. So we may take $k_1 = k(t)^{rc}$. The valuation v_t extends to a valuation of k_1 which is compatible with its ordering. Again by Lemma 8 we may choose a lifting of the ordering of k_1 to K through the valuation w. This ordering on K induces through $v = w \circ v_t$ the same ordering on the residue field k as the ordering on $k(t)^{rc}$ induces through v_t; in particular, we find that the chosen ordering on K is compatible with the valuation v.

Now pick any positive integer n and consider the n-real closure $K^{rc(n)}$ of $(K, <)$ as defined in [Bg]. It is encluded in the real closure K^{rc} of $(K, <)$, so we can extend w to the real closure (cf. Lemma 9) and then restrict it to $K^{rc(n)}$; the valuation so obtained is still compatible with the ordering. As $K^{rc}w = (Kw)^{rc} = k(t)^{rc} = Kw$ by Lemma 9, we have that $K^{rc(n)}w = k(t)^{rc}$.

So $w \circ v_t$ is an extension of v to $K^{\mathrm{rc}(n)}$, and we denote it again by v. As before, we see that it is compatible with the ordering.

Suppose $K^{\mathrm{rc}(n)}$ admits a weak complement R. We proceed as in Example 2, with K replaced by $L := K^{\mathrm{rc}(n)}$. As $L_\Gamma w$ is dense in $(Lw, v_t) = (k(t)^{\mathrm{rc}}, v_t)$ we can infer from Lemma 12 that $v_t(L_\Gamma w) = v_t k(t)^{\mathrm{rc}}$. This in turn is the divisible hull of $v_t k(t) = \mathbb{Z}$ (cf. Lemma 9). Hence, $v_t(L_\Gamma w) = \mathbb{Q}$ and thus also $vL_\Gamma = \mathbb{Q}$. But as in Example 2 one shows that the relative algebraic closure of $k(t')$ in $K(t')$ must be a finite extension E of $k(t')$. Now L_Γ lies in the relative algebraic closure E' of E in L, which is just the n-real closure of E. But the value group of the n-real closure of E is the n-divisible hull of vE, which in turn is a finite extension of $vk(t') = \mathbb{Z}$. So the value group of E' is still isomorphic to the n-divisible hull of $\mathbb{Z}$. This contradicts the fact that its subfield L_Γ has value group $\mathbb{Q}$. This contradiction proves that $K^{\mathrm{rc}(n)}$ does not admit weak complements for its compatible valuation v, and therefore does not admit integer parts.

In order to obtain an example where the valued field (K, v) admits an embedded residue field *and* a cross-section, we modify Example 2 as follows.

EXAMPLE 5. In our basic construction, we take $k = k_0(z)$ where k_0 is any prime field and z is transcendental over k_0. The henselization $k_0(z)^h$ of $k_0(z)$ with respect to the z-adic valuation v_z is a countably generated separable-algebraic extension of $k_0(z)$. Therefore, we may choose k_1 to be a countably generated separable-algebraic extension of $k(t)$ such that $v_t k_1 = \mathbb{Z}$ and $k_1 v_t = k_0(z)^h$ (cf. Theorem 2.14 of [K1]). Then we take

$$v' = v \circ v_z = w \circ v_t \circ v_z.$$

Let Γ be the convex subgroup of $v'K$ such that $v'_\Gamma = w$; now Γ is the minimal convex subgroup containing $v't$. Suppose that (K, v') admits a weak complement R. Then by Lemma 37 the isomorphic image $K_\Gamma w$ of the subfield K_Γ of K is dense in the valued residue field $(k_1, v_t \circ v_z)$. The isomorphism $K_\Gamma \rightarrow K_\Gamma w$ preserves the prime field k_0 of K_Γ. From Lemma 12 we infer that $v_t \circ v_z(K_\Gamma w) = v_t \circ v_z(k_1)$. This value group has two non-trivial convex subgroups, namely, itself and the smallest convex subgroup which contains $v_t \circ v_z(z)$. We choose elements $t', z' \in K_\Gamma$ such that $v_t \circ v_z(t'w) > 0$ lies in the former, but not in the latter, and $v_t \circ v_z(z'w) > 0$ lies in the latter. Then these two values are rationally independent. Thus by Theorem 1 of [Br], Chapter VI, §10.3, $t'w$, $z'w$ are algebraically independent over the trivially valued field k_0. But as $K_\Gamma w \subseteq k_1$, we must have trdeg $K_\Gamma w|k_0 = 2$. Hence $K_\Gamma w|k_0(t'w, z'w)$ is algebraic, and so is $K_\Gamma|k_0(t', z')$.

By Lemma 16, $K_\Gamma w$ is also dense in (k_1, v_t). Hence $(K_\Gamma w)v_t = k_1 v_t = k_0(z)^h$ by Lemma 12. Hence, $z \in (K_\Gamma w)v_t$ and we can in fact choose z' such that $(z'w)v_t = z$. Consequently, $k_0(t'w, z'w)v_t = k_0(z)$ (cf. the already cited Theorem 1 of [Br]). Since $(K_\Gamma w)v_t = k_0(z)^h$ is an infinite extension of $k_0(z)$

by part a) of Proposition 21, it follows from Lemma 10 that $K_\Gamma w$ is an infinite extension of $k_0(t'w, z'w)$. Thus, K_Γ is an infinite extension of $k_0(z', t')$.

Since $\operatorname{trdeg} K|k_0 = 4$ and $\operatorname{trdeg} k_0(z', t')|k_0 = \operatorname{trdeg} k_0(z'w, t'w)|k_0 = 2$, we have that $\operatorname{trdeg} K|k_0(z', t') = 2$. Let $\{x', y'\}$ be a transcendence basis for this extension. Because the algebraic extension $K_\Gamma|k_0(z', t')$ is linearly disjoint from the purely transcendental extension $k_0(z', t', x', y')|k_0(z', t')$, the extension $K_\Gamma(x', y')|k_0(z', t', x', y')$ is infinite. But it is contained in the finite extension $K|k_0(z', t', x', y')$. This contradiction shows that (K, v') cannot admit weak complements.

The value group $v'K$ is the lexicographic product $wK \times v_t k_1 \times v_z k_0(z) \simeq \mathbb{Z} \times \mathbb{Z} \times \mathbb{Z}$ since $v_z k_0(z) = v_z k_0(z)^h = \mathbb{Z}$. This shows that (K, v') admits a cross-section. The residue field $Kv' = k_0$ is embedded in K. Note that K is a rational function field of transcendence degree 4 over its residue field.

EXAMPLE 6. In the foregoing example, take $k_0 = \mathbb{Q}$. By Lemma 8, there is an ordering $<$ on the rational function field $K = k(t, x, y, z)$ which is compatible with the valuation v'. Then $(K, <)$ does not admit an integer part. Nevertheless, the valuation v', which is the natural valuation of the ordering $<$ since $Kv' = \mathbb{Q}$ is archimedean ordered, admits an embedding of its residue field and a cross-section.

Finally, let us note that Proposition 21 shows:

PROPOSITION 38. *None of the valued fields in the above examples are henselian. Also, the natural valuation of the example constructed by Boughattas in* [Bg] *is not henselian.*

PROOF. It follows from part a) of Proposition 21 that the rational function fields of Examples 2 and 3 are not henselian. The fields of Example 4 and Boughattas' example are n-real closures of algebraic function fields. The "n-algebraic closures" of part b) of Proposition 21 are algebraic extensions of the n-real closures. Since they are not henselian, Lemma 15 shows that the same holds for the n-real closures.   ⊣

REFERENCES

[B-K-K] DARKO BILJAKOVIC, MIKHAIL KOCHETOV, and SALMA KUHLMANN, *Primes and irreducibles in truncation integer parts of real closed fields*, **Logic in Tehran** (A. Enayat, I. Kalantari, and M. Moniri, editors), Lecture Notes in Logic, vol. 26, ASL and AK Peters, 2006, this volume, pp. 42–64.

[Bg] S. BOUGHATTAS, *Résultats optimaux sur l'existence d'une partie entière dans les corps ordonnés*, **The Journal of Symbolic Logic**, vol. 58 (1993), no. 1, pp. 326–333.

[Br] N. BOURBAKI, *Commutative Algebra*, Hermann, Paris, 1972.

[E] O. ENDLER, *Valuation Theory*, Springer, New York, 1972.

[F] A. FORNASIERO, *Embedding henselian fields in power series*, preprint.

[Fu] L. FUCHS, *Partially Ordered Algebraic Systems*, Pergamon Press, Oxford, 1963.

226   FRANZ-VIKTOR KUHLMANN

[K1] F.-V. KUHLMANN, *Value groups, residue fields, and bad places of rational function fields*, *Transactions of the American Mathematical Society*, vol. 356 (2004), no. 11, pp. 4559–4600.

[K2] ———, *Book on Valuation Theory (in preparation)*, Preliminary versions of several chapters available at: http://math.usask.ca/~fvk/Fvkbook.htm.

[KS] S. KUHLMANN, *Ordered Exponential Fields*, Fields Institute Monographs, vol. 12, American Mathematical Society, Providence, RI, 2000.

[M-S] S. MACLANE and O. F. G. SCHILLING, *Zero-dimensional branches of rank one on algebraic varieties*, *Annals of Mathematics. Second Series*, vol. 40 (1939), pp. 507–520.

[N] J. NEUKIRCH, *Algebraic Number Theory*, Springer, Berlin, 1999.

[P] A. PRESTEL, *Lectures on Formally Real Fields*, Lecture Notes in Mathematics, vol. 1093, Springer, Berlin, 1984.

[R] P. RIBENBOIM, *Théorie des valuations*, Les Presses de l'Université de Montréal, Montreal, Que., 1968.

[W] S. WARNER, *Topological Fields*, North-Holland Mathematics Studies, vol. 157, North-Holland, Amsterdam, 1989.

[Z-S] O. ZARISKI and P. SAMUEL, *Commutative Algebra. Vol. II*, D. Van Nostrand, New York-Heidelberg-Berlin, 1960.

MATHEMATICAL SCIENCES GROUP
UNIVERSITY OF SASKATCHEWAN
106 WIGGINS ROAD, SASKATOON
SASKATCHEWAN, S7N 5E6, CANADA
E-mail: fvk@math.usask.ca

A RECURSIVE NONSTANDARD MODEL FOR OPEN INDUCTION WITH GCD PROPERTY AND COFINAL PRIMES

SHAHRAM MOHSENIPOUR

Abstract. Berarducci and Otero [1] have constructed a recursive nonstandard model for normal open induction with cofinal primes. We modify their method to construct a recursive nonstandard model for open induction with cofinal primes in which the GCD property also holds.

§1. **Introduction.** Tennenbaum [7] showed that true arithmetic does not have a recursive nonstandard model. This result has been subjected to a number of refinements, culminating in the work of Wilmers [10] where it is shown that IE_1 does not have a recursive nonstandard model (IE_1 is the fragment based on the induction scheme for bounded existential formulas). On the other hand, Shepherdson [5] constructed a recursive model for the fragment *Iopen* of arithmetic based on the induction scheme for quantifier-free formulas. So we can say that from a logical point of view arithmetic can be divided into two areas which we would like to call the *Shepherdson area* and the *Tennenbaum area*. Our work here can be considered as an attempt to reach a better understanding of the extent of the Shepherdson area.

Shepherdson provided an algebraic characterization for models of Iopen as follows: M is a model of open induction iff M is an integer part of its real closure. By using this characterization and Puisseux power series, Shepherdson constructed a recursive nonstandard model of Iopen, a model which also exhibits several independence results such as:

$$\text{Iopen} \nvdash \text{irrationality of } \sqrt{2},$$

$$\text{Iopen} \nvdash \text{cofinality of primes, and}$$

$$\text{Iopen} \nvdash \text{Fermat's last theorem.}$$

This shows that open induction is too weak to prove many true statements of number theory. For this reason a number of algebraic first order properties

2000 *Mathematics Subject Classification.* 03H15, 03D80, 03C57.

Key words and phrases. open induction, recursive nonstandard model, GCD property.

This research was in part supported by a grant from IPM (No. 83030321)

have been suggested to be added to Iopen in order to obtain closer systems to number theory. These properties include: Normality [8], having the GCD property [6], being a Bezout domain [2], cofinality of primes (abbreviated here as cof(prime)) and so on. We mention that GCD is stronger than normality, and Bezout is stronger than GCD. In 1996 Berarducci and Otero [1] constructed a recursive nonstandard model for Iopen + Normality + cof(prime). It is easy to verify that their model doesn't satisfy the GCD property. In this paper we will show that while the Berarducci-Otero model lacks the GCD property, their method of construction is potentially capable of producing a GCD model, provided that we combine it with Smith's theorems [6]. In other words, we can go one step further and establish:

THEOREM 1.1. *There is a recursive nonstandard model of open induction with the GCD property and cofinal primes.*

Our method is the same as Berarducci-Otero's [1] with the exception that the homomorphisms used in Wilkie and $\widehat{\mathbb{Z}}$-constructions are parsimonious (see Definition 2.9). These maps were used by Smith [6] in order to extend Wilkie and $\widehat{\mathbb{Z}}$-constructions to GCD and Bezout domains. Our strategy in presenting this paper is that we move very closely to Berarducci-Otero's paper (in terminology and notation) and bring every theorem of Smith into the context whenever needed. This makes the paper entirely self-contained.

§2. **Preliminaries.** We follow the notation of Berarducci-Otero [1]. Let L be the language of ordered rings based on the symbols $+$, $-$, $\cdot$, 0, 1, $\leq$. We write $\mathbb{N}^*$ for $\mathbb{N} \setminus \{0\}$. We will work with the following set of axioms in L:

DOR: discretely ordered rings, i.e., axioms for ordered rings and

$$\forall x \neg (0 < x < 1).$$

ZR: discretely ordered $\mathbb{Z}$-rings, i.e., DOR and for every $n \in \mathbb{N}^*$

$$\forall x \exists q, r \left(x = nq + r \bigwedge 0 \leqslant r < n \right).$$

Iopen: open induction, i.e., DOR and for every open L-formula $\psi(\overrightarrow{x}, y)$

$$\forall \overrightarrow{x} \left(\psi(\overrightarrow{x}, 0) \bigwedge \forall y \geqslant 0 \, (\psi(\overrightarrow{x}, y) \to \psi(\overrightarrow{x}, y + 1)) \to \forall y \geqslant 0 \psi(\overrightarrow{x}, y) \right).$$

GCD: the existence of greatest common divisor, i.e.,

$$\forall x, y \left((x = y = 0) \bigvee \exists z \left(z|x \bigwedge z|y \bigwedge \left(\forall t \left((t|x \bigwedge t|y) \to t|z \right) \right) \right) \right),$$

where $x|y$ is an abbreviation for $\exists t \, (t \cdot x = y)$.

Let M be an ordered domain, then $\mathbb{Q}M$ will denote the $\mathbb{Q}$-algebra generated by M and $RC(M)$ will be the real closure of its fraction field. Given two ordered domains $I \subset K$ we say that I is an integer part of K if I is discrete and every element of K could be approximated by an element of I at a finite distance. If M is an ordered field, $AC(M)$ will denote $RC(M)[\sqrt{-1}]$.

The following algebraic characterization of Shepherdson is the starting point of all constructions of various models of Iopen:

THEOREM 2.1 (Shepherdson [5]). *Let M be an ordered domain. M is a model of Iopen iff M is an integer part of $RC(M)$.*

We also need some facts from Puisseux series:

DEFINITION 2.2. Let K be a field. The following is the field of Puisseux series in descending powers of x with coefficients in K:

$$K((x^{1/N})) = \left\{ \sum_{k \leq m} a_k x^{k/q} : m \in \mathbb{Z}, q \in \mathbb{N}^*, a_k \in K \right\}.$$

THEOREM 2.3. *If K is real (resp. algebraically) closed field then $K((x^{1/N}))$ is real (resp. algebraically) closed field.*

PROOF. See [9]. ⊣

Let $M(x)$ be an ordered field with $x > a$ for all $a \in M$, then the real closure of $M(x)$ is the subfield of $RC(M)((x^{1/N}))$ of those series which are real algebraic over $M(x)$.

The following subrings of $K((x^{1/N}))$ will play crucial roles in our constructions:

$$K[x^{\mathbb{Q}}] = \left\{ a_l x^{l/q} + \cdots + a_1 x^{1/q} + a_0 : q \neq 0 \text{ and } a_0, \ldots, a_l \in K \right\}$$

$$K[x^{\mathbb{Q}}]^* = \left\{ a_l x^{l/q} + \cdots + a_1 x^{1/q} : q \neq 0 \text{ and } a_1, \ldots, a_l \in K \right\}$$

$$K[x^{\mathbb{Q}}]^* + \mathbb{Z} = \left\{ a + m : a \in K[x^{\mathbb{Q}}]^*, m \in \mathbb{Z} \right\}.$$

Now we can describe the Shepherdson model.

THEOREM 2.4 (Shepherdson [5]). *$RC(\mathbb{Q})[x^{\mathbb{Q}}]^* + \mathbb{Z}$ is a recursive nonstandard model of Iopen.*

As an interesting and easy exercise the reader can deduce the three independence results mentioned in the introduction.

LEMMA 2.5. *Let F be an ordered field. Let $M \subset F$, M a model of ZR and $\beta \in F$. The following are equivalent:*
(1) *β is at infinite distance from M;*
(2) *β is at infinite distance from $\mathbb{Q}M$.*

PROOF. Easy, (See [1, Lemma 2.4. p. 1231]). ⊣

LEMMA 2.6. *Let F be an ordered field contained in $\mathbb{R}$. Let M be a discretely ordered ring. Then $F \cap \mathbb{Q}M = \mathbb{Q}$.*

PROOF. Easy, (See [1, Lemma 2.5. p. 1231]). ⊣

The next two lemmas are due to Wilkie (See [9] and [8]):

LEMMA 2.7. *Let F be an ordered field. Let $M \subset F$ be such that M is a model of ZR. Let $\beta \in F$. If β is not infinitesimally close to $AC(M)$ and β is at infinite distance from M, then $M[\beta]$ is a discretely ordered ring extending M.*

PROOF. See [1, Lemma 2.6. p. 1231].   ⊣

Let $\widehat{\mathbb{Z}}$ denote the product of the rings $\mathbb{Z}_p$ of p-adic integers.

LEMMA 2.8. *Given a ring homomorphism* $\varphi : M \to \widehat{\mathbb{Z}}$ *(M a domain), there is a unique* $\mathbb{Z}$-ring M_φ *with* $M \subset M_\varphi \subset \mathbb{Q}M$ *such that* φ *extends to* $\varphi' : M \to \widehat{\mathbb{Z}}$. *Moreover* $M_\varphi = \{a/k : a \in M, \ k \in \mathbb{Z}^*, \ k | \varphi(a)\}$.

PROOF. See [1, Lemma 2.7. p. 1231].   ⊣

Based on the above lemmas, we shall adopt the following conventions for the rest of the paper: If the ring $M[\beta]$ is built from M via Lemma 2.7, then we say $M[\beta]$ is built by a *Wilkie-construction*, and if the ring M_φ is constructed from M via Lemma 2.8, then we say that M_φ is built by a $\widehat{\mathbb{Z}}$-*construction*. The above two lemmas allow us to form a strictly increasing chain by alternating the applications of Wilkie and $\widehat{\mathbb{Z}}$-constructions with the goal of reaching an integer part of a ground field. Moreover, we intend to achieve the following goals:

1. The model must have the GCD property (by considering the fact that in general GCD property isn't preserved in chains).
2. The model must have cofinal primes.
3. The model must be recursive.

We will show how to achieve these goals by blending the methods of Smith [6] with those of Berarducci-Otero [1]. So far we have been following the strategy of Berarducci-Otero, but now we wish to employ some results of Smith [6]. First we fix some notation from p-adic numbers. We denote each p-adic integer $a \in \mathbb{Z}_p$ by $a = (a_1, a_2, \ldots)$ such that $a_i \in \mathbb{Z}$ and $a_{i+1} \equiv a_i \pmod{p^i}$ (See [4]). Let M be a domain, for every homomorphism $\varphi : M \to \widehat{\mathbb{Z}}$, we denote its p-component by $\varphi^p : M \to \mathbb{Z}_p$ and for every $a \in M$ we set $\varphi^p(a) = (a_1^p, a_2^p, \ldots)$ and $\varphi_i^p(a) = a_i^p$.

The following definition is due to Smith [6, p. 196] but is expressed here in a slightly different form:

DEFINITION 2.9. Let M be a discretely ordered ring with $\varphi : M \to \widehat{\mathbb{Z}}$ a homomorphism. We say that φ is *parsimonious* if for each $a \neq 0$ there are only finitely many pairs (n, p) with $\varphi_n^p(a) = 0$.

The following two theorems say that in the presence of having parsimonious map we can extend the Wilkie-construction to GCD models:

THEOREM 2.10 (Smith [6, Lemma 5.1, p. 201]). *If* $\varphi : M \to \widehat{\mathbb{Z}}$ *is parsimonious, then the extension of* φ *to* M_φ *(as defined in Lemma 2.8) is parsimonious.*

THEOREM 2.11 (Smith [6, Theorem 5.3, p. 202]). *Let* M *be a discretely ordered ring with the GCD property, and let* $\varphi : M \to \widehat{\mathbb{Z}}$ *be a parsimonious map. Then* M_φ *also has the GCD property.*

The next two theorems make $\widehat{\mathbb{Z}}$-constructions available for countable GCD models:

THEOREM 2.12 (Smith [6, Theorem 6.8, p. 206]). *Let M be a GCD domain and suppose x is transcendental over M. Then $M[x]$ is a GCD domain.*

THEOREM 2.13 (Smith [6, Theorem 6.12, p. 207]). *Let M be a countable $\mathbb{Z}$-ring and suppose the remainder homomorphism $\varphi : M \to \widehat{\mathbb{Z}}$ (i.e., $\varphi_n^p(a) :=$ the remainder by dividing over p^n) is parsimonious. Let x be transcendental over M and suppose $M[x]$ is discretely ordered (and this ordering restricts to the original ordering on M). Then φ can be extended to a parsimonious $\varphi' : M[x] \to \widehat{\mathbb{Z}}$ such that $\varphi'(x)$ is a unit of $\widehat{\mathbb{Z}}$ (for the last assertion see the last sentence in the proof of this theorem).*

The next theorem of Smith guarantees the preservation of the GCD property in chains constructed by alternative applications of Wilkie- and $\widehat{\mathbb{Z}}$-constructions via parsimonious maps. We express the theorem in a more restricted form which is adequate for us:

THEOREM 2.14 (Smith [6, Theorem 9.4, p. 226]). *Suppose M_0 is a countable $\mathbb{Z}$-ring with the GCD property and there is a parsimonious $\varphi : M_0 \to \widehat{\mathbb{Z}}$. Let $\{M_i : i \in \mathbb{N}\}$ be the chain of discretely ordered domains such that M_{2i+1} is constructed from M_{2i} by the Wilkie-construction, and M_{2i+2} is constructed from M_{2i+1} by the $\widehat{\mathbb{Z}}$-construction. Then $M = \bigcup_{i \in \mathbb{N}} M_i$ is a discretely ordered domain with the GCD property.*

In the last theorem of this section we see the possibility of having cofinal primes:

THEOREM 2.15 (Macintyre-Marker [2, Lemma 3.3, p. 64]). *If $\varphi : M \to \widehat{\mathbb{Z}}$ is a homomorphism (M a domain), $p \in M$ is prime and $\varphi(p)$ is a unit of $\widehat{\mathbb{Z}}$, then p is prime in M_φ.*

§3. Essential facts from applied recursion theory.

We borrow some definitions and facts from Berarducci-Otero [1]. An ordered ring $R = (R, +, \cdot, <)$ is *recursive* if there is an algorithm which, on inputs $x, y \in R$, computes $x + y$, $x \cdot y$ and decides whether $x \geq y$ or not. Obviously when we talk about an input x, we mean its code.

DEFINITION 3.1. (1) A real number $\gamma \in \mathbb{R}$ is *recursive* if there is an algorithm which, on input $n \in \mathbb{N}$, computes the first n digits of the (non-terminating) decimal expansion of γ.

(2) A sequence of real numbers $\{\pi_i : i \in \mathbb{N}^*\}$ is a *recursive family of recursive real numbers* if there is an algorithm which on input $i \in \mathbb{N}^*$ and $n \in \mathbb{N}$ computes the first n digits of the decimal expansion of π_i.

We fix such a sequence $\{\pi_i : i \in \mathbb{N}^*\}$ for the rest of the paper and define L as the ordered ring $\mathbb{Q}(\{\pi_i : i \in \mathbb{N}^*\})$ in which order is induced by that of $\mathbb{R}$.

LEMMA 3.2 ([1, Lemma 3.2, p. 1232]). *L is a recursive ordered field.*

The following is a classic fact from recursive algebra:

LEMMA 3.3 ([1, Lemma 3.3, p. 1233]). *The real closure K of L is a recursive ordered field.*

For the remainder of this section, we state a recursive version of Theorem 2.13 because we shall need it for keeping track of recursiveness of our model in the process of construction described in the next section. Initially we must define recursiveness for p-adic integers:

DEFINITION 3.4. A p-adic integer $a = (a_1, a_2, \dots)$ is *recursive* if there exists an algorithm which on input $n \in \mathbb{N}^*$, computes a_n.

THEOREM 3.5. *Let M be recursive $\mathbb{Z}$-ring and suppose that the remainder homomorphism $\varphi : M \to \widehat{\mathbb{Z}}$ is parsimonious such that for each $a \in M$ and each prime p, $\varphi^p(a)$ is a recursive p-adic element of $\mathbb{Z}_p$. Let x be transcendental over M and suppose that $M[x]$ is a recursive discretely ordered ring. Then φ can be extended to a parsimonious map $\varphi' : M[x] \to \widehat{\mathbb{Z}}$ such that for every $a \in M[x]$ and every prime p, $\varphi'^p(a)$ is a recursive p-adic element of $\mathbb{Z}_p$ and $\varphi'(x)$ is a unit of $\widehat{\mathbb{Z}}$.*

PROOF. We plan to slightly modify Smith's proof [6, Theorem 6.12, p. 207] in order to make it recursive. At first let us fix some notation. Since φ is the remainder map which is also parsimonious and since for every $a \in M$ and for each prime p, $\varphi^p(a)$ is recursive, then any nonzero element $f \in M[x]$ can be written in the form $f(x) = p^e g(x)$ where $e \in \mathbb{N}$, $p \nmid g(x)$ and $g(x)$ can be deduced recursively. We denote $g(x)$ by $\text{Red}_p f(x)$ and denote e by $\exp(f(x), p)$. We express a *recursive arrangement* of all nonconstant elements of $M[x]$ by $(f_i(x), i \in \omega)$. Take $f_0(x) = x$ and suppose $\deg f_i(x) = d_i$. Clearly it is enough to define $\varphi_n^p(x)$ recursively for each $n \in \mathbb{N}^*$ and each prime p. We will describe an alternating ω-stage construction in which at odd stages we recursively define $\varphi_1^p(x)$ for finitely many primes p and at even stages we recursively define $\varphi_n^p(x)$ for some of the primes for which we had defined $\varphi_1^p(x)$ at earlier stages. More precisely:

(1) At stage $2k + 1$, we define $\varphi_1^{p_j}(x)$ for each prime p_j such that

$$d_0 + \cdots + d_k < p_j \leq d_0 + \cdots + d_{k+1}$$

in such a way that

$$\varphi_{\exp(f_i(x), p_j)+1}^{p_j}(f_i(x)) \not\equiv 0 \ (\text{mod } p_j^{\exp(f_i(x), p_j)+1})$$

for each of these p_j's and for each i, $0 \leq i \leq k$.

(2) At stage $2k + 2$, we consider polynomial $f_k(x)$. For each prime p_j such that

$$p_j \leq d_0 + \cdots + d_k$$

we find an exponent n_{jk} such that we can define $\varphi_{n_{jk}}^{p_j}(x)$ so that

$$\varphi_{n_{jk}}^{p_j}(f_k(x)) \not\equiv 0 \ (\mathrm{mod} \ p_j^{n_{jk}}).$$

Of course we must do this in a consistent way relative to earlier stages.

Now we give the details. Suppose we have done stages $0, 1, \ldots, 2k$. So we continue as follows:

Stage $2k + 1$. Suppose p_j is any prime such that $d_0 + d_1 + \cdots + d_k < p_j \le d_0 + d_1 + \cdots + d_{k+1}$. For each i, $0 \le i \le k$, let $g_i(x) = \mathrm{Red}_{p_j} f_i(x)$ and suppose $\bar{g}_i(x)$ denotes the polynomial obtained from $g_i(x)$ by reducing coefficients module p_j. Then $\bar{g}_i(x)$ is nonzero in $\mathbb{Z}/p_j\mathbb{Z}[x]$ because of the definition of $g_i(x)$. Since $\deg \bar{g}_i \le \deg f_i$ for each i, $0 \le i \le k$, so $\bar{g}_i$ has at most d_i roots in $\mathbb{Z}/p_j\mathbb{Z}$. Then the total number of roots of all the polynomials $\bar{g}_i$ is at most $d_0 + d_1 + \cdots + d_k < p_j$; hence there is at least one $l \in \mathbb{Z}/p_j\mathbb{Z}$ which is not a root of any of them and can be found *recursively*. Note that l is nonzero because $f_0(x) = x$.

Let $\varphi_1^{p_j}(x) = l$. Then $\varphi_1^{p_j}(g_i(x)) = \bar{g}_i(\varphi_1^{p_j}(x)) = \bar{g}_i(l) \ne 0$. So we have

$$\varphi_{e+1}^{p_j}(f_i(x)) = \varphi_{e+1}^{p_j}(p_j^e \cdot g_i(x)) = p_j^e \cdot \varphi_1^{p_j}(g_i(x))$$

where $e = \exp(f_i(x), p_j)$. Hence

$$\varphi_{e+1}^{p_j}(f_i(x)) \not\equiv 0 \ (\mathrm{mod} \ p_j^{e+1}).$$

Stage $2k+2$. Let p_j be any prime such that $p_j \le d_0 + d_1 + \cdots + d_k$. Consider the polynomial $f_k(x)$. Let $l = \varphi_{n_{j,k-1}}^{p_j}(x)$ such that $0 \le l < p_j^{n_{j,k-1}}$. Clearly $\mathrm{Red}_{p_j} f_k = g_k$ has at most d_k roots in M. Let $N \in \mathbb{N}$ be sufficiently large so that $p_j^N > d_k$. In $\mathbb{Z}/p_j^{(n_{j,k-1}+N)}\mathbb{Z}$ there are p_j^N elements which are congruent to l modulo $p_j^{n_{j,k-1}}$, so at least one of them which can be found *recursively* (call it l_1), is not a root of f_k in M. Note that $p_j \nmid l_1$. Since $f_k(l_1) \ne 0$ and $\varphi : M \to \widehat{\mathbb{Z}}$ is parsimonious, there is an $m \in \mathbb{N}$ such that $p_j^m \nmid f_k(l_1)$ in M. Let $n_{jk} = \max(n_{j,k-1} + N, M)$. Then $n_{jk} \ge n_{j,k-1} + N > n_{j,k-1}$. Defining $\varphi_{n_{jk}}^{p_j}(x) = l_1$, we therefore have

$$\varphi_{n_{jk}}^{p_j}(f_k(x)) = \bar{f}_k(l_1) \not\equiv 0 \ (\mathrm{mod} \ p_j^{n_{jk}}).$$

This completes stage $2k + 2$.

It is easily seen that for every prime $p_j \in \mathbb{N}$ and every $n \in \mathbb{N}^*$, $\varphi_n^{p_j}(x)$ is defined and $\varphi_n^{p_j}(x) \ne 0$. It immediately follows that $\varphi(x) \in U(\widehat{\mathbb{Z}})$, the group of units of $\widehat{\mathbb{Z}}$. $\dashv$

§4. **Construction.** Let K be the recursive field defined in the previous section. Let $K_n = RC(\mathbb{Q}(\pi_1, \ldots, \pi_n))$, thus $K = \bigcup_{n \in \omega} K_n = RC(\mathbb{Q}\{\pi_i\})$. We intend to construct an integer part M for $K[x^{\mathbb{Q}}]$. Clearly such an M would be automatically an integer part for $K((x^{1/\mathbb{N}}))$ and so would be a model of

Iopen. The model M will be generated by elements of the form $\beta_i = q_i(x) + \pi_i$ with $q_i(x) \in K_{i-1}[x^Q]^*$. For such β's we have the following lemma:

LEMMA 4.1 ([1, Lemma 4.1, p. 1233]). *Let* $\beta_i = q_i(x) + \pi_i$ *with* $q_i(x) \in K_{i-1}[x^Q]^*$ $(i = 1,\ldots,n)$. *Then* $\beta_1,\ldots,\beta_n$ *are algebraically independent over* $\mathbb{Q}$.

First of all we fix a recursive enumeration of $K[x^Q]$. We construct M inductively in the form $M = \bigcup_{n \in \omega} M_i$ as follows:

At each step n, we construct a pair (M_n, φ_n) with the following properties:

1. $M_n \models \text{ZR} + \text{GCD}$.
2. M_n is a recursive ring.
3. M_n contains previous M_i's.
4. $\varphi : M_n \to \widehat{\mathbb{Z}}$ is parsimonious.
5. For each $a \in M_n$ and each prime p, $(\varphi_n)^p(a)$ is a recursive p-adic element of $\mathbb{Z}_p$.
6. φ_n is an extension of previous φ_i's.

Let $M_0 = \mathbb{Z}$ and $\varphi_0 : \mathbb{Z} \to \widehat{\mathbb{Z}}$ be the remainder map. Clearly (M_0, φ_0) has all the above properties.

Suppose (M_n, φ_n) has been produced. We construct (M_{n+1}, φ_{n+1}). We do this in four steps:

Step 1. Suppose $q_{n+1}(x)$ is the first element in the enumeration of $K[x^Q]^*$ which is in $K_n[x^Q]$ and it is at infinite distance from M_n. The existence of such $q_{n+1}(x)$ is clear since $M_n \subset L[x^Q]$ for some finite algebraic extension L of $K_{n-1}[\pi_n]$. Therefore it is sufficient to get $q_{n+1}(x) \in K_n[x^Q] \setminus L[x^Q]$. In this step for keeping track of recursiveness we must be able to decide whether a given $p(x) \in K_n[x^Q]$ is at finite distance from M_n. We shall prove this in the last part of this section so for the time being we accept it and continue:

Step 2. We put $\beta_{n+1} = q_{n+1}(x) + \pi_{n+1}$. At first we show that $M_n[\beta_{n+1}] \models$ DOR. Since $M_n \models \text{ZR}$ and the distance from β_{n+1} to M_n is infinite, by Lemma 2.7 it suffices to prove that β_{n+1} is not infinitely close to $AC(M_n)$. For this observe that if $\alpha \in AC(M_n)$, $\alpha = \Sigma_{i<k} c_i x^{i/q}$ ($c_i \in AC(K_n)$) say, and $|\alpha - \beta_{n+1}| < m^{-1}$ for each $m \in \mathbb{N}^*$, then we must have $c_0 - \pi_{n+1} = 0$ which is impossible since π_{n+1} is transcendental over K_n. Therefore $M_n[\beta_{n+1}] \models$ DOR. Also since β_{n+1} is transcendental over the quotient field of M_n, $M_n[\beta_{n+1}]$ has the GCD property by Theorem 2.12. Moreover by considering the accepted fact from step 1. we deduce that $M_n[\beta_{n+1}]$ is recursive ring.

Step 3. Now we are in the position to apply Theorem 3.5. Hence we can extend φ_n to a parsimonious map $\varphi'_n : M_n[\beta_{n+1}] \to \widehat{\mathbb{Z}}$ such that for each $a \in M_n[\beta_{n+1}]$ and each prime p, $(\varphi'_n)^p(a)$ is a recursive p-adic element of $\mathbb{Z}_p$ and $\varphi'_n(\beta_{n+1})$ is a unit of $\widehat{\mathbb{Z}}$.

Step 4. By Theorem 2.10, φ'_n is extendable to parsimonious map $\varphi''_n : (M_n[\beta_{n+1}])_{\varphi'_n} \to \widehat{\mathbb{Z}}$. Also by Lemma 2.8 and Theorem 2.11, we have $(M_n[\beta_{n+1}])_{\varphi'_n} \models \text{ZR} + \text{GCD}$. We will show in the last part of this section

$(M_n[\beta_{n+1}])_{\varphi'_n}$ is a recursive ring and for every $a \in (M_n[\beta_{n+1}])_{\varphi'_n}$ and each prime p, $(\varphi''_n)^p(a)$ is a recursive p-adic element. Then we put $M_{n+1} = (M_n[\beta_{n+1}])_{\varphi'_n}$, $\varphi_{n+1} = \varphi''_n$ and $M = \bigcup_{i<\omega} M_i$.

Our construction is complete and clearly (M_{n+1}, φ_{n+1}) satisfies conditions 1 through 6.

THEOREM 4.2. $M \models Iopen + GCD$.

PROOF. Since DOR is $\forall\exists$-theory then it is preserved in chains and consequently $M \models$ DOR. By construction M is an integer part of $K[x^Q]$ so by Shepherdson's theorem $M \models$ Iopen. In order to show $M \models$ GCD, it is sufficient to observe that the conditions of Theorem 2.14 hold for the chain $(N_i)_{i\in\omega}$ when $N_{2i} = M_i$ and $N_{2i+1} = M_i[\beta_{i+1}]$. Thus $M = \bigcup_{i<\omega} M_i = \bigcup_{i<\omega} N_i$ is a model of GCD.   ⊣

THEOREM 4.3. The set of primes of M is cofinal in M.

PROOF. Since β_n is prime in $M_{n-1}[\beta_n]$ by Theorems 2.15 and 3.5 it is also prime in $(M_{n-1}[\beta_n])_{\varphi'_{n-1}} = M_n$. Suppose β_n is prime in $M_k (k > n)$. Then β_n is prime in $M_k[\beta_{k+1}]$ (since β_{k+1} is transcendental over the quotient field of M_k), hence is prime in $M_{k+1} = M_k[\beta_{k+1}]_{\varphi'_k}$ (by Lemma 2.8). So β_n is prime in all M_k's with $k > n$, thus is prime in M.

Since $K[x^Q]$ is cofinal in $K((X^{1/N}))$ clearly $\{\beta_k; k \in \mathbb{N}^*\}$ is a cofinal subset of M.   ⊣

It remains to prove the recursiveness of M, but before that we mention a basic theorem about p-adics.

THEOREM 4.4 ([4, Chapter 2, Proposition 2]).   (a) An element u of $\mathbb{Z}_p$ is a unit if and only if u is not divisible by p.

(b) Every nonzero element of $\mathbb{Z}_p$ can be written uniquely in the form $p^k u$ such that u is unit of $\mathbb{Z}_p$ and $n \geq 0$.

THEOREM 4.5. The model M is recursive.

PROOF. Since we are working inside the recursive ring $K[x^Q]$ in order to show M is recursive it suffices to check that elements of M are recursively recognizable. Namely, the set of codes of elements of M in the recursive enumeration of $K[x^Q]$ is a recursive set. This will be accomplished by verifying the following three facts, which have remained unproved from steps 1 and 4:

Fact 1. At each step n, can recursively find $q_{n+1}(x)$.

Fact 2. At each step n, $(M_n[\beta_{n+1}])_{\varphi'_n}$ is a recursive ring or equivalently, in the process of constructing M_{n+1}, we can decide $m|\varphi'_n(a)$, for each $a \in M_n[\beta_{n+1}]$ and each $m \in \mathbb{Z}^*$.

Fact 3. At each step n, for each $a \in (M_n[\beta_{n+1}])_{\varphi'_n}$ and each prime p, $(\varphi''_n)^p(a)$ is a recursive element of $\mathbb{Z}_p$.

Proof of the Fact 1: We show that we can decide whether a given $p(x) \in K_n[x^Q]$ is at finite distance from M_n. By Lemma 2.5 this is equivalent to deciding if such a $p(x)$ is at finite distance from $\mathbb{Q}[\beta_1, \ldots, \beta_n]$. This essentially reduces to

> finding a recursive bound δ on the total $(\beta_1, \ldots, \beta_n)$-degree of f's in $\mathbb{Q}[\beta_1, \ldots, \beta_n]$ which might be at a finite distance from the given $p(x) \in K_n[x^{\mathbb{Q}}]$.

This is exactly what has been accomplished in the last part of [1, Sections 5 and 6].

Proof of the Fact 2: We will decide for given $a \in M_n[\beta_{n+1}]$ and $m \in \mathbb{Z}^*$ whether $a/m \in (M_n[\beta_{n+1}])_{\varphi'_n}$ or, equivalently, $m | \varphi'_n(a)$. For the sake of simplicity in notation we replace φ'_n by ψ. Clearly $m | \psi(a)$ in $\widehat{\mathbb{Z}}$ iff for each prime p, $m | \psi^p(a)$ in $\mathbb{Z}_p$.

If p is a prime with $\gcd(m, p) = 1$, then by Theorem 4.4, m is a unit of $\mathbb{Z}_p$ and surely $m | \psi^p(a)$. So in this case there is nothing to check.

If $p^k | m$ and $p^{k+1} \nmid m$, then again by Theorem 4.4, it is sufficient to check whether $p^k | \psi^p(a)$ and by assuming $\psi^p(a) = (a_1, a_2, \ldots)$ this is equivalent to $a_k \equiv 0 \pmod{p^k}$ which can be recursively checked because of recursiveness of $\psi^p(a) = (a_1, a_2, \ldots)$ (step 3).

Since there are only finitely many primes p such that $\gcd(m, p) \neq 1$, then it can be recursively checked whether $m | \psi(a)$ in $\widehat{\mathbb{Z}}$, so $(M_n[\beta_{n+1}])_{\varphi'_n}$ is a recursive ring.

Proof of the Fact 3: For given $m \in \mathbb{Z}^*$ and $a \in (M_n[\beta_{n+1}])_{\varphi'_n}$, suppose we have confirmed recursively that $m | \varphi'_n(a)$, so $a/m \in (M_n[\beta_{n+1}])_{\varphi'_n}$ and $\varphi''_n(a) = \varphi'_n(a)/m$. Clearly for every prime p, $(\varphi'_n)^p(a)/m$ is recursive p-adic, since $(\varphi'_n)^p(a)$ is recursive p-adic.

Now here is our decision procedure:

Suppose $k \in \mathbb{N}$ is given. We will decide whether k is the code of an element of M. First we determine whether k is the code of an element $K[x^{\mathbb{Q}}]$. Let k be the code of a $D \in K[x^{\mathbb{Q}}]$. Put $D = D^* + e$ in which e is the constant term of D. Let k' be the code of D^*. We can recursively find m such that $D^* \in K_m[x^{\mathbb{Q}}]$ but $D^* \notin K_{m-1}[x^{\mathbb{Q}}]$. Since by construction, the codes of $q_i(x)$'s are unbounded in $\mathbb{N}$, there exists the least $n \geq m$ such that if s is the code of $q_n(x)$, then $s \geq k'$, where n and s are found effectively. Now we are faced with two separate cases:

(a) $k' = s$,

(b) $k' < s$.

In the first case we have $q_n(x) = D^*$ and then it just remains to check that if $e - \pi_n \in \mathbb{Z}$.

In the second case we deduce that D^* and subsequently D, are at finite distance from M_{n-1}. Since $M_{n-1} \subset M$ are discretely ordered rings, each element of $M \setminus M_{n-1}$ must be at infinite distance from M_{n-1}. Hence $D \in M$ iff $D \in M_{n-1}$ and the latter is decided effectively by recursiveness of M_{n-1}.

Now the proof of the recursiveness of M is complete. $\dashv$

§5. Further results. As the referee has pointed out, we have implicitly obtained a stronger result. The model M is not only a GCD model but it is additionally a *factorial* domain. This is gained simply by noticing that there are analogous versions of Theorems 2.11, 2.12 and 2.14 for DCC (i.e., *divisor chain condition*, See Smith [6, Definition 1.4]). In other words:

1. DCC is preserved in transcendental extensions [6, Theorem 6.10],
2. DCC is preserved in Wilkie-constructions via parsimonious maps [6, Theorem 5.5],
3. DCC is preserved in chains of Theorem 2.14 [6, Theorem 9.8].

The three above facts will guarantee that M has the DCC. Now the factoriality of M is established by considering the following additional fact:

4. M is factorial iff M has both of the GCD property and DCC. [6, Theorem 1.5].

Based on the suggestions of A. Enayat and the referee, we continued our line of thought by further exploring the ramifications of Smith's techniques, with the aim of constructing a recursive nonstandard model for Iopen + Bezout. Recently we managed to achieve this goal by producing a recursive nonstandard model of Iopen + cof(prime) + Bezout (even PID!). This was done by (a) special attention to a recursive version of the so called F-construction in Smith's paper, and (b) the Kronecker algorithm. We mention that this model enables us to take a unified approach to demonstrate non-finite axiomatizability of Iopen, Iopen + Normality, Iopen + GCD, Iopen + Bezout. There are also other applications in end-extension problems, e.g., we can answer Marker's questions in [3]. We leave a detailed exposition of these matters to another paper.

Acknowledgment. We would like to thank Ali Enayat for his guidance and for numerous constructive comments. We would also like to thank the referee for a number of stimulating suggestions and his interest in widening the range of the Shepherdson area. I am grateful to Professor G. B. Khosrovshahi for his kindness during the past years at IPM.

REFERENCES

[1] A. BERARDUCCI and M. OTERO, *A recursive nonstandard model of normal open induction*, *The Journal of Symbolic Logic*, vol. 61 (1996), no. 4, pp. 1228–1241.

[2] ANGUS MACINTYRE and DAVID MARKER, *Primes and their residue rings in models of open induction*, *Annals of Pure and Applied Logic*, vol. 43 (1989), no. 1, pp. 57–77.

[3] DAVID MARKER, *End extensions of normal models of open induction*, *Notre Dame Journal of Formal Logic*, vol. 32 (1991), no. 3, pp. 426–431.

[4] J.-P. SERRE, *A Course in Arithmetic*, Graduate Texts in Mathematics, vol. 7, Springer-Verlag, New York, 1973.

[5] J. C. SHEPHERDSON, *A non-standard model for a free variable fragment of number theory*, *Bulletin de l'Académie Polonaise des Sciences*, vol. 12 (1964), pp. 79–86.

[6] S. SMITH, *Building discretely ordered Bezout domains and GCD domains*, *Journal of Algebra*, vol. 159 (1993), no. 1, pp. 191–239.

[7] S. TENNENBAUM, *Non-archimediam models for arithmetic*, *Notices for American Mathematical Society*, vol. 6 (1959), p. 270.

[8] LOU VAN DEN DRIES, *Some model theory and number theory for models of weak systems of arithmetic*, *Model Theory of Algebra and Arithmetic*, Lecture Notes in Mathematics, vol. 834, Springer, Berlin, 1980, pp. 346–362.

[9] R. WALKER, *Algebraic Curves*, Princeton University Press, New Jersey, 1950.

[10] GEORGE WILMERS, *Bounded existential induction*, *The Journal of Symbolic Logic*, vol. 50 (1985), no. 1, pp. 72–90.

INSTITUTE FOR STUDIES IN
THEORETICAL PHYSICS AND MATHEMATICS (IPM)
P.O. BOX 19395-5746
TEHRAN, IRAN
E-mail: mohseni@ipm.ir

MODEL THEORY OF BOUNDED ARITHMETIC
WITH APPLICATIONS TO INDEPENDENCE RESULTS

MORTEZA MONIRI

Abstract. In this paper we apply some new and some old methods in order to construct classical and intuitionistic models for theories of bounded arithmetic. We use these models to obtain proof theoretic consequences. In particular, we construct an ω-chain of models of BASIC such that the union of its worlds satisfies S_2^1 but none of its worlds satisfies the sentence $\forall x \exists y (x = 0 \lor x = y+1)$. Interpreting this chain as a Kripke model shows that double negation of the above mentioned sentence is not provable in the intuitionistic theory of BASIC plus polynomial induction on coNP formulas.

§1. **Introduction and some backgrounds.** In this paper we are concerned with some well known classical and intuitionistic theories of bounded arithmetic such as S_2^1 and IS_2^1 (see [B1] and [B3]). The language of these theories extends the usual language of arithmetic by adding the function symbols $\lfloor \frac{x}{2} \rfloor$ ($= \frac{x}{2}$ rounded down to the nearest integer), $|x|$ ($=$ the length of binary representation for x) and # ($x \# y = 2^{|x||y|}$). These symbols have clear computational meanings (see [B1]). We also work with a richer language introduced by Stephen Cook containing function symbols for polynomial time computable functions, and with theories such as IPV in this language (see [CU]).

This paper can be considered as a companion to our earlier works [M1] and [M2], but can be read independently. In particular, our results were not based on [M2].

Below, we give some general information concerning the theories mentioned above.

BASIC is a finite set of quantifier-free formulas expressing basic properties of the relation and function symbols.

A *sharply bounded* formula is a bounded formula in which all quantifiers are sharply bounded, i.e. of the form $\exists x \leqslant |t|$ or $\forall x \leqslant |t|$ where t is a term which does not contain x. The class $\Sigma_0^b = \Pi_0^b$ is the class of all sharply

2000 *Mathematics Subject Classification.* 03F30, 03F55, 03F50, 68Q15.

Key words and phrases. Bounded Arithmetic, Intuitionistic Logic, Polynomial Hierarchy, Polynomial Induction, NP, coNP, Kripke Model.

Logic in Tehran
Edited by A. Enayat, I. Kalantari, and M. Moniri
Lecture Notes in Logic, 26

bounded formulas. The syntactic classes Σ_{i+1}^b and Π_{i+1}^b of bounded formulas are defined by counting alternations of bounded quantifiers, ignoring sharply bounded quantifiers (see [B1]).

The Σ_i^b formulas represent exactly the relations in the ith level of the polynomial hierarchy. So, for example, the NP relations in the standard model are exactly the ones that can be defined via Σ_1^b formulas. The same is true for Π_1^b formulas and coNP relations.

The (classical) theory S_2^1 is axiomatized by adding the scheme PIND for Σ_1^b formulas to BASIC, i.e. $[A(0) \wedge \forall x(A(\llcorner \frac{x}{2} \lrcorner) \rightarrow A(x))] \rightarrow \forall x A(x)$, where $A(x)$ is a Σ_1^b formula. Here, $A(x)$ can have more free variables besides x. A function f is said to be Σ_1^b-*definable* in S_2^1 if and only if it is provably total in S_2^1 with a Σ_1^b formula defining the graph of f. Buss proved that a function is Σ_1^b-definable in S_2^1 if and only if it is polynomial time computable.

The theories S_2^i, $i > 1$, are similarly defined as the theories axiomatized by BASIC together with PIND on Σ_i^b formulas. S_2 is the union of all S_2^i, $i \geq 1$.

The theory IS_2^1 is the intuitionistic theory axiomatized by BASIC plus the scheme PIND on positive Σ_1^b formulas (denoted Σ_1^{b+}), i.e. Σ_1^b formulas which do not contain '$\neg$' and '$\rightarrow$'. This theory was introduced and studied by Cook and Urquhart and by Buss (see [CU] and [B3]). A function f is also defined to be Σ_1^{b+}-*definable* in IS_2^1 if it is provably total in IS_2^1 with a Σ_1^{b+} formula defining the graph of f. In [CU] it is proved that f is Σ_1^{b+}-definable in IS_2^1 if and only if it is polynomial time computable. Also, S_2^1 is $\forall \exists \Sigma_1^b$-conservative over IS_2^1, as it follows from 10.7 and 4.2 in [CU], see also [A] for a different proof.

The theory PV is Stephen Cook's equational theory for polynomial time functions and PV_1 is its (conservative) extension to classical first-order logic. PV_1 is a universal theory which proves the polynomial induction on quantifier-free formulas. PV^i denotes the intuitionistic deductive closure of PV. IPV is the intuitionistic theory of PV plus polynomial induction on NP formulas. Here, an NP formula is a quantifier free formula (in the language of PV) prefixed by a block of bounded existential quantifiers. IPV is a conservative extension of IS_2^1. CPV is the classical version of IPV.

It is known that the theory S_2^1 can be axiomatized over the base theory BASIC using either $PIND(\Pi_1^b)$ or $PIND(\Sigma_1^b)$. On the other hand, in [M1], we showed that the same is not true for the corresponding theories based on intuitionistic logic,

FACT 1.1. The intuitionistic theory axiomatized by $BASIC + PIND(\Pi_1^{b+})$ does not imply IS_2^1.

In Section 4, we give more exact results in this direction. The following related results are also proved in [M1].

FACT 1.2. (i) If $IPV \vdash PIND(coNP)$, then $CPV = PV_1$.
(ii) If $IS_2^1 \vdash PIND(\Pi_1^{b+})$, then $CPV = PV_1$.

It is known that, CPV = PV$_1$ implies the collapse of the polynomial hierarchy (see [KPT]).

In the last section we will consider the question of whether

$$PV + PIND(coNP) \vdash_i PIND(NP).$$

We will give a negative answer to this question based on a plausible assumption.

§2. **Basic results on Kripke models of intuitionistic bounded arithmetic.** In this section we briefly describe Kripke models. All theories we will study prove the principle of excluded middle PEM (that is, $\varphi \vee \neg\varphi$) for atomic formulas and so we can use a slightly simpler version of the definition of Kripke models as we have completeness with respect to this restricted class of models (see [B2]).

A Kripke structure K for a language L can be considered as a set of classical structures for L partially ordered by the relation substructure. We can assume, without loss of generality, that this partially ordered set is a rooted tree. For every node α, L_α denotes the expansion of L by adding constants for elements of M_α. The forcing relation $\Vdash$ is defined between nodes and L_α-sentences inductively as follows:

- For atomic φ, $M_\alpha \Vdash \varphi$ if and only if $M_\alpha \vDash \varphi$;
- $M_\alpha \Vdash \varphi \vee \psi$ if and only if $M_\alpha \Vdash \varphi$ or $M_\alpha \Vdash \psi$;
- $M_\alpha \Vdash \varphi \wedge \psi$ if and only if $M_\alpha \Vdash \varphi$ and $M_\alpha \Vdash \psi$;
- $M_\alpha \Vdash \varphi \rightarrow \psi$ if and only if for all $\beta \geq \alpha$, $M_\beta \Vdash \varphi$ implies $M_\beta \Vdash \psi$;
- $M_\alpha \Vdash \forall x \varphi(x)$ if and only if for all $\beta \geq \alpha$ and all $a \in M_\beta$, $M_\beta \Vdash \varphi(a)$; and
- $M_\alpha \Vdash \exists x \varphi(x)$ if and only if there exists $a \in M_\alpha$ such that $M_\alpha \Vdash \varphi(a)$.

A Kripke model K forces a formula $\varphi(\overline{x})$, if each of its nodes (equivalently its root) forces $\forall \overline{x} \varphi(\overline{x})$. A Kripke model is *T-normal*, where T is a set of sentences, if each node (world) of it satisfies T. It decides quantifier free formulas if it forces the axiom PEM restricted to quantifier free formulas. So, for example, any BASIC-normal Kripke model decides atomic formulas (see [B3]).

Below, we mention some facts about Kripke models of bounded arithmetic theories (see [M1]).

PROPOSITION 2.1. (i) *Kripke models of PV^i are exactly PV$_1$-normal Kripke models.*

(ii) *For a Kripke model of PV^i over the frame ω to force PIND(coNP) it is necessary and sufficient that the union of the worlds in it satisfies CPV.*

PROOF. By induction on the complexity of formulas it is easy to see that for each $\exists$-free formula A, each node in such a Kripke model forces A if and only the union of the worlds above it satisfies A. Now apply the definition of forcing.   ⊣

§3. **Constructing models of bounded arithmetic.** In this section we introduce some methods for constructing models for classical and intutionistic bounded arithmetic. First the classical one. This is indeed a variant of the construction given in Johannsen [J1, J2], where a model of the theory S_2^0 (the classical theory axiomatized by BASIC plus PIND on sharply bounded formulas) was constructed to witness a well-known independence result of Takeuti [Ta], i.e. $S_2^0 \nvdash \forall x \exists y (x = 0 \lor x = y + 1)$. Takeuti proved this result through the use of a proof-theoretic method.

Let M and N be two models of BASIC. Let $\mathrm{Log}(M) = \{a \in M : (\exists b \in M)\ a \le |b|\}$. N is called a *weak end-extension* of M and it is written that $M \subseteq_{w.e.} N$, if N extends M and $\mathrm{Log}(N)$ is an end-extension of $\mathrm{Log}(M)$, i.e. for all $a \in \mathrm{Log}(M)$, $b \in \mathrm{Log}(N)$ with $N \vDash b \le a$, we have $b \in \mathrm{Log}(M)$. It is known and easy to see that weak end-extensions are always Σ_0^b-elementary, i.e. if $M \subseteq_{w.e.} N$, then for any Σ_0^b-formula $A(\overline{x})$ and $\overline{a} \in M$, $M \vDash A(\overline{a})$ if and only if $N \vDash A(\overline{a})$. Elements of $\mathrm{Log}(M)$ are called *small* elements of M. The others are *large* elements of M.

Recall that, the axiom exp states that the exponentiation function is total, and the axiom Ω_2 states that the function $x \#_3 y = 2^{|x|\#|y|}$ is total.

It is known that the function $\mathrm{Bit}(x, i)$ which gives the value of the ith bit in the binary expansion of x, and the operation of length bounded counting are definable in S_2^1. Hence the function $\mathrm{count}(a) = \#i < |a|(\mathrm{Bit}(a, i) = 1)$, which gives the number of 1's in the binary expansion of a, is well defined in S_2^1. Now, let M be a countable (nonstandard) model of $S_2^1 + \Omega_2 + \neg\exp$. For a large element $a \in M$, structures of the form

$$M' = \{x \in M : \mathrm{count}(x) < \|a\| \text{ for some } a \in M\}$$

were studied in [J2].

Next we need to modify the definition of M' (in order to prove our independence result).

Fix a large element $a \in M$ and define

$$M^* = \mathrm{Log}(M) \cup \{x \in M : \mathrm{count}(x) < \|a\|^n$$

$$\text{for some non-negative integer } n\}.$$

THEOREM 3.1. $M^* \vDash \mathrm{BASIC}$.

PROOF. First, note that $\mathrm{Log}(M)$ is a model of BASIC (note that since we have assumed M is closed under Ω_2, small elements are closed under #).

Moreover, as it is mentioned in [J1], it can be proved in S_2^1 that

$$\mathrm{count}(a + b) \le \mathrm{count}(a) + \mathrm{count}(b)$$
$$\text{and} \quad \mathrm{count}(a \cdot b) \le \mathrm{count}(a) \cdot \mathrm{count}(b).$$

Furthermore, for any $a, b \in M$, count$(a\#b) = 1$. Therefore, M^* is closed under $+$, $\cdot$, and $\#$. Also, M^* is clearly closed under the function $|x|$ because each small element of M is in M^*. Closure of M^* under the function $\lfloor \frac{x}{2} \rfloor$ can also be easily verified by a simple computation.

Now it should be clear that $M^* \vDash$ BASIC as BASIC is a universal theory. $\dashv$

In fact, since $M^* \subseteq_{w.e.} M$, we have $M^* \vDash L_2^0$, where L_2^0 is the theory axiomatized by BASIC plus length induction on sharply bounded formulas (see [J1, Corollary 2]). The scheme of length induction is $[A(0) \wedge \forall x(A(x) \rightarrow A(x+1))] \rightarrow \forall x A(|x|)$.

THEOREM 3.2. $M^* \nvDash \forall x \exists y(x = 0 \vee x = y + 1)$.

PROOF. The proof is similar to the one given in [J2, Proposition 6].

Note that count$(2^{|a|}) = 1$ and so $2^{|a|} \in M^*$. Consider the element $b = 2^{|a|} - 1 \in M$. Then count$(b) = |a|$. If $b \in M^*$, then $|a| < \|a\|^n$ for some $n \in \mathbb{N}$. As $M \vDash \Omega_2$, there is $a' \in M$ such that $|a| < \|a'\|$ and so $a < 2|a'|$, in contradiction to a being large. $\dashv$

Inductively define $a^{\#0} = 1$, $a^{\#1} = a$, and $a^{\#(n+1)} = a^{\#n}\#a$, for each $a \in M$.

THEOREM 3.3. *There is an ω-chain of models of* BASIC *such that the union of its worlds satisfies* S_2^1 *but none of its worlds satisfies the sentence* $\forall x \exists y(x = 0 \vee x = y + 1)$. *Moreover, for each i, M_{i+1} is a proper weak end-extension of M_i.*

PROOF. As above, let M be a countable (nonstandard) model of $S_2^1 + \Omega_2 + \neg$exp. Consider a cofinal sequence (a_i) of large elements of M such that $a_{i+1} \geqslant (a_i\#_3a_i)$, for all $i \geq 0$.

Now define a sequence of substructures of M as follows:

1) $M_0 = M^*$ (as defined above, with a_0 as a),

2) M_{i+1} is defined as the set

$$\{x \in M : x < a_i^{\#n} \text{ for some } n \in \mathbb{N}\}$$
$$\cup \{x \in M : \text{count}(x) < \|a_{i+1}\|^n \text{ for some } n \in \mathbb{N}\}.$$

Similar proofs as the ones given for M^* above can be applied to show that $M_i \vDash$ BASIC $+ \neg\forall x \exists y(x = 0 \vee x = y + 1)$. Also, clearly, $\bigcup M_i = M \vDash S_2^1$ as $a_i \in M_{i+1}$.

Note that each world contains $\text{Log}(M)$, and so clearly M_{i+1} is a weak end-extension of M_i. Moreover the extension is proper, since for example, $(a_i\#a_i) - 1 \in M_{i+1} \setminus M_i$. $\dashv$

Next we mention an easy fact about Kripke models (see [M1]). Its proof is similar to the one for Proposition 2.1. A Kripke model is a weak end-extension Kripke model if its accessibly relation is a weak end-extension.

PROPOSITION 3.4. *For a weak end-extension Kripke model whose accessibility relation is ω and decides atomic formulas to force* $\text{PIND}(\Pi_1^{b+})$ *it is necessary and sufficient that the union of the worlds in it satisfies* $\text{PIND}(\Pi_1^{b+})$.

PROOF. Note that any such Kripke model is Σ_0^b-elementary.   ⊣

The last two results suggest a way to construct a special Kripke model. Below, we will explicitly define and use it.

§4. **PIND on NP and coNP formulas.** In this section our aim is to apply the models constructed above to show that the sentence $\neg\neg\forall x\exists y(x = 0 \lor x = y + 1)$ is not intuitionistically provable in $\text{BASIC} + \text{PIND}(\Pi_1^{b+})$. A similar method is used in [M2] to show that

$$\text{BASIC} + \text{PIND}(\Pi_1^{b+}) \nvdash_i \neg\neg\forall x, y\exists z \leq y(x \leq |y| \to x = |z|).$$

Each of these results, using $\forall\exists$ conservatively of S_2^1 over IS_2^1, can be easily applied to show that

$$\text{BASIC} + \text{PIND}(\Pi_1^{b+}) \nvdash_i \neg\neg\text{PIND}(\Sigma_1^{b+}).$$

THEOREM 4.1. $\text{BASIC} + \text{PIND}(\Pi_1^{b+}) \nvdash_i \neg\neg\forall x\exists y(x = 0 \lor x = y + 1).$

PROOF. Consider the ω-chain $M_0 \subset_{w.e.} M_1 \subset_{w.e.} M_2 \subset_{w.e.} \cdots$ of Theorem 3.3, and interpret it as an ω-framed Kripke model K. For each i, $M_i \nvDash \forall x\exists y(x = 0 \lor x = y + 1)$.

Therefore, by the definition of forcing, $K \Vdash \neg\forall x\exists y(x = 0 \lor x = y + 1)$. Hence, $K \nVdash \neg\neg\forall x\exists y(x = 0 \lor x = y + 1)$.

On the other hand, since the union of the worlds in this Kripke model is equal to $M \vDash S_2^1$, by Proposition 3.4, we have $K \Vdash \text{PIND}(\Pi_1^{b+})$.   ⊣

Now, we are interested in the question of whether results analogous to the above hold if we work in the language of PV (recall that, by [M1], $\text{IPV} \nvdash \text{PIND}(\text{coNP})$ unless $\text{CPV} = \text{PV}_1$). This question is not easy. The following proposition gives a reason for this claim. By $IS_2(PV)$ we mean the intuitionistic version of S_2 conservatively extended to the language of PV.

PROPOSITION 4.2. *If* $PV^i \vdash P = NP$, *i.e. any NP formula in* PV^i *is equivalent to a quantifier free formula, then* $PV^i \equiv IS_2(PV)$.

PROOF. Let $PV^i \vdash P = NP$. Then, using induction on the complexity of formulas, one can see that, any bounded formula in PV^i would be equivalent to a quantifier free formula. We only examine the $\forall$ case:

Let $PV^i \vdash \varphi(x, \overline{y}) \leftrightarrow \psi(x, \overline{y})$, where ψ is quantifier-free. So, $PV^i \vdash \forall x \leq t\varphi(x, \overline{y}) \leftrightarrow \forall x \leq t\psi(x, \overline{y})$. Using decidability of atomic formulas, we get the following intuitionistic equivalences:

$$\forall x \leq t\psi(x, \overline{y}) \equiv_i \forall x \leq t\neg\neg\psi(x, \overline{y}) \equiv_i \neg\exists x \leq t\neg\psi(x, \overline{y}).$$

Now, using the assumption, we obtain that $\forall x \leq t\psi(x, \overline{y})$ is equivalent to a quantifier free formula. To see $PV^i \equiv IS_2(PV)$, note that PV_1 proves the polynomial induction on quantifier free formulas and so one can see that PV^i does the same, using the negative translation.   ⊣

So proving $PV + PIND(coNP) \nvdash_i PIND(NP)$ would require proving that $PV^i \nvdash P = NP$.

In the theorem below, we give an answer to the above mentioned question under a plausible assumption.

Recall that, the (sharply bounded) replacement (or collection) scheme $BB(\Sigma_0^b)$ is

$$\forall i < |a| \exists x < aA(i, x) \longrightarrow \exists \omega \forall i < |a| A(i, [\omega]_i),$$

where A is a Σ_0^b formula and $[x]_i$ is the ith element of the sequence coded by x. It is known that CPV proves this scheme but, if integer factoring is not possible in probabilistic polynomial time, then PV_1 does not prove the scheme (see [CT]). Also, if $PV_1 + BB(\Sigma_0^b)$ proves CPV, then PV_1 proves CPV (see [Z] and [CT]).

Also, let EF denote an extended Frege proof system. Recall that a Frege proof system is just an ordinary propositional proof system containing finitely many axiom schemes and inference rules, and an extended Frege proof system is a Frege system allowing abbreviations of the form $p_A \equiv A$, where A is a propositional formula and p_A is a new propositional variable. The system EF can be formalized in PV_1 and related results about it can be stated and proved in this theory (see [C] or [K]). For example, it is known that provability of $NP = coNP$ and that of the statement "EF is a complete proof system" in PV_1 are equivalent (see [K, Theorem 15.3.7]).

THEOREM 4.3. *If there exists a model M of $PV + BB(\Sigma_0^b)$ that does not satisfy CPV and in which EF is not a complete proof system, then we have $PV + PIND(coNP) \nvdash_i PIND(NP)$.*

PROOF. Let M be as above. There exists $M' \vDash PV$ such that M embeds in M' and the embedding is not Σ_1^b-elementary (see [K, Corollary 15.3.10]). Now embed M' Σ_1^b-elementarily in a model $M^* \vDash CPV$, see [K, Theorem 7.6.3] for the existence of such a model. Note that the induced embedding between M and M^* is not Σ_1^b-elementary. Putting M^* above M produces a Kripke model which forces $PV + PIND(coNP)$, see Proposition 2.1. We show that it does not force $PIND(NP)$. Suppose the Kripke model forces $PIND(NP)$. We will show that, in this case, the root, i.e. M, (classically) would satisfy $PIND(NP)$ which is a contradiction.

We heavily rely on the easily verifiable fact that forcing and satisfaction of NP formulas in each node of a Kripke model of PV^i are equivalent.

Suppose that $A(y)$ is an NP formula (possibly with parameters from M) such that $M \nvDash A$ but $M^* \vDash A$. Such a formula exists because, modulo $PV_1 + BB(\Sigma_0^b)$, each Σ_1^b formula is equivalent to an NP formula, see [Th] for more detail on this theory.

Let $B(x)$ be an arbitrary NP formula. We are going to show that M satisfies the instance of polynomial induction on $B(x)$. So, let $M \vDash B(0)$ and

$$M \vDash B\left(\lfloor\frac{x}{2}\rfloor\right) \longrightarrow B(x).$$

We will show that $M \vDash \forall x B(x)$. Clearly we have $M \vDash B(0) \vee A$ and

$$M \vDash \forall x\left[\left(B\left(\lfloor\frac{x}{2}\rfloor\right) \vee A\right) \rightarrow (B(x) \vee A)\right].$$

Therefore, $M \Vdash B(0) \vee A$ and, using the assumption $M^* \vDash A$,

$$M \Vdash \forall x\left[\left(B\left(\lfloor\frac{x}{2}\rfloor\right) \vee A\right) \rightarrow (B(x) \vee A)\right].$$

So, we get $M \Vdash \forall x(B(x) \vee A)$ since the Kripke model forces PIND(NP) by our assumption. Hence, $M \vDash \forall x(B(x) \vee A)$. So, $M \vDash \forall x B(x)$. Therefore, $M \vDash$ PIND(NP), which is a contradiction.   ⊣

Acknowledgments. I would like to thank two anonymous referees for useful comments on earlier versions of this paper. I would also like to thank the editor responsible for this paper Iraj Kalantari for his comments improving presentation of the paper, and Chris Pollett for drawing my attention to [CT]. This research was in part supported by a grant from IPM (No. 83030118).

REFERENCES

[A] J. AVIGAD, *Interpreting classical theories in constructive ones*, **The Journal of Symbolic Logic**, vol. 65 (2000), no. 4, pp. 1785–1812.

[B1] S. R. BUSS, **Bounded Arithmetic**, Bibliopolis, Naples, 1986.

[B2] ———, *On model theory for intuitionistic bounded arithmetic with applications to independence results*, **Feasible Mathematics** (S. R. Buss and P. J. Scott, editors), Birkhäuser Boston, Boston, MA, 1990, pp. 27–47.

[B3] ———, *A note on bootstrapping intuitionistic bounded arithmetic*, **Proof Theory (Leeds, 1990)**, Cambridge University Press, Cambridge, 1992, pp. 149–169.

[C] S. A. COOK, *Feasibly constructive proofs and the propositional calculus*, **Seventh Annual ACM Symposium on Theory of Computing**, ACM, New York, 1975, pp. 83–97.

[CT] S. A. COOK and N. THAPEN, *The strength of replacement in weak arithmetic*, **Nineteenth Annual IEEE Symposium on Logic in Computer Science (LICS 2004)**, pp. 256–264.

[CU] S. A. COOK and A. URQUHART, *Functional interpretations of feasibly constructive arithmetic*, **Annals of Pure and Applied Logic**, vol. 63 (1993), no. 2, pp. 103–200.

[J1] J. JOHANNSEN, *A model-theoretic property of sharply bounded formulae, with some applications*, **Mathematical Logic Quarterly**, vol. 44 (1998), no. 2, pp. 205–215.

[J2] ———, *A remark on independence results for sharply bounded arithmetic*, **Mathematical Logic Quarterly**, vol. 44 (1998), no. 4, pp. 568–570.

[K] J. KRAJÍČEK, **Bounded Arithmetic, Propositional Logic, and Complexity Theory**, Cambridge University Press, Cambridge, 1995.

[KPT] J. KRAJÍČEK, P. PUDLÁK, and G. TAKEUTI, *Bounded arithmetic and the polynomial hierarchy*, **Annals of Pure and Applied Logic**, vol. 52 (1991), no. 1-2, pp. 143–153.

[M1] MORTEZA MONIRI, *Comparing constructive arithmetical theories based on NP-PIND and coNP-PIND*, **Journal of Logic and Computation**, vol. 13 (2003), no. 6, pp. 881–888.

[M2] ———, *An independence result for intuitionistic bounded arithmetic*, submitted.

[Ta] G. TAKEUTI, *Sharply bounded arithmetic and the function a$\dot{-}$1*, **Logic and Computation** *(Pittsburgh, PA, 1987)*, Contemporary Mathematics, vol. 106, Amer. Math. Soc., Providence, RI, 1990, pp. 281–288.

[Th] N. THAPEN, **The Weak Pigeonhole Principle in Models of Bounded Arithmetic**, DPhil Thesis, University of Oxford, 2002.

[Z] D. ZAMBELLA, *Notes on polynomially bounded arithmetic*, **The Journal of Symbolic Logic**, vol. 61 (1996), no. 3, pp. 942–966.

INSTITUTE FOR STUDIES IN
    THEORETICAL PHYSICS AND MATHEMATICS (IPM)
        P.O. BOX 19395-5746, TEHRAN, IRAN
and
    DEPARTMENT OF MATHEMATICS
        SHAHID BEHESHTI UNIVERSITY
            P.O. BOX 19839, TEHRAN, IRAN
E-mail: ezmoniri@ipm.ir

IBN-SINA'S ANTICIPATION OF THE FORMULAS
OF BURIDAN AND BARCAN

ZIA MOVAHED

Abstract. In quantified modal logic, there are two theses known as *Barcan formulas*, which have been centers of controversies since the second half of the twentieth century. Also, there are two other famous formulas known as *Buridan formulas*. In this paper, I'll show that Ibn-Sina, the Persian philosopher and logician, anticipated and discussed all of these four formulas, and enunciated the important distinction between *de re* and *de dicto* modality.

§1. Contemporary background. In 1947 Carnap wrote: [2, p. 196]

> Any system of modal logic without quantification is of interest only as a basis for a wider system including quantification. If such a wider system were found to be impossible, logicians would probably abandon modal logic entirely.

Surpisingly Carnap's anticipation turned out to be relevant, but exactly in the opposite direction! What captured mostly the attention of modal logicians was propositional modal logic. Quantified modal logic (QML) has remained under-developed. Of course, there are many systems of QML and many approaches to it (see [3]). But apart from its provability interpretation, other treatments of the subject have mostly been of interest to philosophical logicians. In fact, QMLs have become a battleground for ongoing heated controversies over philosophical problems such as the very ontological nature of possible worlds, necessity, essentialism, transworld identity, and existence.

But among the many existing versions of QML, the one I am interested in, is a system in which all of the following formulas are derivable:

$$BF : \forall x \Box Fx \longrightarrow \Box \forall x Fx$$
$$CBF : \Box \forall x Fx \longrightarrow \forall x \Box Fx$$
$$BUF : \Diamond \forall x Fx \longrightarrow \forall x \Diamond Fx$$

BF and CBF stand respectively for the *Barcan formula* and the *converse Barcan formula*. These names derive from Ruth C. Barcan (later Mrs. J. A. Marcus), who called attention to the first two formulas in her paper of 1946 [1]. Also, BUF stands for the *Buridan formula*, and is attributed by Alvin Plantinga to

Logic in Tehran
Edited by A. Enayat, I. Kalantari, and M. Moniri
Lecture Notes in Logic, 26
© 2006, Association for Symbolic Logic

Jean Buridan, the French philosopher of 14th century [8, p. 58]. BF and CBF, as we shall see, have significant semantical and ontological consequences.

§2. **Introductory informal remarks.** To understand the significance of Ibn-Sina's anticipation of, in particular, BF and CBF, we have to interpret it in the context of modern modal semantics. So I begin with some informal introduction of this semantics. In what follows, I assume familiarity with first-order logic.

Since the second half of the 20th century, what has made modal logic a subject worthy of serious study was the discovery of its semantics based on the notion of *possible worlds*. A possible world, or a possible state of affairs, is a world in which things are different from what they actually are. This notion, so deeply embedded in our use of natural language, goes back to Leibniz. Inspired by the Leibniz' idea, we have the following truth-definitions: a proposition P is *necessarily true* ($\Box P$) iff it is true at all possible worlds; P is *possibly true* ($\Diamond P$) iff it is true at some possible worlds, and P is *impossible* iff it is true at none.

The difficulty with these definitions is that they cannot distinguish between some of the different modal axioms. In fact, all of the following well-known axioms of propositional modal logic become valid by this semantics:

$$T : \Box P \longrightarrow P$$
$$B : P \longrightarrow \Box \Diamond P$$
$$4 : \Box P \longrightarrow \Box \Box P$$
$$5 : \Diamond P \longrightarrow \Box \Diamond P.$$

To see how, take, for example, 5. Suppose $\Diamond P$ is true at the possible world α; then by definition, P is true at some world. But by this definition $\Diamond P$ is true at all possible worlds, and again, by definition $\Diamond P$ is necessarily true; or $\Box \Diamond P$. Therefore, $\Diamond P$ implies $\Box \Diamond P$, and 5 is valid by this semantics. The same, also, holds true for T, B, 4, and many other modal axioms.

To amend this shortcoming, Kripke and, before him but apparently unknown to him, Kanger and Hintikka introduced a binary relation over possible worlds. This simple notion of relative possibility has enabled us to distinguish between many modal axioms and systems. The idea is that for every possible world α, there are worlds *reachable* from α (or *accessible* to α), and there are those which are not. If β is accessible to α, we shall write $\alpha R \beta$. R is called an *accessibility relation*, and may be any sort of binary relation on possible worlds. Now, the truth-definitions of modal propositions are re-defined as follows:

1) $\Box P$ is true at a world α iff P is true at every possible world β that is accessible to α.
2) $\Diamond P$ is true at α iff P is true at some world β accessible to α.

The important aspect of this semantics is that the truth of each of modal axioms mentioned above is naturally connected to a condition on R; that is, for each axiom, R must satisfy a certain condition. For example, T is true if R is reflexive; B is true if R is symmetric; and 4 is true if R is transitive.

Now when we come to quantified modal logic, the situation becomes more complicated. Here, among other questions, a crucial question arises as to whether all possible worlds contain the same objects or different objects; i.e., whether we are dealing with fixed domains or world-relative domains. To show the impact of this question on, specifically, BF and CBF, I begin with two examples related to these formulas. Combining Tarski's truth-definitions for quantified propositions and the truth-definitions for modal propositions outlined above, we get the following conditions for the truth of $\forall x \Box Fx$ and $\Box \forall x Fx$ (Notice that quantifiers in each possible world range only over the objects of that world):

3) $\forall x \Box Fx$ is true at α iff every object of α has F at every possible world β accessible to α.

4) $\Box \forall x Fx$ is true at α iff at every possible world β accessible to α, every object of β has F.

Obviously (3) and (4) do not say the same thing. Particularly if β has objects not contained in α, then $\forall x \Box Fx$ will not imply $\Box \forall x Fx$ at α unless every object in β exists in α. Similarly, it can be shown that $\Box \forall x Fx$ implies $\forall x \Box Fx$ at α if every object of α exists in every β. Consequently, to have both implications, i.e. both BF and CBF, the domains of all possible worlds should have exactly the same objects. So much for our informal discussion.

§3. **Syntax and semantics of QML.** In this section, I shall formally introduce a simple and strong QML respecting our informal semantics; i.e. a QML system containing BF, CBF, and BUF as theorems, and semantically validating all those formulas.

The language of our QML is the result of adding $\Box$ to a standard first-order language without identity, individual constants, and function-letters. The well-formed formulas are inductively constructed out of $\sim$, $\rightarrow$, $\Box$, and $\forall$. While $\Diamond$ and $\exists$ are given their usual definitions. The axioms and rules are:

- All truth functional tautologies;
- $\Box(P \rightarrow Q) \rightarrow (\Box P \rightarrow \Box Q)$;
- $\Box P \rightarrow P$;
- $\Box P \rightarrow \Box\Box P$;
- $P \rightarrow \Box\Diamond P$;
- $\forall x A \rightarrow A(y/x)$, where y is substitutable for x;
- If $A \rightarrow B$ is a theorem, then $A \rightarrow \forall x B$, where x is not free in A, is a theorem;

- If $A \rightarrow B$ and A are theorems, then so is B;
- If A is a theorem, then so is $\Box A$.

In this axiomatization BF, CBF, and BUF are easily derivable [4, pp. 245–247]. The converse of the Buridan formula (CBUF) is not a theorem of any standard QML. This follows, as we shall see, according to Ibn-Sina's counterexample to CBUF, since from the fact that every person can possibly be a writer, it does not follow that possibly all people are writers.

As to the semantics of our QML, let $M = \langle W, R, D, I \rangle$ be a constant domain first-order model in which W is a non-empty set of worlds, R a binary equivalence relation on W, D a non-empty set of objects, and I an interpretation function which assigns to each n-place predicate letter A at each world α a set $I(A)$ of n-tuples $\langle d_1, \ldots, d_n \rangle$, where each d_i is in D. To give the truth-value of wffs containing free variables, we also need an assignment function also. An assignment function in the model M is a mapping μ that assigns to each variable v some member $\mu(v) \in D$. Finally, for giving the truth-value of quantified sentences we need another assignment function μ^*, where μ^* is an assignment identical to μ except perhaps for what it assigns to v. So, we say μ^* is a v-*alternative* of μ iff for every v' except possibly v, $\mu(v') = \mu^*(v')$.

Now the truth-value of every wff at a world α relative to an assignment μ is given as follows:

I(A): $I_\mu(A v_1, \ldots, v_n)$ is true at α iff $\langle \mu(v_1), \ldots, \mu(v_n) \rangle \in I(A)$;

I($\neg$): $I_\mu(\neg \varphi)$ is true at α iff $I_\mu(\varphi)$ is false at α;

I($\rightarrow$): $I_\mu(\varphi \rightarrow \psi)$ is true at α iff if $I_\mu(\varphi)$ is true at α, then $I_\mu(\psi)$ is true at α;

I($\forall$): $I_\mu(\forall v \varphi v)$ is true at α iff $I_{\mu^*}(\varphi v)$ is true at α for every v-alternative μ^* of μ; and

I($\Box$): $I_\mu(\Box \varphi)$ is true at α iff for every β with $\alpha R \beta$, $I_\mu(\varphi)$ is true at β.

A wff is valid in M iff it is true for every $\alpha \in W$ and every assignment μ. A wff is valid on a frame $\langle W, R \rangle$ iff it is valid in every model M based on $\langle W, R \rangle$.

For systems with varying domains, a model is $M' = \langle W, R, D, I \rangle$, where W and R are as before, D is a function assigning to possible worlds α a non-empty set $D(\alpha)$, the set of objects which exist in α. In this semantics, the truth-value of a quantified sentence becomes as listed below, while the truth-values of the others remain as they were[1]:

I($\forall$): $I_\mu(\forall v \varphi v)$ is true at α iff $I_{\mu^*}(\varphi v)$ is true at α for every v-alternative μ^* of μ such that $\mu^*(v) \in D(\alpha)$.

---

[1]There are some intricacies of detail concerning what to do with assignments that give us values that do not exist in the world under consideration. We avoid these intricacies here. Of course, a full treatment would fill in these details.

It is easy to check that BF, CBF, and BUF are all valid in M. Other logically and philosophically interesting aspects of this simple semantics which captures our informal considerations are:

1. In models with varying domains if we have BF, BF is naturally connected to the condition that whenever $\alpha R \beta$, then $D(\beta) \subseteq D(\alpha)$; and if we have CBF, CBF is naturally connected to the condition that whenever $\alpha R \beta$, then $D(\alpha) \subseteq D(\beta)$. So, if we have both BF and CBF, then $\alpha R \beta$ implies $D(\alpha) = D(\beta)$. Therefore, to have both formulas the domains of all worlds should have the same objects. M, as defined above, has this property, so it is called a *model with fixed domain*. Generally, in models with varying domains one can easily find counterexamples to BF, CBF and BUF [7, pp. 67–68].

2. Since R is an equivalence relation, T, B, 4, 5 are valid in M; in fact they are valid in the frame $\langle W, R \rangle$ of M.

3. According to BF and CBF, to move from one world to another, nothing comes into existence and nothing passes out of existence. This has a controversial consequence that existence is a necessary and not contingent attribute of objects.

This is a sort of system that philosophical logicians like R. Barcan, E. N. Zalta, B. Linsky [10], and T. Williamson [9], although with different axiomatizations, argue as best representing our intuitive conception of metaphysical nature of alethic modality.

§4. **Ibn-Sina's anticipations.** According to the received texts of ancient Greek and the early Islamic period (800–950), Ibn-Sina (980–1037) was the first logician who realized the distinction between *de re* and *de dicto* modality, and extensively discussed what are now called the Buridan and the Barcan formulas. The first place where Ibn-Sina introduces the subject is in *Kitab al-Ibara*, the third volume of his major work *al-Shifa* (in Latin known as *Sufficientia*). He writes: [5, pp. 114–115].

> So we say: the proper place for a modal word is to attach to the copula; this is so because it [the modal word] generally qualifies the relation of the predicate to the subject or qualifies the universal or particular quantifier, then [in the latter case] quantifier shows the quantity of predication so modalized. Thus if we say: [∗] *Every human being is possibly a writer*, it is natural and means: every single person is possibly a writer. But if it [the modal word] is attached to the quantifier and it is not misplaced and it is intended to come before the quantifier, then it would not be the mode of predication but the mode of universalization or particularization, then the meaning changes and it becomes possible for all people to be writers at once. and the reason for the changing of meaning is that in the first case no one would doubt [the truth] of it, because

every one knows that there is nothing in one's nature to necessitate one's being a writer or not. But if we say: [**] *possibly every human being is a writer,* and the possibility is the mode of generality and quantifier, then that is doubtful indeed. Because some people would say that it is impossible that all people are writers, or it happens that you would not find any human being but a writer. Therefore, there are differences between these two meanings.

(My translation from Arabic; see the Arabic Text No. 1 at the end of the paper.)

Here, by reading Ibn-Sina's first example, appropriately, as: 'every human being x, possibly x is a writer', we can symbolize it as: $\forall x \Diamond W(x)$; and his second example can be symbolized as: $\Diamond \forall x W(x)$. Furthermore, from what he says, though he does not explicitly say so, one can infer that, according to him, [**]: $\Diamond \forall x W(x)$ implies [*]: $\forall x \Diamond W(x)$, but not vice versa. So, Ibn-Sina accepts the so-called Buridan formula and rejects the converse Buridan formula. This is as it should be within the wisdom of modern QML.

Ibn-Sina, after the passage quoted above, immediately extends his discussion to universal negative propositions and makes the same distinction between $\Diamond \forall x \neg W(x)$ and $\forall x \Diamond \neg W(x)$ as made for universal positive propositions without, again, mentioning the direction of implications. But when he comes to the propositions logically equivalent to the Barcan formulas, he writes [5, 116]:

But to say that: *some people possibly are not writers* is modally the same as saying that: *possibly some people are not writers,* and although one implies the other the meaning of the one may be opposite to the other.

(See the Ibn-Sina's Arabic Text No. 2 at the end of the paper.)

This is, accordingly, translated as:

$$\exists x \Diamond \neg W(x) \longleftrightarrow \Diamond \exists x \neg W(x)$$

which is logically equivalent to:

$$\forall x \Box W(x) \longleftrightarrow \Box \forall x W(x).$$

This is the conjunction of BF and CBF in one biconditional. So Ibn-Sina discovers and endorses both BF and CBF while admitting that there are differences of meaning between the antecedent and the consequent of each conditional. This is an insight of genius into the subject, an insight which has been the center of controversies about the legitimacy of these conditionals, when we consider possible worlds with different objects. A well-known counterexample in this case to Ibn-Sina's version of BF is this: *it may be possible that there should be things of a different species from any actual living organism, but it is not possible of any actual living organism that it should be of different species.* Ibn-Sina's observation that "the meaning of the one may be opposite to the other" might be referring to such cases. However, as we are studying models with constant domains, such cases do not concern our present discussion. I have mentioned it only to show Ibn-Sina's far-reaching reflections on probable cases.

§5. *De re* and *De dicto* **distinction.** In this final section, I would like to mention briefly what I consider as Ibn-Sina's most important and influential innovation in the subject; i.e. his discovery of distinction between two kinds of modality.

In $\Box\forall xFx(\Diamond\forall xFx)$, Ibn-Sina observes that the modal word qualifies the sentence preceded by a quantifier; however, in $\forall x\Box Fx(\forall x\Diamond Fx)$ modal word qualifies the predicate attributed to a thing or to an object. Thus he coins the name "mode of quantifier" (Jahat-e soor) for the former and "mode of predication" (Jahat-e haml) for the latter. Following Ibn-Sina, other Moslem logicians took up the subject and extensively discussed and developed it. This is the distinction that, more than one hundred years later, European medieval logicians, influenced by Ibn-Sina's writings, through Latin translations, called modality *de dicto* (about sentence) and modality *de re* (about thing).

Kneale and Kneale notice that: [6, p. 236]

> Unfortunately medieval philosophers rarely acknowledge borrowings from men of their own age.

But St. Thomas Aquinas (1225-74), the most influential thinker of the medieval period, acknowledges his debts to Ibn-Sina in different places. Interestingly, Kneale and Kneale's account of Aquinas' views on *de re* and *de dicto* modality, based on his little tract *De Modalibus*, has remarkable similarity to Ibn-Sina's wordings quoted above: [6, 237].

> [A] modal proposition *de dicto* is always singular, since it has a *dictum* for its subject, whereas a modal proposition *de re* may be universal or particular according to the sign of quantity.

Since the development of modal logic during the twentieth century, the *de re/de dicto* distinction has become a major issue, particularly, in epistemic logic and belief contexts.

In conclusion, I think, to do justice to the pioneering work of Ibn-Sina, it would be fair to re-name the presently named Buridan and the Barcan formulas as the Ibn-Sina-Buridan (IBUF) and the Ibn-Sina-Barcan (IBF) formulas.

Acknowledgments. I would like to thank Professor W. D. Hart and Professor Timothy Williamson for encouraging me to bring these historical facts to light. I also express my gratitude to the Iranian Institute for Philosophy for supporting my research.

فنقول : إن حق الجهة أن تقرن بالرابطة ، وذلك لأنها تدل على كيفية

الربط للمحمول على شيء،مطلقا أو بسور ،عهم أو مخصص ، فالسور مبين لـكيـة حـل

مكيف الربط . فإذا قلنا : كل إنسان يمكن أن يكون كاتبا ، فهو الطبيعى ، ومعناه : أن

كل واحد من الناس يمكن أن يكون كاتبا ، فإن قرن بالسور ولم يرد به إزالة عن الموضع

الطبيعى على سبيل التوسع ، بل أريد به الدلالة على أن موضها الطبيعى بمجاورة الـسور ،

لم يكن جهة للربط بل جهة للتميم والتخصيص ،وتغير المعنى ، وصار الممكن هو أن كون

كل واحد من الناس كاتهم كاتبا ممكن . والدليل على تغير المعنى أن الأول لا يشك فيه

عند جمهور الناس فإن كل واحد من الناس يعلم أنه لا يجب له فى طبيعته دوام

كتابة أو غير كتابة . وأما قولنا : يمكن أن يكون كل إنسان كاتبا ، على أن الإمكان

جهة الـكلية والسور ، فقد يشك فيه . فإذ من الناس من يقول : محال أن يكون

كل الناس كاتبين أى محال أن يوجد أن كل إنسان هو كاتب ، حتى يكون اتفق أن

لا واحد من الناس إلا وهو كاتب . فإذن بين المعنيين فرقان .

Text No. 1

وأما قولنـا : بعض الناس يمكن أن لايكون كاتبا ، فإنه قد يساوى من جهة

يمكن أن لا يكون بعض الناس كاتبا ، وقد يخالفه وإن لازمه ،

Text No. 2

REFERENCES

[1] R. C. BARCAN, *A functional calculus of first order based on strict implication*, **The Journal of Symbolic Logic**, vol. 11 (1946), pp. 1–16.

[2] R. CARNAP, *Meaning and Necessity*, The University of Chicago Press, Chicago, Ill., 1947.

[3] J. W. GARSON, *Quantification in modal logic*, **Handbook of Philosophical Logic, Vol. II** (D. Gabby and F. Guenthner, editors), Reidel, Dordrecht, 1984, pp. 249–307.

[4] G. E. HUGHES and M. J. CRESSWELL, *A New Introduction to Modal Logic*, Routledge, London, 1996.

[5] IBN-SINA, *Al-Ibara*, Dar-al-Katib-al-Arabi, Cairo, 1970.

[6] W. C. KNEALE and M. KNEALE, *The Development of Logic*, Clarendon Press, Oxford, 1962.

[7] S. KRIPKE, *Semantical considerations on modal logic*, **Acta Philosophica Fennica**, vol. 16 (1963), pp. 83–94.

[8] A. PLANTINGA, *The Nature of Necessity*, Clarendon Press, Oxford, 1974.

[9] T. WILLIAMSON, *Bare possibilia*, **Erkenntnis**, vol. 48 (1998), no. 2-3, pp. 257–279.

[10] E. N. ZALTA and B. LINSKY, *In defense of the simplest modal logic*, **Philosophical Perspectives** (Tomberline, editor), Philosophy of Logic and Language, vol. 8, Ridgeview, 1994.

IRANIAN INSTITUTE FOR PHILOSOPHY
TEHRAN, IRAN
E-mail: szia110@yahoo.com

REMARKS ON ALGEBRAIC D-VARIETIES AND THE MODEL THEORY OF DIFFERENTIAL FIELDS

ANAND PILLAY

Abstract. We continue our investigation of the category of algebraic D-varieties and algebraic D-groups from [5] and [9]. We discuss issues of quantifier-elimination, completeness, as well as the D-variety analogue of "Moishezon spaces".

§1. Introduction and preliminaries. If (K, ∂) is a differentially closed field of characteristic 0, an algebraic D-variety over K is an algebraic variety over K together with an extension of the derivation ∂ to a derivation of the structure sheaf of X. We let $\mathcal{D}$ denote the category of algebraic D-varieties and $\mathcal{G}$ the full subcategory whose objects are algebraic D-groups. $\mathcal{D}$ and $\mathcal{G}$ were essentially introduced by Buium and $\mathcal{G}$ was exhaustively studied in [2]. The category $\mathcal{D}$ is closely related but not identical to the class of sets of finite Morley rank definable in the structure $(K, +, \cdot, \partial)$ (which is essentially the class of "finite-dimensional differential algebraic varieties"). However an object of $\mathcal{D}$ can also be considered as a first order structure in its own right by adjoining predicates for algebraic D-subvarieties of Cartesian powers. In a similar fashion $\mathcal{D}$ can be considered as a many-sorted first order structure. In [5] it was shown that the many-sorted "reduct" $\mathcal{G}$ has quantifier-elimination. We point out in this paper an easy example showing:

PROPOSITION 1.1. $\mathcal{D}$ *does not have quantifier-elimination.*

The notion of a "complete variety" in algebraic geometry is fundamental. Over C these are precisely the varieties which are compact as complex spaces. Kolchin [4] introduced completeness in the context of differential algebraic geometry. This was continued by Pong [12] who obtained some interesting results and examples. We will give a natural definition of completeness for algebraic D-varieties. The exact relationship to the notion studied by Kolchin and Pong is unclear: in particular I do not know whether any of the examples

Supported by NSF grants DMS-0300639 and DMS-0100979

Logic in Tehran
Edited by A. Enayat, I. Kalantari, and M. Moniri
Lecture Notes in Logic, 26
© 2006, ASSOCIATION FOR SYMBOLIC LOGIC

of new "complete" ∂-closed sets in [12] correspond to complete algebraic D-varieties in our sense. In any case we ask whether there are "new" complete algebraic D-varieties (in our sense), and we prove:

PROPOSITION 1.2. *Suppose G is a connected algebraic D-group such that for every n every irreducible algebraic D-subvariety of G^n is a translate of a subgroup. Then G is complete inside the subcategory $\mathcal{G}$ of $\mathcal{D}$.*

Finally we study the algebraic D-variety analogue of "Moishezon spaces" (see [13]) from complex geometry. Among the algebraic D-varieties there are the "trivial" ones, where the underlying variety X is defined over the constants C_K and the derivation acts trivially on the structure sheaf of X over C_K. The subcategory $\mathcal{D}_0$ of $\mathcal{D}$ consisting of the trivial algebraic D-varieties is simply the category of algebraic varieties over the constants. The analogue of a "Moishezon space" is an algebraic D-variety which is a finite cover of a trivial algebraic D-variety.

PROPOSITION 1.3. *There is an algebraic D-variety which is a finite cover of a trivial algebraic D-variety but is not (iso)trivial.*

A model-theoretic restatement of the above proposition is the existence of an "ample" but not "very ample" strongly minimal set definable in a differentially closed field.

On the other hand, we prove that an algebraic D-group which is "D-Moishezon" *will* be isotrivial.

The remainder of this section should be seen as a brief survey of algebraic D-varieties and groups, in which we give precise definitions and recall earlier results. In this section we also discuss quantifier-elimination and prove Proposition 1.1. In Section 2, we discuss completeness, and in Section 3 we introduce the notion of D-Moishezon D-varieties. We will assume acquaintance with model-theoretic notions, in particular the model theory of differentially closed fields (see [6]). For stability theory see [8]. Some of the work on completeness was done while visiting the University of Illinois at Chicago in March 2003. Thanks to Dave Marker and John Baldwin for their hospitality.

We fix a differentially closed field $(\mathcal{U}, \partial)$ of characteristic zero. When we talk about definability in $\mathcal{U}$ we refer to definability in the structure $(\mathcal{U}, +, \cdot, -, 0, 1, \partial)$. It is convenient to assume $\mathcal{U}$ to be a "universal domain" namely to be κ-saturated of cardinality κ. $K, L, \ldots$ will, unless stated otherwise, denote small differential subfields of $\mathcal{U}$. C denotes the field of constants of $\mathcal{U}$. Let $V \subset \mathcal{U}^n$ be an affine algebraic variety which we identify with its set $V(\mathcal{U})$ of $\mathcal{U}$-rational points. The *shifted tangent bundle*, $T_\partial(V)$ of V is the subvariety of $\mathcal{U}^{2n}$ defined by (i) the equations defining V in the the first n coordinates, together with (ii) equations $\sum_{i=1,\ldots,n}(\partial P/\partial x_i)(\bar{x})u_i + P^\partial(\bar{x})$ for P ranging over any set of generators of the ideal $I(V)$ of V. Here P^∂ is the polynomial obtained by applying ∂ to the coefficients of P. The map $\pi : T_\partial(V) \to V$ denotes projection onto the the first n coordinates.

In the case where V is defined over C, $(T_\partial(V), \pi)$ is none other than the tangent bundle of V.

An affine algebraic D-variety is by definition a pair (V, s) where V is an affine algebraic variety and $s : V \to T_\partial(V)$ is a regular (polynomial) section of π.

We say that (V, s) is irreducible if the underlying variety V is irreducible.

If $a \in V$, we can apply ∂ coordinatewise to a and it is rather easy to see that $(a, \partial(a)) \in T_\partial(V)$. We define $(V, s)^\sharp$ to be $\{a \in V : s(a) = (a, \partial(a))\}$, which is visibly a definable set in $\mathcal{U}$ (of finite Morley rank). In fact:

FACT 1.4.   (i) Any definable subset of $\mathcal{U}^n$ of finite Morley rank, is, up to definable bijection and Boolean combination, of the form $(V, s)^\sharp$ for some irreducible affine algebraic D-variety (V, s).
(ii) Let (V, s) be an irreducible affine algebraic D-variety. Then $(V, s)^\sharp$ is Zariski-dense in V.

Explanation. (i) is contained in Proposition 4.2 of [9] and its proof. The main point is that $tp(a/K)$ has finite Morley rank if and only if

$$\text{tr. deg}(K(a, \partial(a), \partial^2(a), \dots)/K)$$

is finite. Replacing the tuple a by $(a, \partial(a), \dots, \partial^n(a))$ for suitable n, we obtain $\partial(a) \in K(a)$, and after adjoining to a something in $K(a)$ we obtain $\partial(a) \in K[a]$, so $(a, \partial(a)) = s(a)$ for some polynomial map s defined over K. Take V to be the algebraic variety over K whose generic point is a. Then (V, s) is an affine algebraic D-variety, and $a \in (V, s)^\sharp$.

More generally we have the notion of an abstract algebraic D-variety (over $\mathcal{U}$), (V, s). V will be an abstract variety: namely V has a covering by affine algebraic varieties $V_1, \dots, V_r$, such that the transition maps between Zariski-open subsets of the V_i are biregular. The $T_\partial(V_i)$ then piece together to form a variety $T_\partial(V)$ together with the projection π to V. $T_\partial(V)$ is the shifted tangent bundle of V. (In previous papers we used the notation $\tau(V)$ in place of $T_\partial(V)$.) If $s : V \to T_\partial(V)$ is a regular section we call (V, s) an (abstract) algebraic D-variety. Again we say that (V, s) is irreducible if the variety V is irreducible. By an algebraic D-subvariety of (V, s) we mean a subvariety W of V such that $s|W : W \to T_\partial(W)$. We also may call W a D-closed subset of (V, s).

If (V, s) and (W, t) are algebraic D-varieties then their "Cartesian product" is $(V \times W, s \times t)$, also an algebraic D-variety.

Note that if V, W are algebraic varieties over $\mathcal{U}$ and $f : V \to W$ is a morphism of varieties (or regular map) then we can modify the differential of f to obtain the "shifted differential" $d_\partial(f) : T_\partial(V) \to T_\partial(W)$. $d_\partial(f)$ is precisely $df + f^\partial$ where again f^∂ is the result of hitting the coefficients of f with ∂.

By a D-morphism (or D-regular map) between (V, s) and (W, t) we mean a morphism $f : V \to W$ of algebraic varieties such that $(d_\partial(f)) \circ s = t \circ f$. Note that the graph of f is then a D-subvariety of $(V \times W, s \times t)$.

If G is an algebraic group over $\mathcal{U}$, then $T_\partial(G)$ has naturally the structure of an algebraic group: if $f : G \times G \to G$ is the group operation, then the group operation on $T_\partial(G)$ is given by $(g, u) \cdot (h, v) = (f(g, h), (df_\partial)_{(g,h)}(u, v))$. By an algebraic D-group we mean an algebraic group G together with a regular section $s : G \to T_\partial(G)$ which is also a homomorphism.

We let $\mathcal{D}$ denote the category of algebraic D-varieties, and $\mathcal{G}$ the full subcategory of $\mathcal{D}$ whose objects are the algebraic D-groups.

As in Fact 1.4(ii), if (V, s) is an algebraic D-variety, then $(V, s)^\sharp = \{x \in V(\mathcal{U}) : \partial(x) = s(x)\}$ is well-defined and is Zariski-dense in V (and is an interpretable set in the structure $\mathcal{U}$). If (G, s) is an algebraic D-group, then $(G, s)^\sharp$ is a group of finite Morley rank definable (or rather interpretable) in $\mathcal{U}$. Moreover every group of finite Morley rank definable in $\mathcal{U}$ is of this form, up to definable isomorphism.

By a D-rational map from (V, s) to (W, t) we mean a rational map f from V to W such that $d_\partial(f) \circ s = t \circ f$ holds on a Zariski-open subset of V.

We say that (V, s) is *trivial* if V is defined over $\mathcal{C}$ and $s = 0$, the 0-section of the tangent bundle of V. Note that if (V, s) is trivial, then the algebraic D-subvarieties of $V, V \times V, \ldots$ are precisely those algebraic subvarieties which are defined over $\mathcal{C}$. A D-morphism between trivial algebraic D-varieties $(V, 0)$ and $(W, 0)$ is precisely a morphism of varieties which is defined over $\mathcal{C}$. We let $\mathcal{D}_0$ denote the full subcategory of $\mathcal{D}$ whose objects are the trivial algebraic D-varieties. Note that $\mathcal{D}_0$ can be thought of as "algebraic geometry over $\mathcal{C}$". We say that (V, s) is *isotrivial* if (V, s) is isomorphic (as an algebraic D-variety) to some trivial algebraic D-variety. In analogy with the category of compact complex spaces (discussed further below) it may be more suitable to use the expression "algebraic" instead of "isotrivial".

We now explain how an algebraic D-variety (V, s) can be viewed as a first order structure. Given (V, s) the underlying set of the associated structure will be $V(\mathcal{U})$. For each algebraic D-subvariety W of $(V \times \cdots \times V, s \times \cdots \times s)$, we give ourselves an n-place predicate whose interpretation is $W(\mathcal{U})$. We call this language $L_{(V,s)}$ and hopefully without ambiguity we write (V, s) for the structure.

We can also make the whole category $\mathcal{D}$ into a many-sorted first order structure. The sorts will be the algebraic D-varieties (up to isomorphism say), and the predicates will correspond to algebraic D-subvarieties W of the various cartesian products $(V_1, s_1) \times \cdots \times (V_n, s_n)$. We call the language $L_{\mathcal{D}}$ and the many sorted structure $\mathcal{D}$, hopefully without ambiguity.

Likewise $\mathcal{G}$ can be considered to be a many-sorted first order structure: the sorts are the algebraic D-groups (or even the algebraic D-subvarieties of algebraic D-groups) and relations are algebraic D-subvarieties of Cartesian products. Again call the structure $\mathcal{G}$ and the language $L_{\mathcal{G}}$.

Note that the category $\mathcal{D}$ belongs, in a natural sense, to algebraic geometry, rather than to Kolchin's differential algebraic geometry. Moreover $\mathcal{D}$ as a many-sorted structure is of course interpretable in the pure algebraically closed field $(\mathcal{U}, +, \cdot)$.

On the other for (V, s) an algebraic D-variety, $(V, s)^{\sharp}$ does belong to Kolchin's differential algebraic geometry. It is a special case of a differential algebraic variety of finite dimension. As a subset of V, $(V, s)^{\sharp}$ inherits an $L_{V,s}$-structure. More generally, let $\mathcal{D}^{\sharp}$ be the substructure of the $L_{\mathcal{D}}$-structure $\mathcal{D}$, which sort-by-sort is $(V, s)^{\sharp}$.

In [9], $(V, s)^{\sharp}$ as an $L_{V,s}$-structure was studied in detail. (Actually we concentrated there on the case where V is affine but everything generalizes.) The following summarizes the situation:

FACT 1.5. Let (V, s) be an irreducible algebraic D-variety. Then

(i) The subsets of $((V, s)^{\sharp})^n$ definable in $(\mathcal{U}, \partial)$ are precisely the subsets definable in the $L_{V,s}$-structure $(V, s)^{\sharp}$.

(ii) $(V, s)^{\sharp}$ has quantifier-elimination as an $L_{V,s}$-structure,

(iii) Moreover, if we define the "closed" subsets of $((V, s)^{\sharp})^n$ to be those given by the basic predicates of $L_{V,s}$ (so the traces on $(V, s)^{\sharp}$ of D-subvarieties of $(V, s)^n$), them $(V, s)^{\sharp}$ is a (possibly incomplete) Zariski structure in the sense of Zilber [14]

Likewise for the many-sorted structure $\mathcal{D}^{\sharp}$.

PROPOSITION 1.6. Let (V, s) be an algebraic D-variety. Then $Th(V)$ in the language $L_{(V,s)}$ has quantifier elimination iff and only $(V, s)^{\sharp}$ is an $L_{(V,s)}$-elementary substructure of V.

PROOF. As by 1.5(ii) $Th((V, s)^{\sharp})$ has quantifier elimination, clearly we get "right implies left".

Conversely, assume the left hand side. Note that any element of $(V, s)^{\sharp}$ is to all intents and purposes named by a constant of the language. So by Tarski-Vaught (and our assumption of quantifier-elimination) we must show that any quantifier-free $L_{(V,s)}$-formula $\phi(x)$ with a solution in V has a solution in $(V, s)^{\sharp}$. We may assume that ϕ is of the form "$x \in X \wedge x \notin Y$" where X, Y are D-subvarieties of V and X is irreducible. But $X^{\sharp} = X \cap (V, s)^{\sharp}$ is Zariski-dense in X. Hence ϕ has a solution in $(V, s)^{\sharp}$. ⊣

Proposition 1.6 has the following consequence: if (V, s) has quantifier-elimination in $L_{V,s}$ then the definable set $(V, s)^{\sharp}$ with all its induced structure from $(\mathcal{U}, \partial)$ is interpretable in a (pure) algebraically closed field.

Let us describe what we know in general about the relationship between the structures $\mathcal{D}$ and $\mathcal{D}^{\sharp}$, and the obstacle to quantifier-elimination for $\mathcal{D}$. The results are either contained in [9] or can be extracted from the proofs there.

First some notation: $X, Y, \ldots$ will denote algebraic D-subvarieties of sorts in D. If X is a D-subvariety of (V, s) then $X^{\natural}$ denotes $X \cap (V, s)^{\natural}$. $f, g, \ldots$ denote D-morphisms between algebraic D-varieties.

FACT 1.7.   (i) X is the Zariski-closure of $X^{\natural}$,

(ii) Suppose X, Y are D-subvarieties of (V, s). Then $X \cap Y$ and $X \cup Y$ are also D-subvarieties of (V, s) and moreover $(X \cap Y)^{\natural} = X^{\natural} \cap Y^{\natural}$ and $(X \cup Y)^{\natural} = X^{\natural} \cup Y^{\natural}$.

(iii) Let $f : (V_1, s_1) \rightarrow (V_2, s_2)$ be a D-morphism, and let $Z \subseteq V_1$ be an irreducible D-subvariety. Then the Zariski-closure of $f(Z)$ is an irreducible D-subvariety W say of (V_2, s_2). Moreover there is a proper D-subvariety W_1 of W such that $f(Z^{\natural}) = W^{\natural} \backslash W_1^{\natural}$, and $f(Z) = W \backslash W_2$ for W_2 some proper subvariety of W which contains W_1. Finally $W_2 = W_1$ if and only if W_2 is a D-subvariety of W.

The obstacle to QE for D is precisely the possibility that W_2 in (iii) above does not equal W_1. The following gives such an example.

Let $t \in U$ be such that $\partial(t) = 1$. Let V be the irreducible subvariety of affine 2-space defined by $(x - t)y = 1$. Then as can be easily checked $T_{\partial}(V)$ is defined by "$(x, y) \in V$" together with $u_2 + (u_1 - 1)y^2 = 0$. So $s : V \rightarrow T_{\partial}(V)$ defined by $s(x, y) = (x, y, 0, y^2)$ is a regular section. Moreover the projection map $p : V \rightarrow A^1$ taking (x, y) to x is a D-variety map from (V, s) to $(A^1, 0)$. However $p(V) = A^1 \setminus \{t\}$, and $\{t\}$ is NOT a D-subvariety of $(A^1, 0)$.

Thus:

PROPOSITION 1.8. D *does not have quantifier-elimination.*

On the other hand the main result of [5] is:

FACT 1.9. G does have quantifier-elimination: $G^{\natural}$ is an elementary substructure of G, hence interpretable in a (pure) algebraically closed field.

The main ingredient in the proof of Fact 1.9 was the "strong socle theorem":

FACT 1.10. Suppose (G, s) is a connected algebraic D-group. Let X be an irreducible D-subvariety of (G, s) which contains the identity and generates G. Then $\text{Stab}(X)$ is an algebraic D-subgroup of (G, s), and $\text{Stab}(X)$ contains a normal algebraic D-subgroup N of (G, s) such that the algebraic D-group $(G/N, s/N)$ is isotrivial. N can be chosen to be $\text{Stab}(X) \cap Z(G)$.

It follows that if (G, s) is an algebraic D-group, and X is a D-subvariety with trivial stabilizer, then $(X, s|X)$ is isotrivial (or "algebraic"). This should be compared to a well-known result of Ueno that if A is a complex torus and X is an analytic subvariety of A with trivial stabilizer, then X is (biholomorphic) to a projective algebraic variety. In any case the D-group statement above also yields a rather direct proof of Mordell-Lang for function fields in characteristic zero (see [10]).

§2. **Completeness.** We will give a rather natural definition of "completeness" for algebraic D-varieties. Kolchin and Pong studied a notion of completeness for differential algebraic varieties. The relation between the two notions is unclear.

DEFINITION 2.1. Let (V, s) be an algebraic D-variety. We say that (V, s) is *complete* if for any algebraic D-variety (V', s') the projection map from $V \times V'$ to V' is D-closed, namely takes D-closed sets to D-closed sets.

REMARK 2.2.   (i) In the definition above it is enough to demand that the image of a D-closed subset of $V \times V'$ under the projection map to V' is Zariski closed, since we know by Fact 1.7(iii) that the Zariski closure of the image of a D-closed set under a D-morphism is D-closed. In particular we see that if the underlying variety V of (V, s) is complete, then (V, s) is complete.
(ii) If (V, s) is complete then it has quantifier-elimination as an $L_{V,s}$-structure.

On the other hand Buium in [1] proves that any algebraic D-variety whose underlying variety is projective is isotrivial. We expect the proof also works if the underlying variety is complete. The main issue that concerns us is whether there exist complete algebraic D-varieties which are NOT isotrivial. We will broach this below, and for now just give some additional remarks on completeness.

The analogy in our context of affine 1-space is $(A^1, 0)$. By a D-*regular function* on an algebraic D-variety (V, s) we mean a D-morphism from (V, s) to $(A^1, 0)$. Likewise by a D-*rational function* on (V, s) we mean a D-rational map from (V, s) to $(A^1, 0)$.

LEMMA 2.3. *Suppose that (V, s) is an irreducible complete algebraic D-variety. Then any D-regular function on (V, s) is constant.*

PROOF. The usual proof that there are no regular functions on a complete algebraic variety works: suppose $f : (V, s) \to (A^1, 0)$ is a D-morphism. We can view f as a D-morphism into $(P^1, 0)$. $f(V)$ is then a Zariski-closed subset of P^1 contained in A^1, hence is finite. If $|f(V)| > 1$ then we contradict the irreducibility of V. So $f(V)$ is a singleton.   ⊣

DEFINITION 2.4. Let (G, s) be a connected algebraic D-group. We call (G, s) modular if every irreducible D-subvariety of $G \times \cdots \times G$ is a (left) translate of an algebraic D-subgroup of $(G \times \cdots \times G, s \times \cdots \times s)$.

REMARK 2.5. (G, s) is modular in the sense of the Definition 3.4 above if and only if G is modular (or 1-based) as a $L_{G,s}$-structure. It is also equivalent to $(G, s)^{\sharp}$ being 1-based as a group definable in (K, ∂). In particular a modular connected algebraic D-group is commutative.

PROOF. This follows from Fact 1.9.   ⊣

The examples of modular algebraic D-groups come from abelian varieties A over $\mathcal{U}$ which have no abelian subvarieties isomorphic to abelian varieties defined over C (see [2] and [7]). Let A be such. Then the smallest definable (in $(\mathcal{U}, \partial)$) subgroup of $A(\mathcal{U})$ which contains the torsion subgroup, is a connected definable group of finite Morley rank. The latter group is definably isomorphic to $(B, s)^{\sharp}$ for some algebraic D-group (B, s) where B is an extension of A by a vector group. Note that the underlying variety of B is NOT complete.

We conjecture that any modular algebraic D-group is complete (as an algebraic D-variety). We have not been able to prove this so far. However we will point out that this IS the case if we restrict ourselves to the category $\mathcal{G}$ of (closed) algebraic D-subvarieties of algebraic D-groups.

PROPOSITION 2.6. *Let (G, s) be a modular algebraic D-group. Then for any algebraic D-group (H, t) and D-closed subset X of $G \times H$, $p_2(X)$ is D-closed in H where p_2 is the second projection. Hence (G, s) is complete in the category $\mathcal{G}$.*

PROOF. Let p_1, p_2 denote the projections from $G \times H$ to G, H respectively. We only have to show that $p_2(X)$ is Zariski-closed in H.

We may clearly assume X to contain the identity (of $G \times H$). Let L be the algebraic subgroup of $G \times H$ generated by X. Then L is a connected algebraic D-subgroup of $G \times H$ and we may clearly assume that $p_1(L) = G$ and $p_2(L) = H$. Let $\text{Stab}(X) = \{(g, h) \in G \times H : (g, h) \cdot X = X\}$. Let $S = \text{Stab}(X) \cap Z(G \times H)$, which is by Fact 1.10 a normal algebraic D-subgroup of $(G \times H, s \times t)$.

CLAIM 1. L/S (with its natural algebraic D-group structure) is isotrivial.

PROOF. By Fact 1.10. ⊣

CLAIM 2. $p_1(S) = G$.

PROOF. Let $p_1(S) = S_1$. The natural map from L/S to G/S_1 is a morphism of algebraic D-groups. But G/S_1 is also modular (as an algebraic D-group). By Claim 1, L/S maps to the identity. As L projects onto G it follows that $S_1 = G$. ⊣

NOTATION. Let $S_0 = \{x \in H : (0, x) \in S\}$, a central D-subgroup of H, For $g \in G$, let $f(g) = \{x \in H : (g, x) \in S\}$, as an element of $p_2(S)/S_0$.

CLAIM 3. $f : G \to p_2(S)/S_0$ is a surjective homomorphism of algebraic D-groups, so the latter is modular.

PROOF. Follows from Claim 2. ⊣

CLAIM 4. The map taking $(g, h)S$ to $f(g)^{-1}(h/S_0)$ is an isomorphism (of algebraic D-groups) between $(G \times H)/S$ and H/S_0.

PROOF. This is immediately verified. ⊣

Now let H_1/S_0 be the image of L/S under the isomorphism in Claim 4, and let X' be the image of X/S. So X' is a D-closed subset of H_1/S_0.

CLAIM 5. The (multiplication) map taking $p_2(S)/S_0 \times H_1/S_0$ onto $p_2(S)H_1/S_0$ has finite kernel.

PROOF. $p_2(S)/S_0$ is modular (by Claim 3) and H_1/S_0 is isotrivial (being an image of L/S), hence the two groups are "orthogonal" which implies the desired conclusion.   ⊣

By Claim 5, $X'H_1/S_0$ is Zariski-closed in H/H_0 (A surjective homomorphism of algebraic groups with finite kernel is a closed map.) But $p_2(X)$ is precisely the preimage in H of $X'H_1/S_0$ so is also Zariski-closed. This completes the proof of Proposition 2.6.   ⊣

§3. **Finite covers of isotrivial algebraic D-varieties.** We work first at the model-theoretic level. It is convenient to work in the universal domain (namely inside the model $(\mathcal{U}, \partial)$ of DCF_0) and then translate the results into the Zariski geometry setting and the algebraic D-variety setting. We recall the notions, almost internal and internal. If $p(x) \in S(A)$ is a stationary type in some saturated model of a stable theory, and Y is some $\emptyset$-definable set, p is said to be almost internal to Y if there is some $B \supseteq A$ and a realizing p independent from B over A such that $a \in acl(B, Y)$. Equivalently there is some set B of parameters containing A such that every realization of $p(x)$ is in $acl(B, Y)$. If we replace algebraic closure by definable closure we get "internality" in place of "almost internality". A definable set X is said to be (almost) internal to Y if there is some set B of parameters over which X is defined such that $X \subseteq dcl(B, Y)$ $(X \subseteq acl(B, Y))$.

We return to the differentially closed field $(\mathcal{U}, \partial)$. If k is a differential subfield and d a tuple, $k\langle d\rangle$ denotes the differential field generated by k and d. We will sometimes work in the underlying algebraically closed field $(\mathcal{U}, +, \cdot)$, in which case we will refer to "types in the field language" or $tp_f(-)$. The first aim is to find a stationary strongly minimal type $p(x)$, such that $p(x)$ is almost internal to C but not internal to C. Let us fix some element $b \in \mathcal{U}$ such that $b \notin C$. Let k be the differential field generated by b. Let $c \in C$, be such that $c \notin acl(k)$. Let d be such that $d^2 = b + c$. Let $p(x) = tp(d/b)$. With this notation:

LEMMA 3.1.   (i) $p(x)$ is stationary. (In fact $tp(dc/k)$ is stationary.)
(ii) $p(x)$ is almost internal to C.
(iii) $p(x)$ is not internal to C.

PROOF. (i) Let $K = acl(k)$. It is enough to show that if d_1, d_2 realize $p(x)$ then they have the same type over K. Let $c_i \in C$ be such that $tp(d_i c_i/k) = tp(dc/k)$. Then $c_1, c_2 \notin K$ so as they are constants they have the same type over K. So (by an automomorphism) we may assume that $c_1 = c_2 = c$.

Now d_1 and d_2 have the same type in the field sense over K (as both are transcendental over K). Thus as $c = d_i^2 - b$ for $i = 1, 2$, (d_1, c) and (d_2, c) have the same type over K in the fields sense. So d_1 and d_2 have same type in the fields sense over $K(c) = K\langle c \rangle$. As $d_i \in acl(K\langle c \rangle)$ it follows that d_1 and d_2 have same type over $K\langle c \rangle$ (in DCF). In particular they have the same type over K. This proof also yields the parenthetical remark.

(ii) In fact $d \in acl(k, c)$, and $c \in C$.

(iii) This is a slight strengthening of part (i). Let L be an arbitrary algebraically closed differential field (saturated if one wants) containing k and independent from d over k (in the sense of DCF). We have to show that $d \notin dcl(L, C)$. But $dcl(L, C)$ is precisely the field compositum LC. So we must prove:

(*) There is no rational function f and tuples e_1 from L and e_2 from C such that $d = f(e_1, e_2)$.

Before proving (*) we make some observations about the set-up.

(1) L and C are algebraically disjoint over their intersection C_L and the latter is algebraically closed.

(2) $b, c \notin C_L$.

PROOF. We know already that b is not a constant. On the other hand, as L is independent (in any sense) from d over k, L is also independent from c over k. As $c \notin acl(k)$ it follows that $c \notin L$. ⊣

So b, c are transcendental over C_L. Let L_1 be the field of Puiseaux series (fractional power series) in b over C_K, and L_2 the same thing with b replaced by c. We may assume L_1, L_2 to be embedded in L, C, respectively over C_L. L_1, L_2 are algebraically closed fields containing $C_L(b), C_L(c)$ respectively.

Now suppose for a contradiction that $d = f(e_1, e_2)$ for a rational function f and $e_1 \in L$, $e_2 \in C$. So $f(e_1, e_2)^2 = b + c$. Now L is algebraically disjoint from $L_1 C$ over L_1, hence we find $e_1' \in L_1$ such that $f(e_1', e_2)^2 = b + c$. Likewise C is algebraically disjoint from $L_1 L_2$ over L_2, hence we find $e_2' \in L_2$ such that $f(e_1', e_2')^2 = b + c$. So $f(e_1', e_2') \in L_1 L_2$ and is a square root of $b + c$. But an easy computation shows this to be impossible. (The field compositum of the fractional power series $C_L((x))^*$ and $C_L((y))^*$ does not contain a square root of $x + y$.) This completes the proof of (iii). ⊣

Let us first interpret Lemma 3.1 in terms of the notions ample/very ample from [3]. These notions were stated for Zariski geometries but make sense for any strongly minimal set.

Let X be a strongly minimal set (defined over a $\emptyset$ say) in some saturated model $\bar{M}$ of a stable theory. X is said to be *ample* (or nonlocally modular), if (after possibly adding parameters) there is a stationary type $p(x_1, x_2, b)$ of Morley rank 1 such that $p(x_1, x_2, b)$ implies $x_1, x_2 \in X$, $b = Cb(p(x_1, x_2, b))$ and $RM(tp(b)) = 2$.

X is said to be *very ample* if there is $p(x_1, x_2, b)$ witnessing ampleness of X such that, in addition, whenever (a_1, a_2), (a_1', a_2') are distinct generic pairs from X, then there is b' realizing $tp(b)$ such that exactly one of (a_1, a_2), (a_1', a_2') realizes $p(x_1, x_2, b')$.

PROPOSITION 3.2. *Let $b \in \mathcal{U}$ be a nonconstant. Then the subset X of $\mathcal{U}$ defined by $\partial(x^2 - b) = 0$ is an ample but not very ample strongly minimal set (in the ambient differentially closed field $\mathcal{U}$).*

PROOF. By Lemma 3.1(i), over any set A containing b there is a unique nonalgebraic complete type over A containing the formula $\partial(x^2 - b) = 0$. Hence X is strongly minimal. As the strongly minimal algebraically closed field $\mathcal{C}$ is contained in $dcl(X)$, it follows that X is ample. If X were very ample, then the proofs in [3] would yield that there is a set A of parameters containing b such that X is contained in $dcl(A, \mathcal{C})$. This contradicts Lemma 3.1(iii).   ⊣

REMARK 3.3.   (i) Even though X is not very ample, it is probably interpretable in a (pure) algebraically closed field, unlike the examples in [3] and [14].

(ii) X with the induced structure is a Zariski geometry, so we have an example of an ample, not very ample Zariski geometry.

We now pass to the interpretation of Lemma 3.1 in the category of algebraic D-varieties. Just before Lemma 2.3 we defined the notion of a D-rational function on an algebraic D-variety (V, s). Note that if V is irreducible then the set of such functions is field of transcendence degree over $\mathcal{C}$ of at most $\dim(V)$.

DEFINITION 3.4. Let (V, s) be an irreducible D-variety, with $\dim(V) = n$. We say that (V, s) is D-*Moishezon* if the field of D-rational functions on (V, s) has transcendence degree n (over $\mathcal{C}$).

The definition is borrowed from compact complex manifolds, where M is said to be Moishezon if the field of meromorphic functions on X has transcendence degree (over the complexes) equal to $\dim_{\mathcal{C}}(M)$. A striking fact about Moishezon (compact complex) manifolds M is that although M need not be biholomorphic with an algebraic variety, there DOES exist a projective algebraic variety V and a generically finite-to-one holomorphic map from V to M. So in the many-sorted structure CCM of compact complex spaces, any Moishezon space is internal to $P^1(\mathcal{C})$. Lemma 3.1 shows that this FAILS in the algebraic D-variety setting.

REMARK 3.5. Suppose (V, s) is an irreducible algebraic D-variety which is D-Moishezon. Then there is a trivial irreducible algebraic D-variety $(W, 0)$ and a generically finite-to-one dominant D-rational map from (V, s) to $(W, 0)$.

Explanation. Let $f_1, \ldots, f_r$ be generate (over C) the field of D-rational functions on (V, s). Then $(f_1, \ldots, f_r)$ is a D-rational map from (V, s) to $(A^r, 0)$. The Zariski-closure of the image is then an irreducible D-subvariety of $(A^r, 0)$ of dimension n, so of the form $(W, 0)$.

Let V be the variety $A^1 \setminus 0$. Let $b \in \mathcal{U}$ be a nonconstant, and let $s : V \to T(V)$ be given by $s(x) = (x, \partial(b)/2x)$. Note that $(V, s)^{\sharp}$ is precisely $\{x \in \mathcal{U} : \partial(x^2 - b) = 0\}$, the strongly minimal set described above.

PROPOSITION 3.6. (V, s) *is D-Moishezon. However there is no D-rational dominant map from a trivial algebraic D-variety to* (V, s).

PROOF. The map $x \to x^2 - b$ is clearly a nonconstant D-regular function on (V, s). As $\dim(V) = 1$, (V, s) is D-Moishezon.

If there were a D-rational dominant map f from a trivial algebraic D-variety $(W, 0)$ to (V, s), then $f(W, 0)^{\sharp}$ would be a cofinite subset of $(V, s)^{\sharp} = X$ (as above). So X is contained in $dcl(C, A)$ for some finite set A of parameters, again contradicting Lemma 3.1 $\dashv$

Finally we observe that algebraic D-groups are better behaved:

PROPOSITION 3.7. *Suppose that (G, s) is an algebraic D-group, which is D-Moishezon. Then (G, s) is isomorphic to a trivial algebraic D-group.*

The proposition will follow from some general results concerning groups of finite Morley rank. We will state and prove these result (using stability-theoretic notions freely) and then point out how they apply.

ASSUMPTIONS. Work in a saturated model $\bar{M}$ of an arbitrary stable theory T. "Definable" means definable (with parameters) in $\bar{M}$.

Remember that a definable group G is said to be connected if G has no proper definable subgroup of finite index. We will say that a definable group G is *abstractly connected* if G has no proper subgroup of finite index.

LEMMA 3.8. *Suppose that every definable connected commutative group of finite Morley rank is abstractly connected. Then every definable connected group of finite Morley rank is abstractly connected.*

PROOF. Let G be a connected definable group of finite Morley rank. We prove by induction on $RM(G)$ that G is abstractly connected. If $RM(G) = 1$ then G is commutative and we can use our hypothesis. Suppose $RM(G) = n > 1$. Suppose for a contradiction that H is a proper (abstract) subgroup of G of finite index. We may assume H to be normal in G. If G is commutative, we have a contradiction by hypothesis. Otherwise $Z(G)$, being definable, has infinite index in G. So pick $a \in H \setminus Z(G)$. Then the set $a^G a^{-1}$ is infinite, "indecomposable" and contains the identity. By Zilber's indecomposability theorem, the normal subgroup N say of G generated by $a^G a^{-1}$ is definable, and is clearly contained in H. As N is infinite, $RM(G/N) < n$, and clearly H/N is a proper subgroup of G/N of finite index. As G/N is connected, we have a contradiction to the induction hypothesis. $\dashv$

LEMMA 3.9. *Let G be a definable group which is abstractly connected. Let N be a finite normal subgroup of G. Suppose G and N are A-definable. Then $G \subseteq dcl(A, G/N)$.*

PROOF. By stability it is enough to show that if α is a group automomorphism of G such that $\alpha(N) = N$ and α induces the identity map on G/N, then α is the identity. For $x \in G$ define $\beta(x) = x^{-1}\alpha(x)$. Our assumptions imply that β is a crossed homomorphism from G to N. Hence $\ker(\beta) = \{x \in G : \alpha(x) = x\}$ has finite index in G, so equals G.   ⊣

We now return to the differential fields context. Our ambient structure is now $(\mathcal{U}, +, \cdot, \partial)$.

COROLLARY 3.10. *Let G be a connected group of finite Morley rank definable in $\mathcal{U}$. Suppose G to be almost internal to $\mathcal{C}$. Then G is definably isomorphic to $H(\mathcal{C})$ for some connected algebraic group H defined over $\mathcal{C}$.*

PROOF. First, it is well-known that any connected commutative group of finite Morley rank definable in $\mathcal{U}$ is divisible, hence abstractly connected (Any connected commutative group of finite Morley rank definable in $\mathcal{U}$ embeds in a connected commutative algebraic group over $\mathcal{U}$, so for every n has only finitely many elements of order n. Hence the map $x \to nx$ has finite kernel so is surjective by connectedness.). So Lemmas 3.8 and 3.9 apply. Now suppose G to be connected definable, of finite Morley rank, and almost internal to $\mathcal{C}$. By [11], there is a finite normal subgroup N of G such that G/N is internal to $\mathcal{C}$. So by Lemma 3.9, G is internal to $\mathcal{C}$. It then follows that G is definably isomorphic to some group definable in $(\mathcal{C}, +, \cdot)$, and by Weil-Hrushovski [11], the latter is of the required form.   ⊣

PROOF OF PROPOSITION 3.7. Let (G, s) be our connected algebraic D-group which is D-Moishezon. It easily follows that $(G, s)^{\sharp}$ is almost internal to $\mathcal{C}$. By Corollary 3.10, $(G, s)^{\sharp}$ is definably isomorphic to $H(\mathcal{C})$ for some connected algebraic group H over $\mathcal{C}$. By [5], this isomorphism lifts to an isomorphism of algebraic D-groups between (G, s) and $(H, 0)$.   ⊣

REFERENCES

[1] A. BUIUM, *Differential Function Fields and Moduli of Algebraic Varieties*, Springer-Verlag, Berlin, 1986.

[2] ———, *Differential Algebraic Groups of Finite Dimension*, Springer-Verlag, Berlin, 1992.

[3] E. HRUSHOVSKI and B. ZILBER, *Zariski geometries*, **Journal of the American Mathematical Society**, vol. 9 (1996), no. 1, pp. 1–56.

[4] E. KOLCHIN, *Differential equations in a projective space and linear dependence over a projective variety*, **Contributions to Analysis**, Academic Press, New York, 1974, pp. 195–214.

[5] P. KOWALSKI and A. PILLAY, *Quantifier elimination for algebraic D-groups*, to appear in Transactions AMS.

[6] D. MARKER, *Model theory of differential fields*, **Model Theory of Fields** (D. Marker, M. Messmer, and A. Pillay, editors), Lecture Notes in Logic, vol. 5, Springer, Berlin, 1996.

[7] A. PILLAY, *Differential algebraic groups and the number of countable models*, **Model Theory of Fields** (D. Marker, M. Messmer, and A. Pillay, editors), Lecture Notes in Logic, vol. 5, Springer, Berlin, 1996.

[8] ——, **Geometric Stability Theory**, Oxford University Press, New York, 1996.

[9] ——, *Two remarks on differential fields*, **Model Theory and Applications**, Quaderni di Matematica, vol. 11, Dept. Mat., 2° Univ. di Napoli, 2002, pp. 325–347.

[10] ——, *Mordell-Lang conjecture for function fields in characteristic zero, revisited*, **Compositio Mathematica**, vol. 140 (2004), no. 1, pp. 64–68.

[11] B. POIZAT, **Stable Groups**, American Mathematical Society, Providence, RI, 2001.

[12] W.-Y. PONG, *Complete sets in differentially closed fields*, **Journal of Algebra**, vol. 224 (2000), no. 2, pp. 454–466.

[13] I. SHAFAREVICH, **Basic Algebraic Geometry**, Springer-Verlag, Berlin, 1977.

[14] B. ZILBER, *Notes on Zariski geometries*, See http://www.maths.ox.ac.uk/~zilber.

DEPARTMENT OF MATHEMATICS
UNIVERSITY OF ILLINOIS
URBANA-CHAMPAIGN, USA
E-mail: pillay@math.uiuc.edu

A SIMPLE POSITIVE ROBINSON THEORY WITH LSTP ≠ STP

MASSOUD POURMAHDIAN AND FRANK WAGNER

Abstract. We construct a simple positive Robinson theory with a bounded type-definable equivalence relation which is not the intersection of definable equivalence relations. Consequently, in this theory strong types are weaker than Lascar strong types, and the Lascar group is non-trivial.

§1. **Introduction.** In his paper [Hr2] Hrushovski introduces the notion of a Robinson theory, and given a compact connected group G and a homogeneous G-space X he constructs a simple Robinson theory whose Lascar group action on its Kim-Pillay space is isomorphic to (G, X). The notion of Robinson theory was generalized by Pillay [Pi], and then by Ben-Yaacov [BY1, BY2], who developed the basic model and stability theory of *positive* Robinson theories. (It should be noted, though, that these categories had already been considered by Shelah in [Sh].)

In this paper we shall use a variant of Hrushovski's amalgamation construction [BS, Hr1, Wa] and obtain our structure as the generic model for an amalgamation with variable weight of the relations. While we cannot first-order axiomatize the class of existentially closed models, we show that the generic model is compact and homogeneous for a certain existentially closed class of formulas, and thus the universal domain of a positive Robinson theory. We hope that the construction method developed will be ultimately useful in constructing a first-order simple theory with prescribed Lascar group.

Notation is standard. In particular, we shall write $\subseteq_\omega$ to denote finite subsets and substructures, and AB for $A \cup B$. Note that $\subset$ will always mean *proper* subset. Lower case letters $a, b, x, y, \ldots$ indicate (possibly infinite) tuples.

2000 *Mathematics Subject Classification.* 03C45.

Key words and phrases. Strong Type, Lascar Strong Type, Lascar Group, Robinson Theory, κ-Generic.

The authors would like to thank Ikuo Yoneda, as well as the two anonymous referees, for valuable suggestions which greatly improved the presentation of the paper. The first author would also like to thank IPM for its partial support, grant no. 820300120.

Logic in Tehran
Edited by A. Enayat, I. Kalantari, and M. Moniri
Lecture Notes in Logic, 26

§2. **Robinson theories and simplicity.** We recall some definitions and results from Ben-Yaacov [BY1] (but see also [Hr2, Po, Sh] for the earlier approaches).

In this section, Δ will be a fixed set of first-order formulas closed under conjunction, disjunction and subformula, and containing all atomic formulas; κ will be an infinite regular cardinal strictly bigger than $|\Delta|$.

DEFINITION 1. A Δ-*formula* is (an instance of) a formula in Δ; a Δ-*existential* (Δ-*universal*, respectively) formula is one of the form $\exists x \varphi(x, y)$ (or $\neg \exists x \varphi(x, y)$, respectively), where $\varphi(x, y) \in \Delta$. The set of Δ-existential (Δ-universal, respectively) formulas will be denoted by Σ (or Π, respectively). A Δ-*morphism* is a morphism preserving Δ-formulas (but not necessarily their negations).

REMARK 2. *Note that a Δ-universal formula is not defined to be a universally quantified Δ-formula: If Δ is not closed under negation, one cannot make the transformation from $\neg \exists$ to $\forall \neg$.*

DEFINITION 3. Let T be a Δ-universal theory. A model M of T is Δ-*existentially closed* if whenever $N \models T$ and $\sigma : M \to N$ is a Δ-morphism, $a \in M$ and $\varphi(x, y) \in \Delta$, if $N \models \exists x \varphi(x, \sigma(a))$, then $M \models \exists x \varphi(x, a)$. A κ-*universal domain* for T is a model $U \models T$ which, with respect to Δ-types, is κ-homogeneous and κ-compact.

Clearly universal domains are Δ-existentially closed.

REMARK 4. *Ben-Yaacov in [BY1] requires all Δ-formulas to be written without the universal quantifier (rewriting $\forall$ as $\neg \exists \neg$ increases the possible subformulas, and thus potentially Δ). This restriction is used only to obtain existence of an existentially closed extension for any model of a Δ-universal theory T; it would be sufficient just to require that the direct limit of a directed Δ-system of models of T is again a model of T. Since we construct directly a universal domain, we shall not need this assumption and still be able to use his results.*

DEFINITION 5. [BY1, Definition 2.1] A Δ-universal theory T is *positive Robinson* if any one of the following equivalent conditions holds:

(1) Whenever $M, N \models T$ are Δ-existentially closed and $a \in M$, $b \in N$ satisfying $\mathrm{tp}_\Delta^M(a) \subseteq \mathrm{tp}_\Delta^N(b)$, then a and b have the same Δ-existential type.

(2) Whenever $M_0, M_1 \models T$ are Δ-existentially closed and $\sigma : M_0 \to M_1$ is a partial Δ-morphism with domain $A \subseteq M_0$, then there is a Δ-existentially closed $N \models T$ and Δ-morphisms $\sigma_i : M_i \to N$ such that $(\sigma_1 \circ \sigma) \upharpoonright_A = \sigma_0 \upharpoonright_A$.

(3) Whenever $M \models T$ is Δ-existentially closed, $a \in M$ and $M \models \neg \exists y \varphi(a, y)$ for some $\varphi \in \Delta$, then there is $\psi(x) \in \Delta$ such that $M \models \psi(a)$ and $T \vdash \forall x y \neg [\varphi(x, y) \wedge \psi(x)]$.

REMARK 6. *If Δ is closed under negation, one obtains the Robinson theories of Hrushovski [Hr2]. If Δ is the set of existential formulas, one obtains the set-up* *of Pillay* [Pi]; *more generally if* $\Delta = \Sigma$, *then the* Π- *theory of any structure is positive Robinson. Finally, if* Δ *is the set of all formulas, one recovers usual first-order logic.*

FACT 7. [BY1, Lemma 2.12, Theorem 2.13] [BY3, Theorem 2.11] Let T be a Δ-universal theory closed under directed Δ-systems. Then T is positive Robinson if and only for every model M there is a κ-universal domain $U \models T$ (for some κ) and a Δ-morphism $\sigma : M \to U$. In particular, if T is *complete* (i.e. the Δ-universal theory of a structure), it is positive Robinson if and only if it has a κ-universal domain for some κ.

DEFINITION 8. Two tuples a and b have the *same Lascar strong type* over a tuple c in a universal domain U if they are in the same class modulo every c-invariant (but not necessarily definable or type-definable) equivalence relation with a bounded number of classes.

We shall now describe simplicity for positive Robinson theories, again following Ben-Yaacov [BY2], who unlike Buechler and Lessmann [BL] or Pillay [Pi] only requires local character (see Fact 11) and proves all the basic results for *extendible* Δ-types (i.e. which have non-forking extensions to arbitrarily large supersets of its domain). As usual, "small" will mean of cardinality strictly less than κ

DEFINITION 9. Let T be a positive Robinson theory with a κ-universal domain U, and a, b, c small tuples from U. We say that $\text{tp}_\Delta(a/b) =: p(x, b)$ *divides* over c if there is an infinite Δ-indiscernible sequence $b = b_0, b_1, \ldots$ over c such that $\bigcup_{i<\omega} p(x, b_i)$ is inconsistent. Otherwise a and b are independent over c, denoted $a \mathop{\smile\hskip-0.9em\vert}\nolimits_c b$.

DEFINITION 10. A positive Robinson theory T is *simple* if for any finite tuple a and any small B in a κ-universal domain U there is $B' \subseteq B$ with $|B'| \leq |T|$ such that $a \mathop{\smile\hskip-0.9em\vert}\nolimits_{B'} B$.

FACT 11. [BY2, Theorem 1.51] [BY3, Theorem 3.9] Let T be a simple positive Robinson theory with κ-universal domain U, and $a, b, c, d \in U$ small.

(1) Invariance: $a \mathop{\smile\hskip-0.9em\vert}\nolimits_c b$ depends only on $\text{tp}_\Delta(abc)$, and is invariant under permutations of each tuple.

(2) Symmetry: $a \mathop{\smile\hskip-0.9em\vert}\nolimits_c b$ if and only if $b \mathop{\smile\hskip-0.9em\vert}\nolimits_c a$.

(3) Transitivity: $a \mathop{\smile\hskip-0.9em\vert}\nolimits_c bd$ if and only if $a \mathop{\smile\hskip-0.9em\vert}\nolimits_c b$ and $a \mathop{\smile\hskip-0.9em\vert}\nolimits_{cb} d$.

(4) Finite character: $a \mathop{\smile\hskip-0.9em\vert}\nolimits_c b$ if and only if $a' \mathop{\smile\hskip-0.9em\vert}\nolimits_c b$ for all finite $a' \subseteq a$.

(5) Existence: $a \mathop{\smile\hskip-0.9em\vert}\nolimits_c c$.

(6) Local character: There is $b' \subseteq b$ with $|b'| \leq |a| + |T|$ and $a \mathop{\smile\hskip-0.9em\vert}\nolimits_{b'} b$.

(7) Saturated extension: If $a \mathop{\smile\hskip-0.9em\vert}\nolimits_c b$ and c is a sufficiently saturated model, then there is $a' \models \text{tp}_\Delta(a/c)$ with $a' \mathop{\smile\hskip-0.9em\vert}\nolimits_c bd$.

(8) Saturated amalgamation: If c is a sufficiently saturated model, $a \mathop{\smile}\limits_{c} b$, $x, y \in U$ have the same Lascar strong type over c, and $a \mathop{\smile}\limits_{c} x$ and $b \mathop{\smile}\limits_{c} y$, then there exists $z \in U$ of the same Lascar strong type as x over ca and as y over cb, with $z \mathop{\smile}\limits_{c} ab$.

Conversely, if a relation $\mathop{\smile}'$ satisfies the axioms above in a positive Robinson theory T, then T is simple and $\mathop{\smile}'$ coincides with $\mathop{\smile}$.

REMARK 12. *"Sufficiently saturated" means "at least* $(2^{|T|})^{+}$*-saturated" (and consequently κ should be chosen even bigger). In fact, to show saturated extension and saturated amalgamation, instead of assuming c to be a sufficiently saturated model, it is sufficient to require* $\mathrm{tp}_{\Delta}(x/c)$ *to be extendible.*

§3. The construction.

Our language $\mathcal{L}$ will consist of binary predicates (R_q : $q \in \mathbb{Q}^{+}$). Let $\mathcal{C}_0$ be the class of structures $\mathfrak{A} = \langle A, \rho \rangle$, such that

(1) A is an $\mathcal{L}$-structure;
(2) ρ maps A to the unit circle S^1;
(3) all R_q are anti-reflexive, symmetric, and disjoint.

We say that $\langle A, \rho \rangle$ is a substructure of $\langle A', \rho' \rangle$ if $A \subseteq A'$ as an $\mathcal{L}$-structure and $\rho = \rho' \upharpoonright_A$. Abusing notation, we shall wite $\mathfrak{A}_1 \mathfrak{A}_2$ for any structure whose universe is $A_1 \cup A_2$ and whose restriction to A_i is $\mathfrak{A}_i$ for $i = 1, 2$.

REMARK 13. $\mathcal{C}_0$ *is not a first-order class of structures. However, instead of ρ one could add predicates* $P_r(x) := \rho^{-1}(r)$ *for all $r \in S^1$ to form the language* $\mathcal{L}_\rho = \mathcal{L} \cup \{P_r : r \in S^1\}$, *and interpret $\mathcal{C}_0$ as the class of $\mathcal{L}_\rho$-structures which omit the type outside all predicates.*

For $\mathfrak{A} \in \mathcal{C}_0$ and $a, a' \in A$ let $d(a, a') \in [0, \pi]$ be the distance of $\rho(a)$ and $\rho(a')$ on the unit circle. Let R_q^A be the set of pairs $(a, b) \in A^2$ satisfying R_q ; for finite $\mathfrak{A}$ put

$$\delta(\mathfrak{A}) = |A| - \sum_{q \in \mathbb{Q}^+} q \sum_{(a,b) \in R_q^A} d(a, b).$$

Let

$$\mathcal{C} = \big\{ \mathfrak{A} \in \mathcal{C}_0 : \delta(\mathfrak{A}_0) \geq 0 \text{ for all } \mathfrak{A}_0 \subseteq_\omega \mathfrak{A} \big\}.$$

For $\mathfrak{A}_1 \mathfrak{A}_2 \in \mathcal{C}$ we say that $\mathfrak{A}_1$ and $\mathfrak{A}_2$ are *freely amalgamated* over $\mathfrak{A}_0 = \mathfrak{A}_1 \cap \mathfrak{A}_2$, denoted $\mathfrak{A}_1 \amalg_{\mathfrak{A}_0} \mathfrak{A}_2$, if whenever a relation $R_q(a_1, a_2)$ in $R_q^{A_1 A_2} \setminus R_q^{A_1} R_q^{A_2}$ holds (i.e. $a_1 \in A_1 \setminus A_0$ and $a_2 \in A_2 \setminus A_0$), then $d(a_1, a_2) = 0$. A free amalgam $\mathfrak{A}_1 \amalg_{\mathfrak{A}_0} \mathfrak{A}_2$ is *canonical* if $R_q^{A_1 A_2} = R_q^{A_1} \cup R_q^{A_2}$ for all $q > 0$, i.e. there are no new relations. Note that $\amalg$ is not an operator, but denotes any free amalgam.

LEMMA 14. *Let* $\mathfrak{A}\mathfrak{B} \in \mathcal{C}_0$ *be finite.*

(1) $\delta(\emptyset) = 0$.
(2) $\delta(\mathfrak{A})$ *depends only on the isomorphism type of* $\mathfrak{A}$.

(3) *If ρ' is an isometry of S^1 and $\mathfrak{A}' = \langle A, \rho' \circ \rho \rangle$, then $\delta(\mathfrak{A}') = \delta(\mathfrak{A})$. In particular $\mathfrak{A}' \in \mathcal{C}_0$.*

(4) *$\delta(\mathfrak{A}\mathfrak{B}) + \delta(\mathfrak{A} \cap \mathfrak{B}) \leq \delta(\mathfrak{A}) + \delta(\mathfrak{B})$; equality holds if and only if $\mathfrak{A}$ and $\mathfrak{B}$ are freely amalgamated over $\mathfrak{A} \cap \mathfrak{B}$.*

PROOF. Standard [BS, Wa].   ⊣

Hence δ is a *predimension* [Wa, Definition 4.1], except that there may be infinite descending chains $\delta(\mathfrak{A}_0) > \delta(\mathfrak{A}_1) > \dots$. For $\mathfrak{A}\mathfrak{B} \in \mathcal{C}$ with $\mathfrak{A}$ finite, define the *relative predimension* of $\mathfrak{A}$ over $\mathfrak{B}$ as

$$\delta(\mathfrak{A}/\mathfrak{B}) = |A \setminus B| - \sum_{q \in \mathbb{Q}^+} q \sum_{(a,a') \in R_q^{AB} \setminus R_q^B} d(a, a').$$

Note that $\delta(\mathfrak{A}/\mathfrak{B})$ may be $-\infty$ if $\mathfrak{B}$ is infinite; if $\mathfrak{B}$ is finite, we have $\delta(\mathfrak{A}/\mathfrak{B}) = \delta(\mathfrak{A}\mathfrak{B}) - \delta(\mathfrak{B})$.

DEFINITION 15. Let $\mathfrak{A} \subseteq \mathfrak{B} \in \mathcal{C}$. We say that $\mathfrak{A}$ is *closed* in $\mathfrak{B}$, denoted $\mathfrak{A} \leq \mathfrak{B}$, if $\delta(\mathfrak{B}_0/\mathfrak{A}) \geq 0$ for all $\mathfrak{B}_0 \subseteq_\omega \mathfrak{B}$.

LEMMA 16. *Suppose $\mathfrak{B} \subseteq \mathfrak{C} \subseteq \mathfrak{D} \in \mathcal{C}$, and $\mathfrak{A}, \mathfrak{A}' \subseteq_\omega \mathfrak{D}$.*

(1) *Monotonicity: If $\mathfrak{A} \cap \mathfrak{B} = \mathfrak{A} \cap \mathfrak{C}$, then $\delta(\mathfrak{A}/\mathfrak{B}) \geq \delta(\mathfrak{A}/\mathfrak{C})$.*

(2) *Transitivity: $\delta(\mathfrak{A}\mathfrak{A}'/\mathfrak{B}) = \delta(\mathfrak{A}'/\mathfrak{A}\mathfrak{B}) + \delta(\mathfrak{A}/\mathfrak{B})$.*

(3) *Submodularity: $\delta(\mathfrak{A}\mathfrak{A}'/\mathfrak{B}) + \delta(\mathfrak{A} \cap \mathfrak{A}'/\mathfrak{B}) \leq \delta(\mathfrak{A}/\mathfrak{B}) + \delta(\mathfrak{A}'/\mathfrak{B})$. If $\delta(\mathfrak{A}\mathfrak{A}'/\mathfrak{B}) > -\infty$, then equality holds if and only if $\mathfrak{A}\mathfrak{B}$ and $\mathfrak{A}'\mathfrak{B}$ are freely amalgamated over $\mathfrak{A}\mathfrak{B} \cap \mathfrak{A}'\mathfrak{B}$.*

(4) *If $\mathfrak{A}_1, \mathfrak{A}_2 \in \mathcal{C}$ and $\mathfrak{A}_1 \cap \mathfrak{A}_2 \leq \mathfrak{A}_1$, then $\mathfrak{A}_1 \amalg_{\mathfrak{A}_1 \cap \mathfrak{A}_2} \mathfrak{A}_2 \in \mathcal{C}$ and $\mathfrak{A}_2 \leq \mathfrak{A}_1 \amalg_{\mathfrak{A}_1 \cap \mathfrak{A}_2} \mathfrak{A}_2$.*

PROOF. Routine [BS, Wa].   ⊣

DEFINITION 17. A structure $\mathfrak{M} \in \mathcal{C}$ is *κ-generic* (in $\mathcal{C}$) if for any $\mathfrak{A} \leq \mathfrak{B} \in \mathcal{C}$ with $\mathfrak{A} \leq \mathfrak{M}$ and $|B| < \kappa$ there is an embedding $f : \mathfrak{B} \hookrightarrow \mathfrak{M}$ which is the identity on $\mathfrak{A}$, such that $f(\mathfrak{B}) \leq \mathfrak{M}$.

REMARK 18. *Recall that a structure is generic if the above holds with $\kappa = \omega$. In the case of infinite closures, uncountable κ is more appropriate.*

LEMMA 19. *Let κ be uncountable. Every structure in $\mathcal{C}$ is contained in a κ-generic model. If $\langle M, \rho \rangle$ is κ-generic and ρ' is an isometry of S^1, then $\langle M, \rho' \circ \rho \rangle$ is again κ-generic, with the same closed subsets.*

PROOF. By Lemma 16(4) and a union of chains argument. The second assertion follows from invariance of δ under isometries.   ⊣

DEFINITION 20. For $\mathfrak{A}\mathfrak{A}' \in \mathcal{C}$ with $\mathfrak{A}'$ finite we say that $\mathfrak{A}\mathfrak{A}'$ is an *intrinsic extension* of $\mathfrak{A}$, written $\mathfrak{A} \subset^i \mathfrak{A}\mathfrak{A}'$, if $\delta(\mathfrak{A}'/\mathfrak{A}\mathfrak{A}'') < 0$ for all $\mathfrak{A}'' \subset \mathfrak{A}'$.

REMARK 21. *Note that $\mathfrak{A} \subset^i \mathfrak{A}\mathfrak{A}'$ implies $\mathfrak{A} \cap \mathfrak{A}' = \emptyset$.*

LEMMA 22. *The union of two intrinsic extensions is intrinsic. An intrinsic extension of an intrinsic extension is intrinsic.* If $\mathfrak{A} \subset^i \mathfrak{A}\mathfrak{A}'$, *there is* $\mathfrak{A}_0 \subseteq_\omega \mathfrak{A}$ *with* $\mathfrak{A}_0 \subset^i \mathfrak{A}_0\mathfrak{A}'$, *and there are only boundedly finitely many isomorphic copies of* $\mathfrak{A}'$ *over* $\mathfrak{A}_0$ *inside any given* $\mathfrak{M} \in C$ *containing* $\mathfrak{A}_0$.

PROOF. Suppose $\mathfrak{A} \subset^i \mathfrak{B}$ and $\mathfrak{A} \subset^i \mathfrak{C}$, and consider $\mathfrak{B}' \subseteq \mathfrak{B}$ and $\mathfrak{C}' \subseteq \mathfrak{C}$ with $\mathfrak{B}'\mathfrak{C}' \subset \mathfrak{B}\mathfrak{C}$. Then $\mathfrak{C} \not\subseteq \mathfrak{B}\mathfrak{C}'$ or $\mathfrak{B} \not\subseteq \mathfrak{B}'\mathfrak{C}'$. By monotonicity

$$\delta(\mathfrak{B}\mathfrak{C}/\mathfrak{A}\mathfrak{B}'\mathfrak{C}') = \delta(\mathfrak{B}\mathfrak{C}/\mathfrak{A}\mathfrak{B}\mathfrak{C}') + \delta(\mathfrak{B}\mathfrak{C}'/\mathfrak{A}\mathfrak{B}'\mathfrak{C}')$$
$$= \delta(\mathfrak{C}/\mathfrak{A}\mathfrak{B}\mathfrak{C}') + \delta(\mathfrak{B}/\mathfrak{A}\mathfrak{B}'\mathfrak{C}')$$
$$\leq \delta(\mathfrak{C}/\mathfrak{A}(\mathfrak{C} \cap \mathfrak{B}\mathfrak{C}')) + \delta(\mathfrak{B}/\mathfrak{A}(\mathfrak{B} \cap \mathfrak{B}'\mathfrak{C}')) < 0.$$

Now suppose $\mathfrak{A} \subset^i \mathfrak{B} \subset^i \mathfrak{C}$, and consider $\mathfrak{C}' \subset \mathfrak{C}$. Then $\mathfrak{B} \cap \mathfrak{C}' \subset \mathfrak{B}$ or $\mathfrak{B}\mathfrak{C}' \subset \mathfrak{C}$. Again by monotonicity

$$\delta(\mathfrak{C}/\mathfrak{A}\mathfrak{C}') = \delta(\mathfrak{C}/\mathfrak{B}\mathfrak{C}') + \delta(\mathfrak{B}\mathfrak{C}'/\mathfrak{A}\mathfrak{C}') \leq \delta(\mathfrak{C}/\mathfrak{B}\mathfrak{C}') + \delta(\mathfrak{B}/\mathfrak{A}\mathfrak{C}')$$
$$\leq \delta(\mathfrak{C}/\mathfrak{B}\mathfrak{C}') + \delta(\mathfrak{B}/\mathfrak{A}(\mathfrak{B} \cap \mathfrak{C}')) < 0.$$

Suppose $\mathfrak{A} \subset^i \mathfrak{A}\mathfrak{A}' \subseteq \mathfrak{M}$. Then there is a finite $\mathfrak{A}_0 \subseteq \mathfrak{A}$ with $\mathfrak{A}_0 \subset^i \mathfrak{A}_0\mathfrak{A}'$ (take $\mathfrak{A}_0 \subseteq_\omega \mathfrak{A}$ big enough so that the relations between $\mathfrak{A}'$ and $\mathfrak{A}_0$ outweigh $|A'|$). Take $\varepsilon > 0$ with $\delta(\mathfrak{A}'/\mathfrak{A}_0\mathfrak{A}'') < -\varepsilon$ for all $\mathfrak{A}'' \subset \mathfrak{A}'$. Suppose $(\mathfrak{A}'_i : i < n) \subseteq \mathfrak{M}$ are isomorphic copies of $\mathfrak{A}'$ over $\mathfrak{A}_0$, with $\mathfrak{A}'_i \not\subseteq \mathfrak{A}_0 \cup \bigcup_{j<i} \mathfrak{A}'_j$ for all $i < n$. Then

$$0 \leq \delta\left(\mathfrak{A}_0 \cup \bigcup_{i<n} \mathfrak{A}'_i\right) = \delta(\mathfrak{A}_0) + \sum_{i<n} \delta\left(\mathfrak{A}'_i/\mathfrak{A}_0 \cup \bigcup_{j<i} \mathfrak{A}'_j\right)$$
$$\leq \delta(\mathfrak{A}_0) + \sum_{i<n} -\varepsilon = \delta(\mathfrak{A}_0) - n\varepsilon,$$

whence $n \leq \delta(\mathfrak{A}_0)/\varepsilon$.   ⊣

DEFINITION 23. Let $\mathfrak{A} \subseteq \mathfrak{M} \in C$. The *closure* of $\mathfrak{A}$ in $\mathfrak{M}$, denoted $\mathrm{cl}_\mathfrak{M}(\mathfrak{A})$ or $\tilde{\mathfrak{A}}$ (if the ambient structure $\mathfrak{M}$ is clear), is the smallest closed subset of $\mathfrak{M}$ containing $\mathfrak{A}$.

PROPOSITION 24. *The closure of* $\mathfrak{A}$ *in* $\mathfrak{M}$ *exists and is unique, and equals the union of all intrinsic extensions of* $\mathfrak{A}$. *In particular, the closure is finitary, and* $|\mathrm{cl}_\mathfrak{M}(\mathfrak{A})| \leq |\mathfrak{A}| + \omega$. *If* $\mathfrak{A} \leq \mathfrak{B} \leq \mathfrak{C} \in C$, *then* $\mathfrak{A} \leq \mathfrak{C}$. *For* $\mathfrak{A} \subseteq \mathfrak{B} \subseteq \mathfrak{M} \subseteq \mathfrak{N} \in C$, *we have*

$$\mathfrak{A} \subseteq \mathrm{cl}_\mathfrak{M}(\mathfrak{A}) \subseteq \mathrm{cl}_\mathfrak{M}(\mathfrak{B}) \subseteq \mathrm{cl}_\mathfrak{N}(\mathfrak{B}) \subseteq \mathfrak{N}.$$

The intersection of two closed sets is closed.

PROOF. By monotonicity any intrinsic extension of $\mathfrak{A}$ in $\mathfrak{M}$ must be contained in any closed superset of $\mathfrak{A}$. Let $\tilde{\mathfrak{A}}$ be the union of all intrinsic extensions of $\mathfrak{A}$ in $\mathfrak{M}$, and suppose that $\tilde{\mathfrak{A}} \not\leq \mathfrak{M}$. So there is $\mathfrak{B} \subseteq_\omega \mathfrak{M}$ with $\delta(\mathfrak{B}/\tilde{\mathfrak{A}}) < 0$; we choose it of minimal size (and hence $\mathfrak{B} \cap \tilde{\mathfrak{A}} = \emptyset$). There is $\mathfrak{A}' \subseteq_\omega \tilde{\mathfrak{A}}$ with

$\delta(\mathfrak{B}/\mathfrak{A}\mathfrak{A}') < 0$; increasing $\mathfrak{A}'$ we may assume that $\mathfrak{A} \subset^i \mathfrak{A}\mathfrak{A}'$ by Lemma 22. Then for all $\mathfrak{B}' \subset \mathfrak{B}$ we have $\delta(\mathfrak{B}'/\mathfrak{A}\mathfrak{A}') \geq \delta(\mathfrak{B}'/\bar{\mathfrak{A}}) \geq 0$, whence

$$\delta(\mathfrak{B}/\mathfrak{A}\mathfrak{A}'\mathfrak{B}') = \delta(\mathfrak{B}/\mathfrak{A}\mathfrak{A}') - \delta(\mathfrak{B}'/\mathfrak{A}\mathfrak{A}') < 0.$$

So $\mathfrak{A} \subset^i \mathfrak{A}\mathfrak{A}' \subset^i \mathfrak{A}\mathfrak{A}'\mathfrak{B}$, and $\mathfrak{A} \subset^i \mathfrak{A}\mathfrak{A}'\mathfrak{B}$ by Lemma 22, whence $\mathfrak{B} \subseteq \bar{\mathfrak{A}}$, a contradiction. So the closure exists uniquely as the union of all intrinsic extensions. By Lemma 22, the closure is finitary, and $|\operatorname{cl}_{\mathfrak{M}}(\mathfrak{A})| \leq |\mathfrak{A}| + \omega$.

If $\mathfrak{A} \leq \mathfrak{B} \leq \mathfrak{C} \in \mathcal{C}$ and $\mathfrak{A} \subset^i \mathfrak{A}' \subseteq \mathfrak{C}$, then

$$\delta(\mathfrak{A}'/\mathfrak{A}' \cap \mathfrak{B}) \geq \delta(\mathfrak{A}'/\mathfrak{B}) \geq 0.$$

Now $\mathfrak{A} \subset^i \mathfrak{A}'$ implies $\mathfrak{A}' \subseteq \mathfrak{B}$, contradicting $\mathfrak{A} \leq \mathfrak{B}$. Hence $\mathfrak{A} \leq \mathfrak{C}$.

Now $\mathfrak{A} \subseteq \operatorname{cl}_{\mathfrak{M}}(\mathfrak{A}) \subseteq \operatorname{cl}_{\mathfrak{M}}(\mathfrak{B})$ follows from the definition, and $\operatorname{cl}_{\mathfrak{M}}(\mathfrak{B}) \subseteq \operatorname{cl}_{\mathfrak{N}}(\mathfrak{B})$ from our construction of the closure. Finally, if $\mathfrak{A}, \mathfrak{B} \leq \mathfrak{M}$, then

$$\operatorname{cl}_{\mathfrak{M}}(\mathfrak{A} \cap \mathfrak{B}) \subseteq \operatorname{cl}_{\mathfrak{M}}(\mathfrak{A}) \cap \operatorname{cl}_{\mathfrak{M}}(\mathfrak{B}) = \mathfrak{A} \cap \mathfrak{B}. \qquad \dashv$$

PROPOSITION 25. *The family of isomorphisms between closed subsets of cardinality $< \kappa$ of two κ-generic models has the back-and-forth property.*

PROOF. Obvious from Proposition 24 and κ-genericity. $\qquad \dashv$

For the rest of the paper we fix an uncountable cardinal κ and a κ-generic structure $\mathfrak{M} = \langle M, \rho \rangle$ in $\mathcal{C}$. We consider a κ-saturated non-principal ultrapower $\mathfrak{M}^* = \langle M^*, (S^1)^*, [0, \pi]^*, \rho^*, d^* \rangle$ of the structure $\langle M, S^1, [0, \pi], \rho, d \rangle$ (so the image $(S^1)^*$ of ρ^* is a non-standard unit circle, and the image $[0, \pi]^*$ of d^* is a non-standard real interval). Let st denote the standard part maps $(S^1)^* \rightarrow S^1$ and $[0, \pi]^* \rightarrow [0, \pi]$, and put $\mathfrak{M}^\# = \langle M^*, \operatorname{st} \circ \rho^* \rangle$. We extend d to $(\mathfrak{M}^\#)^2$ via $d(a, b) = (\operatorname{st} \circ d^*)(a, b)$ (which equals the distance of $(\operatorname{st} \circ \rho^*)(a)$ and $(\operatorname{st} \circ \rho^*)(b)$ on S^1). We shall continue to use German letters to denote structures in $\mathcal{C}$, and Roman letters for the underlying $\mathcal{L}$-structures.

PROPOSITION 26. $\mathfrak{M}^\# \in \mathcal{C}$, and $\mathfrak{M} \leq \mathfrak{M}^\#$.

PROOF. Suppose $\mathfrak{A} \subset_\omega \mathfrak{M}^\#$ with $\delta(\mathfrak{A}) < 0$. So there is $\varepsilon > 0$ such that

$$|A| - \sum_{q \in \mathbb{Q}^+} q \sum_{(a, a') \in R_q^A} [d(a, a') - \varepsilon] < 0.$$

But for almost all i and all $a, a' \in \mathfrak{A}$ we have $d(a_i, a_i') \geq d(a, a') - \varepsilon$, whence $\delta(\mathfrak{A}_i) < 0$, contradicting $\mathfrak{A}_i \in \mathcal{C}$. Equivalently, one could use that $\delta(\mathfrak{A})$ is the standard part of $(\delta(\mathfrak{A}_i))_i$, considered as a non-standard real. Thus $\mathfrak{M}^\# \in \mathcal{C}$.

By Lemma 22, if $\mathfrak{A} \subset^i \mathfrak{A}\mathfrak{A}' \subset_\omega \mathfrak{M}^\#$ with $\mathfrak{A} \subset \mathfrak{M}$, then there are only finitely many isomorphic copies of $\mathfrak{A}'$ over $\mathfrak{A}$ in $\mathfrak{M}^\#$. On the other hand, let $\mathcal{L}_\rho$ be the language from Remark 13 and suppose that every $\mathcal{L}_\rho$-formula in $\operatorname{tp}_{\mathcal{L}_\rho}(\mathfrak{A}'/\mathfrak{A})$ had infinitely many realizations in $\mathfrak{M}$. By ω-saturation we could find infinitely many realizations of $\operatorname{tp}_{\mathcal{L}_\rho}(\mathfrak{A}'/\mathfrak{A})$ in $\mathfrak{M}^*$, and they would still realize that type in $\mathfrak{M}^\#$, a contradiction. Hence $\operatorname{cl}_{\mathfrak{M}^\#}(\mathfrak{A}) \subseteq \operatorname{acl}(\mathfrak{A}) \subseteq \mathfrak{M}$, whence $\mathfrak{M} \leq \mathfrak{M}^\#$. $\qquad \dashv$

DEFINITION 27. For $r \geq 0$ put

$$\pi_r(x, y) := \left\{ \exists^n z \left[R_q(x, z) \wedge R_q(z, y) \right] : n < \omega, \ q < 1/r \right\}.$$

Note that $\pi_r \supset \pi_{r'}$ for $r < r'$; in fact $\pi_r = \bigcup_{r' > r} \pi_{r'}$.

The following lemma relates the $\mathcal{L}$-structure on M^* with the distance function d (and hence the predimension).

LEMMA 28. $M^* \models \pi_r(a, b)$ if and only if $d(a, b) \leq r$.

PROOF. Suppose $d(a, b) \leq r$. Fix $q < 1/r$ and $\varepsilon > 0$ with $q \leq 1/(r + \varepsilon)$. For any $x, y \in M$ with $d(x, y) \leq r + \varepsilon$ and z in between x and y with the only relations on $\{x, y, z\}$ being $R_q(x, z)$ and $R_q(z, y)$ we have

$$\delta(z/xy) = 1 - q \left[d(x, z) + d(z, y) \right] \geq 1 - q \, d(x, y) \geq 1 - (r + \varepsilon)/(r + \varepsilon) = 0.$$

By genericity $(\mathfrak{M}, d)$ satisfies for any n

$$\forall x, y \left\{ d(x, y) \leq r + \varepsilon \rightarrow \exists^n z \left[R_q(x, z) \wedge R_q(z, y) \right] \right\},$$

and the same (with d replaced by d^*) holds for $\mathfrak{M}^*$. But since $d = \mathrm{st} \circ d^*$ and $\varepsilon > 0$ is real, we must have $d^*(a, b) < r + \varepsilon$, so

$$\exists^n z \left[R_q(a, z) \wedge R_q(z, b) \right]$$

is true in M^* (it is an $\mathcal{L}$-formula). Hence $M^* \models \pi_r(a, b)$.

Conversely, if $d(a, b) > r$ choose $q < 1/r$ with $q \, d(a, b) = 1 + \varepsilon > 1$. Then for any witnesses $c_1, \ldots, c_n$ of the existential quantifier

$$0 \leq \delta(abc_1 \cdots c_n) \leq n + 2 - nq \, d(a, b) = 2 - n\varepsilon;$$

now $\mathfrak{M}^{\#} \in \mathcal{C}$ implies $n \leq 2/\varepsilon$, and $\pi_r(a, b)$ fails. $\dashv$

DEFINITION 29. An existential $\mathcal{L}$-formula $\varphi(X, Y)$ is *nintrinsic* if

(1) it is a disjunction of relations R_q, negated relations $\neg R_q$, and formulas from the various π_r;
(2) there is $\varepsilon > 0$ such that $\delta(\mathfrak{B}/\mathfrak{A}\mathfrak{B}') \leq -\varepsilon$ for all $\mathfrak{A}\mathfrak{B} \models_{\mathfrak{M}} \neg\varphi$ and all $\mathfrak{B}' \subset \mathfrak{B}$.

REMARK 30. *Nintrinsic stands for negated intrinsic*: $\neg\varphi(A, B)$ *for some* $\mathfrak{A}\mathfrak{B} \subset_\omega \mathfrak{M}$ *implies* $\mathfrak{A} \sqsubset^i \mathfrak{A}\mathfrak{B}$. *Note that if* $\varphi(X, Y)$ *is nintrinsic and* $Y = Y' Y''$ *with* $Y'' \neq \emptyset$, *then* $\varphi(XY', Y'')$ *is also nintrinsic (the same formula, but with a different partition of variables).*

The predimension of a structure $\mathfrak{A}$ depends not only on its isomorphism type as an $\mathcal{L}$-structure, but also on the function ρ. Nevertheless we are able to describe the notion of an intrinsic extension by an $\mathcal{L}$-formula.

LEMMA 31. *For every* $\mathfrak{A} \sqsubset^i \mathfrak{A}\mathfrak{A}' \subseteq_\omega \mathfrak{M}$ *there is a nintrinsic* φ *such that* $M \models \neg\varphi(A, A')$.

PROOF. For all $\mathfrak{A}'' \subset \mathfrak{A}'$

$$\sum_{q \in \mathbb{Q}^+} q \sum_{(a,a') \in R_q^{AA'} \setminus R_q^{AA''}} d(a,a') > |A' \setminus A''|.$$

Hence there is $\varepsilon > 0$ such that for all $\mathfrak{A}'' \subset \mathfrak{A}'$

$$\sum_{q \in \mathbb{Q}^+} q \sum_{(a,a') \in R_q^{AA'} \setminus R_q^{AA''}} [d(a,a') - \varepsilon] \geq |A' \setminus A''| + \varepsilon.$$

For $(a,a') \in R_q^{AA'} \setminus R_q^A$ put $r = d(a,a') - \varepsilon$. Since $d(a,a') > r$, by Lemma 28 there is a formula $\varphi_{aa'}(x, x') \in \pi_r$ which is false for aa' in M; moreover for any $bb' \in M$ satisfying $\neg\varphi_{aa'}$ we have $d(b,b') > r$. The required nintrinsic formula will then be the disjunction of all the $\varphi_{aa'}(x, x')$, and the negations of all the relations indexing the sum.   ⊣

Let Δ be the set of existential $\mathcal{L}$-formulas, together with all $\mathcal{L}$-formulas $\forall Y\varphi(X, Y)$ for nintrinsic $\varphi(X, Y)$, and closed under positive boolean combinations and existential quantification. Note that Δ is closed under subformulas by Remark 30.

REMARK 32. *A formula* $\forall Y\varphi(X, Y)$ *for nintrinsic* φ *says that a certain intrinsic extension of* X *does not exist. On the other hand, the existence of an intrinsic extension can be expressed by a partial existential type, and is thus also expressible in* Δ.

Let T be the Π-theory of M, our κ-generic model.

THEOREM 33. *T is a positive Robinson theory, and M is an κ-universal domain.*

PROOF. T is positive Robinson by Remark 6, as $\Delta = \Sigma$.

To show κ-compactness of M, take $A \subseteq M$ of cardinality $< \kappa$, and consider $B \subset M^*$, the realization of a partial Δ-type p over A in $< \kappa$ many variables (recall that M^* is κ-saturated). Increasing B by adding variables for witnesses for existential quantifiers, we may suppose that p consists only of quantifier-free formulas and formulas of the form $\forall Y\varphi(X, Y)$ for nintrinsic $\varphi(X, Y)$. Clearly we may assume that $\mathfrak{A} = \mathrm{cl}_{\mathfrak{M}}(\mathfrak{A}) \leq \mathfrak{M}$, so $\mathfrak{A} \leq \mathfrak{M}^\#$ by Proposition 26. For every nintrinsic $\varphi(X, Y)$ true of some tuple in $\mathrm{cl}_{\mathfrak{M}^\#}(\mathfrak{A}\mathfrak{B})$ we may suppose that B contains a witness for the existential quantifier in φ.

Since $\mathfrak{A} \leq \mathrm{cl}_{\mathfrak{M}^\#}(\mathfrak{A}\mathfrak{B}) \in \mathcal{C}$, we can embed $\mathrm{cl}_{\mathfrak{M}^\#}(\mathfrak{A}\mathfrak{B})$ closedly into $\mathfrak{M}$ over $\mathfrak{A}$ by κ-genericity. Let ˜denote the image under this embedding. Since quantifier-free formulas are trivially preserved by the embedding, consider a formula $\forall Y\varphi(X, Y)$ true in M^* of some tuple $B_0 \subseteq_\omega \mathrm{cl}_{\mathfrak{M}^\#}(\mathfrak{A}\mathfrak{B})$. Then it must be true in M of $\tilde{B}_0$: Since $\varphi(X, Y)$ is nintrinsic, any $C \models_M \neg\varphi(\tilde{B}_0, Y)$ must be in $\mathrm{cl}_{\mathfrak{M}}(\tilde{\mathfrak{A}}\tilde{\mathfrak{B}}) = \mathrm{cl}_{\mathfrak{M}^\#}(\mathfrak{A}\mathfrak{B})$, say $C = \tilde{C}_0$ for some $C_0 \in \mathrm{cl}_{\mathfrak{M}^\#}(\mathfrak{A}\mathfrak{B})$. By assumption on φ we have $M^* \models \varphi(B_0, C_0)$ and there is a witness in B for the existential quantifier in φ; its image under ˜will witness $M \models \varphi(\tilde{B}_0, \tilde{C}_0)$.

For κ-homogeneity of M, let A and B be subsets of M satisfying the same Δ-types. By compactness $\mathrm{cl}_{\mathfrak{M}}(\mathfrak{A})$ and $\mathrm{cl}_{\mathfrak{M}}(\mathfrak{B})$ have the same existential $\mathcal{L}$-type (existential formulas tell us what is in the closure and its existential type, formulas $\forall Y \varphi(X, Y)$ with $\varphi(X, Y)$ nintrinsic tell us what is not in the closure); moreover, by Lemma 28 they are isometric via some isometry ρ' of S^1. Since $\mathfrak{M}' := \langle M, \rho' \circ \rho \rangle$ is again κ-generic by Lemma 19, we may consider $\mathrm{cl}_{\mathfrak{M}}(\mathfrak{B})$ as a closed subset of $\mathfrak{M}'$; as such it becomes isomorphic to $\mathrm{cl}_{\mathfrak{M}}(\mathfrak{A})$, and thus has the same $\mathcal{L}$-type by Proposition 25. $\dashv$

DEFINITION 34. Two subsets A and B of M are δ-independent over $C \subseteq M$, denoted $A \underset{C}{\overset{\delta}{\downarrow}} B$, if

$$\mathrm{cl}_{\mathfrak{M}}(\mathfrak{A}\mathfrak{B}\mathfrak{C}) = \mathrm{cl}_{\mathfrak{M}}(\mathfrak{A}\mathfrak{C}) \amalg_{\mathrm{cl}_{\mathfrak{M}}(\mathfrak{C})} \mathrm{cl}_{\mathfrak{M}}(\mathfrak{B}\mathfrak{C}).$$

THEOREM 35. *T is a simple Robinson theory where all types are extendible, and δ-independence coincides with forking independence.*

PROOF. We shall use the characterization given by Fact 11, working in the κ-generic model M. Clearly δ-independence is *symmetric*, and satisfies *existence*.

Isomorphism-invariance: If ABC and $A'B'C'$ have the same Δ-types, then $\mathrm{cl}_{\mathfrak{M}}(\mathfrak{A}\mathfrak{B}\mathfrak{C})$ and $\mathrm{cl}_{\mathfrak{M}}(\mathfrak{A}'\mathfrak{B}'\mathfrak{C}')$ have the same atomic $\mathcal{L}$-types and are isometric; as in the proof of homogeneity above it follows that they have the same $\mathcal{L}$-types. Isomorphism-invariance follows.

Transitivity: We shall write $\bar{\mathfrak{X}}$ for $\mathrm{cl}_{\mathfrak{M}}(\mathfrak{X})$. Clearly

$$\overline{\mathfrak{A}\mathfrak{B}} \amalg_{\bar{\mathfrak{B}}} \overline{\mathfrak{B}\mathfrak{C}\mathfrak{D}} = \left[\overline{\mathfrak{A}\mathfrak{B}} \amalg_{\bar{\mathfrak{B}}} \overline{\mathfrak{B}\mathfrak{C}}\right] \amalg_{\overline{\mathfrak{B}\mathfrak{C}}} \overline{\mathfrak{B}\mathfrak{C}\mathfrak{D}}$$

as structures (with suitable choice of relations between points of distance zero in the free amalgam, but these do not count for δ-dependence anyway). So if $A \underset{B}{\overset{\delta}{\downarrow}} CD$, then $\overline{\mathfrak{A}\mathfrak{B}\mathfrak{C}\mathfrak{D}}$ is equal to that structure. Since $\overline{\mathfrak{B}\mathfrak{C}} \leq \overline{\mathfrak{B}\mathfrak{C}\mathfrak{D}}$ we get

$$\overline{\mathfrak{A}\mathfrak{B}} \amalg_{\bar{\mathfrak{B}}} \overline{\mathfrak{B}\mathfrak{C}} \leq \left[\overline{\mathfrak{A}\mathfrak{B}} \amalg_{\bar{\mathfrak{B}}} \overline{\mathfrak{B}\mathfrak{C}}\right] \amalg_{\overline{\mathfrak{B}\mathfrak{C}}} \overline{\mathfrak{B}\mathfrak{C}\mathfrak{D}} = \overline{\mathfrak{A}\mathfrak{B}\mathfrak{C}\mathfrak{D}} \leq \mathfrak{M}$$

by Lemma 16(4), whence $\overline{\mathfrak{A}\mathfrak{B}} \amalg_{\bar{\mathfrak{B}}} \overline{\mathfrak{B}\mathfrak{C}} = \overline{\mathfrak{A}\mathfrak{B}\mathfrak{C}}$. So $A \underset{B}{\overset{\delta}{\downarrow}} C$ and $A \underset{BC}{\overset{\delta}{\downarrow}} D$.

Conversely, if $A \underset{B}{\overset{\delta}{\downarrow}} C$ and $A \underset{BC}{\overset{\delta}{\downarrow}} D$, then firstly $\overline{\mathfrak{A}\mathfrak{B}\mathfrak{C}} = \overline{\mathfrak{A}\mathfrak{B}} \amalg_{\bar{\mathfrak{B}}} \overline{\mathfrak{B}\mathfrak{C}}$ and secondly $\overline{\mathfrak{A}\mathfrak{B}\mathfrak{C}\mathfrak{D}} = \overline{\mathfrak{A}\mathfrak{B}\mathfrak{C}} \amalg_{\overline{\mathfrak{B}\mathfrak{C}}} \overline{\mathfrak{B}\mathfrak{C}\mathfrak{D}}$, whence $A \underset{B}{\overset{\delta}{\downarrow}} CD$.

Finite character: Follows from finite character of closure and Lemma 16(4) (or one direction of transitivity).

Local character: Fix $A, B \subset M$; by finite character we may assume that A is finite, and by transitivity that B is closed (for any $B' \leq \mathrm{cl}_{\mathfrak{M}}(\mathfrak{B})$ with $A \underset{B'}{\overset{\delta}{\downarrow}} \mathrm{cl}_{\mathfrak{M}}(\mathfrak{B})$ there is $B'' \subseteq B$ with $|B''| \leq [B'] + \omega$ and $B' \subseteq \mathrm{cl}_{\mathfrak{M}}(\mathfrak{B}'')$, whence $A \underset{B''}{\overset{\delta}{\downarrow}} B$).

LEMMA 36. $\mathrm{cl}_{\mathfrak{M}}(\mathfrak{A}\mathfrak{B}) \setminus \mathfrak{A}\mathfrak{B}$ *is countable.*

PROOF. This follows from [VY, Lemma 3.9], but we shall give a quick proof. Consider $\mathfrak{C} \subseteq_\omega \mathrm{cl}_{\mathfrak{M}}(\mathfrak{A}\mathfrak{B}) \setminus \mathfrak{A}\mathfrak{B}$. Then $\delta(\mathfrak{A}\mathfrak{C}/\mathfrak{B}) \geq 0$, whence

$$\delta(\mathfrak{C}/\mathfrak{A}\mathfrak{B}) = \delta(\mathfrak{A}\mathfrak{C}/\mathfrak{B}) - \delta(\mathfrak{A}/\mathfrak{B}) \geq -\delta(\mathfrak{A}/\mathfrak{B}).$$

If $\mathrm{cl}_{\mathfrak{M}}(\mathfrak{A}\mathfrak{B}) \setminus \mathfrak{A}\mathfrak{B}$ were uncountable, we would find $\varepsilon > 0$ such that for uncountably many intrinsic $\mathfrak{A}\mathfrak{B} \subset^i \mathfrak{A}\mathfrak{B}\mathfrak{C}_i$ (with $i < \aleph_1$, and none contained in the union of its predecessors) we had $\delta(\mathfrak{C}_i/\mathfrak{A}\mathfrak{B}\mathfrak{C}'') < -\varepsilon$ for all $\mathfrak{C}'' \subset \mathfrak{C}_i$. But then $\delta(\bigcup_{i<n} \mathfrak{C}_i/\mathfrak{A}\mathfrak{B}) \to -\infty$, a contradiction. ⊣

It follows that there is countable $\mathfrak{B}_0 \subseteq \mathfrak{B}$ with $\mathrm{cl}_{\mathfrak{M}}(\mathfrak{A}\mathfrak{B}) \setminus \mathfrak{A}\mathfrak{B} \subseteq \mathrm{cl}_{\mathfrak{M}}(\mathfrak{A}\mathfrak{B}_0)$. Suppose we have already defined countable $\mathfrak{B}_i \subseteq \mathfrak{B}$. Choose countable $\mathfrak{B}_i' \leq \mathfrak{B}$ containing $\mathrm{cl}_{\mathfrak{M}}(\mathfrak{A}\mathfrak{B}_i) \cap \mathfrak{B}$. For any $a \in \mathrm{cl}_{\mathfrak{M}}(\mathfrak{A}\mathfrak{B}_i') \setminus \mathfrak{B}$ consider the set $\mathfrak{B}_a$ of all $b \in \mathfrak{B}$ of nonzero distance to a which are in some relation R_q to a. As $a \notin \mathfrak{B} = \mathrm{cl}_{\mathfrak{M}}(\mathfrak{B})$, this set is at most countable, and we put $\mathfrak{B}_{i+1} = \mathfrak{B}_i' \cup \bigcup_{a \in \mathrm{cl}_{\mathfrak{M}}(\mathfrak{A}\mathfrak{B}_i')} \mathfrak{B}_a$. Let $\mathfrak{B}' = \bigcup_{i<\omega} \mathfrak{B}_i = \bigcup_{i<\omega} \mathfrak{B}_i'$, clearly a countable closed subset of $\mathfrak{B}$.

As closure is finitary, we get $\mathrm{cl}_{\mathfrak{M}}(\mathfrak{A}\mathfrak{B}') \cap \mathfrak{B} = \mathfrak{B}'$; by construction $\mathrm{cl}_{\mathfrak{M}}(\mathfrak{A}\mathfrak{B}')$ and $\mathfrak{B}$ are freely amalgamated over $\mathfrak{B}'$. Since $\mathrm{cl}_{\mathfrak{M}}(\mathfrak{A}\mathfrak{B}) \subseteq \mathrm{cl}_{\mathfrak{M}}(\mathfrak{A}\mathfrak{B}') \cup \mathfrak{A}\mathfrak{B}$, the free amalgam $\mathrm{cl}_{\mathfrak{M}}(\mathfrak{A}\mathfrak{B}') \amalg_{\mathrm{cl}_{\mathfrak{M}}(\mathfrak{B}')} \mathfrak{B}$ is closed, and $A \underset{B'}{\overset{\delta}{\downarrow}} B$.

Extension: Let $A, B, C \subset M$ be countable. By Lemma 16(4)

$$\mathrm{cl}_{\mathfrak{M}}(\mathfrak{B}\mathfrak{C}) \leq \mathrm{cl}_{\mathfrak{M}}(\mathfrak{A}\mathfrak{B}) \amalg_{\mathrm{cl}_{\mathfrak{M}}(\mathfrak{B})} \mathrm{cl}_{\mathfrak{M}}(\mathfrak{B}\mathfrak{C}) \in \mathcal{C}$$

(we may take any free amalgam here); as $\mathrm{cl}_{\mathfrak{M}}(\mathfrak{B}\mathfrak{C}) \leq \mathfrak{M}$, we may embed $\mathrm{cl}_{\mathfrak{M}}(\mathfrak{A}\mathfrak{B}) \amalg_{\mathrm{cl}_{\mathfrak{M}}(\mathfrak{B})} \mathrm{cl}_{\mathfrak{M}}(\mathfrak{B}\mathfrak{C})$ closedly into $\mathfrak{M}$ over $\mathrm{cl}_{\mathfrak{M}}(\mathfrak{B}\mathfrak{C})$ by genericity; let $\mathfrak{A}'$ be the image of $\mathfrak{A}$. Since $\mathrm{cl}_{\mathfrak{M}}(\mathfrak{A}\mathfrak{B})$ and $\mathrm{cl}_{\mathfrak{M}}(\mathfrak{A}'\mathfrak{B})$ are closed and have the same quantifier-free $(\mathcal{L}, \rho)$-type, they have the same type by Proposition 25. By definition $A' \underset{B}{\overset{\delta}{\downarrow}} C$.

Amalgamation: Let $C \leq M$ contain at least two points of distance $\neq 0, \pi$, and consider $A, B, X, Y \in M$ with $A \underset{C}{\overset{\delta}{\downarrow}} B$, $A \underset{C}{\overset{\delta}{\downarrow}} X$, $B \underset{C}{\overset{\delta}{\downarrow}} Y$ and $\mathrm{cl}_{\mathfrak{M}}(\mathfrak{C}\mathfrak{X}) \equiv_A \mathrm{cl}_{\mathfrak{M}}(\mathfrak{C}\mathfrak{Y})$. Note that by Lemma 28 the Δ-type of xC together with $\rho(\mathfrak{C})$ determines $\rho(x)$ for all $x \in \mathfrak{M}$, so $\mathrm{cl}_{\mathfrak{M}}(\mathfrak{C}\mathfrak{X})$ and $\mathrm{cl}_{\mathfrak{M}}(\mathfrak{C}\mathfrak{Y})$ have the same images under ρ. By extension we may assume that $\mathfrak{A}\mathfrak{C}$ and $\mathfrak{B}\mathfrak{C}$ are sufficiently saturated models closed in $\mathfrak{M}$. Consider the free amalgam $\mathrm{cl}_{\mathfrak{M}}(\mathfrak{C}\mathfrak{X}) \amalg_{\mathfrak{C}} \mathrm{cl}_{\mathfrak{M}}(\mathfrak{A}\mathfrak{B}\mathfrak{C})$, where the relations between $\mathrm{cl}_{\mathfrak{M}}(\mathfrak{C}\mathfrak{X})$ and $\mathfrak{B}\mathfrak{C}$ are those from $\mathrm{cl}_{\mathfrak{M}}(\mathfrak{C}\mathfrak{Y}) \amalg_{\mathfrak{C}} \mathfrak{B}\mathfrak{C}$ (and recall that $\mathrm{cl}_{\mathfrak{M}}(\mathfrak{A}\mathfrak{B}\mathfrak{C}) = \mathfrak{A}\mathfrak{C} \amalg_{\mathfrak{C}} \mathfrak{B}\mathfrak{C}$). Then

$$\mathrm{cl}_{\mathfrak{M}}(\mathfrak{A}\mathfrak{B}\mathfrak{C}) \leq \mathrm{cl}_{\mathfrak{M}}(\mathfrak{C}\mathfrak{X}) \amalg_{\mathfrak{C}} \mathrm{cl}_{\mathfrak{M}}(\mathfrak{A}\mathfrak{B}\mathfrak{C}) \in \mathcal{C}.$$

Hence we can closedly embed $\mathrm{cl}_{\mathfrak{M}}(\mathfrak{C}\mathfrak{X}) \amalg_{\mathfrak{C}} \mathrm{cl}_{\mathfrak{M}}(\mathfrak{A}\mathfrak{B}\mathfrak{C})$ into $\mathfrak{M}$ over $\mathrm{cl}_{\mathfrak{M}}(\mathfrak{A}\mathfrak{B}\mathfrak{C})$; the image Z of X will be δ-independent of AB over C, with $\mathrm{cl}_{\mathfrak{M}}(\mathfrak{A}\mathfrak{C}\mathfrak{X}) \cong \mathrm{cl}_{\mathfrak{M}}(\mathfrak{A}\mathfrak{C}3)$ and $\mathrm{cl}_{\mathfrak{M}}(\mathfrak{B}\mathfrak{C}\mathfrak{Y}) \cong \mathrm{cl}_{\mathfrak{M}}(\mathfrak{B}\mathfrak{C}3)$. Hence $\mathrm{tp}(Z/AC) = \mathrm{tp}(X/AC)$ and $\mathrm{tp}(Z/BC) = \mathrm{tp}(Y/BC)$; since AC and BC were sufficiently saturated models, we even get that Z has the same Lascar strong type as X over AC, and as Y over BC. ⊣

COROLLARY 37. *Closure equals algebraic closure, and M geometrically elimi-nates hyperimaginaries: Every hyperimaginary is interbounded with a real tuple.*

PROOF. Let $\mathfrak{A} \leq \mathfrak{B} \leq \mathfrak{M}$. Then for any $n < \omega$ the n-fold free amalgam of $\mathfrak{B}$ over $\mathfrak{A}$ is in C, and embeds closedly over $\mathfrak{A}$ into $\mathfrak{M}$. It follows that $\mathfrak{B}$ has multiplicity at least n over $\mathfrak{A}$ for all $n < \omega$, and thus cannot be algebraic. Since closure is contained in the algebraic closure, the two closures must coincide.

For geometric elimination of hyperimaginaries, we follow the proof of [VY, Lemma 4.1 and Proposition 4.2]. It is easy to see by intersecting the relevant free amalgams that for $\mathfrak{A}, \mathfrak{B}, \mathfrak{B}_0, \mathfrak{B}_1 \leq \mathfrak{M}$ with $B_i \subseteq B$ and $A \underset{B_i}{\downarrow} B$ for $i < 2$ we get $A \underset{B_0 \cap B_1}{\downarrow} B$. Let now E be a type-definable equivalence relation over $\emptyset$ and consider three tuples a, b_1, b_2 realizing the same E-class, independent over that class, whence $a \underset{b_i}{\downarrow} b_0 b_1$ for $i < 2$. Put $B = \mathrm{acl}(b_0) \cap \mathrm{acl}(b_1)$ (and remember that closure equals algebraic closure). Then $a \underset{B}{\downarrow} b_0 b_1$, so the E-class a_E of a is in $\mathrm{bdd}^{eq}(B)$. On the other hand, since $b_0 \underset{a_E}{\downarrow} b_1$, we have $B = \mathrm{acl}(b_0) \cap \mathrm{acl}(b_1) \subseteq \mathrm{acl}(a_E)$. $\dashv$

REMARK 38. *By Theorem 35 amalgamation holds over a closed set with at least two points of distance $\neq 0, \pi$. Hence Lascar strong type equals type over any such set, which must therefore be boundedly closed. In particular, if B above contains two points of distance $\neq 0, \pi$, then $\mathrm{bdd}^{eq}(B) = B$ and E can be weakly eliminated.*

Note that T cannot be stable, as any relation R_q induces the random graph on the set of points with fixed image under ρ.

REMARK 39. *This partially generalizes results of [VY] to positive Robinson ab initio amalgamation constructions.*

Note that M has a unique type p_0 of a single closed point by Lemma 19 and Proposition 25.

THEOREM 40. *T is a simple Robinson theory which does not eliminate bounded hyperimaginaries over $\emptyset$. In particular Lascar strong type is finer than strong type over $\emptyset$.*

PROOF. The relation π_0 from Definition 27 is a Δ-type-definable equivalence relation over $\emptyset$ on M with boundedly many classes on p_0 (the classes are given by the image under ρ). If $a, a' \in M$ are closed and satisfy π_0, then by genericity there is closed $bb' \in M$ with $d(b, b') \neq 0, \pi$ and such that $\mathrm{cl}(aa'bb') = \mathrm{cl}(aa') \amalg bb'$, with no relations other than those already on $\mathrm{cl}(aa')$. In particular abb' and $a'bb'$ are closed with no relations on them, and have the same type by Proposition 25. So a and a' have the same type over bb', whence the same Lascar strong type by Remark 38; it follows that a and a' have the same Lascar strong type over $\emptyset$, and π_0 is equality of Lascar strong types on p_0.

Suppose E is a finite $\emptyset$-definable equivalence relation, so $\pi_0(x, x') \vdash E(x, x')$ on p_0. Since $\pi_0 = \bigwedge_{r>0} \pi_r$, there must be some $r > 0$ such that $\pi_r(x, x') \vdash E(x, x')$ on p_0. But any two closed points can be linked by a chain of r-close (and hence E-related) closed points, so any two closed points are E-related by transitivity: The only finite equivalence relation on p_0 is the total relation, and p_0 is in fact a strong type.   ⊣

REMARK 41. *Nevertheless, this does not violate the stable forking conjecture: Since any forking is induced by a relation R_q which holds between two points of positive distance ε, it is enough to take a formula $\varphi(x, y) \in \pi_{\varepsilon/2}$; the formula*

$$R_q(x, y) \wedge \neg\varphi(x, y)$$

will be stable and true of those two points.

REMARK 42. *A priori the example constructed depends not only on κ, but on the specific κ-generic model chosen. Note that the question of the existence of a simple Robinson theory, let alone a first-order theory, which does not eliminate (bounded) hyperimaginaries, remains open.*

REFERENCES

[BS] JOHN BALDWIN and NIANDONG SHI, *Stable generic structures, Annals of Pure and Applied Logic*, vol. 79 (1996), no. 1, pp. 1–35.

[BY1] ITAY BEN-YAACOV, *Positive model theory and compact abstract theories, Journal of Mathematical Logic*, vol. 3 (2003), no. 1, pp. 85–118.

[BY2] ——, *Simplicity in compact abstract theories, Journal of Mathematical Logic*, vol. 3 (2003), no. 2, pp. 163–191.

[BY3] ——, *Compactness and independence in non first order frameworks, The Bulletin of Symbolic Logic*, vol. 11 (2005), no. 1, pp. 28–50.

[BL] STEVEN BUECHLER and OLIVIER LESSMANN, *Simple homogeneous models, Journal of the American Mathematical Society*, vol. 16 (2003), no. 1, pp. 91–121.

[Hr1] EHUD HRUSHOVSKI, *A new strongly minimal set, Annals of Pure and Applied Logic*, vol. 46 (1990), pp. 235–264.

[Hr2] ——, *Simplicity and the Lascar group*, Preprint, 1997.

[Pi] ANAND PILLAY, *Forking in the category of existentially closed structures, Connections Between Model Theory and Algebraic and Analytic Geometry* (Angus Macintyre, editor), Quaderni di Matematica, vol. 6, Seconda Univ. Napoli, Caserta, 2000, pp. 23–42.

[Po] MASSOUD POURMAHDIAN, *Simple generic structures*, Ph.D. thesis, Oxford University, 2000.

[Sh] SAHARON SHELAH, *The lazy model-theoretician's guide to stability, Logique et Analyse. Nouvelle Série*, vol. 18 (1975), no. 71-72, pp. 241–308.

[VY] VIKTOR VERBOVSKIY and IKUO YONEDA, *CM-triviality and relational structures, Annals of Pure and Applied Logic*, vol. 122 (2003), no. 1-3, pp. 175–194.

[Wa] FRANK O. WAGNER, *Relational structures and dimensions, Automorphisms of First-Order Structures*, Oxford Sci. Publ., Oxford Univ. Press, New York, 1994, pp. 153–180.

THE INSTITUTE FOR STUDIES IN THEORETICAL
  PHYSICS AND MATHEMATICS (IPM)
    P. O. BOX 19395-5746
      TEHRAN, IRAN
and
  SCHOOL OF MATHEMATICS
    AMIRKABIR UNIVERSITY OF TECHNOLOGY
    TEHRAN, IRAN
E-mail: pourmahd@ipm.ir

INSTITUT CAMILLE JORDAN
  UNIVERSITÉ CLAUDE BERNARD LYON-1
    BÂTIMENT BRACONNIER
      21 AVENUE CLAUDE BERNARD
        69622 VILLEURBANNE-CEDEX
          FRANCE
E-mail: wagner@math.univ-lyon1.fr

CATEGORIES OF THEORIES AND INTERPRETATIONS

ALBERT VISSER

Abstract. In this paper we study categories of theories and interpretations. In these categories, notions of *sameness of theories*, like synonymy, bi-interpretability and mutual interpretability, take the form of isomorphism.

We study the usual notions like monomorphism and product in the various categories. We provide some examples to separate notions across categories. In contrast, we show that, in some cases, notions in different categories *do* coincide. E.g., we can, under such-and-such conditions, infer synonymity of two theories from their being equivalent in the sense of a coarser equivalence relation.

We illustrate that the categories offer an appropriate framework for conceptual analysis of notions. For example, we provide a 'coordinate free' explication of the notion of axiom scheme. Also we give a closer analysis of the object-language/ meta-language distinction.

Our basic category can be enriched with a form of 2-structure. We use this 2-structure to characterize a salient subclass of interpretations, the direct interpretations, and we use the 2-structure to characterize induction. Using this last characterization, we prove a theorem that has as a consequence that, if two extensions of Peano Arithmetic in the arithmetical language are synonymous, then they are identical.

Finally, we study preservation of properties over certain morphisms.

§1. **Introduction.** Interpretations are ubiquitous in mathematics and logic. Some of the greatest achievements of mathematics, like the internal models of non-euclidean geometries are, in essence, interpretations.

Given the importance of interpretations, it would seem that there is some room for a systematic study of interpretations and interpretability as objects in their own right. This paper is an attempt to initiate one such line of enquiry.

2000 *Mathematics Subject Classification.* 00A30, 03B10, 03B30, 18B99.

Key words and phrases. formal theory, relative interpretation, category.

I thank Lev Beklemishev, Harvey Friedman, Spencer Gerhardt, Volker Halbach, Joost Joosten, Benedikt Löwe, Vincent van Oostrom, Darko Sarenac for enlightening conversations. I am grateful to Ieke Moerdijk for giving me some explanation about fibrations. John Corcoran provided historical background in e-mail correspondence, for which I am grateful. I thank Panu Raatikainen for sharing his ideas on interpretations and the definability of truth with me. Panu also provided many pointers to the relevant literature. Finally, I thank the anonymous referee for his or her helpful comments.

It is devoted to the study of the category of interpretations, or, more precisely the study of a sequence of categories of interpretations.

Below, I will briefly address three issues: motivation & desiderata, comparison to some earlier work, and comparison to boolean morphisms.

1.1. Motivation & Desiderata. The fact that interpretations play an important role in mathematics does not ipso facto mean that we should study them in a systematic way. Perhaps, as a totality, they are too diverse to make systematic study sensible. Perhaps, the only general insights are trite and not very useful. The work in this paper certainly does not bring the subject far enough to exclude such pessimistic expectations. The paper should be viewed in an experimental spirit: it is at least worth the effort to pursue such a study up to some level.

Interpretations have several uses. First, *comparing theories* is a philosophical need of human beings. E.g., we want to explicate the notion of strength of a theory. One possible explication is in terms of degrees of interpretability.[1] Or, we just want to say of certain theories that they are essentially the same. Consider, for example, the theory of partial order involving the weak ordering relation versus a formulation involving the strong ordering. Many people would judge these versions to embody the same theory as a matter of course. But what does sameness mean here? One possible explication is *synonymy* (see [dB65a], [dB65b], [Cor80], [Kan72]), which, as will be shown in this paper, can be best viewed as isomorphism in a certain category of interpretations.

Secondly, there are mathematico-logical applications. Interpretations can be used to prove properties of theories. E.g., Tarski proves that group theory is undecidable because true arithmetic can be interpreted in an extension of group theory. (See [TMR53].) Certain interpretations preserve important properties. E.g., consistency is preserved by interpretations in the reverse direction: from interpreting theory to interpreted theory. Decidability is preserved by *faithful* interpretations in the reverse direction.[2]

---

[1]Of course, there are other explications. The most succesful one is Π^0_2-conservativity: a theory is stronger if it proves more Π^0_2-sentences. Note, however, that conservativity implicitly uses interpretations: in many examples the designated class of sentences (e.g. Π^0_2) is not really a class of sentences of the language (e.g. the language of set-theory), but is embedded via an interpretation. Thus, the 'objects' to which we ascribe conservativity with respect to a class of sentences Γ are really pairs $\langle T, \tau \rangle$, where τ is an translation of the language containing Γ into the language of T.

Conservativity and interpretability diverge in some cases: GB is conservative with respect to the full language of set theory over ZF. However, since GB proves the consistency of ZF on a definable cut, GB is not interpretable in ZF.

[2]Note however, that we do not necessarily want to uncritically maximize preservation of properties. For certain purposes, we might wish to have an equivalence relation over which the properties from a given class of properties $\mathcal{P}$ are preserved, but such that every equivalence class contains an element having a designated desirable property Q. Clearly, we do not wish Q to be

Thirdly, interpretations are an essential ingredient of other notions. For example, the notion of conservativity (implicitly) employs interpretations. Also notions of axiom scheme and rule employ interpretations, e.g. when we say that set theory enjoys full induction. See Section 5. A third example is the relation of object-theory and meta-theory. See Section 7.

A good theory of interpretations should do some justice to the various uses of interpretations. So, we want to be able to analyze notions of strength and notions of sameness. We also want to be able to analyze notions like axiom scheme and object-theory/meta-theory in our setting. Moreover, we want to develop a theory of preservation of properties along interpretations. In this paper, we will pay special attention to preservation over retractions (in a certain category). (See Section 10.) In connection with questions of preservation, it is important that we are able to distinguish *sorts of interpretation*. We often want to know, not just that there is an interpretation, but that *a certain kind of interpretation* (like an isomorphism or a retraction) holds between two theories. Further information will follow from the fact that the interpretation is of such-and-such a kind.[3]

One further desideratum is simply that (known) salient reasoning about interpretations can be reconstructed within the categorical framework. We did some work in this direction, e.g. in Sections 8 and 9.

We opted for developing *a category of interpretations*. Two clear internal desiderata are the following. (a) As many notions as possible should receive categorical definitions. (b) There should be interesting uses of category theory. With regard to desideratum (a), we followed a middle road. A number of important concepts like *extension of theories in the same language* are treated as *enrichments* of the category, rather than somehow defined. A nice example of a characterization in categorical terms is provided by Theorem 6.4, where we characterize *direct interpretations* in terms of discrete fibrations (defined in a certain 2-category). With regard to (b), certainly more should be done. We only use quite elementary category theory.

Some caveats with respect to the present project are in order. First there is the question how the enterprise relates to model theory. It is certainly true that one can think of lots of connections. However, it is important to realize that the present approach is too narrow to constitute something like a framework for applications of interpretations in model theory. The reader is referred to the discussion of interpretations in [Hod93] where this point is made very clear. A second caveat is that I am neither a model theorist or a category theorist. Undoubtedly, I will have missed some questions and methods that would have

---

preserved. For an example, see [Per97], where Q is *finite axiomatizability* and the equivalence relation is *semantical similarity with respect to* $\mathcal{P}$.

[3] An important question in this connection, which is left unresolved in this paper, is formulated in Appendix A, item (8).

come naturally for a specialist in one of these subjects. For example, since most of my work is concerned arithmetical theories, there is more attention for the wild world in this paper than for the tame stuff so dear to model theorists.

1.2. Comparison to earlier work. There is a considerable literature on the systematic study of interpretations. There are two main approaches, which will be presented in the next to subsections.

1.2.1. *Interpretability logic.* There is the study of modal logics for interpretability. These logics are extensions of provability logic. There are two survey articles [JdJ98] and [Vis98]. Since these survey articles are in part complimentary they can very well be read together.

The main focus of the study of interpretability logics is to characterize schematic principles of interpretability that can be verified in theories that contain a sufficient amount of arithmetic. This line of research has yielded some beautiful results and some great techniques. It is however clear that it only treats a rather restricted range of theories. Moreover, it offers no possibility to talk explicitly about specific interpretations. Also there are some questions that it simply cannot address. E.g., it does not provide tools to make *distinctions* among interpretations.

1.2.2. *Degrees of interpretability.* Degree structures of interpretability have been extensively studied. See, for example, [Sve78], [Ben86], [Lin97], [MPS90] and [Vis05].[4] This study yielded many remarkable results. As in the case of interpretability logic, many results were proven for theories with a lot of coding machinery. However, this limitation is an accident of interest of the researchers, not intrinsic to the subject. Apart from orderings of interpretability, also orderings of faithful interpretability and of local interpretability[5] are studied. For local interpretability, see [MPS90].

Even if, for many purposes, interpretability is a rather crude relation, still there are some very useful preservation properties.

- Interpretability preserves inconsistency (from interpreted theory to interpreting theory), and, hence, preserves consistency in the reverse direction.
- A theory U is *reflexive* if it interprets a suitable weak arithmetical theory, like Buss's S_2^1, plus all statements of the form $con_n(U)$, where $con_n(U)$ expresses the consistency of the first n axioms of U for provability involving formulas of complexity below n. Here the measure of complexity is depth of quantifier changes. One can show that reflexive theories are not finitely axiomatizable. We have: mutual interpretability preserves reflexivity.

---

[4]The paper [MPS90] has an excellent introduction that in some respects supplements our introduction.

[5]A theory U is locally interpretable in a theory V iff every finite subtheory of U is interpretable in V.

- A theory is *essentially undecidable* iff every consistent extension of it is undecidable. Interpretability preserves essential undecidability from interpreted theory to interpreting theory.
- Interpretability does *not* preserve decidability from interpreting theory to interpreted theory. Not even mutual interpretability preserves decidability. However, faithful interpretability does in the direction from interpreting theory to interpreted theory.

On the other hand, many properties are not preserved, or are at least not obviously preserved. Moreover, degree theory gives us no tool to make distinctions between different interpretations of a theory U in a theory V.

In this paper, we will consider degree theory as a 'limiting case' of a category of interpretations. It will appear as our category INT_4.

1.3. Recursive boolean morphisms. Recursive boolean morphisms are recursive morphisms of the Lindenbaum algebras of theories considered as numerated objects. Recursive boolean morphisms are extensively and deeply studied. See [Han65], [PEK67], [Per97] and [Haj70][6]. We highlight two of the main results. Pour-El and Kripke show that all theories into which Q is interpretable are recursively boolean isomorphic. (See [PEK67].) The result continues to hold when we replace *recursive* by *primitive recursive* or even *elementary*. Peretyat'kin proves a theorem that implies that every recursively axiomatizable theory without finite models is recursively boolean isomorphic with a finitely axiomatized theory. (See [Per97]).

Every relative interpretation gives rise to a recursive boolean morphism, but not vice versa. Isomorpism in each of our categories INT_i, for $i = 0, 1, 2, 3$, will be a refinement of recursive boolean isomorphism. However, mutual relative interpretability and recursive boolean isomorphism are incomparable. E.g., Q and PA are, by the result of Pour-El and Kripke, recursively boolean isomorphic, but not mutually relatively interpretable. Conversely, predicate logic with only unary predicate symbols and predicate logic with at least one more-than-unary predicate symbol (distinct from identity) are mutually interpretable, but not recursively boolean isomorphic. (Recursive boolean isomorphisms preserve decidability.) We will give an example of two theories that are mutually faithfully interpretable, but not recursively boolean isomorphic. See Subsubsection 4.8.4.

Peretyat'kin, in his book [Per97], studies an equivalence relation, *semantical similarity* with respect to a list of model-theoretic properties, that is finer than recursive boolean isomorphism. Roughly, his notion is recursive boolean isomorphism plus the demand that the model-theoretic properties from the list are preserved. None of our notions of isomorphism can fulfil the desiderata for Peretyat'kin's notion: all our notions are refinements of mutual relative

---

[6]Hájek studies a category with morphisms which are a sort of generalized boolean morphisms.

interpretability. However, PA cannot be mutually interpretable with a finitely axiomatized theory.

§2. **Translations and interpretations.** In this section, we sketch the basic framework.[7]

In some respects, the framework introduced here is too limited. We will briefly discuss three natural extensions of the framework in Appendix B: multidimensional interpretations, interpretations for many-sorted predicate logic and interpretations with parameters. We will not pursue these extensions in this paper, even if, in many cases, it is rather obvious how results from the paper can be lifted to the more general context.[8]

We will consider interpretations between *theories* rather than between *models*. Since we insist on theories with simple axiom sets and finite signatures, there is no direct comparison between our categories and categories of models and interpretations. However, most of our results will work if we allow arbitrary axiom sets and arbitrary languages. If we extend our framework like this, we can view a category of models as a subcategory of a category of theories via the identification of a model and its full diagram.

2.1. Predicate logic. Officially, we will consider only relational languages for predicate logic of finite signature. Unofficially, we will also consider languages with terms. These languages can be translated using a standard algorithm to corresponding relational languages.

A signature Σ is a triple $\langle \text{Pred}, \text{ar}, E \rangle$, where Pred is a finite set of predicate symbols, where ar : Pred $\rightarrow \omega$ is the arity function. and where E is a binary predicate representing the identity. We will often write '=' for: E.

We assume we are given a fixed ω-ordered sequence of variables $v_0, v_1, \ldots$. We will use $x, y, x_0, \ldots$ as metavariables ranging over variables. (We will follow the usual convention that, if e.g. "v_3", "x" and "y" are used in one formula, they are supposed to be distinct.) Formulas and sentences based on Σ and $v_0, v_1, \ldots$ are defined in the usual way.

A theory of signature Σ will be given by a set of sentences of the signature, the axioms. Derivability from the theory employs the axioms and the rules of predicate logic including the identity axioms and rules for E. We will assume that the axiom set of our theories is appropriately simple, say p-time decidable. However, most of the results of this paper will hold for more complicated axiom sets too.

---

[7]A much better treatment of the predicate logical language is possible, using sharing graphs. Under this alternative treatment, a lot of *ad hoc* choices and unpleasant details concerning α-conversion would simply disappear. However, setting things up in the alternative way would take a lot of space and would detract from the proper subject of this paper.

[8]I feel that each of these extensions should definitely be investigated. E.g. interpretations with parameters occur often in mathematics. Moreover, our analysis of schemes clearly asks for generalization to the case with parameters.

2.2. Relative translations. Let Σ and Θ be signatures. A *relative translation* $\tau : \Sigma \to \Theta$ is given by a pair $\langle \delta, F \rangle$. Here δ is a Θ-formula representing the *domain* of the translation. We demand that δ contains at most v_0 free. The mapping F associates to each relation symbol R of Σ with arity n an Θ-formula $F(R)$ with variables among $v_0, \ldots, v_{n-1}$.

We translate Σ-formulas to Θ-formulas as follows:

- $(R(y_0, \ldots, y_{n-1}))^\tau := F(R)(y_0, \ldots, y_{n-1})$;
  here $F(R)(y_0, \ldots, y_{n-1})$ is our sloppy notation for:

  $$F(R)[v_0 := y_0, \ldots, v_{n-1} := y_{n-1}],$$

  the result of substituting the y_i for the v_i; we assume that some mechanism for α-conversion is built into our definition of substitution to avoid variable-clashes;
- $(\cdot)^\tau$ commutes with the propositional connectives;
- $(\forall y\, A)^\tau := \forall y\, (\delta(y) \to A^\tau)$;
- $(\exists y\, A)^\tau := \exists y\, (\delta(y) \wedge A^\tau)$.

Suppose τ is $\langle \delta, F \rangle$. Here are some convenient notations.

- We write δ_τ for δ and F_τ for F.
- We write R_τ for $F_\tau(R)$.
- We write $\vec{x} : \delta$ for: $\delta(x_0) \wedge \cdots \wedge \delta(x_{n-1})$.
- We write $\forall \vec{x} : \delta\, A$ for: $\forall x_0 \ldots \forall x_{n-1}\, (\vec{x} : \delta \to A)$.
- We write $\exists \vec{x} : \delta\, A$ for: $\exists x_0 \ldots \exists x_{n-1}\, (\vec{x} : \delta \wedge A)$.

We can compose relative translations as follows:

- $\delta_{\tau\nu} := (\delta_\nu \wedge (\delta_\tau)^\nu)$,
- $R_{\tau\nu} = (R_\tau)^\nu$.

We write $\nu \circ \tau := \tau\nu$. Note that $(A^\tau)^\nu$ is provably equivalent in predicate logic to $A^{\tau\nu}$. The identity translation $\mathrm{id} := \mathrm{id}_\Theta$ is defined by:

- $\delta_{\mathrm{id}} := (v_0 E v_0)$,
- $R_{\mathrm{id}} := R(v_0, \ldots, v_{n-1})$.

Note that translations as defined here only have good properties modulo provable equivalence. E.g., $\delta_{\mathrm{idoid}} = (v_0 E v_0 \wedge v_0 E v_0)$, which is not strictly identical to δ_{id}.

Consider a relative translation $\tau : \Sigma \to \Theta$. Let $\mathcal{M} = \langle M, I \rangle$ be a model of signature Θ. Suppose $\mathcal{M}$ satisfies $\exists v_0\, \delta$ and the τ-translations of the identity axioms in the Σ-language. We can define a model $\mathcal{N} := \tau^{\mathcal{M}}$ as follows.

- Clearly, E_Σ^τ defines an equivalence relation, say $\simeq$, in $\mathcal{M}$ on $N_0 := \{m \in M \mid \mathcal{M} \models \delta_\tau(m)\}$. The domain N of $\mathcal{N}$ is $N_0/\simeq$.
- The relation $\simeq$ is a congruence with respect to the relations R, given by:

  $$R(\vec{m}) :\Longleftrightarrow \mathcal{M} \models P^\tau(\vec{m}).$$

Thus, it makes sense to define, for $\vec{n}$ a sequence of elements of N,

$$P^{\mathcal{N}}(\vec{n}) :\Longleftrightarrow \exists m_0 \in n_0 \ldots \exists m_{k-1} \in n_{k-1} \; R(m_0, \ldots, m_{k-1}).$$

Suppose $\vec{n}$ is a sequence of elements of N, the domain of $\mathcal{N} := \tau^{\mathcal{M}}$. Let $\vec{m}$ be a sequence of elements of M, the domain of $\mathcal{M}$, such that $m_i \in n_i$. One can show:

$$\mathcal{N} \models A(\vec{n}) \Longleftrightarrow \mathcal{M} \models A^\tau(\vec{m}).$$

Further, one can show that $\tau^{\nu^{\mathcal{M}}}$ is isomorphic to $(\tau\nu)^{\mathcal{M}}$ (if defined). We have, e.g., for sentences A of the appropriate signature:

$$\begin{aligned}
\tau^{\nu^{\mathcal{M}}} \models A &\Longleftrightarrow \nu^{\mathcal{M}} \models A^\tau \\
&\Longleftrightarrow \mathcal{M} \models (A^\tau)^\nu \\
&\Longleftrightarrow \mathcal{M} \models A^{\tau\nu} \\
&\Longleftrightarrow (\tau\nu)^{\mathcal{M}} \models A.
\end{aligned}$$

2.3. Relative interpretations. A translation τ supports a *relative interpretation* of a theory U in a theory V, if, for all U-sentences A, $U \vdash A \Rightarrow V \vdash A^\tau$. (Note that this automatically takes care of the theory of identity. Moreover, it follows that $V \vdash \exists v_0 \, \delta_\tau$.) We will write $K = \langle U, \tau, V \rangle$ for the interpretation supported by τ. We write $K : U \to V$ for: K is an interpretation of the form $\langle U, \tau, V \rangle$. If M is an interpretation, τ_M will be its second component, so $M = \langle U, \tau_M, V \rangle$, for some U and V.

Suppose T has signature Σ and $K : U \to V$, $M : V \to W$. We define:

- $\mathrm{id}_T : T \to T$ is $\langle T, \mathrm{id}_\Sigma, T \rangle$,
- $M \circ K : U \to W$ is $\langle U, \tau_M \circ \tau_K, W \rangle$.

We identify two interpretations $K, K' : U \to V$ if:

- $V \vdash \delta_K \leftrightarrow \delta_{K'}$,
- $V, \vec{v} : \delta \vdash P^K \leftrightarrow P^M$, where $\mathrm{ar}(P) = n$ and $\vec{v} = v_0, \ldots, v_{n-1}$.

One can show that modulo this identification, the above operations give rise to a category of interpretations that we call INT.

If $K = \langle U, \tau, V \rangle$, and $\mathcal{M} \models V$, we will often write $K^{\mathcal{M}}$ for $\tau^{\mathcal{M}}$.

We assign to a theory T its class of models $\mathrm{MOD}(T)$. Consider $K : U \to V$. This interpretation gives rise to the mapping $\mathrm{MOD}(K) : \mathrm{MOD}(V) \to \mathrm{MOD}(U)$ given by $\mathcal{M} \mapsto K^{\mathcal{M}}$. We have:

- $\mathrm{MOD}(\mathrm{id}_T) := \mathrm{id}_{\mathrm{MOD}(T)}$,
- $\mathrm{MOD}(M \circ K) = \mathrm{MOD}(K) \circ \mathrm{MOD}(M)$.

Thus, MOD gives rise to a contravariant functor from INT to the category of sets and classes.

§3. **Categories of interpretations.** In this section, we introduce various categories of interpretations. Our basic category is INT. We call two interpretations that implement the same morphism in the sense of INT: *equal*. We call theories that are isomorphic in this category: *synonymous* or *definitionally equivalent*.[9]

3.1. Definable maps between interpretations. We extend INT with extra structure. In this enriched category, $\mathsf{INT}^{\mathrm{morph}}$, the arrows between two objects play themselves the role of objects in a category as follows. Consider $K, M : U \to V$. An arrow $F : K \Rightarrow M$ is a V-definable, V-provable morphism from K to M considered as 'parametrized internal models'. Specifically, this means that a morphism from K to M is given as a triple $\langle K, F, M \rangle$, where F is a formula with the following properties.

- The free variables of F are among v_0, v_1. We write $F(x, y)$ or xFy, for: $F[v_0 := x, v_1 := y]$.
- $V \vdash xFy \to (x : \delta_K \wedge y : \delta_M)$.
- Writing E for the identity relation:

$$V \vdash (x : \delta_K \wedge y : \delta_M \wedge xE_K x' Fy' E_M y) \longrightarrow xFy.$$

- $V \vdash \forall x : \delta_K \exists y : \delta_M \, xFy$.
- $V \vdash (xFy \wedge xFy') \to yE_M y'$.
- $V \vdash \vec{x}F\vec{y} \to (P_K \vec{x} \to P_M \vec{y})$.[10]

Here '$\vec{x}F\vec{y}$' abbreviates $x_0 Fy_0 \wedge \cdots \wedge x_{n-1}Fy_{n-1}$, for appropriate n.

We will call the arrows between interpretations: *i-maps*. We consider $F, G : K \Rightarrow M$ as *equal* when they are V-provably the same. The identity $\mathsf{ID}_K : K \Rightarrow K$ is given by: $v_0(\mathsf{ID}_K)v_1 :\leftrightarrow v_0, v_1 : \delta_K \wedge v_0 E_K v_1$. If $A : K \Rightarrow L$ and $B : L \Rightarrow M$, then $B \cdot A : K \Rightarrow M$ is the obvious composition of B and A.

An isomorphism of interpretations is easily seen to be a morphism with the following extra properties.

- $V \vdash \forall y : \delta_M \exists x : \delta_K \, xFy$,
- $V \vdash (xFy \wedge x'Fy) \to xE_K x'$,
- $V \vdash \vec{x}F\vec{y} \to (P_M \vec{y} \to P_K \vec{x})$.

Suppose we have the constellation depicted in the diagram below.

$$U \underset{K}{\overset{L}{\rightrightarrows}} {\Uparrow F} \, V \xrightarrow{M} W$$

[9]Karel de Bouvère uses *synonymy* in his [dB65a] and [dB65b]. Stig Kanger, John Corcoran and Wilfrid Hodges use *definitional equivalence*. See [Kan72], [Cor80] or [Hod93]. Mikhail Peretyat'kin uses *isomorphism*. See [Per97].

[10]Note that if P represents a function in U, then, by elementary reasoning, we have: $V \vdash \vec{x}F\vec{y} \to (P_K \vec{x} \leftrightarrow P_M \vec{y})$.

We define $M \circ F : M \circ K \Rightarrow M \circ L$ as follows:

- $v_0(M \circ F)v_1 :\leftrightarrow v_0, v_1 : \delta_M \wedge v_0 F^M v_1$.

We may show that, indeed, $M \circ F : M \circ K \Rightarrow M \circ L$.

The corresponding idea of composing F with an interpretation to the right does not generally make sense: an i-map is analogous to a morphism between structures. Such morphisms do not generally induce morphisms between 'substructures with defined operations'. It does only make sense if we put suitable constraints on the interpretations or if we restrict ourselves to i-isomorphisms. We will follow this last strategy in the present paper. Suppose we have the situation depicted in the diagram below. Let G be an i-isomorphism.

$$U \xrightarrow{\quad K \quad} V \begin{array}{c} M \\ \xrightarrow{\quad\quad} \\ \Uparrow G \\ \xrightarrow{\quad\quad} \\ L \end{array} W.$$

We define $G \circ K : L \circ K \Rightarrow M \circ K$ as follows:

- $v_0(G \circ K)v_1 :\leftrightarrow v_0 : \delta_{L \circ K} \wedge v_1 : \delta_{M \circ K} \wedge \exists x, y \; v_0 E_{L \circ K} x G y E_{M \circ K} v_1$.

We can easily show that: $G \circ K : L \circ K \Rightarrow M \circ K$, where $G \circ K$ is an i-isomorphism. Suppose we have the following situation, where G is an i-isomorphism.

$$U \begin{array}{c} L \\ \xrightarrow{\quad\quad} \\ \Uparrow F \\ \xrightarrow{\quad\quad} \\ K \end{array} V \begin{array}{c} N \\ \xrightarrow{\quad\quad} \\ \Uparrow G \\ \xrightarrow{\quad\quad} \\ M \end{array} W.$$

We define $G \circ F : M \circ K \Rightarrow N \circ L$ as follows:

- $G \circ F := (N \circ F) \cdot (G \circ K)$.

One may show that: $G \circ F = (G \circ L) \cdot (M \circ F)$.

We will call the restriction of $\mathsf{INT}^{\mathsf{morph}}$ to i-isomorphisms: $\mathsf{INT}^{\mathsf{iso}}$. We can show that $\mathsf{INT}^{\mathsf{iso}}$ is 2-category with the defined operations. See [Mac71] and [Bor94] for an explanation.

Here is a pleasant way to look at the properties of our enriched category $\mathsf{INT}^{\mathsf{morph}}$. Let us fix some theory, say $\mho$. We define a 2-functor $[\![\cdot]\!]$ (or, more explicitly $[\![\cdot]\!]_\mho$) from $\mathsf{INT}^{\mathsf{iso}}$ to the 2-category of categories, functors and natural transformations as follows.

- $[\![U]\!]$ is the category with as objects the interpretations $M : \mho \to U$ and as arrows the i-maps between the objects.
- Suppose $K : U \to V$, we define $[\![K]\!] : [\![U]\!] \to [\![V]\!]$, by $[\![K]\!](M) := K \circ M$, $[\![K]\!](F) := K \circ F$.
- Suppose $G : K \Rightarrow K'$ is an i-isomorphism. Then, $[\![G]\!] : [\![K]\!] \Rightarrow [\![K']\!]$ is given by: $[\![G]\!](M) := G \circ M$.

The verification that we have indeed defined a 2-functor holds no surprises. Similarly, we can define a contravariant 2-functor $[\![\cdot]\!]$ from $\mathsf{INT}^{\mathsf{iso}}$ to the

2-category of categories, functors and natural transformations, as follows. Again we fix a theory $\mho$.

- $[\![U]\!]$ is the category with as objects the interpretations $M : U \to \mho$ and as arrows the i-isomorphisms between the objects.
- Suppose $K : U \to V$, we define $[\![K]\!] : [\![V]\!] \to [\![U]\!]$, by $[\![K]\!](M) := M \circ K$, $[\![K]\!](F) := F \circ K$.
- Suppose $G : K \Rightarrow K'$ is an i-isomorphism. Then, $[\![G]\!] : [\![K]\!] \Rightarrow [\![K']\!]$ is given by: $[\![G]\!](M) := M \circ G$.

Note that the definition of $[\![\,\cdot\,]\!]$ also makes sense on $\mathsf{INT}^{\mathrm{morph}}$. Only we get no 2-functor since $\mathsf{INT}^{\mathrm{morph}}$ is not a 2-category.

3.2. Subcategories. We may wish to restrict our interpretations to certain subclasses. E.g., we restrict ourselves to unrelativized interpretations, obtaining the category $\mathsf{INT}_{\mathrm{unr}}$. Or we restrict ourselves to interpretations that interpret identity by itself, obtaining the category $\mathsf{INT}_{=}$. An important case is the restriction to unrelativized interpretations that preserve identity. Such interpretations (and the corresponding morphisms) we call: *direct*. The associated category is $\mathsf{INT}_{\mathrm{unr},=}$.

3.3. When are two interpretations the same? The most important variations on INT are obtained by considering cruder notions of equality on interpretations.

1. The finest identification that we will consider is equality of interpretations as defined above. For reasons of systematicity we will also call this equality: equal$_0$ or $=_0$. Similarly, we sometimes call the category INT: INT_0.

2. The next level of identification is i-isomorphism of interpretations in our extended category $\mathsf{INT}^{\mathrm{morph}}$. We will call i-isomorphism also: equal$_1$ or $=_1$. Wilfrid Hodges, in his [Hod93], uses *homotopy* for i-isomorphism and calls i-isomorphic interpretations *homotopic*. We will call the category of theories with the morphisms so obtained: INT_1 or hINT.

3. We may also consider two interpretations $K : U \to V$ and $M : U \to V$ as the same iff, for all models $\mathcal{M} \in \mathrm{MOD}(V)$, we have that $\mathrm{MOD}(K)(\mathcal{M})$ is isomorphic to $\mathrm{MOD}(M)(\mathcal{M})$. We call this equality: equal$_2$ or $=_2$. We will call the category of theories with the morphisms so obtained: INT_2 or whINT.

   There are all kinds of variants of this category which can be obtained by restricting the class of models.

4. We may take two interpretations $K : U \to V$ and $M : U \to V$ as the same iff, for all U-sentences A, we have $V \vdash A^K \leftrightarrow A^M$. We will say that these interpretations are equivalent, equal$_3$ or $=_3$. We will call the category of theories with the morphisms so obtained: INT_3.

5. We can simply identify all interpretations between U and V. In this case we obtain the preorder associated with the degrees of interpretability. We will call the category of theories with the morphisms so obtained: INT_4 or DEG.

REMARK 3.1. There are two choices on how to talk about the morphisms in our various categories. If we divide out the equality relation introduced for the identification of interpretations in the category at hand, we will be speaking about morphisms as equivalence classes of interpretations. A morphism of INT_i would only occur in INT_{i+1} via a standard embedding. I didn't follow this way of doing things, but opted for not explicitly dividing out equality of morphisms. Thus, we will speak about a morphism K in INT_i also occurring in INT_{i+1} and we will simply extract the underlying translation τ_K from K. We will write things like: 'K is a monomorphism in INT_0, but not in INT_1'.

We note a coincidence of categories.

THEOREM 3.2. *The category* INT_3 *coincides with* INT_2^{rs}, *i.e. the category* INT_2 *with the class of models restricted to recursively saturated models.*

PROOF. The proof is immediate from two well-known facts. Definable internal models of recursively saturated models are isomorphic iff they are elementary equivalent. Moreover, any consistent theory has a recursively saturated model. ⊣

Our various categories induce different notions of sameness on theories:

1. two theories are *synonymous* or *definitionally equivalent* if they are isomorphic in INT;
2. two theories are *bi-interpretable*, if they are isomorphic in INT_1;
3. two theories are *weakly bi-interpretable* if they are isomorphic in INT_2;
4. two theories are *sententially equivalent* if they are isomorphic in INT_3;
5. two theories are *mutually interpretable* if they are isomorphic in INT_4.

Note that, due to the simple form of the definition of isomorphism, the more interpretations we identify, the more theories are isomorphic. So, each subsequent notion of sameness of theories is cruder.

Clearly, there cannot be a uniform answer to the question *which category is the best one*? E.g. INT_4 can already be very useful if we want to prove relative consistency or to establish undecidability. As we will see, in INT_3 notions like *faithful interpetation* and *surjective interpretation* can be characterized as *monomorphism*, resp. *epimorphism*. The category INT_1 has good preservation properties, as illustrated in Section 10. Finally, INT^{morph} supplies the most convenient framework to work in for theoretical purposes. In fact, to reason about INT_1, one naturally uses INT^{morph}.

3.4. Generalizing the MOD-functor.

In Subsection 2.3, we introduced the contravariant MOD-functor on INT. Consider $K =_i M : U \to V$ for $i = 1, 2$.

Note that, for any model $\mathcal{M}$ of V, we have that the model $\text{MOD}(K)(\mathcal{M})$ need not be the same as the model $\text{MOD}(M)(\mathcal{M})$, however these models will be isomorphic. Thus, to make sense of the MOD-functor on INT_1 and INT_2, we should consider it as a functor to models modulo isomorphism.

Clearly, there is a cardinality problem here, since isomorphism classes will not be sets. There are all kinds of ways to get around that problem. E.g., we can stipulate that we only consider models in V_κ for some cardinal κ, or we can refrain from dividing out the equivalence relation, or we can demand the elements of the domains of the models are chosen from the cardinal that is the cardinality of the domain, etc. We pretend that we have settled on some such solution.

We will call the new functors MOD_i, for $i = 1, 2$, or, if no confusion is possible, simply MOD. For uniformity, sometimes we will call our original MOD-functor: MOD_0.

In the case of INT_3, we only get, for $K =_3 M : U \to V$, that $\text{MOD}(K)(\mathcal{M})$ is elementarily equivalent to $\text{MOD}(M)(\mathcal{M})$. Thus, in the case of INT_3, we take the MOD_3-functor to give us models modulo elementary equivalence, or alternatively: complete theories.

For INT_4 we will not have a MOD-functor.

3.5. Factorization. A slightly odd aspect of interpretations and the corresponding morphisms in our categories is the fact that there is a certain favouritism towards the target theory, i.e. the interpreting theory.[11] The interpreted theory often plays a minor role. E.g. if you look at the question whether two interpretations between U and V are the same, the answer only depends on the underlying translations and the target theory. Similarly, i-morphisms are defined in terms of the underlying translations and target theories. In this subsection, we spell out some consequences of this lopsidedness.

We work in $\text{INT}^{\text{morph}}$. Consider an interpretation $K = \langle U, \tau, V \rangle$. We define the theory $\tau^{-1}[V]$ or $K^{-1}[V]$ as the theory in the signature of U given by $\{A \in \mathcal{S}_U \mid V \vdash A^\tau\}$. (We may find an efficient axiomatization using Craig's Trick.) The interpretation K is *faithful* iff $U = K^{-1}[V]$.

We write $\mathcal{E}_{TZ}$ for the identical interpretation witnessing that T is a subtheory of Z in the same language. Note that restricting the morphisms to the $\mathcal{E}_{TZ}$ will give us a subcategory. In Subsection 4.6, we will see that $\mathcal{E}_{UV}$ is an epimorphism in each of our categories.

Let $\check{K} := \langle K^{-1}[V], \tau, V \rangle$. Then, we obviously have the following factorization of K:

$$ U \xrightarrow{\mathcal{E}_{U,K^{-1}[V]}} K^{-1}[V] \xrightarrow{\check{K}} V. $$

---

[11] The first point where we use the contents of this section is in Subsubsection 4.8.2. The reader could postpone reading the section until that point.

Trivially, $\check{K}$ is faithful. In Subsection 4.5, we will see that faithful interpretations are monomorphisms in categories INT_0 and INT_3. Ergo, in INT_0 and INT_3, our factorization is an epi-mono factorization. We have the following obvious theorem.

THEOREM 3.3. *Suppose we have* $K : U \to V$, $M : W \to V$, $U \subseteq W$ *and* $K = M \circ \mathcal{E}_{UW}$. *Then,* $K^{-1}[V] = M^{-1}[V]$ *and* $\check{K} = \check{M}$.

We omit the proof.

COROLLARY 3.4. *Suppose* $K : U \to V$ *and* $M : V \to W$. *Then,* $K^{-1}[V] \subseteq (M \circ K)^{-1}[W]$.

PROOF. Suppose $K : U \to V$ and $M : V \to W$. We have $(M \circ K) : U \to W$, $M \circ \check{K} : K^{-1}[V] \to W$ and $U \subseteq K^{-1}[V]$. Hence, by Theorem 3.3,

$$K^{-1}[V] \subseteq (M \circ \check{K})^{-1}[W] = (M \circ K)^{-1}[W]. \qquad \dashv$$

More generally, we can consider the interaction of an $\mathcal{E}$-arrow and two subsequent arrows. We get the following theorem.

THEOREM 3.5. *Suppose the following diagram obtains:*

$$U \xrightarrow{\;\mathcal{E}_{UV}\;} V \begin{array}{c} K \\ \Uparrow F \\ L \end{array} W.$$

We define $F \circ \mathcal{E}_{UV} : (L \circ \mathcal{E}_{UV}) \Rightarrow (K \circ \mathcal{E}_{UV})$, *where* $F \circ \mathcal{E}_{UV}$ *is given by the same formula as* F.[12] *We have:*

- *$(\cdot) \circ \mathcal{E}_{UV}$ is an injective and full functor from the category of arrows from V to W to the category of arrows from U to W;*
- *$(F \circ \mathcal{E}_{UV}) \circ \mathcal{E}_{TU} = F \circ \mathcal{E}_{TV}$;*
- *suppose $M : W \to Z$, then $(M \circ F) \circ \mathcal{E}_{UV} = M \circ (F \circ \mathcal{E}_{UV})$.*

Conversely, suppose we have the following situation.

$$U \begin{array}{c} K \circ \mathcal{E}_{UV} \\ \Uparrow G \\ L \circ \mathcal{E}_{UV} \end{array} W.$$

Then, we can find a unique $\tilde{G} : L \Rightarrow K$ *such that* $G = \tilde{G} \circ \mathcal{E}_{UV}$. *Moreover* $(\tilde{\cdot})$ *is a functor from the subcategory of arrows of the form* $P \circ \mathcal{E}_{UV} : U \to W$ *to the category of arrows from V to W.*

§4. **A closer look at familiar concepts.** In this section, we treat some evident 'homework questions'. If you have a category, you want to know whether there are, e.g., products and, if so, what they are. We do not try to be exhaustive here: we just treat some nice bits.

---

[12] Remember that, previously, we only admitted $F \circ M$ in case F was an i-isomorphism.

4.1. Initial objects. The theory 1 is the theory in the language with just identity, which states that there is precisely one object. It is the initial object in INT_i with $i \in \{1, 2, 3, 4\}$, i.e. it has precisely one interpretation (modulo the chosen notion of equivalence) in any other theory.

The situation is different for INT_0. Consider the theory U with unary predicate symbol P and axiom:

- $\exists x, y \, (P(x) \wedge \neg P(y) \wedge \forall z \, (z = x \vee z = y))$.

We can interpret 1 into U via K_0 given by:

- $\delta_{K_0} :\hookrightarrow P(v_0)$,
- $E_{K_0} :\hookrightarrow v_0 = v_1$.

We can also employ K_1 given by:

- $\delta_{K_1} :\hookrightarrow \neg P(v_0)$,
- $E_{K_1} :\hookrightarrow v_0 = v_1$.

Suppose J would be an initial object of INT_0 and $M : J \to 1$. Then, it is easily seen that $K_0 \circ M : J \to U$ cannot be equal to $K_1 \circ M : J \to U$. A contradiction. Of course there are many *weak* initial objects in INT.

When we restrict INT_i with $i \in \{1, 2, 3, 4\}$ to interpretations that send identity to identity. the situation also changes. Consider $\mathfrak{I} := \mathsf{FOL}_{\mathsf{id}}$, predicate logic with just identity. Suppose J is an initial object in one of the $\mathsf{INT}_{i,=}$, for $i = 1, 2, 3, 4$. Then, there is a unique morphism $M : J \to \mathsf{FOL}_{\mathsf{id}}$. Note that, by a simple model theoretical argument, δ_M must be equivalent to $v_0 = v_0$. Now we can consider identity preserving interpretations K_0 and K_1 of $\mathsf{FOL}_{\mathsf{id}}$ into, say, PA such that $\mathsf{PA} \vdash 1^{K_0}$ and $\mathsf{PA} \vdash 2^{K_1}$. Here 2 is the theory in the langue of pure identity with as axiom the statement that there are precisely two objects. It is easy to see that $K_0 \circ M : J \to \mathsf{PA}$ cannot be equal to $K_1 \circ M : J \to \mathsf{PA}$.

For any theory T, we define the morphism $\mathcal{I}_T : \mathfrak{I} \to T$ by: $\delta_{\mathcal{I}_T} :\hookrightarrow v_0 = v_0$ and $v_0 E^{\mathcal{I}_T} v_1 :\hookrightarrow v_0 = v_1$. It follows that $\mathfrak{I}$ is a weak initial object in any of our categories. It is easy to see that $\mathfrak{I}$ is the initial object in $\mathsf{INT}_{i,\mathsf{unr},=}$, for $i = 0, 1, 2, 3, 4$.

4.2. End objects. The inconsistent theory of any signature is the end object in any of our categories. Any theory has precisely one interpretation into the inconsistent theory.

If we restrict our categories to faithful interpretations, the situation changes dramatically: there is not even a *weak* end object.

4.3. The cartesian product. Surprisingly, there is a stable notion of product that works for all our categories. The easiest way to introduce the product is as follows. First we consider theories U and V of the same signature such that $U \cup V$ is inconsistent. By a simple compactness argument, we find that there is an A such that $U \vdash A$ and $V \vdash \neg A$. We will call A, a *separating* formula for U and V. (Note: a separating formula is assigned to the *ordered* pair U, V.)

LEMMA 4.1. *Suppose U and V have the same signature and suppose $U \cup V$ is inconsistent. Let A be a separating formula for U and V. The theory $W := U \cap V$ can be axiomatized by the following sets of axioms: axioms of the form $A \to B$, where B is an axiom of U, and axioms of the form $\neg A \to C$, where C is an axiom of V.*

Note that it follows that we can write U as $W + A$ and V as $W + \neg A$.

PROOF. It is clear that the axioms $(A \to B)$ and $(\neg A \to C)$ are in the intersection. Conversely, suppose $W \vdash D$. Then some conjuction β of U-axioms B proves D and some conjunction γ of V-axioms C proves D. We claim that $\delta := ((A \to \beta) \wedge (\neg A \to \gamma))$ proves D. This is immediate, since, ex hypothesi, δ implies $(A \to D) \wedge (\neg A \to D)$. Finally, δ can be rewritten to a conjunction of axioms of the form $(A \to B)$ and $(\neg A \to C)$. ⊣

We claim that $W := U \cap V$ is the cartesian product of U and V in any of our categories. The projections are the standard embeddings $\mathcal{E}_{WU}$ and $\mathcal{E}_{WV}$.

To show that $W = U \times V$ with the stated projections, we have to uniquely provide the dotted arrow that makes the following diagram commute.

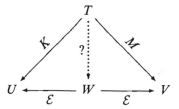

Suppose the translations associated with K, M are τ, ρ. Let A be the chosen separating sentence for U and V. We define a new translation $\tau\langle A \rangle \rho$ as follows.

- $\delta_{\tau\langle A\rangle\rho} :\leftrightarrow ((A \wedge \delta_\tau) \vee (\neg A \wedge \delta_\rho))$.
- $P_{\tau\langle A\rangle\rho} :\leftrightarrow ((A \wedge P_\tau) \vee (\neg A \wedge P_\rho))$.

The interpretation $N := K\langle A \rangle M : T \to W$ is the interpretation corresponding to $\tau\langle A \rangle \rho$. We take:

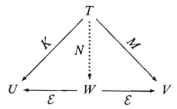

In each of our categories we may easily verify that $K\langle A \rangle M$ supports the unique morphism that makes our diagram commute. This also shows that equality is a congruence for $(\cdot)\langle A\rangle(\cdot)$. Thus, $(\cdot)\langle A\rangle(\cdot)$ can be viewed as an operation on arrows. Note that at the level of the interpretations as mappings from models to models, the set of models of W is the disjoint union of the set of models of U and the set of models of V. We have: $\mathrm{Mod}(K\langle A\rangle M) = \mathrm{Mod}(K) \cup \mathrm{Mod}(M)$. (The union concerns the functions considered as sets of pairs.)

Here are some alternative notations for $K\langle A\rangle M$.

- My notation in earlier papers: $K[A]M$.
- Pseudo-code: if A then K else M.
- Hoare style: $K \lhd A \rhd M$.

We defined the cartesian product for a special case. We extend the definition to arbitrary theories as follows. Let U and V be arbitrary with signatures Σ and Θ. Let $\Sigma \cup \Theta$ be the union of our signatures. To make sense of taking the union, it is best to consider the arities to be built in into the symbols, so that, say, a binary predicate P and a ternary predicate P are automatically counted as different predicates.

Consider a theory T with signature Σ. Let T' be a theory with signature Σ'. We say that T' is a *definitional extension* of T iff Σ' extends Σ and the axioms of T' are the axioms of T plus, for each predicate symbol P of $\Sigma' \setminus \Sigma$, an axiom of the form: $\vdash P\vec{x} \leftrightarrow A\vec{x}$, where A is in the language of T with at most $\vec{x}$ free. Remember that synonymy is isomorphism in INT. We have:

THEOREM 4.2. *Any definitional extension of a theory is synonymous to that theory.*

The proof of Theorem 4.2 is easy. Consider definitional extensions U' and V' of U and V in the signature $\Sigma \cup \Theta$. In case $U' \cup V'$ is inconsistent, we take, as $U \times V$, the intersection $W' := U' \cap V'$. Our first projection becomes $\pi_0 := J \circ \mathcal{E}_{W'U'}$, where $J : U' \to U$ is the standard isomorphism associated with definitional extensions. Similarly, $\pi_1 := L \circ \mathcal{E}_{W'V'}$, where $L : V' \to V$ is an isomorphism. It is easy to see that we have defined a product in this way.

If $U' \cup V'$ is consistent, we extend the signature $\Sigma \cup \Theta$ with a fresh 0-ary predicate symbol P, and replace U' be the definitional extension $U' + P$ and V' by the definitional extension $V' + \neg P$. Now we may proceed as before.

Inspecting the construction, we see that it even works in the strict world of direct interpretations, i.e. $\mathsf{INT}_{\mathrm{unr},=}$.

REMARK 4.3. Sequentiality of a theory T is the existence of a morphism in $\mathsf{INT}_{\mathrm{unr},=}$ from a certain theory of sequences SEQ to T. See Section 10. Thus, it follows that sequentiality is preserved by products, as witnessed by the following commutative diagram in $\mathsf{INT}_{\mathrm{unr},=}$.

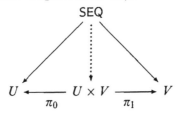

It is an interesting exercise to compute $T^n := \overbrace{T \times \cdots \times T}^{n \times}$. The result of the exercise is the following theorem. We say that a finite boolean algebra B is *presented* by a formula $\phi(\vec{P})$ over propositional variables $\vec{P}$ if the Lindenbaum algebra of the theory ϕ in the propositional language over $\vec{P}$ is (isomorphic to) B. Alternatively, we can say that B is presented by $\phi(\vec{P})$ over $\vec{P}$ if B is (isomorphic to) $\mathcal{F}_{\vec{P}}/[\phi]$, where $\mathcal{F}_{\vec{P}}$ is the free algebra over generators $\vec{P}$ and where $[\phi]$ is the prime filter generated by ϕ.

THEOREM 4.4. *We work in any of our categories. Let $\vec{P}$ be a finite set of 0-ary predicates. Suppose $\phi(\vec{P})$ is a presentation over $\vec{P}$ of a finite boolean algebra B with n atoms on generators $\vec{P}$.*

Consider a theory T with signature Σ and suppose that the symbols $\vec{P}$ do not occur in Σ. Then, we can take as T^n the theory in the signature $\Sigma + \vec{P}$ such that $T^n := T + \phi(\vec{P})$.

Remember that there is precisely one finite boolean algebra with n atoms. We allow the degenerated case, where $n = 0$. This is simply the one point algebra where top en bottom coincide.

PROOF. Suppose $P_0, P_1, \ldots$ is a fixed infinite sequence of 0-ary predicates. We write $\vec{P}_{[n]}$ for $P_0, \ldots, P_{n-1}$. We first prove, by induction on n, that, for each n, we can find a specific ψ_n such that $T + \psi_n(\vec{P}_{[n]})$, in signature $\Sigma + \vec{P}_{[n]}$, is T^n and such that $\psi_n(\vec{P}_{[n]})$ is a presentation of the n-atom finite boolean algebra on generators $\vec{P}_{[n]}$. Then, we show that any ϕ presenting the n-atom finite boolean algebra on some finite set of generators $\vec{R}$ will do the job.

We take:

- $\psi_0 := \bot$.

Suppose T^n is given as $T + \psi_n(\vec{P}_{[n]})$ in signature $\Sigma + \vec{P}_{[n]}$. We compute $T^{n+1} := T^n \times T$. Clearly, T^{n+1} can be taken to be the theory in the signature $\Sigma + \vec{P}_{[n]}$, which is the intersection of $T + \psi_n(\vec{P}_{[n]}) + P_n$ and $T + \bigwedge \vec{P}_{[n]} + \neg P_n$. So we take:

- $\psi_{n+1}(\vec{P}_{[n+1]}) := (\psi_n(\vec{P}_{[n]}) \wedge P_n) \vee (\bigwedge \vec{P}_{[n]} \wedge \neg P_n)$.

It is easy to see that the algebra presented by ψ_{n+1} on generators $\vec{P}_{[n]}$ has precisely $n + 1$ atoms.

Suppose $\phi(\vec{R})$ is a presentation of the n-atom finite boolean algebra over generators $\vec{R}$. This algebra is isomorphic to the algebra defined by by $\psi_n(\vec{P}_{[n-1]})$. Let σ be the isomorphism these algebras. We define unrelativized interpretations K, M between $T + \phi$, of signature $\Sigma + \vec{R}$, and $T + \psi_n$ of signature $\Sigma + \vec{P}_{[n]}$ and back as follows. K and M are the identity on Σ and $R^K := \sigma(R)$ (to be precise: some $\vec{P}_{[n]}$-formula defining $\sigma(R)$) and $P^M := \sigma^{-1}(P)$. It is easy to see that K and M specify a synonymy of theories. $\dashv$

Of course, we can find far more efficient sequences of propositional formulas that do the job. E.g., we could use the sequence χ_n with:

- $e(0) := 0$, $e(2k + 1) := e(2k)$, $e(2k + 2) := e(k + 1) + 1$,
- $\chi_0 := \bot$, in signature Σ,
- $\chi_1 := \top$, in signature Σ,
- $\chi_{2k+2} := \chi_{k+1}$, in signature $\Sigma + \vec{P}_{[e(2k+2)]}$,
- $\chi_{2k+3} := \left(\chi_{2k+2} \wedge P_{e(2k+3)}\right) \vee \left(\bigwedge \vec{P}_{[e(2k+3)]} \wedge \neg P_{e(2k+3)}\right)$, in signature $\Sigma + \vec{P}_{[e(2k+2)]}$.

EXAMPLE 4.5. One may show that PA^2 is not bi-interpretable with PA. See Corollary 9.9.

We provide an example, due to Lev Beklemishev, of a theory T such that T^2 is synonymous to T. Take T to be PA expanded with a unary relation symbol Q. We interpret T^2 in T via K, taking $P_K := Q(0)$ and $Q_K(v_0) := Q(Sv_0)$. For the rest of the signature K is the identity interpretation. Conversely we interpret T in T^2 via M, taking $Q_M(v_0) := ((v_0 = 0 \wedge P) \vee \exists y\, (v_0 = Sy \wedge Qy))$. For the rest of the signature, M is the identity interpretation.

4.4. The sum. In the present subsection we provide a partial treatment of the occurrence of sums in our categories.[13] Consider two theories U and V. Say, U has signature Σ_U and V has signature Σ_V. The sum $U \oplus V$ is given as a theory W of signature Σ_W, where Σ_W is given as the disjoint union of Σ_U and Σ_V plus two additional fresh unary predicate symbols Δ_U and Δ_V and a new binary identity symbol E_W.[14] Let τ_U and τ_V be the obvious translations of the languages of U, respectively V into the language of W, where we relativize to Δ_U in the first case and to Δ_V in the second case. We take W to be axiomatized by the following axioms.

- $\vdash P_{\tau_U}\vec{v} \to \vec{v} : \Delta_U$,
- $\vdash P_{\tau_V}\vec{v} \to \vec{v} : \Delta_V$,
- $\vdash A^{\tau_U}$, for A a U-axiom,
- $\vdash A^{\tau_V}$, for A a V-axiom,
- $\vdash \forall x\, (x : \Delta_U \vee x : \Delta_V)$,
- $\vdash x E_W y \leftrightarrow \forall z\, ((x E_U z \leftrightarrow y E_U z) \wedge (x E_V z \leftrightarrow y E_V z))$.

Note that, in the presence of the other axioms, the last axiom says that E_W is the crudest congruence relation with respect to all predicates of W.

It is easy to check that $\oplus$ is a sum in the sense of category theory for the category INT. It follows that $\oplus$ yields a *weak* sum in the categories INT_1, INT_2, INT_3. Moreover, $\oplus$ is again a sum in INT_4. I have not thought about the

---

[13] A number of details of the sum definition were studied by Spencer Gerhardt in the context of a small project. Specifically, he formulated the definition of sum in INT.

[14] Of course, if $U = V$, we take the appropriate measures to make the Δ disjoint.

situation in INT_2 and INT_3, but it is not difficult to see that the situation with respect to the categorical sum is markedly different in INT_1, i.e. hINT.

Let's call the categorical sum of hINT, whenever it exists: $+$. First, consider 1, the theory in the language with just identity stating that there is precisely on element. The theory 1 is an initial object of hINT. Hence we have, for any A, $A + 1 = A$. Hence, e.g. $1 + 1$ is not bi-interpretable with $1 \oplus 1$.

Now consider the theory 2. This is the theory in the language with just identity stating that there are precisely two elements. We claim that $2 + 2$ does not exist. Suppose, to get a contradiction, that $2 + 2$ does exist. We take α_0 and α_1 to be the in-arrows associated with $2 + 2$. We will call the formulas Δ of $2 \oplus 2$: Δ_0 and Δ_1 Let $2 \boxplus 2$ be $2 \oplus 2$ extended with the axiom stating that the intersection of Δ_0 and Δ_1 is empty. Let $\text{in}_i^* := \mathcal{E}_{2\oplus 2, 2\boxplus 2} \circ \text{in}_i$. Let K be the unique arrow making the following diagram commutative.

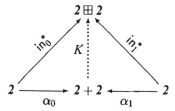

Note that, in $2 \boxplus 2$, we have $E_i := E_U \restriction \Delta_i$. We find that $2 \boxplus 2$ is categorical. Let $\mathcal{A}$ be a model of $2 \boxplus 2$ with domain $\{a, b, c, d\}$ and with $\Delta_0^{\mathcal{A}} = \{a, b\}$ and $\Delta_1^{\mathcal{A}} = \{c, d\}$. Since definable sets are closed under automorphisms, the only definable sets of $\mathcal{A}$ are $\emptyset$, $\{a, b\}$, $\{c, d\}$ and $\{a, b, c, d\}$. Similarly, there are just three interpretations of 2 in $\mathcal{A}$. Moreover, there are no definable isomorphisms between any two different interpretations of 2. Thus, we may conclude that:

$$2 \boxplus 2 \vdash x : \Delta_i \longleftrightarrow x : \delta_{K \circ \alpha_i}.$$

Thus, $2 \boxplus 2 \vdash (\forall x \, \neg (\delta_{\alpha_0}(x) \wedge \delta_{\alpha_1}(x)))^K$. Now compare the following commutative diagrams.

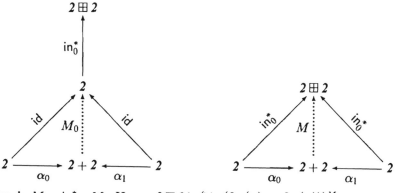

Clearly $M = \text{in}_0^* \circ M_0$. Hence, $2 \boxplus 2 \vdash (\forall x \, (\delta_{\alpha_0}(x) \leftrightarrow \delta_{\alpha_1}(x)))^M$.

Let V be $2 \boxplus 2$ extended with a new predicate Φ and an axiom stating that Φ is a bijection between Δ_0 and Δ_1. Let e be the obvious embedding of U in V. We compare the commutative following diagrams.

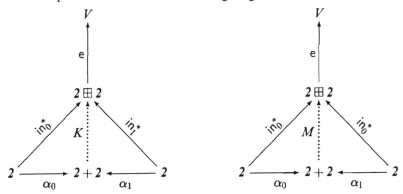

Note that $e \circ \text{in}_0^*$ is equal to $e \circ \text{in}_1^*$ in hINT. By the uniqueness property of $+$, it follows that $e \circ K$ is equal to $e \circ M$. On the other hand,

1. $V \vdash (\forall x \, \neg (\delta_{\alpha_0}(x) \wedge \delta_{\alpha_1}(x)))^{e \circ K}$,
2. $V \vdash (\forall x \, (\delta_{\alpha_0}(x) \leftrightarrow \delta_{\alpha_1}(x)))^{e \circ M}$.

This gives us a contradiction.

4.5. Monomorphisms. The arrow $K : U \to V$ in INT_i is a monomorphism if, for all W and for all $M_0 : W \to U$ and $M_1 : W \to U$, we have $K \circ M_0 =_i K \circ M_1$ implies $M_0 =_i M_1$. So, whenever the diagram

$$W \xrightarrow[\substack{M_0 \\ \dashv \\ M_1}]{} U \xrightarrow{K} V$$

commutes, then so does:

$$W \underset{M_1}{\overset{M_0}{\rightrightarrows}} U$$

Since equality in INT_i occurs both positively and negatively in the definition of *monomorphism*, the property of being a monomorphism need not be preserved when i increases. We have the following theorem.

THEOREM 4.6. (a) *Monomorphisms in* INT_i *for* $i = 0, 1, 2, 3$ *are faithful. This also holds when we relativize* INT_2 *to a class of models in which we have the completeness theorem.* (b) *The faithful interpretations of* INT_j *for* $j \in \{0, 3\}$ *are monomorphisms.* (c) *All interpretations of* INT_4 *are monomorphisms,*

PROOF. We prove (a). Let $i \in \{0, 1, 2, 3\}$. We work in INT_i. Suppose that $K : U \to V$ is a monomorphism. Suppose that $V \vdash B^K$. Let W be U extended with a 0-ary predicate symbol P. Let $M_0 : W \to U$ be the interpretation that is the identity when restricted to the signature of U and that sends P to $\top$. Let $M_1 : W \to U$ be the interpretation that is the identity when restricted to the signature of U and that sends P to B. Then, clearly, $K \circ M_0$ is equal$_0$, and, hence, equal$_i$, to $K \circ M_1$. So, since K is a monomorphism, we have that M_0 is equal$_i$, and, hence equal$_3$, to M_1. Ergo $V \vdash P^{M_0} \leftrightarrow P^{M_1}$. I.o.w. $V \vdash B$. Note that this last step fails for $i = 4$.

We prove (b). Let $i = 0, 3$. Suppose that $K : U \to V$ is faithful. We show that K is a monomorphism in INT_i. Suppose that $M_j : W \to U$, for $j = 0, 1$, and that $K \circ M_0$ is equal$_i$ to $K \circ M_1$. We have to show that M_0 is equal$_i$ to M_1.

We first treat $i = 0$. By our assumption V proves that $\tau_K \circ \tau_{M_0}$ and $\tau_K \circ \tau_{M_1}$ are identical. So:

- $V \vdash \forall x \, ((x : \delta_K \wedge x : \delta^K_{M_0}) \leftrightarrow (x : \delta_K \wedge x : \delta^K_{M_1}))$.
- $V \vdash \forall \vec{x} \, ((\vec{x} : \delta_K \wedge \vec{x} : \delta^K_{M_0}) \to (P^K_{M_0}(\vec{x}) \leftrightarrow P^K_{M_1}(\vec{x})))$.

But this just another way of saying:

- $V \vdash (\forall x \, (x : \delta_{M_0} \leftrightarrow x : \delta_{M_1}))^K$.
- $V \vdash (\forall \vec{x} : \delta_{M_0} \, (P_{M_0}(\vec{x}) \leftrightarrow P_{M_1}(\vec{x})))^K$.

By faithfulness, we find that U proves that τ_{M_0} and τ_{M_1} are identical.

We treat $i = 3$. Consider any W-sentence A. We have $V \vdash A^{M_0 K} \leftrightarrow A^{M_1 K}$. Hence: $V \vdash (A^{M_0} \leftrightarrow A^{M_1})^K$. By the faithfulness of K, we find $U \vdash A^{M_0} \leftrightarrow A^{M_1}$. Ergo M_0 is equal$_3$ to M_1.

(c) is trivial. $\dashv$

In the next theorem, we give some connections between *being a monomorphism* and the behaviour of the MOD-functor.

THEOREM 4.7. (a) For $i = 0, 2, 3$, if $\mathsf{MOD}_i(K)$ is surjective, then $K : U \to V$ is a monomorphism. (b) If $K : U \to V$ is a monomorphism in INT_3, then $\mathsf{MOD}_3(K)$ is surjective.

In (a), we can improve the cases (0) and (3): these work also if we have the completeness theorem in the range of $\mathsf{MOD}_i(K)$.

PROOF. We prove (a). Let $i \in \{0, 2, 3\}$. We will say that two models are *equal$_i$* if they are the same if $i = 0$, isomorphic if $i = 2$ and elementarily equivalent if $i = 3$.

Suppose that $\mathsf{Mod}_i(K)$ is surjective. Suppose $K \circ M_0 =_i K \circ M_1$, for $M_j : W \to U$. Consider any model $\mathcal{M} \models U$. By surjectivity, there is a model $\mathcal{N} \models V$ such that $\mathcal{M}$ is equal$_i$ to $K^{\mathcal{N}}$. It follows that $M_j^{\mathcal{M}}$ is equal$_i$ to $M_j^{K^{\mathcal{N}}}$ which is, in its turn, equal$_i$ to $(M_j K)^{\mathcal{N}}$. Since also $(M_0 K)^{\mathcal{N}}$ and $(M_1 K)^{\mathcal{N}}$ are equal$_i$ to each other, we find that $M_0^{\mathcal{M}}$ and $M_1^{\mathcal{M}}$ are equal$_i$. Ergo $M_0 =_i M_1$. (For the cases $i = 0, 3$, we only use, in the last step, that '$\mathcal{M}$' ranges over a class of models in which we have the completeness theorem.)

We prove (b). Let U^* be any consistent complete extension of U. If $V + U^{*K}$, were inconsistent, then for some A in U^*, $V \vdash \neg A^K$, and so

$V \vdash (\neg A)^K$. Ergo: $U \vdash \neg A$. Quod non. Take V^* any complete extension of $V + U^{*K}$. Clearly, $\mathrm{Mod}_3(K)(V^*) = U^*$. $\dashv$

We consider a number of examples of morphisms and consider the question whether they are monomorphisms in our various categories.

a: id : PA → PA.

b: ext_0 : PA → $\mathrm{PA}^c + \{c \neq \underline{0}, c \neq \underline{1}, \dots\}$. Here PA^c is simply PA with its signature extended with constant c. ext_0 is the standard 'identical' interpretation.

c: Let $2 \boxplus 2$ and V be as defined in Subsection 4.4. Let ext_1 be the standard 'identical' interpretation of $2 \boxplus 2$ in V. (This interpretation was called e in Subsection 4.4.)

d: ext_2 : PA → ACA_0. Here ext_2 is the standard 'identical' interpretation.

e: ext_3 : PA → $\mathrm{PA}_{\mathrm{ns}}$. Here $\mathrm{PA}_{\mathrm{ns}}$ is PA plus a non-standard satisfaction predicate and ext_3 is the standard identical interpretation.

f: $\mathcal{E}$: PA → PA + con(PA). Here $\mathcal{E}$ witnesses the subtheory relation.

In the diagram below $i \in 0, 1, 2, 3, 4$ indicates INT_i. $2'$ indicates $\mathrm{INT}_2^{\mathrm{count}}$, i.e. INT_2, where we restrict ourselves in the definition of equality of interpretations to countable models. We first sum up our results in a diagram and then run through the proofs.

	0	1	2	2′	3	4
a	+	+	+	+	+	+
b	+	+	?	−	+	+
c	+	−	+	+	+	+
d	+	−	+	+	+	+
e	+	−	?	?	+	+
f	−	−	−	−	−	+

The a-row is trivial and so is the 4-column. the 0- and 3-columns are easy because monomorphisms in INT_0 and INT_3 are precisely the faithful interpretations. The f-row is immediate since monomorphisms in INT_i, for $i = 0, 1, 2, 3$ are all faithful. We treat the remaining cases.

Case **b1:** Suppose $\mathrm{ext}_0 \circ M_0 =_1 \mathrm{ext}_0 \circ M_1$. So $\mathrm{PA}^c + \{c \neq \underline{0}, c \neq \underline{1}, \dots\}$ proves that τ_{M_0} and τ_{M_1} are isomorphic via some isomorphism represented by $J[x := c]$. Here J is a formula of the language of PA having only x, v_0, v_1 free. The fact that $J[x := c]$ presents an isomorphism can be stated in a single sentence. Hence, by compactness there must be some number N, such that PA proves that $J[x := \underline{N}]$ is an isomorphism between τ_{M_0} and τ_{M_1}.

Case **b2′:** Let Z_0, Z_1 be the obvious interpretations of the order theory of $\mathbb{Z}$ and $\mathbb{Z} \times \mathbb{Q}$ in PA. In $\mathbb{N}$ these interpretations give us precisely the orderings of $\mathbb{Z}$ and $\mathbb{Z} \times \mathbb{Q}$. In a countable non-standard model, both will give us the ordering of $\mathbb{Z} \times \mathbb{Q}$.

Case c0,2,3: Every model of $2 \boxplus 2$ can be ext_1-extended to a model of V. This makes $\text{Mod}_i(\text{ext}_1)$ surjective, for $i \in \{0, 2, 3\}$. (Note that we already knew c0 and c3.)

Case c1: By a simple model theoretical argument, in_0^* and in_1^* are distinct in INT_1. However, $\text{ext}_1 \circ \text{in}_0^* =_1 \text{ext}_1 \circ \text{in}_1^*$.

Case d0,2,3: Every model of PA can be ext_2-extended to a model of ACA_0. This makes $\text{Mod}_i(\text{ext}_2)$ surjective, for $i \in \{0, 2, 3\}$. (Note that we already knew d0 and d3.)

Case d1: This beautiful example is due to Harvey Friedman (in conversation). Let W be the theory of linear order. We take:

- $\delta_{M_0} :\leftrightarrow \forall y \, (\text{proof}_{\text{PA}}(y, \perp) \to v_0 < y)$,
- $\delta_{M_1} :\leftrightarrow \forall y \, (\text{proof}_{\text{PA}}(y, \perp) \to v_0 \le y)$.
- $\le_{M_i}$ will be $\le$ restricted to δ_{M_i}.

In ACA_0 we may produce an isomorphism between $\text{ext}_2 \circ M_0$ and $\text{ext}_2 \circ M_1$ as follows. There is a definable cut I in ACA_0 such that $\text{ACA}_0 \vdash \text{con}^I(\text{PA})$. Thus $\delta_{M_i} \supseteq I$. Define:

$$v_0 G v_1 :\longleftrightarrow v_0 : \delta_{M_0} \wedge v_1 : \delta_{M_1} \wedge$$
$$((v_0 : I \wedge v_1 = v_0) \vee (\neg (v_0 : I) \wedge v_1 = S v_0)).$$

It is easy to see that, indeed, G provides the desired isomorphism. Finally, note that the existence of a definable isomorphism in PA would imply, in $\text{PA} + \text{incon}(\text{PA})$, the existence of a definable order isomorphism between p and $p - 1$, where p is the smallest proof of inconsistency. Quod impossibile.

Case e0,3: Lachlan's theorem tells us that a countable model of PA can be extended to a model of PA_{ns} iff it is recursively saturated. See [Kay91]. Since we have the completeness theorem for countable recursively saturated models we find that ext_3 is faithful. A proof-theoretical argument for faithfulness is given in [Hal99].

Case e1: The argument is the same as the one for d1.

4.6. Epimorphisms. Just as in the case of monomorphisms, *being an epimorphism* is not necessarily preserved when we move from INT_i to INT_{i+1}.

An important class of epimorphisms (in all our categories) is constituted by the extension morphisms $\mathcal{E}_{UV}$. Suppose we have $M_0, M_1 : V \to W$ and $M_0 \circ \mathcal{E}_{UV} = M_1 \circ \mathcal{E}_{UV}$. Then, we must have, in any of our categories, that $M_0 = M_1$, because equality only depends the theory W and the underlying translations, which are not affected by composing an $\mathcal{E}$-morphism.

An interpretation $K : U \to V$ is *surjective* iff, for all $B \in \mathcal{S}_V$, there is an $A \in \mathcal{S}_U$ such that $V \vdash B \leftrightarrow A^K$. Note that equality of interpretations in the categories INT_i, for $i = 0, 1, 2, 3$, preserves surjectivity. So in these categories we can sensibly consider surjectivity to be a property of morphisms.

THEOREM 4.8. *The epimorphisms of INT_3 are precisely the surjective morphisms.*

Before giving the proof we define an auxiliary sum-like operation — which we already met in the special case of $2 \boxplus 2$. Consider theories T and Z. We define the theory $T \boxplus Z$ as the extension of $T \oplus Z$ with the axiom stating that the intersection of Δ_T and Δ_Z is empty. We define $\mathsf{in}_i^* := \mathcal{E}_{T \oplus Z, T \boxplus Z} \circ \mathsf{in}_i$.[15]

Similarly, we build a disjoint sum of models $\mathcal{M} \boxplus \mathcal{N}$ with domain the disjoint union of the domains of $\mathcal{M}$ and $\mathcal{N}$.

PROOF. Consider a morphism $K : U \to V$. Suppose that, for some V-sentence B, there is no U-sentence A such that $V \vdash B \leftrightarrow A^K$. We construct a complete extension U^* of U such that B is independent over $V + U^{*K}$. The theory U^* will be the union of an increasing sequence of theories U_n, where B is is not equivalent with any A^K over $V + U_n^*$. We fix some enumeration C_n of the U-sentences.

- Let $U_0 := U$.
- Suppose $V + U_n^K + C_n^K \vdash B \leftrightarrow A_0^K$ and $V + U_n^K + \neg C_n^K \vdash B \leftrightarrow A_1^K$. Then, $V + U_n^K \vdash B \leftrightarrow ((C_n \land A_0) \lor (\neg C_n \land A_1))^K$. Quod non. In case, for no A_0, $V + U_n^K + C_n^K \vdash B \leftrightarrow A_0^K$, we take $U_{n+1} := U_n + C_n$. Otherwise, we take $U_{n+1} := U_n + \neg C_n$.

It is easy to see that this construction does the trick. Now consider the following theory:

$$W := (V \boxplus V) + \{A^{K \mathsf{in}_0} \leftrightarrow A^{K \mathsf{in}_1} \mid A \in \mathcal{S}_U\} + B^{\mathsf{in}_0} + \neg B^{\mathsf{in}_1}.$$

(In the definition of W we confuse K and the in_i with their underlying translations.) If we take models $\mathcal{M}_0 \models V + U^{*K} + B$ and $\mathcal{M}_1 \models V + U^{*K} + \neg B$, then $\mathcal{M}_0 \boxplus \mathcal{M}_1 \models W$. Hence, W is consistent.

Let $M_i := \mathcal{E}_{V \boxplus V, W} \circ \mathsf{in}_i^*$. Clearly, $M_0 \circ K$ and $M_1 \circ K$ are the same in INT_3. However, $W \vdash B^{M_0}$ and $W \vdash \neg B^{M_1}$, so M_0 and M_1 are different. Hence, K is not an epimorphism.

Conversely, suppose that K is surjective. Let $M_0, M_1 : V \to W$ and suppose that $M_0 \circ K$ is equal to $M_1 \circ K$ in INT_3. Consider any V-sentence B. By surjectivity, we can find a U-sentence A such that $V \vdash B \leftrightarrow A^K$.

---

[15]The operation $\boxplus$ gives the sum in the category of (recursive) boolean morphisms. It does not generally give the sum in any of our categories INT_i. We have the remarkable property that every sentence of $T \boxplus Z$ is a boolean combination of sentences of the form $A^{\mathsf{in}_0^*}$, where A is a T-sentence, and $B^{\mathsf{in}_1^*}$, where B is a Z-sentence.

We have:

$$W \vdash B^{M_0} \longleftrightarrow A^{KM_0}$$
$$\longleftrightarrow A^{KM_1}$$
$$\longleftrightarrow B^{M_1}.$$

Ergo, M_0 is equal to M_1.   ⊣

4.7. Split monomorphisms. Suppose $K : U \to V$, $M : V \to U$ and $M \circ K = \mathrm{id}_U$. In these circumstances, we call K a *split monomorphism* or *co-retraction*. We call M a *split epimorphism* or *retraction*. We say that U *is a retract of* V. Note that if K is a split monomorphism in INT_i and $i < j$, then K is a split monomorphism in INT_j. Note also that the contravariant MOD_i-functor sends split monomorphisms to split epimorphisms and that it sends split epimorphisms to split monomorphisms. It is easily seen that split monomorphisms are monomorphisms and split epimorphisms are epimorphisms.

It seems that many interesting interpretations are split monomorphisms. Here are two examples.

EXAMPLE 4.9. Gödel's interpretation gödel : $(\mathsf{ZF} + \mathsf{V} = \mathsf{L}) \to \mathsf{ZF}$ is a split monomorphism in INT, as can be seen by inspecting the construction. The corresponding split epimorphism is simply $\mathcal{E}_{\mathsf{ZF},\mathsf{ZF}+\mathsf{V}=\mathsf{L}}$. Note that it follows that gödel is faithful.

EXAMPLE 4.10. Suppose $N : \mathsf{S}_2^1 \to T$. Here S_2^1 is Buss's arithmetic (see [Bus86] or [HP91]). However, any sufficiently rich arithmetical theory would do as well. We will use an arithmetization of metamathematics in T that is implicitly relativized to N. Consider the following commutative diagram in hINT:

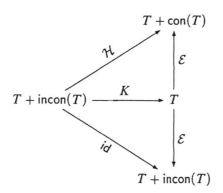

Note that $T \vdash \mathrm{con}(T) \to \mathrm{con}(T + \mathrm{incon}(T))$, by the Second Incompletness theorem. The interpretation $\mathcal{H} : (T + \mathrm{incon}(T)) \to (T + \mathrm{con}(T))$ is the Henkin interpretation based on $con(T + incon(T))$. See e.g. [Vis91], for an explanation. We take: $K := \mathcal{H}\langle con(T)\rangle id$. Clearly, K is a split monomorphism.

In Section 7, we will show that the interpretations involved in the Orey phenomenon, to wit the interpretability of both $T + O$ and $T + \neg O$ in T, for certain T and O, are split monomorphisms. We will prove further several results on (co)retractions in this paper.

4.8. Isomorphisms. In this subsection, we discuss isomorphisms in our various categories.

4.8.1. *Bisimulation.* Here is a useful insight concerning isomorphisms.

THEOREM 4.11. *Being isomorphic in* INT_i *for* $i = 0, 1, 2, 3$ *is a bisimulation with respect to theory extension in the same language.*

PROOF. Suppose $L : U \to V$ and $M : V \to U$ witness the fact that U and V are isomorphic in one of INT_i, for $i = 0, 1, 2, 3$. Suppose $U \subseteq U'$. Take $V' := \{A \in \mathcal{S}_V \mid U' \vdash A^M\}$. Clearly, M lifts to an interpretation $\tilde{M} : V' \to U'$ with the same underlying translation. Moreover,

$$U' \vdash A \implies U' \vdash A^{LM}$$
$$\implies V' \vdash A^L.$$

So L lifts to an interpretation of $\tilde{L} : V' \to U'$. The fact that this pair of interpretations witness isomorphisms only depends on the underlying translations and the fact that we have at least U and V available.   ⊣

Note that our argument even establishes that recursive boolean isomorphism is a bisimulation with respect to theory extension. Our theorem does not hold for INT_4 or even for $INT_{4,faith}$, which is INT_4 restricted to faithful interpretations. See Subsubsection 4.8.4.

4.8.2. *Synonymy.* In this subsubsection, we study the classical notion of synonymy.

THEOREM 4.12. *If two theories are synonymous, then they are isomorphic in* $INT_{unr,=}$

This is an immediate consequence of Theorem 6.1, which tells us that, if $M \circ K$ in INT is direct, then M is direct.

We remind the reader of the notion of *definitional extension*. Consider a theory T with signature Σ. Let T' be a theory with signature Σ'. We say that T' is a *definitional extension* of T iff Σ' extends Σ and the axioms of T' are the axioms of T plus, for each predicate symbol P of $\Sigma' \setminus \Sigma$, an axiom of the form: $\vdash P\vec{x} \leftrightarrow A\vec{x}$, where A is in the language of T with at most $\vec{x}$ free. Theorem 4.2 tells us that any definitional extension of a theory is synonymous to the given theory.

One immediate consequence of Theorems 4.12 and 4.2 is that Karel de Bouvère's notion of synonymity coincides with ours. See [dB65a] and [dB65b].

This notion has also been called *definitional equivalence*. See [Kan72] or [Cor80] or [Hod93]. It has also been called *isomorphism*. See [Per97].

Since MOD_i is a contravariant functor, it follows that $MOD_i(K)$ is an isomorphism if K is an isomorphism in INT_i. The following theorem shows that, if K is direct and $i = 0$, the converse is also true. In the proof of the theorem, we will use the factorization of $K : U \to V$ introduced in Subsection 3.5:

$$U \xrightarrow{\mathcal{E}_{U,K^{-1}[V]}} K^{-1}[V] \xrightarrow{\check{K}} V.$$

THEOREM 4.13. *Suppose $K : U \to V$ is direct. We have*:

1. $MOD(K)$ *is injective iff* $\check{K} : K^{-1}[V] \to V$ *is an isomorphism in* INT;
2. $MOD(K)$ *is bijective iff K is an isomorphism in* INT.

PROOF. The proof is an adaptation of the proof of Theorem 2 of [dB65b]. Suppose K is direct.

We first prove (1). Suppose that $MOD(K)$ is injective. Suppose U has signature Σ and V has signature Θ. Without loss of generality, we may assume that the identity symbol of Σ is the same as the one of Θ and that, for all other predicate symbols, Σ and Θ are disjoint.

Let $\tilde{\Theta}$ be a disjoint copy of Θ. This means that we replace every predicate symbol P of Σ, except the identity symbol E, by a disjoint copy $\tilde{P}$. Again, we may assume that, for all non-identity symbols, Σ and $\tilde{\Theta}$ are disjoint. So, we end up with three signatures that are pairwise disjoint, except for the shared identity symbol.

We take $\tilde{V}$ the obvious copy of V in the signature $\tilde{\Theta}$. Let V^+ be the theory in the signature $\Sigma + \Theta$ obtained by extending the axioms of V with axioms $\forall \vec{v} \, (P(\vec{v}) \leftrightarrow P_K(\vec{v}))$, for each predicate symbol P of Σ. The theory $\tilde{V}^+$ is similarly defined. Let W be the theory in the signature $\Sigma + \Theta + \tilde{\Theta}$ axiomatized by $V^+ + \tilde{V}^+$.

We claim that, for every predicate symbol Q of Θ, $W \vdash \forall \vec{v} \, (Q(\vec{v}) \leftrightarrow \tilde{Q}(\vec{v}))$. Suppose not. By symmetry, we may conclude that, for some Q, there is a model $\mathcal{M}$ of $W + \exists \vec{v} \, (Q(\vec{v}) \wedge \neg \tilde{Q}(\vec{v}))$. We define $\mathcal{M}_0$ as the reduct of $\mathcal{M}$ to Θ, $\mathcal{N}$ as the reduct of $\mathcal{M}$ to Σ, and $\mathcal{M}_1$ as the result of first restricting $\mathcal{M}$ to $\tilde{\Theta}$ and, then, replacing the predicates of $\tilde{\Theta}$ by the corresponding predicates of Θ. As is easily seen:

$$MOD(K)(\mathcal{M}_0) = \mathcal{N} = MOD(K)(\mathcal{M}_1).$$

Ergo, by injectivity, $\mathcal{M}_0 = \mathcal{M}_1$. A contradiction.

By Beth's Theorem, we find that (†) $V^+ \vdash \forall \vec{v} \, (Q(\vec{v}) \leftrightarrow B_Q(\vec{v}))$, where B_Q is some formula having only $\vec{v}$ free, in the language of Σ. We define a direct translation μ by setting $Q_\mu :\leftrightarrow B_Q$. We have, for $B \in \mathcal{S}_\Theta$,

$$K^{-1}[V] \vdash B^\mu \Longleftrightarrow V^+ \vdash B^\mu$$
$$\Longleftrightarrow V^+ \vdash B$$
$$\Longleftrightarrow V \vdash B.$$

So μ lifts to an interpretation $M : V \to K^{-1}[V]$. Also $V^+ \vdash \forall \vec{v} \, (Q(\vec{v}) \leftrightarrow Q_M^K(\vec{v}))$, and so $V \vdash \forall \vec{v} \, (Q(\vec{v}) \leftrightarrow Q_M^K(\vec{v}))$. Moreover, $V^+ \vdash \forall \vec{v} \, (P(\vec{v}) \leftrightarrow P_K^M(\vec{v}))$. Hence, $K^{-1}[V] \vdash \forall \vec{v} \, (P(\vec{v}) \leftrightarrow P_K^M(\vec{v}))$. We may conclude that $\check{K}$ and M are inverses.

Conversely, suppose $\check{K}$ is an isomorphism in INT between $K^{-1}[V]$ and V. since MOD is a contravariant functor, it follows that MOD($\check{K}$) is a bijection. Clearly, MOD($\mathcal{E}_{U,K^{-1}[V]}$) is injective. Hence, since

$$\mathsf{MOD}(K) = \mathsf{MOD}\big(\check{K} \circ \mathcal{E}_{U,K^{-1}[V]}\big) = \mathsf{MOD}\big(\mathcal{E}_{U,K^{-1}[V]}\big) \circ \mathsf{MOD}(\check{K}),$$

we find that MOD(K) is injective.

We prove (2). Suppose MOD(K) is a bijection. By Theorem 4.7, it follows that K is a monomorphism in INT. Hence, by Theorem 4.6, K is faithful. We may conclude that $U = K^{-1}[V]$ and $K = \check{K}$. So, by (1), we are done. The converse is easy. $\dashv$

REMARK 4.14. Theorem 4.13 can be considered as a generalization of Beth's Theorem. We sketch how to obtain Beth's Theorem from Theorem 4.13. Suppose Σ is a signature and Q is a predicate symbol, not in Σ. Let $\Sigma(Q)$ be Σ extended with Q. Suppose $W(Q)$ is a theory of signature $\Sigma(Q)$ in which Q is implicitly definable, i.e. $W(Q) + W(Q') \vdash \forall \vec{v} \, (Q(\vec{v}) \leftrightarrow Q'(\vec{v}))$. Let $\hat{W}$ be the theory of the consequences of W in the language of Σ. We have the obvious extension interpretation ext : $\hat{W} \to W$. Evidently, MOD(ext) maps a model of $W(Q)$ to its restriction to Σ. Using the implicit definability of Q, we easily see that MOD(ext) is injective. Clearly, ext$^{-1}[W] = \hat{W}$. By Theorem 4.13, it follows that ext has a direct inverse K. We find that Q_K is the desired explicit definition of Q.

4.8.3. *Bi-interpretability.* We prove a modest preservation result for i-limits and i-colimits.

THEOREM 4.15. *Suppose that $K : U \to V$ is an isomorphism in* hINT. *Then $[\![K]\!]$ preserves limits and colimits from $[\![U]\!]$ to $[\![V]\!]$.*

PROOF. Clearly, $[\![K]\!]$ is an equivalence of $[\![U]\!]$ and $[\![V]\!]$. Since equivalences are both left and right adjoints, $[\![K]\!]$ preserves limits and colimits. $\dashv$

4.8.4. *Some examples.* Consider FOL$_{\text{arith}}$, predicate logic in the language of arithmetic and $\mathbf{1}$ the theory in the language of pure identity that states that there is precisely one element. FOL$_{\text{arith}}$ and $\mathbf{1}$ are mutually interpretable. The example shows the following points:

- Mutual interpretability does not preserve decidability, nor does it preserve categoricity.
- Mutual interpretability is not a bisimulation with respect to the subtheory relation, since FOL_{arith} extends e.g. to Robinson's Arithmetic Q, but there is no matching extension of I.
- Mutual interpretability is not the same as isomorphism in INT_i, for $i = 0, 1, 2, 3$. In fact, our two theories are not even isomorphic in the boolean sense.

Here is a second example. Predicate Logic with just unary and 0-ary predicate symbols is mutually directly interpretable with Predicate logic with at least one n-ary predicate symbol, for some $n > 1$. However, the first theory is decidable and the second is not. Since, isomorphism in INT, for $i = 0, 1, 2, 3$ preserves decidability, it follows that our theories are not isomorphic in INT_i, for $i = 0, 1, 2, 3$.

In our third example, we separate mutual *faithful* interpretability from isomorphism in INT_i, for $i = 0, 1, 2, 3$. Consider the theory Q^- axiomatized by $(\exists x, y \; x \neq y \rightarrow \bigwedge Q)$ in the language of arithmetic. Clearly, the theories Q^- and FOL_{arith} are mutually interpretable. (We can interpret the statement 'there is at most one element' in FOL_{arith}. This implies Q^-.) Since both theories are trustworthy (see [Vis05]), they are mutually faithfully interpretable. We extend FOL_{arith} by adding the axiom 'there are precisely two elements'. Say the resulting theory is T. Suppose there is an extension U of Q^-, which is mutually faithfully interpretable with T. Since T is decidable, it follows that U is decidable. Hence the extension of U with Q is inconsistent. Since U contains the Q^--axiom, we may conclude that U proves: *there is at most one element*. But then it is impossible that U interprets T.

It follows that isomorphism in INT_i, for $i = 0, 1, 2, 3$, is not the same as mutual faithful interpretability.[16]

Note that the example also makes clear that mutual faithful interpretability is not a bisimulation with respect to extension of theories in the same language.

In example 9.5, we will separate mutual faithful direct interpretability from isomorphism in INT_i, for $i = 0, 1$.

There are several examples of sameness of theories, that prima facie belong in hINT. One such example is as follows. Consider ZF. We expand the signature with a unary predicate symbol U and with a binary relation symbol F. Let T be the theory in the expanded language given by the usual axioms for

---

[16]Our example also works for recursive boolean isomorphism in the place of isomorphism in one of the categories INT_i, for $i = 0, 1, 2, 3$. The argument for non-isomorphism of the theory T and any theory U over FOL_{arith} that implies 'there is at most one element', is a simple count of the propositions in the associated Lindenbaum algebra. The theory U has at most 2^{2^3} propositions, where T has strictly more.

ZF with ur-elements, where the class of ur-elements is given by U, plus an axiom that says that F is a bijection between U and ω. We can interpret T in ZF by representing ur-elements as pairs $\langle 0, n \rangle$, for $n \in \omega$, and sets as pairs $\langle 1, x \rangle$, where x is a set of representations of sets and ur-elements. To get things to work we need $\in$-induction and $\in$-recursion. We can show that the interpretation so constructed is an isomorphism in hINT. So one might wonder: can we improve this result to show that ZF and T are synonymous? Benedikt Löwe produced an argument to show that indeed one can do this. So we are left with the following question.

OPEN QUESTION 4.16. Give an example of two theories that are bi-interpretable but not synonymous.

OPEN QUESTION 4.17. Is there an interesting class of theories on which mutual interpretability and isomorphism in one of INT_i, for $i = 0, 1, 2, 3$, always coincide? What is the situation for the finitely axiomatizable sequential theories?

§5. Axiom schemes. Consider the principle of complete induction over numbers in PA and in ZF. We would like to say that this is *the same scheme* only realized in different languages. Moreover, in the case of ZF, the scheme uses numbers which are not reflected in the signature of the theory. The machinery of relative interpretations provides a simple and convenient way to define such schemes. Consider the following three theories.

- Q, i.e. Robinson's Arithmetic;
- Q^X, i.e. Robinson's Arithmetic with an extra unary predicate symbol X n the signature, but with no further axioms;
- $Q^{ind} := Q^X + IND(X)$, where $IND(X)$ is the principle of induction over X.

We have the obvious embeddings emb $:= emb_{Q,Q^X}$ of Q in Q^X and $\mathcal{E} := \mathcal{E}_{Q^X,Q^{ind}}$ of Q^X into Q^{ind}. We can now say that an interpretation $K : Q \to U$ satisfies full induction iff, for all interpretations $K^X : Q^X \to U$, such that $K^X \circ emb = K$, there is an interpretation $K^{ind} : Q^{ind} \to U$, such that $K^{ind} \circ \mathcal{E} = K^X$.

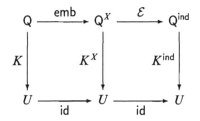

Reflecting on this example leads us to the following definition. An *ae-scheme* is a pair of composable arrows $\langle K, L \rangle$. To emphasize a pair is used in the role of ae-scheme we will use $K \twoheadrightarrow L$. We define, for $K : S_0 \to S_1$, $L : S_1 \to S_2$,

$M_0 : S_0 \to U$, that $M_0 \models K \twoheadrightarrow L$ iff, for all interpretations $M_1 : S_1 \to U$, such that $M_1 \circ K = M_0$, there is an interpretation $M_2 : S_2 \to U$, such that $M_2 \circ L = M_1$.

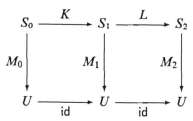

In other words, $M_0 \models (K \twoheadrightarrow L)$ iff $[\![K]\!]_U^{-1}(M_0) \subseteq \mathrm{range}[\![L]\!]_U$. Note that, in our set-up, ae-schemes are not ascribed to theories but to interpretations. Of course, the most standard kind of ae-schemes are ascribed to the identity interpretation. Thus, id : PA $\to$ PA satisfies $\mathrm{emb}_{Q,Q^x} \twoheadrightarrow \mathcal{E}_{Q^x,Q^{emb}}$, the induction scheme. The definition of satisfaction of an ae-scheme can be used in all our categories. The default is INT. If we employ other categories INT_i we will write: $M \models^i K \twoheadrightarrow L$. Here are some further notations. Suppose $K : U \to V$.

- $[\![\sigma]\!] := \{M \mid M \models \sigma\}$;
  We write $[\![\sigma]\!]_i$ if we want to indicate that we consider the satisfiers in INT_i. No index corresponds with $i = 0$.
- $?K := (\mathrm{id}_U \twoheadrightarrow K)$.
  We will also represent $?K$ as $\langle K \rangle$. We will call a scheme of the form $?K$ an *e-scheme*.
- $\sim K := (K \twoheadrightarrow \bot_V)$.
  Here $\bot$ is the unique arrow from V to the inconsistent theory in the signature with only identity.

The following theorem is obvious.

THEOREM 5.1. *Consider any ae-scheme σ. Consider $M : S_0 \to U$, such that $M \models^i \sigma$. Suppose that $N : U \to V$ is an isomorphism of theories in INT_i. Then, $N \circ M \models^i \sigma$. In slogan: ae-schemes are preserved under isomorphism of theories.*

A ae-scheme $(K \twoheadrightarrow L)$ is *direct* iff both K and L are direct. Thus, induction is a direct scheme. We have the following theorem.

THEOREM 5.2. *Suppose the ae-scheme σ is direct. Then, $[\![\sigma]\!]_0 = [\![\sigma]\!]_1$.*[17]

The proof uses that direct interpretations are forward looking. See Section 6. It follows that induction is satisfied in hINT whenever it is satisfied in INT. So, induction in the ordinary sense (the 0-sense) is preserved over bi-interpretability.

---

[17]Strictly speaking we should say something like: M is in $[\![\sigma]\!]_0$ iff its standard embedding into hINT is in $[\![\sigma]\!]_1$.

A severe restriction of our present approach is that we can only treat schemes without parameters. This is due to the fact that we only consider parameter-free interpretations. To get around the restriction, we have to extend our category. Note that, in the case of full induction, the restriction does not matter: parameter-free full induction and full induction with parameters happen to be equivalent.

We provide some extra information about e-schemes. Note that $M \models ?K$ iff we can find an N that makes the following diagram commute.

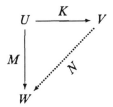

The interpretation $M : U \to W$ satisfies the e-scheme $?K$, for $K : U \to V$, if U^M can be 'expanded' in W to V^N, for some via $N : V \to W$. So satisfaction of an e-scheme monitors uniform expandability. The next theorem is obvious.

THEOREM 5.3. *Let* $K : U \to V$, $M : U \to W$, $N : W \to Z$. *Suppose* $M \models ?K$. *then* $N \circ M \models ?K$.

In the next theorem, we study how e-schemes are preserved upwards and downwards iff we change categories. Remember our convention that the variables $K, \ldots$ really run over the tuples $\langle U, \tau, V \rangle$.

THEOREM 5.4.   1. *Suppose* $i \leq j$ *and* $M \models^i ?K$. *Then,* $M \models^j ?K$.
2. *Suppose* $i \leq j$ *and* $M \models^j ?K$. *Suppose further that, for any* L, L' *with* $L =_j L'$, *we have* $L \models^i ?K$ *iff* $L' \models^i ?K$. *Then,* $M \models^i ?K$.

PROOF. We prove (2). Suppose $i \leq j$ and (a): $M \models^j ?K$. Suppose further that (b): for any L, L' with $L =_j L'$, we have $L \models^i ?K$ iff $L' \models^i ?K$.

By (a), we have that, for some N, $M =_j N \circ K$. Clearly, $N \circ K \models^i ?K$. Hence, by (b), $M \models^i ?K$.   ⊣

Let ϕ be a function from theories U to interpretations K with $\operatorname{dom}(K) = U$. We say that $?\phi := \langle \phi \rangle$ is a *uniform* e-scheme if, for any $K : U \to V$, there is a K' such that $K' \circ \phi_U = \phi_V \circ K$. Define:

- $M \models ?\phi :\Leftrightarrow M \models ?\phi_{\operatorname{dom}(M)}$,
- $\operatorname{SAT}_\phi := \{M \mid M \models ?\phi\}$.

It would be nice to define the notion of uniform e-scheme for ϕ as a natural transformation. However, our application in Section 7 asks for our present less restrictive definition. The following theorems are easy.

THEOREM 5.5. *Suppose* $M \circ N$ *exists. Then* $M \circ N$ *is in* SAT_ϕ *if* M *is in* SAT_ϕ *or* N *is in* SAT_ϕ.

THEOREM 5.6. *If $?\phi$ is a uniform e-scheme in* INT_i *and* $i \leq j$, *then* $?\phi$ *is a uniform e-scheme in* INT_j. *Moreover, if* $K \models^i ?\phi$, *then* $K \models^j ?\phi$.

§6. **i-Isomorphisms.** In this section we develop our knowledge of the 2-category $\mathsf{INT}^{\mathrm{iso}}$ a bit further. Subsection 6.1 is devoted to a characterization of direct interpretations. In Subsection 6.2, we provide some sufficient conditions for the transfer of certain properties of morphisms in hINT to corresponding morphisms in INT and vice versa. In Subsection 6.3, we show how in some circumstances one may replace an interpretation by an i-isomorphic one with 'better' properties. Finally, in Subsection 6.4, we consider a specific example, the Ackermann interpretation in some detail.

6.1. Direct interpretations and discrete fibrations. An interpretation K is *direct* iff it is unrelativized and preserves identity. Thus, direct interpretations are the morphisms of $\mathsf{INT}^{\mathrm{unr},=}$. Here is an immediate insight.

THEOREM 6.1. *If* $M \circ K$ *in* INT *is direct, then* M *is direct.*

Here is a first characterization of direct interpretations. Let $\mathfrak{S}$ be predicate logic with just the identity symbol. For any theory T we define the morphism $\mathcal{I}_T : \mathfrak{S} \to T$ by: $\delta_{\mathcal{I}_T} :\leftrightarrow v_0 = v_0$ and $v_0 E_{\mathcal{I}_T} v_1 :\leftrightarrow v_0 = v_1$. Thus, $\mathfrak{S}$ is a weak initial object. The following theorem is obvious.

THEOREM 6.2. $K : U \to V$ *is direct iff* $K \circ \mathcal{I}_U = \mathcal{I}_V$.

It turns out that we can characterize direct interpretations fully in terms of $\mathsf{INT}^{\mathrm{iso}}$. We need some preliminary definitions to do this. Consider a functor $\phi : \mathcal{C} \to \mathcal{D}$. The functor ϕ is *a discrete fibration* if, for every c in $\mathcal{C}$ and for every $g : d \to \phi(c)$, there is a unique f in $\mathcal{C}$ with $\phi(f) = g$ and $\mathrm{cod}(f) = c$. We will write $\overline{g}(c) : g^*(c) \to c$ for the unique f corresponding to g and c. If we want to make the role of ϕ explicit, we will write, e.g., $\overline{g}_\phi(c)$.[18]

A morphism K in INT is *forward looking* iff, for every $\mathfrak{V}$, the contravariant functor $[\![K]\!]_{\mathfrak{V}}$ is a discrete fibration. Setting $L := M \circ K$, $L' := M' \circ K$, $M' := F^*_{[\![K]\!]_{\mathfrak{V}}}(M)$ and $G := \overline{F}_{[\![K]\!]_{\mathfrak{V}}}(M)$, we have the following diagram.

$$
\begin{array}{ccc}
U & \xrightarrow{\quad K \quad} & V \\[2mm]
L' \Downarrow{\scriptstyle F} \Big| L & \quad M' \Downarrow{\scriptstyle G} \Big| M \\[2mm]
\mathfrak{V} & \xrightarrow[\mathrm{id}]{\quad\quad} & \mathfrak{V}
\end{array}
$$

Here F and G are i-isomorphisms. The next theorem shows what happens if we move via a morphism to another 'base theory'.

---

[18]For an extensive presentation of the theory on fibrations, see [Jac99]. In this paper, we will only use a few trivial facts.

THEOREM 6.3. *Suppose K is forward looking and that $N : \mho_0 \to \mho_1$. We have:*

- $(N \circ F)^*_{\overline{[K]}_{\mho_1}} (N \circ M) = N \circ F^*_{[K]_{\mho_0}} (M),$
- $\overline{(N \circ F)}_{[K]_{\mho_1}} (N \circ M) = N \circ \overline{F}_{[K]_{\mho_0}} (M).$

The proof of the theorem is easily seen by contemplating the following diagram and using the uniqueness clause in the definition of discrete fibration.

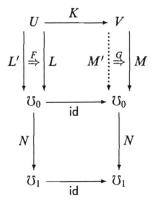

We prove the promised characterization of direct interpretations.

THEOREM 6.4. *An interpretation is direct iff it is forward looking.*

PROOF. We first prove the left-to-right direction. Suppose we are given a theory $\mho$, a direct interpretation $K : U \to V$, interpretations $M : V \to \mho$, $L' : U \to \mho$, and an i-isomorphism $F : L' \Rightarrow L$, where $L := M \circ K$. We have to show that there is a unique pair $M' : V \to \mho$, $G : M' \Rightarrow M$ such that $L' = M' \circ K$, $G = F \circ K$.

$$
\begin{array}{ccc}
U & \xrightarrow{\quad K \quad} & V \\
L' \Downarrow^{F} L & & M' \Downarrow^{G} M \\
\mho & \xrightarrow{\text{id}} & \mho
\end{array}
$$

Note that $\mho$ will prove that δ_L is δ_M and E_L is E_M. We define M' and G as follows.

- $\delta_{M'} :\leftrightarrow \delta_{L'},$
- $P_{M'}(\vec{v}) :\leftrightarrow \exists \vec{w} \, (\vec{v} F \vec{w} \land P_M(\vec{w})).$
- $G :\leftrightarrow F.$

The verification that M' and G indeed uniquely have the desired property holds no surprises.

We turn to the right-to-left direction. Suppose $K : U \to V$ is forward looking. We define a theory W as follows. The signature of W is the signature of V plus a fresh unary predicate symbol Δ and a fresh binary predicate symbol G. Par abus de langage we will use Δ for the translation of V-formulas by their relativizations to Δ in the language of W and also for the interpretation from V to W carried by the translation Δ. The theory W will be axiomatized by:

- A^{Δ}, for all axioms A of V,
- an axiom expressing that G is a permutation of the domain of W,
- two axioms expressing that G 'leaves $\Delta \circ K$ fixed', i.e.:
  - $\vdash (xGy \wedge x : \Delta \wedge x : \delta_K^{\Delta}) \to (y : \Delta \wedge y : \delta_K^{\Delta} \wedge xE_K^{\Delta}y)$,
  - $\vdash (xGy \wedge y : \Delta \wedge y : \delta_K^{\Delta}) \to (x : \Delta \wedge x : \delta_K^{\Delta} \wedge xE_K^{\Delta}y)$.

We define an interpretation $M : V \to W$ as follows.

- $x : \delta_M :\leftrightarrow \exists y \, (xGy \wedge y : \Delta)$,
- $P_M(\vec{x}) \leftrightarrow \exists \vec{y} \, (\vec{x}G\vec{y} \wedge P(\vec{y}))$.

As is easily seen, we indeed have: $M : V \to W$ and $G : M \Rightarrow \Delta$. Let $L := \Delta \circ K$ and $F := G \circ K$. Since G 'leaves L fixed', we find that F is ID_L. Note that also $\mathsf{ID}_\Delta \circ K = \mathsf{ID}_L$. Ergo, since K is forward looking, we find that $G = \mathsf{ID}_\Delta$ and $M = \Delta$.

To arrive at a contradiction, suppose K were not direct. Then, there is a model $\mathcal{M}$ of V in which either $\delta_K^{\mathcal{M}}$ is not the full domain of $\mathcal{M}$, or where $E_K^{\mathcal{M}}$ has a non-trivial equivalence class. In the first case, we can extend $\mathcal{M}$ to a model $\mathcal{N}$ of W as follows.

- The domain of $\mathcal{N}$ is the domain of $\mathcal{M}$ plus one new element a.
- $\Delta^{\mathcal{N}}$ is the domain of $\mathcal{M}$.
- The $R^{\mathcal{N}}$ restricted to $\Delta^{\mathcal{N}}$ are the $R^{\mathcal{M}}$; the other choices are don't care, except in the case of identity.
- $G^{\mathcal{N}}$ is the transposition of one element of $\Delta^{\mathcal{N}} \setminus \delta_K^{\Delta\mathcal{N}}$ and the new element a.

Clearly, $G^{\mathcal{N}}$ is not $\mathsf{ID}_\Delta^{\mathcal{M}}$. In the second case, the construction is similar. We need not extend the domain and take $G^{\mathcal{N}}$ a transposition of two elements of a non-trivial equivalence class of $E_K^{\mathcal{N}}$.

We arrive at a contradiction. Hence, K must be direct. $\dashv$

REMARK 6.5. Note that, in the proof of the right-to-left direction of Theorem 6.4, we only use the uniqueness clause in the definition of discrete fibration. Thus, in reality, we prove a stronger theorem.

6.2. hINT meets INT. In this subsection, we illustrate that, for direct arrows, we can often transfer properties of morphisms from hINT to INT or from INT to hINT.

THEOREM 6.6. *Direct epimorphisms in* INT *are epimorphisms in* hINT.

Proof. Let $K : U \to V$ and $M, M' : V \to W$. Suppose K is a direct epimorphism in INT. Suppose further that we have an isomorphism $F : M' \circ K \Rightarrow M \circ K$. Since K is forward looking, we can find an $M'' : V \to W$ and a $G : M'' \Rightarrow M$ such that $M' \circ K = M'' \circ K$ and $G \circ K = F$. Since K is an epimorphism in INT, it follows that $M' = M''$. Hence M' is i-isomorphic to M.   ⊣

THEOREM 6.7. *Consider $K : U \to V$. Suppose K is direct. Then, K is a split monomorphism in INT iff K is a split monomorphism in hINT.*

Proof. Suppose K is direct. It is clear that, if K is a split monomorphism in INT, then K is a split monomorphism in hINT. We prove the converse. Suppose K is a split monomorphism in hINT. Let M be the corresponding split epimorphism in hINT. So, we have an isomorphism $F : \mathrm{id}_U \Rightarrow M \circ K$. Let $L := M \circ K$. We have:

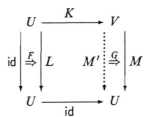

So we can take $M' := F^\star_{[\![K]\!]_U}(M)$, as the split epimorphism corresponding to K in INT.   ⊣

A morphism in $\mathsf{INT}^{\mathrm{iso}}$ is *rigid* iff it has no non-trivial i-automorphisms.

LEMMA 6.8. *Consider $K : U \to V$ and $M, M' : V \to W$. Suppose $G : M' \Rightarrow M$ is an isomorphism. Suppose further that $L := M \circ K = M' \circ K$. Suppose L is rigid and K is direct. Then, $M = M'$, and $G = \mathrm{ID}_M$. In other words, for direct K, the $[\![K]\!]_W$-fiber over a rigid L consists of a set of disconnected rigid M's.*

Proof. Let $F := G \circ K$. The corresponding i-arrows in the following three diagrams must be the same.

$$
\begin{array}{ccc}
U \xrightarrow{\ K\ } V & U \xrightarrow{\ K\ } V & U \xrightarrow{\ K\ } V \\
L \overset{F}{\Rightarrow} L \quad M' \overset{G}{\Rightarrow} M & L \overset{\mathrm{ID}}{\Rightarrow} L \quad M' \overset{G}{\Rightarrow} M & L \overset{\mathrm{ID}}{\Rightarrow} L \quad M \overset{\mathrm{ID}}{\Rightarrow} M \\
W \xrightarrow{\ \mathrm{id}\ } W & W \xrightarrow{\ \mathrm{id}\ } W & W \xrightarrow{\ \mathrm{id}\ } W
\end{array}
$$

By rigidity, $F = \mathrm{ID}_U$. By the discreteness of $[\![K]\!]_W$, $G = \mathrm{ID}_V$.   ⊣

THEOREM 6.9. *Suppose K is rigid and direct. Then K is an isomorphism in* INT *iff K is an isomorphism in* hINT.

PROOF. Suppose K is rigid and direct. It is clear that, if K is an isomorphism in INT, then K is an isomorphism in hINT. For the converse, suppose that K is an isomorphism in hINT. Suppose $M : V \to U$ and F and H are isomorphisms such that $F : \mathrm{id}_U \Rightarrow M \circ K$ and $H : \mathrm{id}_V \Rightarrow K \circ M$. As in the proof of Theorem 6.7, we can find $M' : V \to U$ and an isomorphism $G : M' \Rightarrow M$ such that $\mathrm{id}_U = M' \circ K$ and $F = G \circ K$. So we have:

$$\mathrm{id}_V \xRightarrow{\ H\ } K \circ M \xRightarrow{\ K \circ G^{-1}\ } K \circ M'.$$

Thus, $J := (K \circ G^{-1}) \cdot H$ is an isomorphism between id_V and $N := K \circ M'$. Thus, we have:

$$
\begin{array}{ccc}
U & \xrightarrow{\ K\ } & V \\
\Big\downarrow{\scriptstyle K} & \mathrm{id} \overset{J}{\Rightarrow} N & \Big\downarrow \\
V & \xrightarrow[\ \mathrm{id}\ ]{} & V
\end{array}
$$

Moreover, $N \circ K = K \circ M' \circ K = K \circ \mathrm{id}_U = K$. So we may apply Lemma 6.8, to obtain $M' \circ K = N = \mathrm{id}_V$.   ⊣

In Subsection 6.4, we apply Theorem 6.9 to the Ackermann Interpretation.

6.3. Improving interpretations. In this subsection, we will treat some well-known constructions to replace interpretations by i-isomorphic counterparts having some extra desired properties like directness.

Consider $K : U \to V$. Let U^c be U in the language of U extended with a constant c. Let $\mathrm{emb}_{U,U^c} : U \to U^c$ be the standard embedding of U in U^c. We say that K admits a constant iff $K \models \text{?emb}_{U,U^c}$.

THEOREM 6.10. *Suppose K admits a constant. Then, there is an unrelativized interpretation $K^\star$ i-isomorphic to K.*

PROOF. Suppose A defines the promised constant. We define the equivalence relation $E^\star$ in V as follows:

$$
\begin{aligned}
v_0 E^\star v_1 :&\longleftrightarrow (\neg\,(v_0 : \delta_K) \wedge \neg\,(v_1 : \delta_K)) \vee \\
&\quad (\neg\,(v_0 : \delta_K) \wedge v_1 : \delta_K \wedge A v_1) \vee \\
&\quad (v_0 : \delta_K \wedge \neg\,(v_1 : \delta_K) \wedge A v_0) \vee \\
&\quad (v_0 : \delta_K \wedge v_1 : \delta_K \wedge v_0 E_K v_1).
\end{aligned}
$$

We define $M' : \Im \to V$ and $F : M' \Rightarrow K \circ \mathcal{I}_U$ by:

- $\delta_{M'} :\leftrightarrow v_0 = v_0$,
- $E_{M'} :\leftrightarrow E^\star$,

- $v_0 F v_1 :\leftrightarrow \delta_K(v_1) \wedge v_0 E^\star v_1$.

Let $M := K \circ \mathcal{I}_U$. Now consider the following diagram, noting that $\mathcal{I}_U$ is direct and, hence, forward looking.

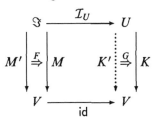

We can take $K' := F_\phi^\star(K)$, where $\phi = [\![\mathcal{I}_U]\!]_V$. Clearly, K' is unrelativized. Moreover, it is i-isomorphic to K via $G := \overline{F}_\phi(K)$.   ⊣

EXAMPLE 6.11. Consider the theory $T = 3 \boxplus 1$.[19] We have: $\mathrm{in}_0^\star : 3 \to T$. However, by a simple modeltheoretic argument, there is no unrelativized interpretation of 3 in T.

REMARK 6.12. If we would consider interpretations *with parameters* (see Subsection B.3), we do not need the assumption of a 'constant' in Theorem 6.10. We can always 'unrelativize' using a parameter. See [MPS90].

THEOREM 6.13. *Suppose V is an extension of* PA *in the language of* PA. *Consider $K : U \to V$. Suppose V proves that the domain of K modulo E_K is infinite. Then, K is i-isomorphic to a direct interpretation $K^\star$.*

PROOF. Let V be an extension of PA in the language of PA. Consider $K : U \to V$. Suppose V proves that the domain of K modulo E^K is infinite. Define:

- $fx := \mu y : \delta_K \cdot \forall x' < x \ fx' \neq^K y$.
- $v_0 F v_1 :\leftrightarrow v_1 : \delta_K \wedge f v_0 E_K v_1$.

From this point on the proof proceeds like the proof of Theorem 6.10.   ⊣

OPEN QUESTION 6.14. Is there a direct interpretation of $T + \mathrm{incon}(T)$ into T, where $T \in \{Q, S_2^1, EA, I\Sigma_1, ACA_0, GB\}$? Here $\mathrm{incon}(T)$ is given some standard arithmetization and, in ACA_0 and GB we employ the usual interpretation of arithmetic.

6.4. The Ackermann interpretation. Let ZF^- be set theory without the axiom of infinity. Let HF be $ZF^- + \neg \mathsf{Inf}$. We show that PA is a retract (in INT) of ZF^- and that PA is synonymous with HF.

First we interpret ZF^- in arithmetic. This employs an interpretation ackermann, or, in short, A, first found by Ackermann. A number in binary is

---

[19]The operation $\boxplus$ is defined in Subsection 4.6. The theory 3 is the obvious theory in the language of pure identity stating that there are precisely three objects.

read, from right to left, as the characteristic function of a finite set of numbers which in their turn again are taken to code sets. Thus, A is given as follows.

- $\delta_A :\leftrightarrow v_0 = v_0$.
- $v_0 E_A v_1 :\leftrightarrow v_0 = v_1$.
- $v_0 \in_A v_1 :\leftrightarrow \exists x, y \, (v_1 = (2x + 1)2^{v_0} + y \wedge y < 2^{v_0})$.

When no confusion is possible, we will simply write $\in$ for $\in_A$.

The interpretation A is not faithful, since it also interprets the negation of Inf, the axiom of infinity. Clearly, A is direct. Also, A is rigid because, PA-verifiably, if an i-automorphism F of A is the identity on the $\in_A$-elements of x, then $F(x)$ must be x by extensionality. In the converse direction, we proceed as follows. We work in ZF^-, using some well-established abbreviations. Define:

- $0 := \emptyset$,
- $Sv_0 := v_0 \cup \{v_0\}$,
- $\mathsf{prenum}^{v_0}(v_1) :\leftrightarrow v_1 = 0 \vee \exists x \in v_0 \, v_1 = Sx$,
- $v_0 : \omega :\leftrightarrow \forall y \in Sv_0 \, \mathsf{prenum}^{v_0}(y)$.

We run through a series of lemmas.

LEMMA 6.15. [ZF^-] 0 *is in* ω, 0 *is not a successor.*

LEMMA 6.16. [ZF^-] *The successor function is injective.*

LEMMA 6.17. [ZF^-] ω *is closed under predecessor, i.o.w. if* $Su : \omega$, *then* $u : \omega$.

PROOF. Suppose $Su : \omega$ and $v \in Su$. It is sufficient to show: $\mathsf{prenum}^u(v)$.

We certainly have $v \in SSu$. Hence, since $Su : \omega$, we have $\mathsf{prenum}^{Su}(v)$. This means that $v = 0$ or, for some $w \in Su$, $v = Sw$. In the first case, we are immediately done. So suppose for some $w \in Su$, $v = Sw$. We have either $w \in u$ or $w = u$. In the first case we are done. In the second case, we have $v = Su \in Su$. A contradiction. $\dashv$

LEMMA 6.18. [ZF^-] $z : \omega \leftrightarrow (z = 0 \vee \exists u : \omega z = Su)$.

PROOF. Suppose $z : \omega$. Then, since $z \in Sz$, we have $\mathsf{prenum}^z(z)$. Hence, $z = 0$ or $z = S(u)$, for some u. In the first case, we are done by Lemma 6.15. In the second case we are done by Lemma 6.17.

For the converse direction, suppose $z = 0$ or $z = Su$, for some $u \in \omega$. In the first case we are done by Lemma 6.15. In the second case we have to show that, for all $v \in Sz$, we have $\mathsf{prenum}^z(v)$. Suppose $v \in S(z)$. In case $v \in z$, we have $v \in Su$, and, hence, $\mathsf{prenum}^u(v)$. Thus, a fortiori, $\mathsf{prenum}^z(v)$. In case $v = z$, we clearly have $\mathsf{prenum}^z(z)$, since $z = Su$. $\dashv$

LEMMA 6.19. [ZF^-] *We have induction on* ω.

PROOF. We prove induction on ω. Reason in ZF^-. Suppose:

$$A0 \quad \text{and} \quad \forall x \in \omega \, (Ax \rightarrow ASx).$$

Suppose further that, for some z, we have in $z \in \omega$ and $\neg Az$. Let z^* be $\in$-minimal with the property. Since $z^* \in \omega$ and $z^* \neq 0$, we have $z^* = Su$

for some $u \in z^*$. Since, by Lemma 6.17, u is in ω it follows that Au. But, then, Az. $\dashv$

Now we can develop the theory of plus and times in the usual way, defining sequences as functions from elements of ω to sets. Thus we have an interpretation, neumann, or, in short, N, of PA in ZF^-.

It is now easy to see that $N \circ A : PA \to PA$ is i-isomorphic to the identity interpretation on PA. The mapping, on which the i-isomorphism, say $\mathcal{I}$, is based, is given by the following recursion:

- $\mathcal{I}0 := 0$,
- $\mathcal{I}Sx := 2^{\mathcal{I}x} + \mathcal{I}x$.

Thus, we have shown that $A : ZF^- \to PA$ is a split monomorphism in hINT. Thus, this interpretation is faithful, since, by Theorem 4.6, monomorphisms in hINT are faithful. It follows, by Theorem 6.7, that A is a split monomorphism in INT and, hence, by Theorem 6.1, A is a split monomorphism in $INT_{unr,=}$.

Our two translations also support morphisms $A^+ : HF \to PA$ and $N^+ : PA \to HF$. By our preceding result $A^+ \circ N^+$ is i-isomorphic to id_{PA}. We show that $N^+ \circ A^+$ is i-isomorphic to id_{HF}. Consider the mapping $\mathcal{J}$ from ω to sets given by:

- $\mathcal{J}x := \{\mathcal{J}y \mid y \in^{A^+N^+} x\}$.

Note that $\mathcal{J}x$ is indeed a set by Replacement, since $y \in^{A^+N^+} n$ implies $y <^{N^+} x$, which in its turn implies $y \in x$. We can show that $\mathcal{J}$ is injective by $\in$-induction. Now we want to show that the image of $\mathcal{J}$ is all sets. Suppose not. Let z^* be an $\in$-minimal element not in the image of $\mathcal{J}$. If $\mathcal{J}^{-1}[z^*]$ is bounded by a number x, we can easily construct a number y with $\mathcal{J}y = z^*$. So $\mathcal{J}^{-1}[z^*]$ is unbounded in ω. We find that $\mathcal{J}^{-1}[z^*]$ is a set, since $\mathcal{J}^{-1}$ is a function. The union of this set is ω, which will also be a set. Quod non. The further verification that $\mathcal{J} : (N^+ \circ A^+) \Rightarrow id_{HF}$ is an i-isomorphism holds no surprises.

We may conclude that HF and PA are bi-interpretable and, hence, by Theorem 6.9, synonymous.

REMARK 6.20. The Ackermann translation is for many purposes a good translation, but it has as disadvantage that e.g. the singleton function is exponential: $\{x\}$ is coded as 2^x. This makes it unsuitable for working in weak theories. There are other translations, lacking many of the good properties of the Ackermann translation, for which the code $\{x\}$ is of order x^2 and for which union is of the order of multiplication.

§7. Restricted interpretations. In this section we explore what Tarski's theorem means modulo interpretation.[20] A first question is: what precisely

---

[20]The need to connect the theory of the definability of truth with the study of interpretations was clearly seen by Panu Raatikainen. An containing some of his ideas can be found at ⟨http://www.math.helsinki.fi/logic/LC2003/abstracts/⟩ Alternatively, see: Panu

is meant by *object-language* and *meta-language*. This question becomes more interesting if we do not demand that the object-language is part of the meta-language, but just that the object-language is translatable into the meta-language. A first point is that the talk about 'language' is misleading: these 'languages' are really *theories*: object-theory and meta-theory. Secondly, I submit, we do not want to existentially away from the translation. Thus, the right explication of *candidate object-language/meta-language pair* is *morphism in* INT_i, for a chosen $i = 0, 1, 2, 3$. The next step is to define *(succesful) object-language/meta-language pair*. This will be explained as the satisfaction of a certain uniform e-scheme.

Let Q^{true}, be Robinson's arithmetic extended with a unary predicate true (and no further axioms concerning true). Consider any theory U. We extend U to U^{meta} which is $U \oplus \mathsf{Q}^{\text{true}}$ plus the axioms $\text{true}^{\text{in}_1}(\underline{\#A}) \leftrightarrow A^{\text{in}_0}$, for all U-sentences A. We will notationally suppress the superscript in_i, writing the T-scheme simply as $\text{true}(\underline{\#A}) \leftrightarrow A$. Let $\text{om}_U := \mathcal{E} \circ \text{in}_0$. Here 'om' stands for the object-meta embedding. Suppose $K : U \to V$. We can represent the mapping $A \mapsto A^K$ in arithmetic. We will call this arithmetization $(\cdot)^\kappa$.

We may extend K to K^{meta} by interpreting the U-predicates of U^{meta} via K, by interpreting the Q-predicates via the identity, and by interpreting true of U^{meta} by $\text{true}(v_0^\kappa)$ in V^{meta}. We will have, for any U-sentence A,

$$
\begin{aligned}
V^{\text{meta}} \vdash \text{true}^{K^{\text{meta}}}(\underline{\#A}) &\longleftrightarrow \text{true}((\underline{\#A})^\kappa) \\
&\longleftrightarrow \text{true}(\#(A^K)) \\
&\longleftrightarrow A^K \\
&\longleftrightarrow A^{K^{\text{meta}}}.
\end{aligned}
$$

So, $K^{\text{meta}} : U^{\text{meta}} \to V^{\text{meta}}$. Moreover, we will clearly have $\text{om}_V \circ K = K^{\text{meta}} \circ \text{om}_U$. Ergo, ?om will be a uniform e-scheme (in each of our categories).

REMARK 7.1. One might hope that $U \mapsto U^{\text{meta}}$, $K \mapsto K^{\text{meta}}$ would be a functor and that om would be a natural transformation from ID to $(\cdot)^{\text{meta}}$. However, regrettably, this fails. The mapping $K \mapsto K^{\text{meta}}$ does not necessarily preserve equality of arrows in our categories. So, on the level of interpretations, it is not necessarily a mapping.

An interpretation $K : U \to V$ is *restricted* if $K \models ?\text{om}$.

LEMMA 7.2. *Suppose* $L =_3 L'$ *and* $L \models^0 ?\text{om}$. *Then* $L' \models^0 ?\text{om}$.

PROOF. Consider $L, L' : U \to V$. Suppose $L =_3 L'$ and $L \models^0 ?\text{om}$. Say M witnesses $L \models^0 ?\text{om}$, i.e. $M \circ \text{om}_U = L$. Consider $M_0' : U \oplus \mathsf{Q}^{\text{true}} \to V$

Raatikainen: 'Translation and the definability of truth (Abstract)' in: Logic Colloquium 2003, Helsinki Finland, August 14-20, Abstracts. Yliopistopaino, Helsinki, 2003.

defined by $M_0' := L' \oplus (M \circ \text{in}_1)$. We get:

$$V \vdash \text{true}^{M_0'}(\underline{\#A}) \longleftrightarrow \text{true}^M(\underline{\#A})$$
$$\longleftrightarrow A^L$$
$$\longleftrightarrow A^{L'}$$
$$\longleftrightarrow A^{M_0'}. \qquad\qquad \dashv$$

Thus M_0' 'lifts' to an interpretation $M' : U^{\text{meta}} \to V^{\text{meta}}$. It is easy to see that M' witnesses the fact that $L' \models^0$?om. Thus, by Theorem 5.4, we find that K is restricted in INT_i iff it is restricted in INT_j, for any $i, j \in \{0, 1, 2, 3\}$.

THEOREM 7.3. *We have in* INT_i, *where* $i = 0, 1, 2, 3$, *that* id_U *is not restricted.*

PROOF. Suppose M witnesses the fact that id_U were restricted. By the Gödel Fixed Point Lemma, we can find a sentence L such that $U \vdash L \leftrightarrow \neg\,\text{true}^M(\underline{\#L})$. This leads immediately to a contradiction with the T-scheme.
$\dashv$

THEOREM 7.4. *Suppose* $K : U \to V$, $M : V \to W$ *in* INT_i, *for* $i = 0, 1, 2, 3$. *We have:*

1. *If K is restricted, then so is $M \circ K$.*
2. *If M is restricted, then so is $M \circ K$.*
3. *If $M \circ K$ is restricted and K is surjective, then M is restricted.*

PROOF. Suppose $K : U \to V$, $M : V \to W$. Items (1) and (2) are special cases of Theorem 5.5. We prove (3). Suppose that P witnesses that $M \circ K$ is restricted. Suppose further that K is surjective. Clearly, there is a recursive function f such that $f(\#B)$ is some A such that $V \vdash B \leftrightarrow A^K$. Let ν be a bi-representation of f in Q. We construct Q witnessing the fact that M is restricted as follows.

- We interpret Q according to P.
- $\text{true}^Q(v_0) :\leftrightarrow \text{true}^P(\nu(v_0))$.
- We interpret V according to M.

Suppose $f(\#B) = \#A$. We have:

$$W \vdash \text{true}^Q(\underline{\#B}) \longleftrightarrow \text{true}^P(\nu(\underline{\#B}))$$
$$\longleftrightarrow \text{true}^P(\underline{\#A})$$
$$\longleftrightarrow A^{KM}$$
$$\longleftrightarrow B^M$$
$$\longleftrightarrow B^Q.$$

So $Q : V^{\text{meta}} \to W$. It is easy to see that $Q \circ \text{om}_V = M$. $\dashv$

COROLLARY 7.5. *In each of our categories* INT_i, *for* $i = 0, 1, 2, 3$, *we have: no split monomorphism is restricted. Similarly, no split epimorphism is restricted.*

From the philosophical point of view, we think, that the statement that no split monomorphism is restricted, is a good statement of Tarski's Theorem of the undefinability of truth, where we take into account the fact that the objectlanguage is translated into the metalanguage. The reasonable condition on this translation is that it is a split monomorphism or co-retraction.

COROLLARY 7.6. *No surjective morphism is restricted (in* INT_i, *for* $i = 0, 1, 2, 3$).

PROOF. Suppose $K : U \to V$ is surjective and restricted. It follows, by Theorem 7.4, that id_V is restricted. Quod non, by Theorem 7.3.   ⊣

Restricted interpretations $K : U \to U$ give rise to the Orey phenomenon. We have restricted $K : U \to U$ e.g., for reflexive theories, like PRA, PA and ZF, and, for finitely axiomatized sequential theories, like Q, S_2^1, EA, $I\Sigma_1$, ACA$_0$ and GB. (All these examples are constructed via the Henkin construction.)

Suppose $K : U \to U$ is restricted. Let O satisfy: $U \vdash O \leftrightarrow \neg\mathrm{true}^{K^{\mathrm{meta}}}$ (#O). We find: $U \vdash O \leftrightarrow \neg O^K$. So there are interpretations K_0 and K_1 with the same underlying translation as K, such that $K_0 : U + \neg O \to U + O$ and $K_1 : U + O \to U + \neg O$. Thus, in each of our categories the following diagrams commute:

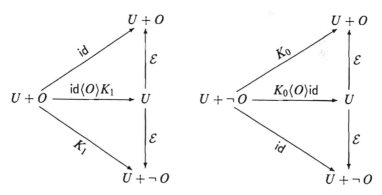

So, there are split monomorphisms both from $U + O$ and from $U + \neg O$ to U. Note that it follows that $U + O$ and $U + \neg O$ are both faithfully interpretable in U. I feel that it is remarkable that the Orey phenomenon occurs even for such a strict notion as split monomorphism.

OPEN QUESTION 7.7. Do we have a restricted $K : T \to T$, for all sequential T?

§8. i-Initial arrows.
In this section, we study the meaning of the existence of i-initial arrows in $\mathsf{INT}^{\mathrm{morph}}$. An arrow $K : U \to V$ is *i-initial* iff it is initial in the category of arrows $M : U \to V$ with the i-morphisms $F : M \Rightarrow M'$.

We fix a weak arithmetic F. We could choose e.g. Robinson's Arithmetic Q or Buss's Arithmetic S_2^1 (or, rather, an appropriate variant of S_2^1 in the arithmetical language) or $I\Delta_0 + \Omega_1$ (aka S_2). Consider the following theories.

- F^X, i.e. Robinson's Arithmetic with an extra unary predicate symbol X n the signature, but with no further axioms;
- $F^{ind} := Q^X + IND(X)$, where $IND(X)$ is the principle of induction over X.

We have the obvious embeddings emb $:=$ emb$_{F,F^X}$ of F in F^X and $\mathcal{E} := \mathcal{E}_{F,F^{ind}}$ of F^X into F^{ind}. We can now say that an interpretation $K : F \rightarrow U$ satisfies full induction iff $K \models (\text{emb} \rightarrow \mathcal{E})$.[21]

THEOREM 8.1. *Suppose $\iota : F \rightarrow U$ is i-initial. Then, ι satisfies induction.*

PROOF. Suppose $\iota^X : F^X \rightarrow U$ and $\iota = \iota^X \circ \text{emb}$. We write:

- $\Omega :\leftrightarrow \delta_\iota$.
- $PROG(X) :\leftrightarrow (X0 \wedge \forall x\,(Xx \rightarrow XSx))$.

Let:

- $B(v_0) :\leftrightarrow (v_0 : \Omega \wedge (PROG(X) \rightarrow Xv_0)^{\iota^X})$.

Let $K : F^X \rightarrow U$ be the extension of ι that interprets X as B. Clearly, $U \vdash (PROG(X))^K$. Now —and here we use the fact that F is a weak theory— we apply Solovay's method of shortening cuts (see e.g. [HP91]) to obtain a formula $C(v_0)$ such that $U \vdash C \rightarrow B$ and such that we relativizing to C yields an interpretation of F. Specifically, C is such that we have $R_C : F \rightarrow U$, where R_C is defined as follows.

- $\delta_{R_C} :\leftrightarrow C(v_0)$,
- $P_{R_C}(v_0, \ldots, v_{n-1}) :\leftrightarrow P_\iota(v_0, \ldots, v_{n-1})$, for any arithmetical predicate P of F.

We have an i-morphism $e_C : R_C \rightarrow \iota$ given by:

- $v_0(e_C)v_1 :\leftrightarrow (C(v_0) \wedge v_0 E_\iota v_1)$.

By i-initiality, there is an arrow $F : \iota \Rightarrow R_C$. We find: $e_C \circ F : \iota \Rightarrow \iota$. By the uniqueness clause of i-initiality, we must have: $(e_C \circ F) = ID_\iota$. So, we have in U:

$$v_0 E_\iota v_1 \longleftrightarrow v_0(e_C \circ F)v_1$$
$$\longleftrightarrow \exists y\,(v_0 F y \wedge y(e_C)v_1)$$
$$\longleftrightarrow \exists y\,(v_0 F y \wedge C(y) \wedge y E_\iota v_1)$$
$$\longleftrightarrow v_0 F v_1.$$

It follows that $U \vdash \forall v_0 : \Omega\ C(v_0)$, and hence $U \vdash \forall v_0 : \Omega\ B(v_0)$, which is equivalent to the induction principle for ι^X.   ⊣

---

[21]Strictly speaking this defines only induction without parameters, but, as is well-known, *full* induction without parameters implies full induction with parameters.

OPEN QUESTION 8.2. Is there an example of an arrow $\iota : \mathsf{F} \to U$ that is *weakly i-initial*, but that does not satisfy full induction?

In case U is sequential, we have a converse of Theorem 8.1. The notion of sequentiality is due to Pavel Pudlák. See, e.g., [HP91] or our Section 10. The idea is that U 'has sequences of all objects of the domain (including the sequences)'. These sequences are not extensional: two sequences may be different even if they have the same projections. The numbers with respect to which we project are given by some interpretation, say N, of Q in U. We will need three important properties of sequences.

- There is an empty sequence.
- Given a sequence σ and an object a, we may form $\sigma * \langle a \rangle$.
- Given a sequence σ of length n, and a number $k \leq n$, there is a sequence τ of length k, such that $(\sigma)_i = (\tau)_i$, for all $i < k$.[22]

THEOREM 8.3. *Suppose $\iota : \mathsf{F} \to U$ satisfies full induction and $K : \mathsf{F} \to U$.*

1. *Suppose $F, G : \iota \Rightarrow K$. then $F = G$.*
2. *Suppose U is sequential, then there is an $F : \iota \Rightarrow K$.*

It follows that, for sequential U, ι is i-initial.

PROOF. The proof of (1) is easy. We sketch the proof of (2). Suppose U is sequential and that $\iota : \mathsf{F} \to U$ satisfies full induction. Consider $K : \mathsf{F} \to U$. We have to produce $F : \iota \Rightarrow K$. We work in U.

Note that we have to work with three number systems. There are the N-numbers that are used in taking projections from sequences. There are the ι-numbers that satisfy full induction. And there are the K-numbers about which we do not know much. We will write Ω for δ_ι.

We define approximations of the desired F as follows. An approximation is a sequence of pairs where the first components are from Ω and the second components are from δ_K. The first elements of approximations are pairs of zeros. Successor elements of approximations are pairs of successors (in the respective number systems) of the preceding pair. We take $v_0 F v_1$ iff there is an approximation ending in the pair $\langle v_0, v_1 \rangle$. Now we may prove, by induction, that F is a total relation. We briefly look at the argument for functionality. We prove by induction on x in Ω that:

$$\forall x' : \Omega \; \forall y, y' : \delta_K \left((xFy \wedge x'Fy' \wedge xE_\iota x') \to yE_K y' \right).$$

Suppose first $Z_\iota(x)$, xFy, $x'Fy'$, $xE_\iota x'$. Consider an approximating sequence σ for xFy. The N-length of this sequence is either 1 or bigger than 1. In the first case, we find $Z_K(y)$. In the second case, we would have that x is a ι-successor. Quod non. Since $xE_\iota x'$, we find $Z_\iota(x')$. By copying the above reasoning, it follows that $Z_K(y')$. We may conclude that $yE_K y'$.

---

[22] This property is not among the properties stipulated in Section 10. However, we can obtain it by switching to a 'better' set of numbers for the projections.

Next suppose x is a ι-successor and xFy, $x'Fy'$, $xE_\iota x'$. The Induction Hypothesis, tells us that we have the desired property for and ι-predecessor of x. Consider an approximating sequence σ for xFy. The N-length n of this sequence is either 1 or bigger than 1. In the first case, we find $Z_\iota(x)$. Quod non. In the second case, we find $(\sigma)_{n-2} = \langle u, w \rangle$ and $uS_\iota x$ and $wS_K y$. By restricting σ to the first $(n-1)$ elements we obtain a witness for uFw. Since $xE_\iota x'$, we find that x' is a ι-successor. Thus, reasoning as before, we find u', w' with $u'S_\iota x'$, $w'S_K y'$ and $u'Fw'$. Since, uSx, $u'S_\iota x'$, $xE_\iota x'$, we find $uE_\iota u'$. Applying the induction hypothesis, we get $wE_K w'$, and, hence, $yE_K y'$.

We leave the proof that F commutes with successor, plus and times to the reader.   ⊣

We end this section with a theorem that will be useful in Section 9.

THEOREM 8.4. *Suppose ι : F $\to$ U satisfies full induction. Suppose K : F $\to$ U. Then, there is at most one F : K $\Rightarrow$ ι. Moreover, if such an F exists, it is an i-isomorphism.*

We omit the easy proof.

§9. On comparing arithmetic and set theory.

In this section, we prove some results on arithmetics and set theories. We first prove that retractions in hINT preserve weakly i-initial arrows.

THEOREM 9.1. *Let $\mho$ be any theory. Let $[\![\cdot]\!] := [\![\cdot]\!]_\mho$. Suppose that $L : V \to U$ is a retraction in hINT. Then $[\![L]\!]$ preserves weakly i-initial arrows in $[\![V]\!]$. (In the last statement, we consider L as a morphism in INT.)*

PROOF. We reason in $\mathsf{INT}^{\mathrm{morph}}$. We assume the conditions of the theorem. Let $N : \mho \to V$ be weakly i-initial in $[\![V]\!]$. We want to show that $L \circ N$ is weakly i-initial in $[\![U]\!]$.

Let $K : U \to V$ be the split monomorphism in hINT corresponding to L. Let M be any arrow from $\mho$ to U. By the weak i-initiality of N, we have, for some G:

$$\mho \xrightarrow[\quad N \quad]{\overset{K \circ M}{\underset{\Uparrow G}{}}} V \xrightarrow{\quad L \quad} U.$$

By the defining property of retractions, we have, for some i-isomorphism F:

$$\mho \xrightarrow{\quad M \quad} U \xrightarrow[\quad L \circ K \quad]{\overset{\mathsf{id}}{\underset{\Uparrow F}{}}} U.$$

Ergo:

1. $L \circ G : L \circ N \Rightarrow L \circ K \circ M$,
2. $F \circ M : L \circ K \circ M \Rightarrow M$.

($F \circ M$ exists, since F is an i-isomorphism.) So we have:

$$H := (F \circ M) \cdot (L \circ G) : L \circ N \Longrightarrow M.$$

Since M was arbitrary, it follows that $L \circ N$ is weakly i-initial.   ⊣

THEOREM 9.2. *Suppose $\iota_U : \mathsf{F} \to U$ and $\iota_V : \mathsf{F} \to V$ are i-initial arrows. Suppose also that U is a retract of V in* hINT. *Let $L : V \to U$ be the retraction (split epimorphism). Then the following diagram commutes in* hINT.

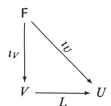

It follows that $\iota_V^{-1}[V] \subseteq \iota_U^{-1}[U]$.

PROOF. We assume the conditions of the theorem. Since ι_V is, a fortiori, weakly i-initial, we may conclude, by Theorem 9.1, that $L \circ \iota_V$ is weakly i-initial. So we have an arrow $H : L \circ \iota_V \Rightarrow \iota_U$. Since ι_U is i-initial, it satisfies full induction. Thus, we may conclude, by Theorem 8.4, that H is an i-isomorphism.

It follows, by Corollary 3.4, that $\iota_V^{-1}[V] \subseteq \iota_U^{-1}[U]$.   ⊣

OPEN QUESTION 9.3. Prove or refute the analogue of Theorem 9.2 for INT$_3$.

COROLLARY 9.4. *Suppose U and V are extensions of* PA *in the language of* PA. *Suppose U is a retract of V in* hINT. *Then $V \subseteq U$.*

In Subsection 4.7, we showed that PA + incon(PA) is a retract of PA. This illustrates the fact that, in Corollary 9.4, we cannot replace the subset relation by identity.

EXAMPLE 9.5. It is well-known that there is a restricted interpretation from PA to PA. We have seen in Section 7, that it follows that there is an arithmetical sentence O such that both PA + O and PA + $\neg O$ are retracts of PA. Clearly one of O, $\neg O$ must be true. Say it is O. Then, PA is interpretable in PA + O, and, since PA + O is Σ_1^0-sound, PA is faithfully interpretable in PA + O, by the results of Lindström, see e.g. [Lin94]. (Alternatively, see [Vis05].) It follows that PA and PA + O are mutually faithfully interpretable. By Theorem 6.13, it follows that PA and PA + O are mutually faithfully directly interpretable. On the other hand, by Corollary 9.4, they are not bi-interpretable.

COROLLARY 9.6. *Suppose U is an extension of* PA *in the language of* PA *and V is an extension of* ZF *in the language of* ZF. *Then U is not a retract of V in* hINT.

PROOF. Suppose U is an extension of PA in the language of PA and V is an extension of ZF in the language of ZF. Moreover, suppose that $K : U \to V$ and $L : V \to U$ witness that U is a retract of V. By Theorem 9.2, the following diagram commutes in hINT.

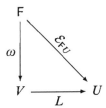

Here $\omega := \mathcal{E}_{ZF,V} \circ$ neumann, where neumann is the von Neumann interpretation of the natural numbers in ZF. Since neumann is restricted, we find, by Theorem 7.4, that $L \circ \omega$ is restricted. Hence, $\mathcal{E}_{F,U}$ is restricted. This gives a contradiction with Corollary 7.6, since $\mathcal{E}_{F,U}$ is surjective.   ⊣

COROLLARY 9.7. ZF^{-}, *i.e.* ZF *minus the axiom of infinity, is not isomorphic in* hINT *to any extension of* PA *in the language of* PA.

PROOF. Let U be an extension of PA in the language of PA. Suppose that ZF^{-} is isomorphic in hINT to U. It follows, by the bisimulation property of isomorphisms, that ZF is isomorphic in hINT to some extension W of U in the language of PA. But this contradicts Corollary 9.6.   ⊣

COROLLARY 9.8. *Suppose* U *is an extension of* ZF *in the language of* ZF *and* V *is an extension of* PA *in the language of* PA. *Then* U *is not a retract of* V *in* hINT.

PROOF. Suppose U is an extension of ZF in the language of ZF and V is an extension of PA in the language of PA. Moreover, suppose that $K : U \to V$ and $L : V \to U$ witness that U is a retract of V. By Theorem 9.2, the following diagram commutes in hINT.

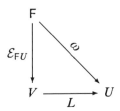

Here $\omega := \mathcal{E}_{ZF,V} \circ$ neumann, where neumann is the von Neumann interpretation of the natural numbers in ZF. Note that $\mathcal{E}_{FU}$ is surjective. Moreover, L is a split epimorphism in hINT. Since split epimorphisms are preserved by functors, L also yields a split epimorphism in INT_3. Hence, L is surjective. It follows that ω is surjective. But ω is also restricted, contradicting Corollary 7.6.

(Alternatively, we can reason as follows. Since $\mathcal{E}_{FU}$ is surjective and $L \circ \mathcal{E}_{FU} = \omega$ is restricted, it follows that L is restricted. Quid impossibile, since L is a retraction.)   ⊣

COROLLARY 9.9. PA^2 *is not a retract of* PA *in* hINT.

PROOF. It is easily seen that PA^2 can be taken to be PA in the language expanded with a 0-ary predicate symbol P. We reason analogously to the first part of the proof of Corollary 9.8. This gives us that the standard embedding of F into PA^2 is surjective. But then P would be provably equivalent to an arithmetical sentence, quod non.   ⊣

§10. **Preservation over retractions.** We write $U \sqsubseteq_i V$ iff U is a retract of V in INT_i. Clearly $\sqsubseteq_i$ is a pre-order. The induced equivalence relation of $\sqsubseteq_i$ will be $\equiv_i$.

A property $\mathcal{P}$ of theories is *preserved over retractions* in INT_i, or *preserved to retracts in* INT_i iff whenever $\mathcal{P}(V)$ and $U \sqsubseteq_i V$, then $\mathcal{P}(U)$. Note that iff $\mathcal{P}$ is preserved under retractions in INT_i and $j \leq i$, then $\mathcal{P}$ is preserved under retractions in INT_j. We treat some examples of preservation.

THEOREM 10.1. *κ-categoricity is preserved under retractions in* whINT.

PROOF. Suppose $K : U \to V$ is a co-retraction and $M : V \to U$ is the corresponding retraction, both in whINT. Suppose V is κ-categorical and that $\mathcal{M} \models U$ and $|\mathcal{M}| = \kappa$. We have $M^{\mathcal{M}} \models V$ and $K^{M^{\mathcal{M}}} \models U$. Moreover, $K^{M^{\mathcal{M}}}$ is isomorphic to $\mathcal{M}$. Hence, $|K^{M^{\mathcal{M}}}| = \kappa$. Since, the cardinality of $M^{\mathcal{M}}$ must be between the cardinalities of $K^{M^{\mathcal{M}}}$ and $\mathcal{M}$, we find $|M^{\mathcal{M}}| = \kappa$. Consider any other model with $\mathcal{N} \models U$ and $|\mathcal{N}| = \kappa$. Again we find $|M^{\mathcal{N}}| = \kappa$. Since, $M^{\mathcal{M}}$ and $M^{\mathcal{N}}$ are models of V, we see that $M^{\mathcal{M}}$ and $M^{\mathcal{N}}$ are isomorphic. Hence, $K^{M^{\mathcal{M}}}$ and $K^{M^{\mathcal{N}}}$ are isomorphic and, so, $\mathcal{M}$ and $\mathcal{N}$ are isomorphic. We may conclude that U is κ-categorical.   ⊣

THEOREM 10.2. *Finite axiomatizability is preserved under retractions in* hINT

PROOF. Suppose $K : U \to V$ is a co-retraction and $M : V \to U$ is the corresponding retraction, both in hINT. Let $F : \mathrm{id} \Rightarrow M \circ K$ be an i-isomorphism.

Suppose that V is finitely axiomatized and that A is the conjunction of a finite set of axioms of V. We claim that the following axioms form an axiomatization of U.

- The statement witnessing that $F : \mathrm{id} \Rightarrow M \circ K$ is an i-isomorphism.
- A^M.

Call the theory given by these axioms: $U^\star$. Clearly, $U^\star$ is a subtheory of U. Conversely, we have:

$$U \vdash B \Longrightarrow A \vdash B^K$$
$$\Longrightarrow \exists x\, \delta_M(x), A^M \vdash B^{KM}$$
$$\Longrightarrow U^\star \vdash B^{KM}$$
$$\Longrightarrow U^\star \vdash B.$$

The last step is proved by verifying, by induction on formulas $C(\vec{x})$, that

$$U^\star \vdash \vec{x} F \vec{y} \longrightarrow \left(C(\vec{x}) \leftrightarrow C^{KM}(\vec{y}) \right). \qquad \dashv$$

OPEN QUESTION 10.3. It is a great open problem whether S_2, aka $I\Delta_0 + \Omega_1$, is finitely axiomatizable. We do know that S_2 is interpretable in Q. By Theorem 10.2, we know that, if S_2 were an hINT-retraction of Q, then S_2 would be finitely axiomatizable. Hence, our question: prove or refute that S_2 is an hINT-retraction of Q.

Some properties of theories are associated with other theories. Consider any theory W. The theory U *has the property* $\mathcal{D}_W$ iff there is a direct interpretation from W to U.

THEOREM 10.4. $\mathcal{D}_U$ *is extensionally the same as* $\mathcal{D}_V$ *iff* U *and* V *are mutually directly interpretable.*

PROOF. Suppose $\mathcal{D}_U$ is extensionally the same as $\mathcal{D}_V$. Clearly, we have $\mathcal{D}_U(U)$. Hence, we have $\mathcal{D}_U(V)$ and, thus, there is a direct interpretation of U in V. Similarly, there is a direct interpretation of U in V. Hence, U and V are mutually directly interpretable.

Conversely, suppose U and V are mutually directly interpretable. Suppose further that $\mathcal{P}_U(W)$. Say, $L : U \to W$ is a direct interpretation witnessing this fact. Also, there is a direct interpretation $P : V \to U$. Hence, $L \circ P : V \to W$ is direct and, so, $\mathcal{P}(V)$. Similarly, $\mathcal{P}(U)$ follows from $\mathcal{P}(V)$, $\qquad \dashv$

An important example of a property associated with a theory that is preserved over retractions in hINT is *sequentiality*.[23]

The notion of sequentiality is due to Pavel Pudlák. See, e.g., [HP91]. A theory is sequential iff it satisfies $\mathcal{D}_{SEQ}$, where SEQ is the following theory.

- We have predicates num, E, Z, S, A, M and axioms to the effect that Q^N, where N is the obvious translation of arithmetic to these predicates. The axioms include axioms to the effect that E is an equivalence relation on num, etc.

---

[23]After this paper was finished, I realized that there is an alternative elegant treatment of sequentiality. It can be treated as a uniform e-scheme. In this way, sequentiality becomes primarily a property of interpretations. A theory U is sequential iff id_U is sequential. We will explore this line of thought in a later paper.

- We have predicates seq and proj and the following axioms:
  1. $\vdash \mathsf{seq}(s,u) \to (\mathsf{num}(u) \land \forall v <^N u \, \exists! x \, \mathsf{proj}(x,v,s))$,
  2. $\vdash \exists e, z \, (\mathsf{Z}(z) \land \mathsf{seq}(e,z))$,
  3. $\vdash \mathsf{seq}(s,u) \to \forall y \, \exists s', u' \, (\mathsf{S}(u,u') \land \mathsf{seq}(s',u') \land$
     $\forall v <^N u \, \forall x \, (\mathsf{proj}(x,v,s) \leftrightarrow \mathsf{proj}(x,v,s')) \land \mathsf{proj}(y,u,s'))$.

Sequentiality is a very robust notion. We can work with much stronger variants of SEQ. E.g., we could take, instead of Q, the theory $I\Delta_0 + \Omega_1$. Stronger versions are more pleasant for applications. For the purpose of verifying that a theory is sequential, (seemingly) weaker definitions are better. Here is one such weaker version, due to Pudlák. (See [MPS90].) Let WSET be the following theory.

- The language of WSET has one binary relation symbol $\in$ (in addition to the identity symbol),
- We have the following axioms:
  1. $\exists y \, \forall x \, x \notin y$,
  2. $\forall x, y \, \exists z \, \forall u \, (u \in z \leftrightarrow (u \in y \lor u = x))$.

THEOREM 10.5 (Pudlák). WSET *is mutually directly interpretable in* SEQ. *Hence, a theory is sequential iff it directly interprets* WSET.

We provide a slight variant of WSET, that is even more convenient, let's call it WSET′.

- The language of WSET′ has one binary relation symbol $\in$ and one unary symbol set (in addition to the identity symbol),
- We have the following axioms:
  1. $\exists y : \mathsf{set} \, \forall x \, x \notin y$,
  2. $\forall x \, \forall y : \mathsf{set} \, \exists z : \mathsf{set} \, \forall u \, (u \in z \leftrightarrow (u \in y \lor u = x))$.

THEOREM 10.6. WSET′ *is mutually directly interpretable in* WSET. *Hence, a theory is sequential iff it directly interprets* WSET′.

PROOF. The direct interpretation of WSET′ in WSET sends set(x) to $x = x$ and leaves the rest unchanged. The direct interpretation of WSET in WSET′ sends $x \in y$ to $x \in y \land \mathsf{set}(y)$ and leaves the rest unchanged. For the verification of the second WSET-axiom we distinguish the cases where y : set and where not y : set. In the first case, we are immediately done. In the second case, y functions as an empty set. We are guaranteed an empty set $y^\star$ in set. We use $y^\star$ to find the desired z.   ⊣

We show that sequentiality is preserved over retractions in hINT.

THEOREM 10.7. *Suppose V is sequential and U is a retract of V in* hINT. *Then, U is sequential.*

PROOF. Suppose $K : U \to V$ and $M : V \to U$ and suppose that the isomorphism $F : \mathrm{id}_U \Rightarrow (M \circ K)$ witness the fact that U is a retract of V in hINT. Suppose V is sequential. Let $L : \mathsf{WSET}' \to V$ be a direct interpretation witnessing the sequentiality of V. We define a direct interpretation P of WSET' in U. We take:

- $\delta_P :\leftrightarrow v_0 = v_0,$
- $E_P :\leftrightarrow v_0 = v_1,$
- $\text{set}_P :\leftrightarrow \delta_M(v_0) \wedge \text{set}_L^M(v_0),$
- $\in_P :\leftrightarrow \text{set}_P(v_1) \wedge \exists x \, (v_0 F x \wedge x \in_L^M v_1).$

We verify the axioms of WSET' under P. Reason in U.

First we verify the empty-set axiom. We have:

(1) $\qquad\qquad\qquad (\exists y : \text{set } \forall x \; x \notin y)^{LM}.$

So, we find:

(2) $\qquad\qquad\qquad \exists y' : \text{set}_P \, \forall x' : \delta_M \; x' \notin_L^M y'.$

Pick $y := y'$ and suppose $x \in_P y$. We have, for some x', xFx' and $x' \in_L^M y$. Also, from xFx', we get: $x' : \delta_M$. A contradiction with equation (2). So y witnesses the empty-set axiom for $\in_P$.

Next we verify the addition-of-an-element-axiom. We have:

(3) $\qquad (\forall x \, \forall y : \text{set } \exists z : \text{set } \forall u \, (u \in z \leftrightarrow (u \in y \vee u = x)))^{LM}.$

Hence,

(4) $\qquad\quad \forall x' : \delta_M \, \forall y' : \text{set}_P \, \exists z' : \text{set}_P \, \forall u' : \delta_M$
$\qquad\qquad (u' \in_L^M z' \leftrightarrow (u' \in_L^M y' \vee u' E_M x')).$

Now consider any x and $y : \text{set}_P$. Pick x' such that xFx', By equation (4), we can find $z' : \text{set}_P$ such that (†) $\forall u' : \delta_M \, (u' \in_L^M z' \leftrightarrow (u' \in_L^M y \vee u' E_M x'))$. We take $z := z'$. Consider any u.

First suppose $u \in_P z$. It follows that, for some u', uFu' and $u' \in_L^M z$. From uFu', it follows that $u' : \delta_M$. Hence, by (†), $u' \in_L^M y$ or $u' E_M x'$.

From the first case, to wit $u' \in_L^M y$, we get $u \in_P y$. From the second case, $u' E_M x'$, we have $u' E_{M \circ K} x'$, since $u', x' : \delta_{M \circ K}$ and, on $\delta_{M \circ K}$, the relation $E_{M \circ K}$ is a coarser equivalence relation than E_M. Since xFx' and uFu', we may conclude that $u = x$.

Next suppose $u \in_P y$ or $u = x$. In the first case, there is an u', such that uFu' and $u' \in_L^M y$. We find $u' : \delta_M$ and, hence, by (†), $u' \in_L^M z$. Ergo $u \in_P z$. In the second case, we have, from (†), $x' \in_L^M z$, and, hence $u = x \in_P z$. ⊣

Appendix A. Questions.

1. Provide separating examples concerning various notions, like monomorphism and isomorphism, across our categories.
2. Treat the notion of sum in INT_2 and INT_3.
3. Prove that equality of interpretations is complete for its prima facie complexity class, for INT_i, $i = 0, 1, 3, 4$. What is the complexity in the case of INT_2?

4. Which important properties of theories are preserved or antipreserved by morphisms? monomorphisms? isomorphisms? etc.
5. Which properties of theories and interpretations have natural formulations in terms of our categories (including DEG)? E.g., given two specific theories (e.g., PA and ZF) are there functorial automorphisms that interchange them?
6. Give an example of two theories that are bi-interpretable but not synonymous. (This is Question 4.16.)
7. Is there an interesting class of theories on which mutual interpretability and isomorphism in one of INT_i, for $i = 0, 1, 2, 3$, always coincide? What is the situation for the finitely axiomatizable sequential theories? (This is Question 4.17.)
8. There are many intimately connected pairs like PRA and $I\Sigma_1$, PA and ACA_0, PA and PA plus a non-standard satisfaction predicate, ZF and GB. For each pair we have strong conservativity results. On the other hand each the second theory of the pair proves the consistency of the first component on a definable cut. One consequence is a superexponential speed-up in the second theory. (See e.g. [Pud85], [Pud86].)

    Is there anything illuminating to say about the nature of the interpretation which embeds the first of the pair into the second?
9. Is there a direct interpretation of $T + \mathrm{incon}(T)$ into T, where

$$T \in \{ \mathsf{Q}, \mathsf{S}_2^1, \mathsf{EA}, I\Sigma_1, \mathsf{ACA}_0, \mathsf{GB} \}?$$

    Here $\mathrm{incon}(T)$ is given some standard arithmetization and, in ACA_0 and GB we employ the usual interpretation of arithmetic. (This is Question 6.14.)
10. Do we have a restricted $K : T \to T$, for all sequential T? (This is Question 7.7.)
11. Is there an example of an arrow $\iota : \mathsf{F} \to U$ that is *weakly initial*, but that does not satisfy full induction? (This is Question 8.2.)
12. Prove or refute the analogue of Theorem 9.2 for INT_3. (This is Question 9.3.)
13. It is a great open problem whether S_2, aka $I\Delta_0 + \Omega_1$, is finitely axiomatizable. We do know that S_2 is interpretable in Q. By Theorem 10.2, we know that, if S_2 were an hINT-retraction of Q, then S_2 would be finitely axiomatizable. Hence, our question: prove or refute that S_2 is an hINT-retraction of Q. (This is Question 10.3.)

Appendix B. More general notions. In this appendix we sketch some possible natural extensions of the notions of translation and interpretation as discussed in this paper.

B.1. Multidimensional interpretations. We could employ δ, containing variables $v_0, \ldots, v_{n-1}$, where we use *several* variables to represent *one* object. Suppose $\mathrm{ar}(P) = m$. Then, P would τ-translate to a formula with variables among $v_0, \ldots, v_{nm-1}$. Each subsequent block of m variables would stand for one object. Such translations are called *multidimensional*. They give rise in the obvious way to multidimensional interpretations.

B.2. Many-sorted predicate logic. Translations and interpretatons generalize immediately to the many-sorted case. For many purposes, this is a very natural generalization. I suspect that the categories so obtained would have slightly better properties.

B.3. Interpretations with parameters. Many famous interpretations like 'the Klein model' and Tarski's interpretation of true arithmetic into an extension of the theory of groups are interpretations with parameters. This extension is sufficiently important to at least state the basic definitions.[24]

Let Σ and Θ be signatures. A *relative translation with parameters* $\tau : \Sigma \to \Theta$ is given by a triple $\langle p, \delta, F \rangle$. The variables $v_0, \ldots, v_{p-1}$ will have the role of parameters. The formula δ is a Θ-formula, which contains at most $v_0, \ldots v_p$ free. The mapping F associates to each relation symbol R of Σ with arity n a Θ-formula $F(R)$ with variables among $v_0, \ldots, v_{n+p-1}$. Here the $v_p, \ldots v_{n+p-1}$ represent the argument places.

We translate Σ-formulas of to Θ-formulas of as follows:

- $(R(v_{i_0}, \ldots, v_{i_{n-1}}))^\tau := F(R)[v_p := v_{i_0+p}, \ldots, v_{p+n-1} := v_{i_{n-1}+p}]$; We have to 'shift' the v_{i_j} to v_{i_j+p} to avoid confusion with the parameters.
- $(\cdot)^\tau$ commutes with the propositional connectives;
- $(\forall v_j \, A)^\tau := \forall v_j \, (\delta[v_p := v_{j+p}] \to A^\tau)$;
- $(\exists v_j \, A)^\tau := \exists v_j \, (\delta[v_p := v_{j+p}] \wedge A^\tau)$.

We can compose relative translations with parameters as follows.

- $p_{\tau v} = p_\tau + p_v$,
- $\delta_{\tau v} := (\delta_v \wedge (\delta_\tau)^v)$,
- $R_{\tau v} = (R_\tau)^v$.

The identity translation $\mathrm{id} := \mathrm{id}_\Theta$ is defined in the obvious way, setting $p_{\mathrm{id}} := 0$.

Consider a translation with parameters $\tau : \Sigma \to \Theta$. Let $\mathcal{M} = \langle M, I \rangle$ be a model of signature Θ. Let f be an assignment for $\mathcal{M}$. Let

$$\Delta := \{ m \in M \mid \mathcal{M}, f[v_p := m] \models \delta_\tau \}.$$

Suppose $\mathcal{M}, f$ satisfies $\exists v_p \, \delta$ and the τ-translations of the identity axioms in the Σ-language.

---

[24]I am not aware of a statement of this definition anywhere in the literature. One reason for this defect, may be that Tarski in [TMR53] opted for a different style of treatment, to wit: adding constants.

Suppose that $f(v_\ell) \in \Delta$, for all $\ell \geq p$. We can define a new model assignment pair $\langle \mathcal{N}, g \rangle := \tau^{\langle \mathcal{M}, f \rangle}$ as follows.

- We define, for m, m' in Δ, $m \simeq m' :\Leftrightarrow \mathcal{M}, f[v_p := m, v_{p+1} := m'] \models E_\tau$. Clearly $\simeq$ is an equivalence relation. We write $[m]$ for the $\simeq$-equivalence class of m. We take N, the domain of $\mathcal{N}$, to be $\Delta/\simeq$.
- $g(v_i) := [f(v_{i+p})]$.
- Suppose $\mathrm{ar}(P) = k$. We find that $\simeq$ is a congruence (on Δ) with respect to the relation R_0, given by:

$$R_0(\vec{m}) :\Longleftrightarrow \mathcal{M}, f[v_p := m_0, \ldots, v_{k+p-1} := m_{k-1}] \models P^\tau.$$

Thus, it makes sense to define, for $\vec{n}$ a sequence of elements of N,

$$P^{\mathcal{N}}(\vec{n}) :\Longleftrightarrow \exists m_0 \in n_0 \ldots \exists m_{k-1} \in n_{k-1} \ R_0(m_0, \ldots, m_{k-1}).$$

One can show, for $\langle \mathcal{N}, g \rangle = \langle \mathcal{M}, f \rangle^\tau$, that:

$$\mathcal{N}, g \models A \Longleftrightarrow \mathcal{M}, f \models A^\tau.$$

Further, one can show that $\tau^{v^{\mathcal{M},f}}$ exists iff $(\tau v)^{\mathcal{M},f}$ exists and that if both exist, they are isomorphic.

A *relative interpretation with parameters* is a quadruple $\langle U, \tau, C, V \rangle$, where U is a theory of signature Σ, V is a theory of signature Θ, where $\tau : \Sigma \to \Theta$ is a translation with parameters and where C is a Θ-formula containing at most $v_0, \ldots, v_{p_\tau - 1}$ free. The formula C provides a constraint on the parameters.[25] We demand:

- $V \vdash \exists v_0 \ldots \exists v_{p_\tau - 1} \ C$,
- for any Σ-sentence A, if $U \vdash A$, then $V \vdash \forall v_0 \ldots \forall v_{p_\tau - 1} (C \to A^\tau)$.

The identity interpretation is defined in the obvious way. The only new aspect of composition of interpretations is the transformation of the constraining formula C. This works as follows:

- $C_{M \circ K} := C_M \wedge (v_{p_M}, \ldots, v_{p_M + p_K - 1} : \delta_M) \wedge C_K^M$.

There are several choices for the MOD functor. We can make $\mathrm{MOD}(T)$ the set of models of T and simply say that, for $K : U \to V$, $\mathrm{MOD}(K)$ is a total binary relation from models of V to models of U. We can also take $\mathrm{MOD}(T)$ to be a set of model/assignment pairs $\langle \mathcal{M}, f \rangle$ such that $\mathcal{M} \models T$. In this case, $\mathrm{MOD}(K)$ is a *partial* map from $\mathrm{MOD}(V)$ to $\mathrm{MOD}(U)$, to wit,

$$\mathrm{MOD}(K)(\langle \mathcal{M}, f \rangle) = \langle \mathcal{M}, f \rangle^{\tau_K}.$$

We still have some freedom in stipulating to which category the MOD-functor is a functor, since we can put further conditions on partial maps Φ from $\mathrm{MOD}(V)$ to $\mathrm{MOD}(U)$. E.g., we can demand that for every $\mathcal{M} \models V$, there be an f, such that $\Phi(\langle \mathcal{M}, f \rangle)$ is defined.

---

[25] In the paper [MPS90], no constraining formula is used. The reason that this can be avoided is the fact that the authors study *local* interpretability.

REFERENCES

[Ben86] C. BENNET, *On Some Orderings of Extensions of Arithmetic*, Department of Philosophy, University of Göteborg, 1986.

[Bor94] F. BORCEUX, *Handbook of Categorical Algebra 1, Basic Category Theory*, Encyclopedia of Mathematics and its Applications, Cambridge University Press, Cambridge, 1994.

[Bus86] S. BUSS, *Bounded Arithmetic*, Bibliopolis, Napoli, 1986.

[Cor80] J. CORCORAN, *Notes and queries, History and Philosophy of Logic*, vol. 1 (1980), pp. 231–234.

[dB65a] K. L. DE BOUVÈRE, *Logical synonymy, Indagationes Mathematicae*, vol. 27 (1965), pp. 622–629.

[dB65b] ———, *Synonymous Theories, The Theory of Models, Proceedings of the 1963 International Symposium at Berkeley* (J. W. Addison, L. Henkin, and A. Tarski, editors), North Holland, Amsterdam, 1965, pp. 402–406.

[Haj70] P. HÁJEK, *Logische Kategorien, Archiv für Mathematische Logik und Grundlagenforschung*, vol. 13 (1970), pp. 168–193.

[HP91] P. HÁJEK and P. PUDLÁK, *Metamathematics of First-Order Arithmetic*, Perspectives in Mathematical Logic, Springer, Berlin, 1991.

[Hal99] VOLKER HALBACH, *Conservative theories of classical truth, Studia Logica*, vol. 62 (1999), pp. 353–370.

[Han65] W. HANF, *Model-theoretic methods in the study of elementary logic, The Theory of Models, Proceedings of the 1963 International Symposium at Berkeley* (J. W. Addison, L. Henkin, and A. Tarski, editors), North Holland, Amsterdam, 1965, pp. 132–145.

[Hod93] W. HODGES, *Model Theory*, Encyclopedia of Mathematics and its Applications, vol. 42, Cambridge University Press, Cambridge, 1993.

[Jac99] B. JACOBS, *Categorical Logic and Type Theory*, Studies in Logic and the Foundations of Mathematics, no. 141, North Holland, Amsterdam, 1999.

[JdJ98] G. JAPARIDZE and D. DE JONGH, *The logic of provability, Handbook of Proof Theory* (S. Buss, editor), North-Holland Publishing Co., Amsterdam edition, 1998, pp. 475–546.

[Kan72] S. KANGER, *Equivalent theories, Theoria*, vol. 38 (1972), pp. 1–6.

[Kay91] RICHARD KAYE, *Models of Peano Arithmetic*, Oxford Logic Guides, Oxford University Press, 1991.

[Lin94] P. LINDSTRÖM, *The Arithmetization of Metamathematics*, vol. 15, Filosofiska meddelanden, blå serien, Institutionen för filosofi, Göteborgs universitet, Göteborg, 1994.

[Lin97] ———, *Aspects of Incompleteness*, Lecture Notes in Logic, vol. 10, Springer, Berlin, 1997.

[Mac71] S. MACLANE, *Categories for the Working Mathematician*, Graduate Texts in Mathematics, no. 5, Springer, New York, 1971.

[MPS90] J. MYCIELSKI, P. PUDLÁK, and A. S. STERN, *A Lattice of Chapters of Mathematics (Interpretations between Theorems)*, Memoirs of the American Mathematical Society, vol. 426, AMS, Providence, Rhode Island, 1990.

[Per97] M. G. PERETYAT'KIN, *Finitely Axiomatizable Theories*, Consultants Bureau, New York, 1997.

[PEK67] M. B. POUR-EL and S. KRIPKE, *Deduction-preserving "recursive isomorphisms" between theories, Fundamenta Mathematicae*, vol. 61 (1967), pp. 141–163.

[Pud85] P. PUDLÁK, *Cuts, consistency statements and interpretations, The Journal of Symbolic Logic*, vol. 50 (1985), pp. 423–441.

[Pud86] ———, *On the length of proofs of finitistic consistency statements in finitistic theories, Logic Colloquium '84* (J. B. Paris et al., editors), North-Holland, 1986, pp. 165–196.

[Sve78] V. ŠVEJDAR, *Degrees of interpretability, Commentationes Mathematicae Universitatis Carolinae*, vol. 19 (1978), pp. 789–813.

[TMR53] A. TARSKI, A. MOSTOWSKI, and R. M. ROBINSON, *Undecidable Theories*, North–Holland, Amsterdam, 1953.

[Vis91] A. VISSER, *The formalization of interpretability*, *Studia Logica*, vol. 51 (1991), pp. 81–105.

[Vis98] ――――, *An Overview of Interpretability Logic*, *Advances in Modal Logic, vol 1* (M. Kracht, M. de Rijke, H. Wansing, and M. Zakharyaschev, editors), CSLI Lecture Notes, no. 87, Center for the Study of Language and Information, Stanford, 1998, pp. 307–359.

[Vis05] ――――, *Faith & Falsity: a study of faithful interpretations and false Σ_1^0-sentences*, *Annals of Pure and Applied Logic*, vol. 131 (2005), pp. 103–131.

SUBFACULTY OF PHILOSOPHY
HEIDELBERGLAAN 8
3584CS UTRECHT
THE NETHERLANDS
E-mail: albert.visser@phil.uu.nl

LECTURE NOTES IN LOGIC
General Remarks

This series is intended to serve researchers, teachers, and students in the field of symbolic logic, broadly interpreted. The aim of the series is to bring publications to the logic community with the least possible delay and to provide rapid dissemination of the latest research. Scientific quality is the overriding criterion by which submissions are evaluated.

Books in the Lecture Notes in Logic series are printed by photo-offset from master copy prepared using LaTeX and the ASL style files. For this purpose the Association for Symbolic Logic provides technical instructions to authors. Careful preparation of manuscripts will help keep production time short, reduce costs, and ensure quality of appearance of the finished book. Authors receive 50 free copies of their book. No royalty is paid on LNL volumes.

Commitment to publish may be made by letter of intent rather than by signing a formal contract, at the discretion of the ASL Publisher. The Association for Symbolic Logic secures the copyright for each volume.

The editors prefer email contact and encourage electronic submissions.

Editorial Board

David Marker, Managing Editor
Dept. of Mathematics, Statistics,
   and Computer Science (M/C 249)
University of Illinois at Chicago
851 S. Morgan St.
Chicago, IL 60607-7045
marker@math.uic.edu

Vladimir Kanovei
Lab 6
Institute for Information
   Transmission Problems
Bol. Karetnyj Per. 19
Moscow 127994 Russia
kanovei@mccme.ru

Steffen Lempp
Department of Mathematics
University of Wisconsin
480 Lincoln Avenue
Madison, Wisconsin 53706-1388
lempp@math.wisc.edu

Lance Fortnow
Department of Computer Science
University of Chicago
1100 East 58th Street
Chicago, Illinois 60637
fortnow@cs.uchicago.edu

Shaughan Lavine
Department of Philosophy
The University of Arizona
P.O. Box 210027
Tuscon, Arizona 85721-0027
shaughan@ns.arizona.edu

Anand Pillay
Department of Mathematics
University of Illinois
1409 West Green Street
Urbana, Illinois 61801
pillay@math.uiuc.edu

Editorial Policy

1. Submissions are invited in the following categories:

i) Research monographs   iii) Reports of meetings

ii) Lecture and seminar notes   iv) Texts which are out of print

Those considering a project which might be suitable for the series are strongly advised to contact the publisher or the series editors at an early stage.

2. Categories i) and ii). These categories will be emphasized by Lecture Notes in Logic and are normally reserved for works written by one or two authors. The goal is to report new developments quickly, informally, and in a way that will make them accessible to non-specialists. Books in these categories should include

– at least 100 pages of text;

– a table of contents and a subject index;

– an informative introduction, perhaps with some historical remarks, which should be accessible to readers unfamiliar with the topic treated;

In the evaluation of submissions, timeliness of the work is an important criterion. Texts should be well-rounded and reasonably self-contained. In most cases the work will contain results of others as well as those of the authors. In each case, the author(s) should provide sufficient motivation, examples, and applications. Ph.D. theses will be suitable for this series only when they are of exceptional interest and of high expository quality.

Proposals in these categories should be submitted (preferably in duplicate) to one of the series editors, and will be refereed. A provisional judgment on the acceptability of a project can be based on partial information about the work: a first draft, or a detailed outline describing the contents of each chapter, the estimated length, a bibliography, and one or two sample chapters. A final decision whether to accept will rest on an evaluation of the completed work.

3. Category iii). Reports of meetings will be considered for publication provided that they are of lasting interest. In exceptional cases, other multi-authored volumes may be considered in this category. One or more expert participant(s) will act as the scientific editor(s) of the volume. They select the papers which are suitable for inclusion and have them individually refereed as for a journal. Organizers should contact the Managing Editor of Lecture Notes in Logic in the early planning stages.

4. Category iv). This category provides an avenue to provide out-of-print books that are still in demand to a new generation of logicians.

5. Format. Works in English are preferred. After the manuscript is accepted in its final form, an electronic copy in LaTeX format will be appreciated and will advance considerably the publication date of the book. Authors are strongly urged to seek typesetting instructions from the Association for Symbolic Logic at an early stage of manuscript preparation.

Lecture Notes in Logic

From 1993 to 1999 this series was published under an agreement between the Association for Symbolic Logic and Springer-Verlag. Since 1999 the ASL is Publisher and A K Peters, Ltd. is Co-publisher. The ASL is committed to keeping all books in the series in print.

Current information may be found at http://www.aslonline.org, the ASL Web site. Editorial and submission policies and the list of Editors may also be found above.

Previously published books in the *Lecture Notes in Logic* are:

1. *Recursion Theory.* J. R. Shoenfield. (1993, reprinted 2001; 84 pp.)

2. *Logic Colloquium '90; Proceedings of the Annual European Summer Meeting of the Association for Symbolic Logic, held in Helsinki, Finland, July 15–22, 1990.* Eds. J. Oikkonen and J. Väänänen. (1993, reprinted 2001; 305 pp.)

3. *Fine Structure and Iteration Trees.* W. Mitchell and J. Steel. (1994; 130 pp.)

4. *Descriptive Set Theory and Forcing: How to Prove Theorems about Borel Sets the Hard Way.* A. W. Miller. (1995; 130 pp.)

5. *Model Theory of Fields.* D. Marker, M. Messmer, and A. Pillay. (First edition, 1996, 154 pp. Second edition, 2006, 155 pp.)

6. *Gödel '96; Logical Foundations of Mathematics, Computer Science and Physics; Kurt Gödel's Legacy. Brno, Czech Republic, August 1996, Proceedings.* Ed. P. Hajek. (1996, reprinted 2001; 322 pp.)

7. *A General Algebraic Semantics for Sentential Objects.* J. M. Font and R. Jansana. (1996; 135 pp.)

8. *The Core Model Iterability Problem.* J. Steel. (1997; 112 pp.)

9. *Bounded Variable Logics and Counting.* M. Otto. (1997; 183 pp.)

10. *Aspects of Incompleteness.* P. Lindstrom. (First edition, 1997. Second edition, 2003, 163 pp.)

11. *Logic Colloquium '95; Proceedings of the Annual European Summer Meeting of the Association for Symbolic Logic, held in Haifa, Israel, August 9–18, 1995.* Eds. J. A. Makowsky and E. V. Ravve. (1998; 364 pp.)

12. *Logic Colloquium '96; Proceedings of the Colloquium held in San Sebastian, Spain, July 9–15, 1996.* Eds. J. M. Larrazabal, D. Lascar, and G. Mints. (1998; 268 pp.)

13. *Logic Colloquium '98; Proceedings of the Annual European Summer Meeting of the Association for Symbolic Logic, held in Prague, Czech Republic, August 9–15, 1998.* Eds. S. R. Buss, P. Hájek, and P. Pudlák. (2000; 541 pp.)

14. *Model Theory of Stochastic Processes.* S. Fajardo and H. J. Keisler. (2002; 136 pp.)

15. *Reflections on the Foundations of Mathematics; Essays in Honor of Solomon Feferman.* Eds. W. Seig, R. Sommer, and C. Talcott. (2002; 444 pp.)

16. *Inexhaustibility; A Non-exhaustive Treatment.* T. Franzén. (2004; 255 pp.)

17. *Logic Colloquium '99; Proceedings of the Annual European Summer Meeting of the Association for Symbolic Logic, held in Utrecht, Netherlands, August 1–6, 1999.* Eds. J. van Eijck, V. van Oostrom, and A. Visser. (2004; 208 pp.)

18. *The Notre Dame Lectures.* Ed. P. Cholak. (2005, 185 pp.)

19. *Logic Colloquium 2000; Proceedings of the Annual European Summer Meeting of the Association for Symbolic Logic, held in Paris, France, July 23–31, 2000.* Eds. R. Cori, A. Razborov, S. Todorčević, and C. Wood. (2005; 408 pp.)

20. *Logic Colloquium '01; Proceedings of the Annual European Summer Meeting of the Association for Symbolic Logic, held in Vienna, Austria, August 1–6, 2001.* Eds. M. Baaz, S. Friedman, and J. Krajiček. (2005, 486 pp.)

21. *Reverse Mathematics 2001.* Ed. S. Simpson. (2005, 401 pp.)

22. *Intensionality.* Ed. R. Kahle. (2005, 265 pp.)

23. *Logicism Renewed: Logical Foundations for Mathematics and Computer Science.* P. Gilmore. (2005, 230 pp.)

24. *Logic Colloquium '03; Proceedings of the Annual European Summer Meeting of the Association for Symbolic Logic, held in Helsinki, Finland, August 14–20, 2003.* Eds. V. Stoltenberg-Hansen and J. Väänänen. (2006; 407 pp.)

25. *Nonstandard Methods and Applications in Mathematics.* Eds. N.J. Cutland, M. Di Nasso, and D. Ross. (2006; 248 pp.)

26. *Logic in Tehran: Proceedings of the Workshop and Conference on Logic, Algebra, and Arithmetic, held October 18–22, 2003.* Eds. A. Enayat, I. Kalantari, M. Moniri. (2006; 341 pp.)

Printed and bound by CPI Group (UK) Ltd, Croydon, CR0 4YY

23/10/2024

01777672-0006